AQUACULTURE SCIENCE

**Delmar Publishers is proud
to support FFA activities**

Aquaculture
Science

Rick Parker

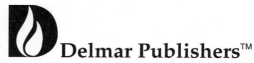

I⟨T⟩P™ An International Thomson Publishing Company

Albany • Bonn • Boston • Cincinnati • Detroit • London • Madrid
Melbourne • Mexico City • New York • Pacific Grove • Paris • San Francisco
Singapore • Tokyo • Toronto • Washington

NOTICE TO THE READER

Cover credit: CJ Hepburn Associates, Inc.

Delmar Staff:
Publisher: Tim O'Leary
Administrative Editor: Cathy Esperti
Senior Project Editor: Andrea Edwards Myers
Production Manager: Wendy Troeger

COPYRIGHT © 1995
By Delmar Publishers
a division of International Thomson Publishing Inc.

The ITP logo is a trademark under license

Printed in the United States of America

For more information, contact:

Delmar Publishers
3 Columbia Circle, Box 15015
Albany, New York 12212-5015

International Thomson Publishing Europe
Berkshire House 168-173
High Holborn
London, WC1V7AA
England

Thomas Nelson Australia
102 Dodds Street
South Melbourne, 3205
Victoria, Australia

Nelson Canada
1120 Birchmont Road
Scarborough, Ontario
Canada M1K 5G4

International Thomson Editores
Campos Eliseos 385, Piso 7
Col Polanco
11560 Mexico D F Mexico

International Thomson Publishing GmbH
Königswinterer Strasse 418
53227 Bonn
Germany

International Thomson Publishing Asia
221 Henderson Road
#05-10 Henderson Building
Singapore 0315

International Thomson Publishing - Japan
Hirakawacho Kyowa Building, 3F
2-2-1 Hirakawacho
Chiyoda-ku, Tokyo 102
Japan

1 2 3 4 5 6 7 8 9 10 XXX 01 00 99 98 97 96 95

Library of Congress Cataloging-in-Publication Data

Parker, Rick, 1949-
  Aquaculture science / Rick Parker.
    p. cm.
  Includes index.
  ISBN 0-8273-6454-7
  1. Aquaculture—Juvenile literature. 2. Aquaculture industry—
Juvenile literature. [1. Aquaculture.] I. Title.
SH135.P37 1994   94-34435
639'.8—dc20   CIP
                                        AC

Dedicated to
My father, Dick Parker, who taught me to love agriculture
And to
My mother, Louise Parker, who taught me to love life.

Contents

Preface

Like other aspects of agriculture, humans have engaged in some form of aquaculture for several thousand years. Also, like agriculture, humans perfected aquaculture to produce more food. Yet, ask any ten people to explain aquaculture and each person will give a different explanation. Each person's concept of aquaculture centers on only one small aspect because the practice of aquaculture is so broad.

Writing a textbook on aquaculture is a lot like writing a textbook on all of agriculture. If the textbook included all of aquaculture, it would be several volumes. Deciding what to include and how much was a challenge. With the explosion of information worldwide, perhaps the fate of those who inherit the future is to learn how to find information and how to use it for their circumstances. Indeed, learning to find and use information should be a goal of all formal education.

Anyone who attempts to learn of aquaculture soon realizes how much science is in aquaculture. Aquaculture demands a reasonable understanding of water chemistry. Next, a species cannot be cultured until its biology is known. Being able to reproduce significant numbers of an aquatic animal for culture requires a thorough understanding of reproductive life cycles. To recognize healthy animals and prevent diseases, an understanding of anatomy and physiology is necessary. Feed costs represent a significant share of the cost of production so an understanding of the science of nutrition is essential. Finally, science uses the metric system. Throughout the book, metric equivalents are given in parenthesis.

This book is packed with information. Several features of the book make the information more useable for the student and the instructor. These features include the chapter objectives, key terms, tables, charts, graphs, illustrations, pictures, teaching aids, exercises, a glossary, and an appendix. Each chapter also provides at least one sidebar to help stimulate students' interest. Each chapter and each feature must be used as a whole. Each part complements the others.

An education prepares students for a productive life. Preparation is difficult without knowing what is required. Each chapter in this book starts with a list of learning objectives. These help the student identify what is really important from all of the information in the chapter.

Also, the beginning of each chapter contains a list of key words. Every field of study uses unique words or words with unique meanings. Aquaculture has its own language. Just to read and discuss any topic in aquaculture requires learning the words and their meanings. Many of the words are defined within the text, but the reader should not forget the glossary. It is thorough and contains many words common in aquaculture but not necessarily used in the text. For the beginner, the glossary can be used like a traveler uses a foreign language dictionary.

Often, presenting information in text form is the least effective. This book relies on tables, pictures, charts, graphs, and illustrations to present information. Tables, charts, and graphs provide quick access to information without wading through excess words. Students need to learn how to read these and use the information they supply. Also, pictures, charts, and graphs are used since people often remember more of what they see than what they hear or read. The appendix contains helpful tables for converting units of measure, making contact with the industry and agencies affecting the industry, and information to supplement the text of the book. Unless noted otherwise, pictures in the text were taken by the author.

Knowledge and information alone are useless unless they can be applied. The end of each chapter provides a selection of ideas the student or the instructor can use to apply the content of the chapter—an opportunity to learn by doing. Also, for the serious student and the great teacher, the end of each chapter provides a complete list of learning/teaching aids. These are listed because this book could not contain everything and because in different areas of the United States, aquaculture has a different emphasis.

Aquaculture is a bright spot with great potential. Many areas need to be explored and developed. Its future depends on the preparation of those who choose aquaculture as their vocation. Chapter 9, Aquaculture Business, and Chapter 11, Career Opportunities in Aquaculture, set the stage for those who make this choice.

Finally, this book would still be a dream or idea without the help of great friends. Marilyn Parker is not only a wife to the author and mother of eight beautiful children, she is a friend who critiques the author's ideas and writing. She typed some of the manuscript, kept track of all the artwork for the book, and checked the format of the manuscript. She also settled for less of the author so he could write. Rosemary Vaughn is a coworker, but more than that, she is another friend who contributed endless hours to collecting information for the book and typing the manuscript. Each day the author thanked the Great Spirit of Words that Rosemary could type as fast as he could think! Two other friends and coworkers, Liane Taylor and Leann Bywater, typed many of the tables. Also, the author appreciates the help and encouragement of Cathy Esperti and the team at Delmar.

Aquaculture Basics and History

A quaculture is the only way to satisfy an increasing demand for fish and seafood products. Aquaculture can continue as the fastest growing agricultural industry in the United States by successfully meeting challenges as it has in the past.

Learning Objectives

After completing this chapter, the student should be able to:

In Aquaculture

- Explain the development of aquaculture as a part of agriculture
- Name three civilizations practicing aquaculture more than 200 years ago
- Define aquaculture
- Compare traditional farming to aquaculture
- Discuss why aquaculture evolved from fishing practices
- Discuss how the catfish industry developed and why Mississippi leads in catfish production
- Explain why Idaho leads in trout production

■ List five main activities that are a part of aquaculture but often become a separate industry

■ Discuss how aquaculture is expanding and what the future holds for aquaculture

In Science

■ Identify significant scientific events or people that contributed to the development of aquaculture

■ Explain the National Sea Grant Program and its role in scientific research

■ Discuss the role of science and technology in the development of aquaculture

■ Indicate the role of scientific research in the future of aquaculture

Understanding of this chapter will be enhanced if the following terms are known. Many are defined in the text and others are defined in the glossary.

KEY TERMS		
Agriculture	Freshwater	Polyculture
Aquaculture	Grow-out	Processing
Aquifer	Harvesting	Salinity
Brackish	Hatchery	Seed
Broodstock	Husbandry	Self-feeders
Coldwater	Incubate	Spawn
Culture	Larvae	Warmwater
Eggs	Mariculture	
Fingerlings	Monoculture	

INTRODUCTION AND DEFINITIONS

From prehistoric times to the present, two primary needs of humans remain—food and shelter. Through time, only the means of obtaining food and shelter change. As societies moved from hunting and gathering to the culturing of plants and animals for food, their needs for shelter changed from temporary to permanent. Also, as societies learned to culture plants and animals for food, they generated surpluses of food that allowed members of the society to pursue other interests and stimulated the need for preserving and marketing the surpluses.

FIGURE 1-1 Shelter and food have always been primary human needs. Bannock Shoshoni Indians photographed near the turn of the century. (Photo courtesy Idaho State Historical Society)

Agriculture is the art, science, and business—the culture—of producing every kind of plant and animal useful to humans. Agriculture is the oldest and the most important of all the industries. It continues to evolve with the knowledge and needs of civilization. Typically, agriculture evolves through four stages—

1. a hunting-gathering activity
2. an object of husbandry
3. a craft
4. a science and business.

Agriculture includes not only the cultivation of the land but dairy production, beef production, sheep production, pig production, and all other farming activities, including aquaculture. Examples of aquaculture include catfish farming, crawfish farming, trout farming, salmon ranching, and oyster culture.

Aquaculture is a relatively new word used to describe the art, science, and business of producing aquatic plants and animals useful to humans. Aquaculture is a type of agriculture. It is farming in water instead of on land. Often agriculture and aquaculture include all of the activities involved in producing plants and animals, the supplies and services needed, the processing and marketing, and other steps that deliver products to the consumer in the desired form.

Aquaculture and farming share some similarities and some differences. Table 1-1 compares traditional farming to aquaculture.

TABLE 1-1 Comparison of Traditional Farming to Aquaculture

Farming	Aquaculture
Occurs on land	Occurs in water
Limited by water supply	Limited by oxygen dissolved in water
Many plant and animal crops	Many plant and animal crops
Domesticated plants and animals	Wild and/or domesticated plants and animals

Aquaculture occurs in these general environments—

■ **warmwater** aquaculture

■ **coldwater** aquaculture

■ **mariculture** or marine culture (saltwater).

Warmwater aquaculture is the commercial raising of stock that thrives in warm, often turbid, **freshwater** with temperatures between 70° and 90°F (21.1° and 32.2°C). Examples of warmwater species include catfish, crayfish (crawfish),

FIGURE 1-2 Three-fourths of the Earth's surface is covered with water. People still enjoy fishing the ocean for a meal.

baitfish, and many sport fish. Coldwater aquaculture involves the commercial production of stock that thrives in cool, clear freshwater with temperatures of 65°F (18.3°C) and under. Trout and salmon are examples of coldwater aquaculture. Warmwater and coldwater are also generally considered freshwater—no salinity. Shrimp, oysters, and seaweed cultures are examples of mariculture (marine culture) where the crop thrives in saltwater of various temperatures. The salinity of saltwater ranges from 30 to 35 parts per thousand (ppt) and the salinity of brackish water is 1 to 10 ppt.

Aquaculture, like agriculture, involves controlled culture and an individual or individuals who own the crop. Fisheries are different from aquaculture but are involved in aquaculture. Fisheries involve hunting and the general public accesses to the crops—fish—being hunted. Aquaculture enhances fisheries by providing fish to restock streams, lakes, and oceans. This role of aquaculture in fisheries makes sport fishing more enjoyable and stable and helps ensure the economic success of commercial fisheries.

Historical events that made aquaculture a viable, growing, and profitable enterprise may not be easily identified. Aquaculture likely evolved through

FIGURE 1-3 Aquaculture provides stock to make sport fishing more enjoyable. (Photo courtesy Chuck Weirich, Delta Research and Extension Center, Stoneville, MS)

observation and serendipity in several areas of the world at different times. Perhaps aquaculture developed from fishing practices that involved trapping fish and holding them for freshness, which lead to trapping, holding, and feeding to maintain a food supply for a longer time. Once people saw that fish could be fed and held, they refined techniques to ensure a more constant supply of fish. Possibly cage culture developed when fishers realized that their surplus catch could be held in baskets in the water. Pond culture likely developed when some fishers observed fish trapped in pools of water formed by a flood. Some aquaculture likely developed in conjunction with farming and irrigation, since irrigation provided structures and a source of water.

In the United States, this relatively new business grew more than 15 percent annually from 1980 to 1990. Catfish production, which dominates the U.S. aquaculture output, accounting for about half the total production, soared from 1980 to 1990.

As the demand for aquaculture products increases and technology is developed for different species, aquaculture will grow worldwide.

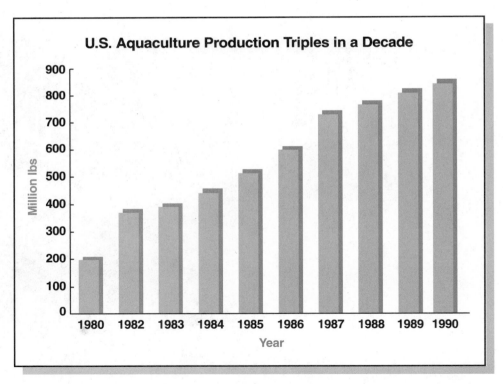

FIGURE 1-4 U.S. aquaculture increased rapidly in the decade from 1980 to 1990. No data available for 1981. (Source: *Agricultural Outlook,* 1991)

HISTORICAL PATTERNS AND PRACTICES

Aquaculture seems like a fairly new agricultural endeavor. But many civilizations actually developed some form of aquaculture to satisfy the needs of the people.

Chinese Aquaculture

Aquaculture in China began around 3500 B.C. with culture of the common carp. These carp were grown in ponds on silkworm farms. The silkworm pupae and feces provided supplemental food for the fish. Carp are hardy and easy to raise in freshwater ponds, and because fish were an important part of life in ancient China, their culture developed very early. In Chinese, the word for fish means surplus. Indeed, fish were equated with a bountiful harvest.

In 475 B.C., Fan-Li wrote the oldest document on fish culture. Fan-Li was a politician and administrator, renowned for his self-taught expertise in carp culture. Fan-Li's document described methods for pond construction, broodstock selection, stocking, and managing ponds.

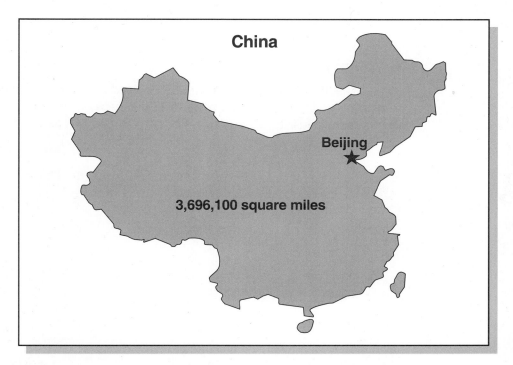

China

Beijing

3,696,100 square miles

FIGURE 1-5 China is home to 1.2 billion people, and aquaculture is still important in helping feed them all.

Emperor Li of the Tang Dynasty (618 to 906 A.D.) banned the culture of common carp since the word for carp was Li. Emperors were considered sacred. Apparently, anything that shared the emperor's name was sacred and should not be eaten. This ban lead the Chinese to develop polyculture—growing more than one species in the same water. Realizing that water is a three-dimensional habitat, all of the productive portions of a pond will not be used by just one species. Different species occupy different locations in the pond and feed on different food. Using polyculture, the Chinese cultured four species of carp, as Table 1-2 indicates.

China is generally considered the cradle of aquaculture. Chinese aquaculture evolved from providing food for the elite to providing a staple food for the common people.

In the Zhujiang Delta of South China, a dike-pond system of agriculture still exists after more than 500 years. Mulberry, sugarcane, fruit, forage crops, vegetables, silkworm breeding, and pig raising integrate with fish rearing. Crops and crop residues are fed directly to bighead carp, grass carp, silver carp, mud carp, common carp, black carp, bream, and tilapia.

Common carp raised in the United States since the 1870s never gained a broad acceptance as a food fish. Since 1963, three species of Chinese carp—the grass carp, silver carp, and bighead carp—have been introduced into the United States for testing as farm fish.

Egyptian Aquaculture

Early aquaculture for Egyptians likely evolved as a part of the irrigation systems they developed. Aquaculture in Egypt focused on tilapia, and developments seem consistent with those of carp in China. No written documents

TABLE 1-2 Polyculture with Four Species of Carp

Species	Location	Feed
Grass carp	Topwater	Large vegetation near shore
Bighead carp	Midwater	Minute animals known as zooplankton
Silver carp	Midwater	Minute plants or algae known as phytoplankton
Mud carp	Bottom	Wide variety of plants and animals

from early Egyptian aquaculture exist, but drawings in tombs dated about 2000 B.C. show tilapia.

Roman Aquaculture

During the days of the Roman Empire, fish were kept in ponds called stews next to the manors of the wealthy. Roman aquaculture focused on mullet and trout. Pliny the Elder recorded that saltwater and freshwater fish culture was practiced in Rome the 1st century B.C.

During the Middle Ages, stew ponds became important for the monks and lay people, providing a source of fresh fish.

English and European Aquaculture

In central Europe the history of pond fish culture began at the close of the 11th century and the beginning of the 12th. Pond management in Bohemia, a part of the Czech Republic, peaked in the 14th century, and Bohemia had about 185,000 A (74,867 ha) of ponds for carp. From spawning to marketing required four to six years. Each acre was stocked with about 120 two-year-old carp. The 16th century was the golden era of Bohemian pond culture.

Dom Pinchon, a 14th century French monk, possibly was the first person to artificially fertilize trout eggs. At the very least, he is the first person to collect natural spawn and incubate them in a hatching box. In 1600, John Taverner of England presented the first known comprehensive paper on the management of carp, bream, trench, and perch in ponds. From his experiences, he provided very accurate details that mesh with today's practices. Also, Gervais Markham, in 1613, detailed raising carp in ponds. In 1713, Sir Roger North presented a treatise on fish husbandry techniques reflecting the ideas of Taverner but without acknowledgement.

In southwestern Germany, Stephan Jacobi published a series of articles in English and German describing his results in propagating several species of freshwater fish. He perfected the technique of spawning trout, incubating the fertilized eggs, and raising the fish. His methods did not become widespread, and little was written about fish husbandry for the next 100 years.

For a couple of reasons, many consider France as the birthplace of modern aquaculture. Two commercial fisherman, Joseph Remy and Antoine Gehin, became concerned with the decline of trout in French streams. Using their observations of trout, natural habits, and with the help of two well-known scientists, M. Miline Edwards and M. Coste, the first fish hatchery was established in 1852 in Huningue. With Coste as the director, the hatchery became well known and supplied trout eggs for most of central Europe.

A BRIEF HISTORY OF FISHING

Fishing is one of the oldest and most important activities of humankind. Ancient remains of spears, hooks, and fishnets have been found in ruins of the Stone Age. The people of early civilizations drew pictures of nets and fishing lines in their art. Through the ages, people wrote about fishing, used fish in exchange for services, and even learned to fish farm.

Early hooks were made from the upper bills of eagles and from bones, shells, horns, and thorn plants. Spears were tipped with the same materials, or sometimes with flints. Lines and nets were made from leaves, plant stalks, and cocoon silk. Ancient fishing nets were rough in design and material, but they were amazingly like some now in use.

Many examples of fishing remain in art or are mentioned in writing. An Egyptian tomb more than 4,000 years old contains a picture of fishermen. An old Chinese proverb recognized the value of fishing: "Give a man a fish and he will live for a day, teach him to fish and he will have food for life."

Fish were often used as a medium of exchange or as payment for services rendered. In the twenty-ninth year of Ramses III, the Union of Grave Diggers in Egypt filed a petition with the royal authorities for higher wages. As part of their wages, these workers received large amounts of fish four times each month. The petition requested a pay increase, pointing out that the petitioners came to the authorities without clothes and ointments—and even without fish.

The herring industry grew up around the Baltic Sea in the 12th century and was controlled by the Hanseatic League, a group of German cities whose merchants traded all over northern Europe in fish, timber, cloth, salt, and many other goods.

A 14th century discovery by a Dutchman named Beukelszoon helped the Dutch fishing industry. Buekelszoon pickled herring in brine instead of preserving them in dry salt.

In the 15th century, the herring mysteriously disappeared from the Baltic Sea. Fishermen had to seek their herring in the North Sea and the Atlantic Ocean. The Dutch took over the herring fisheries and led commercial fishing of all kinds until the end of the 17th century.

Commercial fishing on the North American continent started over 300 years ago with the first colonists. So many fish were close to shore that the colonists did not have to build large sailing vessels as the Europeans did. Instead, the colonists followed the Indians' example and fished from small boats. Some fish were caught in traps and weirs (brush fences) set in the mouths of rivers and harbors. Shore fishermen used nets or, when the tide had gone out, searched the rocks and sand for shellfish.

As colonization progressed, fishers began sailing farther out to sea to find enough fish for a good catch. They sailed for months as they worked the fishing banks off Canada and northeastern United States. Many of the early houses along the coast in colonial America had a walk around the roof so the family could watch for returning ships. Many ships did not return, and this walk became known as the widow's walk.

As ships grew larger and fishing methods were developed and refined, the success of fishing voyages and the types of fish and seafood increased. Like other commercial operations, the fishing industry became mechanized. With new technology, ships sailed to new fishing grounds by sailing farther from port and returning safely, loaded with fish. During the years between 1900 and the late 1960s, the world fish catch increased by 27 times.

Fishing rights of a country have been for some time, a source of concern and agitation. As early as 1377, records indicate lawsuits against fishermen who used a net called the wondrychoun, a large net that fishermen dragged through the water. The net caught little fish as well as big, and some people were afraid that soon there would be no fish left.

In the 1860s, individuals and groups acknowledged that fishery resources were limited, and that they must be managed through international agreements. In 1902, the International Council for the Exploration of the Sea (ICES) was formed by the major European fishing countries. Other nations joined, including the United States, in the mid-1960s. ICES led to several conventions for the regulation of fisheries by mesh size of nets and by quota in order to obtain the highest yields consistent with the maintenance of fish stocks.

During the 1800s, one more contribution added to the body of knowledge of aquaculture. In 1856, V. P. Vrasski developed the dry or Russian method of fertilizing trout eggs. Unfortunately, this was not published until 1871, the same year G. C. Atkins perfected the American method of dry fertilization.

Native Americans and Aquaculture

In America, almost every young student is introduced to the story of how Indians showed the white man how to plant a fish with corn seed to improve corn harvest. Native Americans knew much more aquaculture, but much of what they knew was not recorded. Now we must surmise from things they left behind.

Hawaii. Hawaiian society centered around the ocean, agriculture, and aquaculture. By 400 A.D., an organized system of aquaculture existed in Hawaii. Extensive pond systems were developed, and the chiefs controlled aquaculture by leasing tracts of land to governors, who ensured that the ponds produced fish and were maintained.

Four types of agriculture/aquaculture existed in Hawaii—

1. Freshwater fish ponds fed by canals from streams
2. Taro ponds that irrigated agricultural plots
3. Brackish water fish ponds located near the shoreline
4. Seawall fish ponds along the shoreline walled off from the ocean by human-made walls.

Hawaiian integrated aquaculture existed until 1778, when Europeans arrived, destroying the ancient religion and removing the chiefs from ruling the ponds. A few of the old ponds are still used, and Hawaii maintains a prominent role in modern aquaculture.

America. In southern California near the Salton Sea, the Cahuilla people built fish ponds.

The Maya developed irrigation systems. Some evidence suggests that the Maya trapped or cultured fish in ponds or canals somewhere between 500 and 800 B.C.

U.S. Aquaculture Development

Theodatus Garlick collaborated with H. A. Ackley in working with brook trout. Their work inspired pioneers like S. H. Ainsworth, T. Norris, Seth Green, and Livingston Stone. Supposedly, Seth Green established the first public hatchery at Mumford, New York, in 1864. Early emphasis in Europe and America was on restocking depleted streams and lakes, not the culture of food crops.

Soon, several New England states established fish and game commissions and private facilities increased. The combined interest of the state and private agencies lead to the formation of the American Fish Cultural Society in 1870. In 1885, this organization changed its name to the American Fisheries Society. The formation of this society is credited with providing the push to establish the U.S. Commission of Fish and Fisheries, which eventually became the U.S. Fish and Wildlife Service.

While aquaculture developed in several areas in the United States, two species dominate U.S. aquaculture development—catfish and trout. Two states dominate—Mississippi and Idaho.

Catfish and Mississippi. Commercial warmwater fish farming began in the late 1920s and early 1930s by a few individuals who raised minnows to supply the growing demand for baitfish for sport fishing. Shortly after World War II, the demand for minnows increased as the result of the boom in farm pond and reservoir construction and the many water conservation projects inspired by the dust bowl years of the 1930s. By the early 1950s, the number of producers increased enormously, and farmers also began to raise food fish such as buffaloes, bass, and crappies. Many of these early attempts at fish husbandry failed because the operators were not experienced in fish culture, ponds were not properly constructed, and low-value species were being raised.

FIGURE 1-6   Roadside sign along highway 49W in Mississippi proclaiming Humphreys County as the catfish capital of the world.

From 1955 to 1959, the U.S. Fish and Wildlife Service, with funds from the Saltonstall-Kennedy Act for Commercial Fisheries, sponsored research on channel catfish at the University of Oklahoma. The purpose of the research was to develop better production methods in national fish hatcheries and to develop a basis for commercial fish farming. Other agencies and universities also became interested in channel catfish as a commercial and sport species. During the next few years, the Service established three warmwater fish cultural research facilities: the Southeastern Fish Cultural Laboratory, Marian, Alabama, in 1959; the Fish Farming Experimental Station, Stuttgart, Arkansas, in 1960; and the Fish Farming Development Station, Rohwer, Arkansas, in 1963. These stations began research mainly with buffaloes, catfishes, and baitfishes, but other species were added later.

Land grant universities, in cooperation with the U.S. Department of Agriculture, made substantial contributions to warmwater aquaculture research through Agricultural Experiment Stations in Alabama, Arkansas, California, Georgia, Hawaii, Louisiana, Mississippi, Puerto Rico, South Carolina, Tennessee, Texas, and the Virgin Islands.

Farmers received the technology through intensive extension efforts involving the Fish Farming Experimental Station, State cooperative extension services, and the Extension Service of the U.S. Department of Agriculture. Various research laboratories, National Fish Hatcheries, and Fish Cultural Development Centers of the U.S. Fish and Wildlife Services throughout the country also provided technical assistance to fish farmers, as have the agricultural experiment stations, Sea Grant Programs, the U.S. Soil Conservation Service, universities, Tennessee Valley Authority, National Marine Fisheries Service, state departments of conservation and various private foundations. (Chapter 10, Federal, State, and International Agencies and Regulations, provides more detail on the agencies and services that supported and continue to support the development of aquaculture.)

The National Sea Grant Program, established in 1966, provides grants to U.S. universities that are designated sea grant colleges. The program, administered by the National Oceanic and Atmospheric Administration (NOAA) in the Department of Commerce, encourages those schools to provide education, research, and advisory programs in such areas as ocean engineering, aquaculture, pollution studies, environmental studies, seafood processing, coastal management, and mineral resources. In its concern for the marine environment, the program parallels the one that established the land grant colleges to develop the agricultural environment.

Warmwater aquaculture blossomed during the 1960s. The channel catfish industry was originally limited to south-central Arkansas but now it is centered in the delta region of northwestern Mississippi. Mississippi accounts for over 70 percent of U.S. catfish sales.

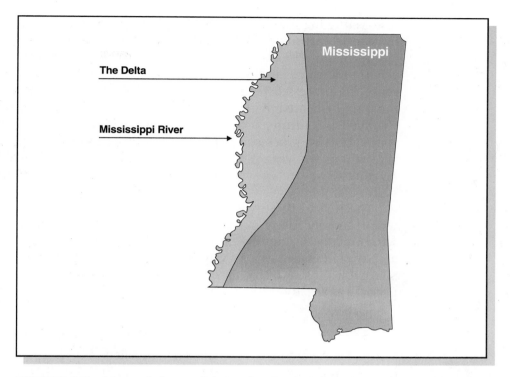

FIGURE 1-7   Most of the catfish production in Mississippi is located on the Delta.

The development of the catfish industry moved through three identifiable phases. Passing through each of these phases, Mississippi emerged as the leader in catfish production.

The first phase or pioneering phase saw relatively high production costs that resulted from low yields and inefficiency. High processing costs added more. High processing costs resulted from a chronic underuse of processing capacity. Farm-raised catfish faced severe competition from channel catfish caught in rivers by commercial fishers and from imported fish. The markets were fragmented. Low product acceptance outside the principal market areas of the deep South and the lack of an effective marketing strategy limited expansion.

During the second phase, from 1971 to 1976, production improved and unit costs declined. Average annual yields increased from 1,500 to 2,000 lbs per A to 3,000 to 4,000 or more lbs per A (1,700 to 2,268 kg per ha to 3,403 to 4,535 kg per ha). Processing, typically limited to the fall, became less seasonal. Unprofitable and marginal producers quit the business when feed costs rose as a result of a scarcity of fish meal. Competition from river fish and imports continued, but supplies of these fish became stabilized while the total demand for catfish rose. Marketing strategy improved, and Mississippi emerged as the clear leader in channel catfish production, processing, and related activities.

In the third phase, from 1977 to 1982, productivity continued improving, acreage increased, and production costs declined. The major sales outlet became the processed fish market. These developments were coupled with a more sophisticated marketing approach that lead to single companies being involved in culture, processing, and marketing. This vertical integration started because processors needed to handle a nearly constant volume of fish throughout the year. A fall production peak is a built-in feature of catfish farming because most of the fish stocked as fingerlings in spring reach harvest size in fall.

The 1973 to 1975 shakeout period provides a lesson for all agribusinesses. Unfortunately, many farmers constructed ponds and started producing fish without considering two critical factors—management expertise and identifiable, dependable markets. Even when catfish farmers in areas like Georgia and South Carolina produced fish, they often had no ready markets. Local oversupply was especially critical when high feed prices reduced profit margins. In

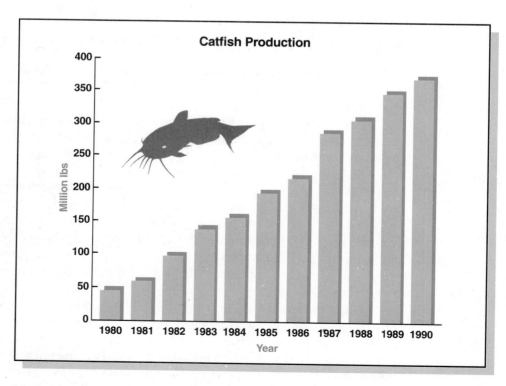

FIGURE 1-8 Catfish production soared from 1980 to 1990. (Source: *Agricultural Outlook*, 1991)

Mississippi, processing technology grew with the industry to provide a market for the crop.

Trout and Idaho. Rainbow trout were introduced into commercial fish farming in the early 1900s. Beginning in 1906 and continuing to 1947, the State of Idaho built 14 hatcheries located mainly in the southern part of the state. These hatcheries produced mostly rainbow trout to maintain productive fishing in rivers, lakes, and reservoirs. From the early 1920s until the end of World War II (1945), private trout hatchery development proceeded slowly because of the easy availability of sport-caught fish and low demand.

The first commercial trout farm began operation in 1909 at the Devil's Corral Spring near Shoshone Falls in the Snake River Canyon. By 1914, Warren Meader started broodstock production, and by 1940 he supplied up to 60 million eggs to public and private hatcheries around the country. Another early innovator was Jack Tingey, the former commissioner of the Utah Fish and Game agency. In 1928, he and his wife, Selma, started the first commercial hatchery near Buhl, Idaho. In the late 1940s, the trout industry began to grow.

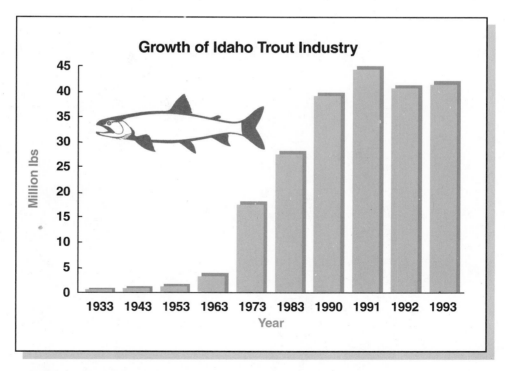

FIGURE 1-9 Trout production began slowly but increased rapidly from the 1960s to the 1980s. Then production plateaued. (Source: *Northwest Environmental Journal*, 1989)

From the 1960s through the 1980s, trout production rapidly increased. Most of the growth occurred in the processed fish segment of the industry.

Today the industry in Idaho is dominated by the world's largest trout production facility—Clear Springs Trout Company in Buhl, Idaho. The trout industry in Idaho is complete with feed mills and processing plants and a training facility at the College of Southern Idaho in Twin Falls. Recent surveys indicate that Idaho trout producers—only about 30—produce about 76 percent of the commercial trout in the United States, a farm value of about $30 million. For now with current water supplies, the Idaho production of trout will probably stay at about 40 million lbs (18,000 metric tons) per year.

While trout is cultured in 45 of the 48 contiguous states, Idaho leads in trout production. Idaho leads because of an abundant supply of water at an ideal year-round temperature, and entrepreneurs who use knowledge from various agencies and their own observations.

History credits the Idaho trout industry with the development of many technological advances. Dry diets developed in the 1950s allowed the industry to expand. In 1956, the Snake River Trout Company started the first processing plant allowing the opportunity for product diversity and distribution. By 1970, selective breeding for off-season spawn provided a year-round supply of eggs.

FIGURE 1-10 Water coming out of the Southern Idaho Aquifer into the Snake River Canyon near Buhl, Idaho.

The idea for self-feeders (demand feeders) developed at the College of Southern Idaho fish technology training facility. This idea was modified into many types of self-feeders and spread throughout the industry.

Water in Idaho comes from the Southern Idaho Aquifer. The water in this aquifer enters the vast and extremely porous lava plain in southern Idaho. Eventually, water emerges from the aquifer at the ideal temperature of 55° to 58°F (12.8° to 14.4°C).

Some of the water in Idaho comes from warm and hot water wells. This water is being used to produce catfish, tilapia, and prawns on a limited basis. Some individuals are even trying to raise alligators.

Oysters, crawfish, clams, and shrimp are also important to the history of American aquaculture.

Oysters. About 43 A.D. Roman settlers in England harvested oysters along the seacoasts. In the winter, they packed the oysters in cloth bags and sent them to Rome. Eventually, to satisfy their taste for oysters, ancient Romans learned to farm oysters in the water off the Italian coast.

When Europeans first came to North America they found Indian tribes along the coast who depended upon oysters as part of their diet. Large piles of oyster shells existed around Indian settlements. The new Americans developed a taste for oysters and harvested the natural supplies. In 1894 the harvest of Chesapeake Bay oysters peaked at 15 million bushels and then began to decline. Soon, Americans developed culture methods to supplement the natural supply. The culture of oysters in the United States is over 100 years old. Worldwide, South Korea, Japan, the United States, and France lead in the culture of oysters.

Crayfish (Crawfish). Culture of crawfish developed as a simulation of their natural life cycle in ponds. Now, some crawfish culture is tied to agricultural practices such as rice fields in the south.

Clams. Shortages and increasing prices for clams are creating more interest in aquaculture, either as an investment venture or to replace over-harvested stocks in public areas. Methods of spawning and growing hard clam larvae were described as early as 1927 and patented in 1929. Interest in culturing clams remained low until the early 1950s. The first commercial aquaculture operation, including a hatchery, began in 1957 near the town of Atlantic, Virginia. A short time later, another project was started in Sayville, New York, by Joe Glancey, who later patented a new method of growing clams.

During this period, a number of other companies formed, tried various methods of growing clams, and generally failed. The major problems involved in raising clams from seed to market size is the ability to culture large enough numbers and to culture them at a reasonable cost. A few companies managed to survive as producers of clam seed for experimental planting and replenishment programs carried out by various state agencies. By 1970, new technology and new materials contributed to the development of several promising methods for aquaculture of hard clams.

Shrimp. Shrimp are widely cultured in Asia where historically the culture occurred almost by accident in brackish water ponds. Culturing shrimp alone, or monoculture, is a fairly recent occurrence in the United States and even Asia.

The National Aquaculture Act of 1980 established aquaculture as a national priority. It consolidated the federal support for aquaculture and the development of national planning for policy and cooperation by federal and state governments, universities, and industry. The purpose of the act was to enhance aquaculture as an industry making major contributions to the nation.

AQUACULTURE ACTIVITIES

When any industry such as aquaculture develops, the functions or activities performed to produce the product become identified in groups. Often, these become separate industries. In aquaculture, five main activities are performed: hatchery, grow-out, harvesting, marketing, and processing.

Hatcheries produce the seed or young fish used to stock growing facilities. Seed are obtained by capturing wild seed or raising from broodstock—adults kept for reproduction.

Grow-out facilities produce crops (fish) from the seed. Like any agriculture venture, these can be intensive or extensive production systems. Intensive systems involve a very dense population of fish in relatively small spaces and require careful management. Extensive systems involve lower populations and less stringent management. Grow-out facilities may be land-based like ponds, tanks, and runways. Or, they may be water-based like pens, cages, and ranching.

Harvesting involves the gathering or capturing of the fish for marketing and processing. Aquaculture harvesting is typically topping (partial) or total harvesting.

Marketing connects producers with consumers. The purpose of marketing is to provide a consumer with desired products and provide the producer with a price to cover production and make a profit. Fish are the major aquacrop in the United States. Five markets, depending on the reason for production, are associated with fish: (1) food for human consumption, (2) bait for sport fishing,

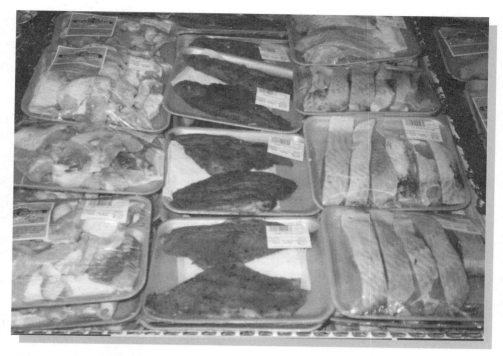

FIGURE 1-11   Marketing is an important activity of aquaculture.

(3) pets or ornamentals for home or office aquaria, (4) sport fish for release into lakes and steams, and (5) fish for feed ingredients.

Processing changes the form of the product into something more desirable to consumers. Processing occurs in three forms—minimal, medium, and value-added.

All of the activities and functions of aquaculture are covered in more detail in other chapters.

THE FUTURE OF AQUACULTURE

In the United States most traditional fisheries are being harvested at or near maximum sustainable yields. More than 60 percent of the fishery products consumed in this country are imported to meet the high demand. Thriving aquaculture industries improve the balance of trade, increase the stability of seafood industries and markets, and provide more jobs for American workers.

At present, world aquacultural production represents about 10 percent of the world aquatic food production by fisheries. That figure is expected to reach 20 percent. Worldwide, aquacultural production in the 43 countries that have such industries produce more than 9 billion lbs (almost 4 million metric tons) of fish and fish products. Exclusive of the aquaculture of sport, bait, ornamental organisms, and pearls, this production includes more than 152 species including finfish, species of shrimp and prawns, crawfish, diverse marine plants, oysters, clams, and other mollusks.

Several technological breakthroughs have increased the potential of aquaculture in the United States—

- Development of net/pen culture and ocean ranching in the Pacific Northwest

- Establishment of abalone culture in California

- Introduction of Malaysian prawn culture to Hawaii and South Carolina

- Improvement of raft culture of blue mussels and oysters in New England

- Development of oyster hatcheries in the Pacific Northwest and the Atlantic States

- Establishment of marine shrimp farms in Central America by U.S. firms.

Table 1-3 lists technological development of the aquaculture of species included in the Sea Grant plan.

Aquaculture throughout the world exists in different levels of development for a variety of reasons. Levels of development include commercial aquaculture, infant industries, pilot scale or partially developed technology, and major lack of technology.

- Commercial aquaculture represents enterprises with established production facilities, profitable markets, and continuity of sales. Research needs are similar to those that support established agricultural enterprises. These include product improvement, increased production efficiency, and effective marketing

- Infant industries may require research on several aspects of production, marketing, and creation of an acceptable institutional framework.

- Pilot scale includes promising organisms for which proof of concept is established and basic breakthroughs in production technology have been achieved. Pilot scale aquaculture requires refinements to solve scale-up problems and ensure reasonable prospects for making money.

- Major lack of technology represents those species of high market potential for which many major problems (such as reproduction, larval survival, domestication, strain selection, nutrition, and production systems) must still be solved.

TABLE 1-3 Organisms Grouped by Level of U.S. Commercial Aquacultural Development

Level of Commercial Development	Species
Commercial Industry	Baitfish Channel Catfish Crawfish (Crayfish) Rainbow Trout
Infant Industry	Penaeid shrimp Prawns (Hawaii) Salmon (net/pen rearing and ocean ranching) Yellow perch Oyster (hatchery/nursery production) Mussels Abalone Striped bass
Technology Developed to Pilot Scale or Partial Level	Prawns (continental United States) Scallops (bay and rock) Seaweeds Clams Eels Bait leech Channel bass Scallops (other than bay and rock) Red drum Sturgeon
Major Lack of Technology	Southern flounder Speckled trout Red snapper Pompano Milkfish Lobster

Aquaculture is now considered a significant part of U.S. agricultural food production. Several factors suggest the role of aquaculture will continue to grow: increased demand, new marketing and processing, and the culture of new species. Continual research on the problems facing aquaculture will ensure its future.

Demand

Aquaculture is the only mode of increasing domestic fish production. The world's capture fisheries—wild-caught fish—are harvested at close to the maximum sustainable level. U.S. demand for fish increases in several ways. A more health-conscious public consumes more fish each year. Recent marketing breakthroughs in several national fast food and restaurant businesses have extended sales of the southern tradition, catfish, into nontraditional regions.

Every 1-lb increase in per capita consumption requires 700 million more lbs (317 million kg) of fresh fish. Over the past 30 years, per capita consumption of fish increased from 11 lbs to about 15 lbs (5.0 kg to about 6.8 kg). Experts predict the per capita consumption of fish and shellfish to reach 25 lbs (11.3 kg) by the year 2025. Even if the per capita consumption remained constant, the U.S. population continues to increase. Fish raised to replenish dwindling wild stock also increases the demand for fish.

Marketing and Processing

The success of aquaculture depends on how the product meets the demands of the market—different products for different markets. Food service, retailers, and food processors market fish. The trend is toward more value-added, fresh refrigerated products including bone fillets, seasoned and marinated products, smoked products, and vacuum-packed prepared fresh products that are ready to bake or broil.

Techniques and Technology

New techniques and technology continue to improve the profitability of aquaculture. Feeding represents 40 to 50 percent of the costs associated with aquaculture production. New feeding techniques and technology will improve feed conversion and use. Biotechnology, genetics, and selective breeding will increase aquaculture production. New rearing methods such as cage culture and closed systems will open the door for more people to try aquaculture.

New Species

Scientists recognize about 21,000 kinds of fish but only a few of these are widely used as food. U.S. aquaculture is dominated by catfish production. This will continue but the culture, technology, and marketing for many other species are being developed.

Some of these include carp, tilapia, hybrid striped bass, alligators, buffalofish, red drum and shrimp, prawns, and some aquatic plants. Chapter 2 discusses the potential species for aquaculture.

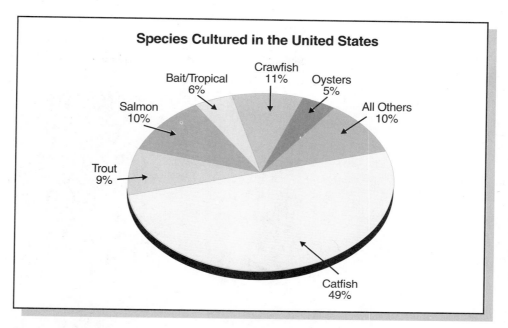

FIGURE 1-12 With thousands of species to culture, catfish make up the majority of finfish and shellfish cultured in the United States. (Source: *Feedstuffs*, 1990)

Research and Problems

Although aquaculture is generally successful, it is still several decades behind traditional livestock husbandry in research and development. Virtually every aspect of aquaculture can still be improved. Hundreds of thousands of acres of land are still available for expansion of fish farming. The water supply, if properly used, is adequate. The cooperative efforts of federal and state governments, private agencies, universities, and industry will be necessary to overcome the barriers that prevent the development of that acreage.

Research needs identified by members of the aquaculture industry touch every aspect of aquaculture. General topics needing research include—

■ Life history and biology

■ Genetics and reproduction

■ Nutrition and diet

■ Environmental requirements

■ Effluent (waste) control and water availability

■ Control of diseases and parasites

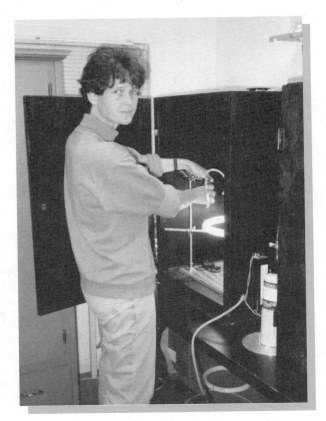

FIGURE 1-13   Research continues to be important to aquaculture. Here, Dr. Martine van de Ploeg develops a method to culture algae.

■ Predation and competition

■ Harvesting, processing, and distribution

■ Transportation

■ Introduction of non-native species

■ Drug and chemical registration

■ Production of rare seed stocks.

Environmental issues will continue to receive attention for all aspects of agriculture. This will present some problems—challenges—to the aquaculture industry. For aquaculture, these environmental issues include waste feed and excretory products, reduced water resources, endangered species, multiple uses of water, and water pollution from other sources.

SUMMARY

A thriving and developing aquaculture industry is important for several reasons. Aquaculture supplies a quality, healthy food source to a growing human population and does so through the efficient use of resources. Aquaculture creates jobs and stimulates economic activity. It provides valuable nonfood items such as eel skins, alligator hides, and by-products from the processing of finfish and shellfish. The feed demand of aquaculture increases the demand for other agricultural products such as corn, soybeans, wheat, oats, and barley. Finally, aquaculture contributes to recreation by providing fish to stock lakes, steams, and ponds for sport and fee-fishing.

For U.S. aquaculture producers, the future markets will grow, but producers will be faced with increased environmental regulations, the need for new and better technology through research, and competition from foreign producers as aquaculture expands worldwide.

STUDY/REVIEW

Success in any career requires knowledge. Test your knowledge of this chapter by answering these questions or solving these problems.

True or False

1. Aquaculture is a form of agriculture.

2. Modern Americans were the first to practice aquaculture.

3. Trout were one of the first fish involved in polyculture.

4. Aquaculture helps fisheries by providing fish to restock streams, rivers, oceans, and lakes.

5. Aquaculture is a minor part of U.S. food production.

6. Aquaculture developed rapidly in the United States from the 1960s to the 1980s.

Short Answer

7. Scientific _____ of the problems in aquaculture helps the industry grow.

8. _____ connects the aquaculture producer with consumers.

9. The state of _____ leads in trout production while the state of _____ leads in catfish production.

10. _____ and _____ are examples of coldwater fish.

11. _____ _____ in water limits aquaculture.

12. List four stages of evolution for all agricultural activities.

13. What are the significant aquatic species cultured in the United States?

14. List five activities that are a part of aquaculture and often become separate industries.

Essay

15. Define aquaculture.

16. Explain how aquaculture may help maintain a traditional fishing industry and sport fishing.

17. Compare farming the water to farming the land.

18. Define freshwater, saltwater, warmwater, and coldwater.

19. Describe the importance of aquaculture to two ancient civilizations.

20. Why is the history of aquaculture in England and Europe important to American aquaculture?

21. How did Mississippi emerge as the leader in catfish production and Idaho emerge as the leader in trout production?

22. Why is the National Sea Grant Program important to U.S. aquaculture?

23. List and describe five areas that will determine the future of aquaculture.

KNOWLEDGE APPLIED

1. Visit local grocery stores and survey the fish and seafood sold in the store. How much is a freshwater product? How much is a saltwater (marine) product? How are the products sold—fresh, frozen, canned? Are any of the products produced locally?

2. The history and development of aquaculture can be used to teach geography. Obtain a world map and identify the locations discussed in the chapter. The history of agriculture may be used as a springboard to other history lessons.

3. Visit a local aquaculture production site. Have questions ready to ask about the operation. For example—

    How did the production facility get started?

    What problems have they encountered with environmental concerns, diseases, feeding, processing, and marketing?

    How are prices for the product established?

    What type of training is required to be successful?

    What legal regulations are involved?

    If a local aquaculture production facility is not available, arrange a teleconference using a speaker phone with one in another part of the state or a different state.

4. Using the list of Regional Aquaculture Centers listed in the Appendix Table A-11, write letters requesting information on one of the species listed in Table 1-3. When requesting information, make sure the information is requested from a center that is apt to represent the species. For example, request trout information from centers in the northwest or north and marine species information from centers near the coasts.

5. While carp farming never really became important in the United States, it has been very important in China and parts of Europe. Develop a series of reports on the culture of carp in either China or Europe. Include in the reports the development of carp culture, polyculture with other species, integration with other forms of agriculture, and any recipes for carp or ways of serving carp. Several books in the Learning/Teaching Aids section will be helpful.

LEARNING/TEACHING AIDS

Books

Bardach, J. E., Ryther, J. H., & McLarney, W. O. (1972). *Aquaculture: The farming and husbandry of freshwater and marine organisms.* New York: John Wiley & Sons.

Council for Agricultural Education. (1992). *Aquaculture.* Module I: Discovering the origins and opportunities in aquaculture. Alexandria, VA: Council for Agricultural Education.

Kirk, R. (1987). *A history of marine fish culture in Europe and North America.* Surrey, England: Fishing News Books, Ltd.

Michaels, V. K. (1988). *Carp farming.* Surrey, England: Fishing News Books, Ltd.

Walker, S. S. (1990). *Aquaculture*. Stillwater, OK: Mid-America Vocational Curriculum Consortium, Inc.

Wittwer, S., Youtai, Y., Hans, S., & Lianzheng, W. (1987). *Feeding a billion: Frontiers of Chinese agriculture*. East Lansing, MI: Michigan State University Press.

Articles

Behrendt, A. (1988, September/October). The debt we owe to the early experimenters. *Fish Farmer*, 40–41.

Brannon, E., & Klontz, G. (1989). The Idaho aquaculture industry. *Northwest Environmental Journal, 5,* 23–35.

Deakman, E. (1987). The D. C. Booth Historic Fish Hatchery. *Fisheries, 12,* (2), 23–24.

Harvey, D. J (1991, October–December). Aquaculture: A diverse industry poised for growth. *FoodReview,* 18–23.

Leffler, M. (1986, December). Bringing up oysters. *Oceans,* 38–43.

Videos

Aquaculture: Farming the waters. (Available from American Association for Vocational Instructional Materials (AAVIM), 745 Gaines School Road, Athens, GA 30602.)

Aquatic Plants and Animals

A quaculture includes the art, science, and business of cultivating plants and animals in water. U.S. aquaculture involves only a few successful species of fish and plants. This could change as aquaculture evolves. Future aquaculturists need an awareness of potential plants and animals and a basic knowledge of aquatic plants and animals.

Learning Objectives

After completing this chapter, the student should be able to:

In Aquaculture

- Name the major aquatic species in the United States
- Name five aquatic animals that hold potential for aquaculture in the United States
- Explain why aquatic crops may be more productive than terrestrial crops
- Briefly describe the general water and feeding characteristics of five aquatic animals
- List three aquatic plants that potentially could be cultured in the United States

■ List three other uses for aquatic plants besides human food
■ Give examples of aquatic animals and plants that could be used in polyculture

In Science

■ Recognize the scientific names for some common aquatic species
■ List and describe important biological characteristics in selecting a species for aquaculture
■ Explain how aquatic species save energy when compared to terrestrial species
■ List and describe the major characteristics of aquatic plants and animals
■ Discuss the morphology, anatomy, and physiology of common aquatic animals
■ Name and describe the nine body systems of aquatic animals
■ Identify and describe the internal and external anatomy of a fish
■ Identify and describe the basic structure and internal anatomy of crustaceans
■ Identify and describe the basic structure and internal anatomy of an oyster or mussel
■ Describe the basic morphology of aquatic plants

Understanding of this chapter will be enhanced if the following terms are known. Many are defined in the text and others are defined in the glossary.

KEY TERMS		
Adductor	Biomass	Diffusion
Anatomy	Bivalve	Digestive system
Antennae	Bloom	Dorsal
Anterior	Calcareous	Esophagus
Anus	Carnivores	Exoskeleton
Appendages	Chitin	Feedstuffs
Arteries	Chlorophyll	Fertilization
Asexually	Cilia	Food chain
Assimilation	Copepods	Fusiform
Barbels	Decapods	Gametes

KEY TERMS

Gastropods	Nerve fibers	Siphon
Gills	Omnivores	Spinal cord
Gonads	Ovaries	Spores
Herbivores	Phycocolloid	Stimuli
Hermaphroditic	Physiology	Swimmerets
Heterocercal	Phytoplankton	Terrestrial
Homocercal	Polysaccharide	Testes
Incubation	Posterior	Thorax
Inorganic	Primary producers	Urinary bladder
Kidneys	Protandrous	Urinary ducts
Lateral lines	Regeneration	Urinary opening
Macrophytes	Respiration	Uropod
Mandibles	Rotifers	Veins
Mantle	Scales	Ventral
Maxillae	Semipermeable	Zooplankton
Molting	Sensory receptors	Zygote
Morphology	Sinuses	

FIGURE 2-1 Aquaculture requires a good supply of clean water.

U.S. AQUATIC PLANT SPECIES

Aquatic plants are important components of aquaculture in other parts of the world, particularly in Asia. Europe and North America rank dead last in aquatic plant production worldwide. Some interest exists for cultivating aquatic plants for (1) production of food, feed, and chemical products; (2) wastewater treatment, and (3) biomass production for conversion to energy. Table 2-1 lists aquatic plants and their potential use.

TABLE 2-1 Aquatic Plants for Aquaculture

Common Name	Scientific Name	Water[1]	Uses	Notes
Spirulina	Spirulina spp.	F	Food	Protein content of some species 70 percent; collected and dried into patties for human consumption in some Asian countries and Mexico; nutritious supplement; distinct taste.
Brown algae or Kelp	Undaria pinnatifida macro Macrocystis pyrifera Macrocystis integrifolia	S	Food Mulch Fertilizer Phyococolloids	Called wakame in Japan; dried, chopped, and used in salads; brownish color comes from xanthophyll; giant kelp may grow to 200 feet (61.0 m).
Green algae	Monostroma macro Enteromorpha Chlorella	S,F	Food Mulch Fertilizer	Least cultured of three macroalgae; called aonori in Japan.
Red algae or Laver	Porphyra spp. Gelidium spp. Gracilaria spp.	S,B	Food Feed Mulch Fertilizer Phycocolloids	Cultured in Japan back to 1570; dried and high in protein; some harvested for livestock feed; United States leads in carrageenan production— a phycocolloid.
Duckweed	Lemna spp. Spirodela spp. Wolffia spp. Wolfiella spp.	F	Feed Waste water treatment	Favorite food of herbivorous fish and water fowl; harvested and used for livestock feed; one of least expensive to produce.

(continued)

[1] Freshwater (F), saltwater (S), brackish (B).

TABLE 2-1 Aquatic Plants for Aquaculture *(concluded)*

Common Name	Scientific Name	Water[1]	Uses	Notes
Water spinach	*Ipomoea reptans*	F	Feed Food	Commonly cultured in Thailand, Malaysia, and Singapore; often in polyculture; low protein and carbohydrate content.
Water hyacinth	*Eichhornia crassipes*	F	Waste water treatment Fuel source	Effectively removes waste from water and easy to harvest; possibly used for methane gas production.
Chinese waterchestnut	*Eleocharis dulcis*	F	Food	Small-scale production in the United States compared to Asia; corm consumed; each corm produces about 20 lbs (9.1 kg) of new corms in about 220 days; labor intensive; useful in polyculture.
Watercress	*Nasturtium officinale*	F	Food	Primary freshwater aquatic plant produced in the United States; requires abundant continuous flowing water; many people harvest wild crop.
Cattail	*Typha latifolia* *T. angustifolia*	F	Ornamental	Grown in aquatic gardens and used in dried flower arrangements; edible parts but not cultured for food.
Arrowhead	*Sagittaria sp.*	F	Ornamental	Grown in aquatic gardens; edible parts but not cultured for food.

[1] Freshwater (F), saltwater (S), brackish (B).

Of the chemicals or products obtained from aquatic plants, a phyco-colloid called carrageenan is most widespread. Carrageenan is used in foods for gelling, thickening, and stabilizing. It is a polysaccharide.

Phytoplankton. A list of potential aquatic plants for culture should not overshadow the important role of phytoplankton in aquaculture. They are the primary producers, forming the first link in the aquatic food chain. Through photosynthesis, phytoplankton use sunlight to produce food energy and contribute oxygen to the water. Phytoplankton serves as a food source for zooplankton and some fish and produces a bloom that helps shade out unwanted rooted aquatic plants. Pond fertilization encourages the production of phytoplankton.

U.S. AQUATIC ANIMAL SPECIES

Catfish and salmonids (trout and salmon) dominate U.S. aquaculture. Other species of freshwater finfish, marine finfish, mollusks, and crustaceans hold promise for the future of U.S. aquaculture. Tables 2-2 and 2-3 categorize the important points of U.S. aquaculture species.

TABLE 2-2 Finfish for Aquaculture

Common Name	Scientific Name	Water Temp.[1]	Water Type[2]	Diet[3]	Notes
Atlantic Salmon	Salmo salar	C	A	C	Important as rod catalyst, sport fish, and commercial netting; fishing regulated by national, international, and local laws.
Bighead Carp	Aristichthys nobilis	W	F	C	Excellent food animal; suited for polyculture; acceptance increasing in the United States.
Black Bullhead	Ictalurus melas	W	F	O	Susceptible to disease; tolerant of adverse water conditions; demand low.
Blue Catfish	Ictalurus furcatus	W	F	C	Some culture work; silvery white to light blue color.
Brook Trout	Salvelinus fontinalis	C	F	C	Used in hybrid crosses with Lake Trout—splake.
Brown Trout	Salmo trutta	C	F (A)	C	Naturalized populations on every continent except Antarctica.
Buffalofish (Largemouth)	Ictiobus cyprinellus	W	F	C	Technology for spawning and rearing available; possible polyculture species.
Channel Catfish	Ictalurus punctatus	W	F	O	Principle farm-raised species in the United States; oxygen depletion major problem.

(continued)

[1] Warmwater (W) temperature of 70° to 90°F (21.1° to 32.2°C) or coldwater (C) temperature of 65°F (18.3°C) and under.
[2] Freshwater (F), saltwater (S), brackish water (B), or anadromous (A).
[3] Herbivorous (H), carnivorous (C), or omnivorous (O).

TABLE 2-2 Finfish for Aquaculture *(continued)*

Common Name	Scientific Name	Water Temp.[1]	Water Type[2]	Diet[3]	Notes
Chinook Salmon (King)	Oncorhynchus tshawytscha	C	A (F)	C	Coastal species; researched and cultured in New Zealand; may live in fresh water.
Chum Salmon	Oncorhynchus keta	C	A	C	Most cold tolerant of Pacific salmon; widest distribution; hatchery techniques developed in Japan.
Coho Salmon	Oncorhynchus kisutch	C	A (F)	C	Grow rapidly second year when feeding on other fish; introduced into Great Lakes to feed on alewife, smelts, and sea lampreys.
Common Carp	Cyprinus carpio	W	F	O	Deep yellow body; member of minnow family.
Crappie	Pomoxis spp.	W	F	C	Member of sunfish family, *centrachidae*; spawn readily.
Cutthroat Trout	Salmo clarki	C	F	C	Possible to propagate artificially; hybrid potential.
Fathead Minnow	Pimephales promelas	W	F	O	Baitfish; short-lived; seldom reach 3 inches (7.6 cm) or 3 years.
Flathead Catfish	Pylodictis olivaris	W	F	C	Predator species; not economical to raise on large scale.
Golden Shiner	Notemigonus crysoleucas	W	F	C	Baitfish; large member of minnow family; grows to over 8 inches (20.3 cm).
Goldfish	Carassius auratus	W	F	H	Baitfish; very hardy; used as feeder fish or forage fish.
Grass Carp	Ctenopharyn- godon idella	W	F	H	Slim carp; feeds on aquatic plants but accepts pelleted feed when cultured; cultured in Asia.

(continued)

[1] Warmwater (W) temperature of 70° to 90°F (21.1° to 32.2°C) or coldwater (C) temperature of 65°F (18.3°C) and under.
[2] Freshwater (F), saltwater (S), brackish water (B), or anadromous (A).
[3] Herbivorous (H), carnivorous (C), or omnivorous (O).

TABLE 2-2 Finfish for Aquaculture *(continued)*

Common Name	Scientific Name	Water Temp.[1]	Water Type[2]	Diet[3]	Notes
Lake Trout	*Salvelinus namaychus*	C	F	C	Used in hybrid crosses with Brook Trout—Splake.
Largemouth Bass	*Micropterus salmoides*	W-C	F	C	Large bass eat small ones; spawn in gravel nest; jaw extends beyond eye.
Milkfish	*Chanos chanos*	W	S-B	H	Very disease resistant; popular in tropical Pacific; will not spawn in captivity.
Mullet, Striped	*Mugil cephalus*	W-C	F-B-S	H	Commonly cultured; tropical and semi-tropical; possible polyculture.
Muskellunge	*Esox masquinongy*	C	F	C	Some cannibalism; prefer temperatures warmer than trout but cooler than catfish.
Northern Pike	*Esox lucius*	C	F	C	Wild stock usually captured for egg-taking; requires forage fish.
Pink Salmon	*Oncorhynchus gorbuscha*	C	A	C	Attempts to extend range not very successful; ranched in Alaska.
Pompano	*Trachinotus carolinus*	W	S	C	Naturally not very abundant; commercial production expensive.
Rainbow Trout	*Oncorhynchus mykiss*	C	F	C	Tolerant to relatively high water temperatures and low oxygen levels; fast growth.
Red Drum	*Sciaenops ocellata*	W	S-B	O	Popular in Cajun-style restaurants; popular sport fish; some successful culture.
Smallmouth Bass	*Micropterus dolomieui*	W	F	C	Special equipment and techniques to collect fry.

(continued)

[1] Warmwater (W) temperature of 70° to 90°F (21.1° to 32.2°C) or coldwater (C) temperature of 65°F (18.3°C) and under.
[2] Freshwater (F), saltwater (S), brackish water (B), or anadromous (A).
[3] Herbivorous (H), carnivorous (C), or omnivorous (O).

TABLE 2-2 Finfish for Aquaculture *(concluded)*

Common Name	Scientific Name	Water Temp.[1]	Water Type[2]	Diet[3]	Notes
Sockeye Salmon	*Oncorhynchus nerka*	C	A (F)	C	Landlocked form called kohanec; crustaceans diet pigments flesh red.
Steelhead	*Oncorhynchus mykiss*	C	A	C	Anadromous form of Rainbow Trout.
Striped Bass, Hybrid	*Morone saxatilis x Morone chrysops*	W	F	C	Cross of female striped bass and male white bass; approved for aquaculture late 1970s.
Sturgeon	*Acinpenseridae spp.*	C	F	O	Cultured to increase numbers; some culture for roe.
Sunfish (Green, Bluegill, Redear)	*Lepomis spp.*	W	F	C	Spawn readily; hybridize easily; female drab.
Tilapia	*Tilapia spp.*	W	F	H	Controlling reproduction a major problem to culture; feed on algae, detritus, and waste feed.
Walleye	*Stizostedion vitreum vitreum*	W-C	F	C	Wild stock captured for egg-taking; requires long, slender forage fish.
White Catfish	*Ictalurus catus*	W	F	C	Determined inferior to channel catfish for aquaculture; hardy; stocked for fee-fishing ponds.
White Sucker	*Catostomus commersoni*	C	F	C	Forage fish; adapt to formulated feed as a supplemental diet.
Yellow Perch	*Perca flavescens*	C	F	C	Famous in the Midwest; cultured in Holland; some culture trials in the United States.

[1] Warmwater (W) temperature of 70° to 90°F (21.1° to 32.2°C) or coldwater (C) temperature of 65°F (18.3°C) and under.
[2] Freshwater (F), saltwater (S), brackish water (B), or anadromous (A).
[3] Herbivorous (H), carnivorous (C), or omnivorous (O).

TABLE 2-3 Mollusks and Crustaceans for Aquaculture

Common Name	Scientific Name	Water Temp.[1]	Notes
Abalone, red	Haliotis rugescens	S	The only gastropod (snail) of significance cultured in the United States; largest hatchery in California; prolific spawners
Clams Hard clam Soft clam	Mercenaria mercenaris (hard clam)	S	More culture of hard clam; not widely cultured around the world; the United States has most advanced culture; two to seven years to market size depending on location
Crabs Blue crab	Callinectes Spidus	S	Primarily a fisheries product; aquaculture techniques produce soft-shelled crabs
Crawfish (crayfish)	Procambarus clarkii P. blandingi acutus	F	About 300 species in the United States; harvested from wild and cultured; found on every continent except Africa and Antarctica; six to fourteen months to reach market size
Mussels	Mytilus edulis	S	New to U.S. culture; easy to raise; grow faster than other shell fish
Lobster	Homarus americanus	S	Farming from egg to market size not profitable; minimum of five years to reach market size
Prawns (Malaysian prawn)	Macrobrachium rosenbergii	F-B	High demand; started in Hawaii
Oysters (American oyster)	Crassostrea virginica	S	Culture over 100 years old in the United States; larvae swim free then attach to something for rest of life
Shrimp	Penaeus spp.	S	Widely cultured in Asia but new to the United States; great demand for shrimp

[1] Freshwater (F), saltwater (S), brackish water (B), or anadromous (A).

Hobby Fish

Hobby, tropical, and aquarium fish represent several families and over 100 species of small, colorful, and unique fish. These fish occur naturally in tropical, semitropical freshwater, saltwater, or brackish water. The major hobby fish industry is located in central Florida, but hobby fish are raised in most of the other states. Water temperature management is a prime concern since hobby fish are sensitive to cool temperatures. Culturalists specialize in the production of colorful varieties that are easy to propagate. Some common hobby fish include sailfin mollies, guppies, clown barbs, black tetras, angelfish, and blue gouramies.

Bullfrogs

In the United States, most bullfrogs (*Rana catesbeiana*) for consumption come from the wild. Demand for food frogs and live frogs for biological research is greater than the supply and the availability is seasonal. All of this makes the possibility of commercial production appealing. Unfortunately, frog culture is very complex because of the complicated life cycle and demanding feeding habits of the bullfrog. The Japanese and Taiwanese practice open pond culture of bullfrogs from eggs to adults.

Alligators

Alligators (*Alligator Mississippienis*) are large aquatic reptiles valued for their meat and hide. They were once abundant in the lower South before overhunting and habitat destruction reduced their numbers. Extensive conservation efforts restored alligators where the habitat permitted and led to the development of alligator culture techniques. Presently, alligators are commercially cultured in Texas, Georgia, South Carolina, Louisiana, and Florida. Some producers in Idaho with access to warmwater wells are even considering raising alligators. The demand for alligator meat and hide keeps prices high and production profitable. Strict regulations govern intra- and interstate commerce in alligators and alligator products.

Eels

Eels are considered a gourmet food in Japan, Taiwan, and most European countries. The commercial production of food-sized eels for export captured the interest of some U.S. aquaculturists. The life cycle of eels is complicated. They spawn at sea, and seed stock must be captured from the wild when the elvers—small eels—migrate upstream from the sea. Captured eels are raised in ponds and need to be trained to eat artificial feed. Eel culture is risky business without a stable supply of elvers, and there are few markets.

Zooplankton. Discussing the potential of aquatic animals tends to overshadow the minute animals important to aquaculture. Zooplankton, primarily copepods (very small crustaceans) and rotifers serve as a vital food source for all fish fry, and they feed on the phytoplankton. They are primary consumers in the food chain, as Figure 2-2 indicates.

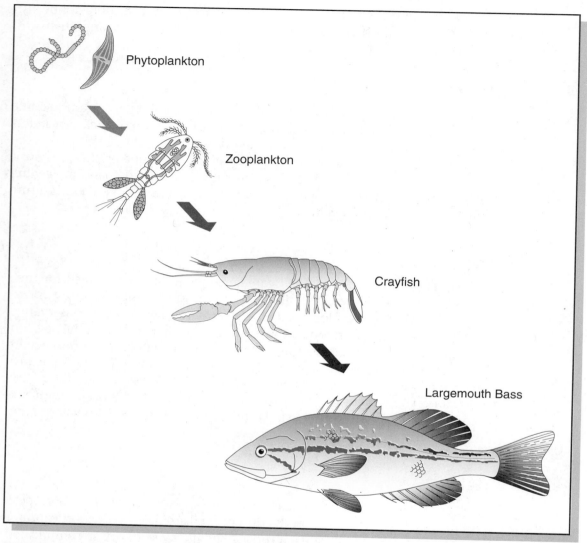

FIGURE 2-2 An example of the food chain. Zooplankton feed on the plankton. Crawfish eat the zooplankton and a bass eats the crawfish.

COMMON CHARACTERISTICS OF AQUATIC SPECIES

Aquatic plants and animals hold a greater productive potential than terrestrial plants and animals. Reasons for this include—

■ Body temperature about same as environment (ectothermic)

■ Body density similar to habitat

■ Reduced energy required for getting food

■ Efficient feed conversion

■ Rapid growth

■ Live in multidimensional environment.

Since the body temperature of aquatic animals is near that of their environment, energy normally required to regulate body temperatures can be directed toward growth. Since their body density is near that of their habitat, energy normally used to overcome gravity can be used for growth.

In land animals, the search for food requires energy. In aquatic species, the search can be minimized. For example, filter feeders like clams filter surrounding water through their bodies for particles of suspended food.

Compared to livestock such as beef cattle and hogs, some aquatic species efficiently convert feed to growth. For example, beef cattle and hogs require 4 to 8 lbs (1.8 to 3.6 kg) of feed for 1 lb of gain. Catfish and trout produce 1 lb of gain from 1.5 to 2 lbs (0.7 to 0.7 kg) of feed. The less feed used, the more profit made.

Some aquatic organisms known as primary producers can grow rapidly. Some species of algae represent the best example of this, growing at a rate of almost 10 percent per day. Figure 2-3 shows some examples of algae.

Different species inhabit various space and positions within the aquatic environment. This expands the aquaculture options at one site. Polyculture with different species of carp is a good example. Fish, crustaceans, and mollusks all occupy different space. Structures such as floating cages, pens attached to the bottom extending above the surface, and strings on poles extending into the water create a dimensional variety (see Figure 2-4).

Choosing a species for aquaculture is similar to choosing any crop or livestock for culture. Successful culture means considering—

■ Reproductive habits

■ Egg and larvae requirements

■ Nutritional needs and feeding habits

■ Polyculture possibilities

■ Adaptability to crowding

■ Disease resistance

■ Market demand.

The ability to reproduce easily is a primary requirement. For successful culture, a stable supply of seed (young) must be available. Also, the reproductive processes of the species must be understood and genetic selection and improvement must be possible. Reproduction should produce massive quantities and occur frequently. Eggs and larvae need to be large, hardy, and easy to

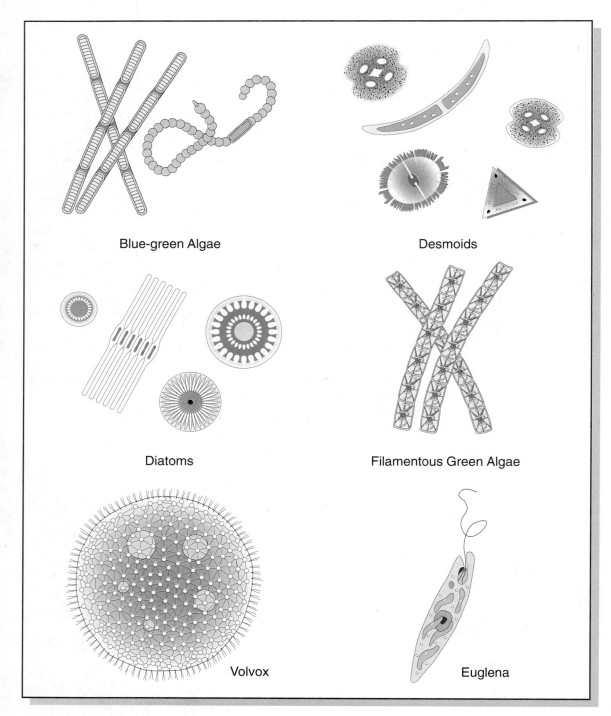

FIGURE 2-3a Types of algae.

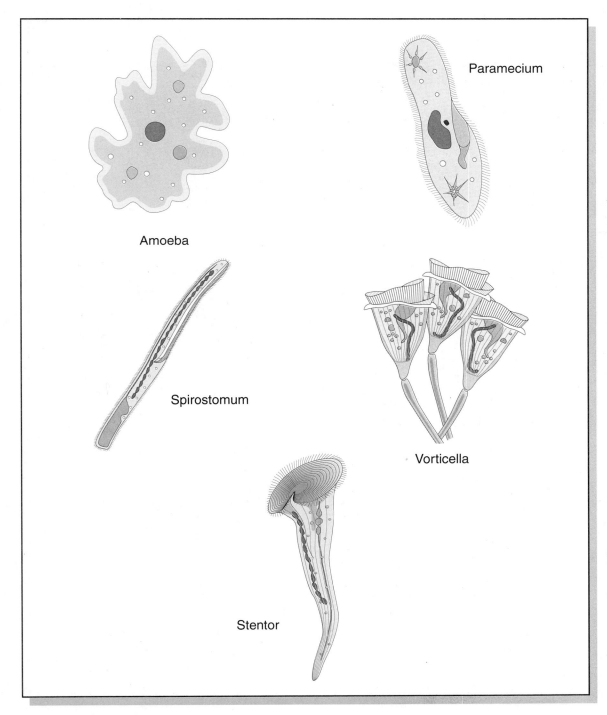

FIGURE 2-3b Pond plankton.

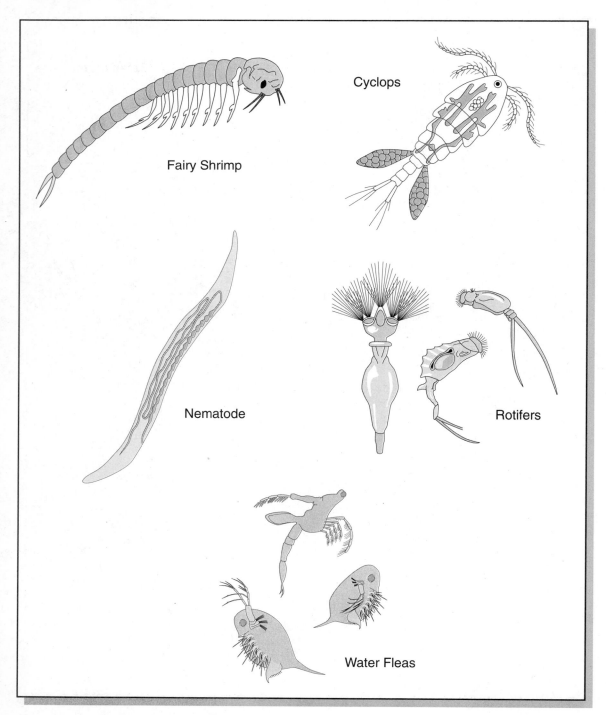

FIGURE 2-3c Pond plankton metazoa.

culture. Table 2-4 indicates the wide variation in finfish and their ability to produce eggs.

One female oyster may produce 500 million eggs per year. In the crustaceans, crayfish produce 100 to 500 eggs and shrimp produce 500,000 to 1 million eggs.

After reproductive ability, the next factor to consider when selecting an aqua species to culture is nutritional needs and feeding habits. For feeding habits, a species can be selected low on the food chain or high on the food chain. An aquaculture species low on the food chain consumes low-cost vegetable matter or by indirectly consuming primary foods within the pond. Examples of species low on the food chain include carp, tilapia and crayfish. Species high on the food chain include shrimp, trout, and bass. These species require a more expensive high-protein diet.

Aquatic animals, like terrestrial animals, require protein, carbohydrates, fat, vitamins, and minerals. Unlike terrestrial animals, some of the nutritional needs are met directly from the aquatic environment. Research on the optimal

TABLE 2-4 Spawning Frequency and Egg Production in Various Finfish[1]

Species	Spawning Frequency	Eggs Per Pound[2] of Fish
Chinook salmon	Once per life span	350
Coho salmon	Once per life span	400
Sockeye salmon	Once per life span	500
Atlantic salmon	Annual-Biennial	800
Trout	Annual	1,000–1,200
Northern pike	Annual	9,100
Walleye	Annual	25,000
Striped bass	Annual	100,000
Channel catfish	Annual	3,750
Largemouth bass	Annual	13,000
Smallmouth bass	Annual	8,000
Bluegill	Intermittent	50,000
Golden shiner	Intermittent	75,000
Goldfish	Intermittent	50,000
Common carp	Intermittent	60,000

[1] Source: *Fish Hatchery Management.*
[2] 1 lb = 0.45 kg

amounts and forms for each species continues. The more completely nutritional needs are understood, the more efficiently other aquatic animals can be produced. Complete information about nutritional needs is covered in Chapter 5, Fundamentals of Nutrition in Aquaculture.

Selecting polyculture as a criterion for determining which species to produce depends on the type of production system. In an intensive culture system, such as trout or catfish production, growth rate could be more a concern than the efficiency in use of water space and nutrients. Polyculture increases the total aquatic production in a volume of water by using species that occupy different dimensions of the water and feed on different **feedstuffs**.

Aquaculture crowds species that are not used to crowding. Crowding increases productivity of a space while increasing management for the space. Aquatic species selected for culture exhibit adaptability to withstand crowding.

Species vary widely in their ability to resist disease. Aquaculturalists select species for disease resistance based on the conditions at their production site.

Production of an aqua crop can be successful and efficient, but without a market, production efforts are wasted. A market for a product consists of—

- Desire by consumers
- Price consumers can afford
- Prepared, easy-to-use forms of the product
- Storage to reach consumer
- Desired flavor.

Chapter 3, Marketing Aquaculture, contains more complete information about marketing.

FIGURE 2-4 Structures such as cages create dimensional variety. (Photo courtesy Chuck Weirich, Delta Research and Extension Center, Stoneville, MS)

STRUCTURE AND FUNCTIONS OF AQUATIC ANIMALS AND PLANTS

A study of aquaculture requires some information of the structure and form or morphology and anatomy of aquatic animals and plants, and the function of aquatic animals and plants or their physiology. The suitability of an organism for culture depends on its morphology, anatomy, and physiology.

Animal Surfaces

Any discussion of the structure and function of animals begins with an understanding of dorsal, ventral, anterior, and posterior. Dorsal pertains to the upper surface of an animal. Ventral relates to the lower or abdominal surface. Anterior applies to the front or head of an animal. Posterior pertains to the tail or rear of an animal. These are easy to understand in many species, but in species like clams and oysters, these positions can be a little confusing. Figure 2-5 shows the dorsal, ventral, posterior, and anterior of fish, crayfish, and clams.

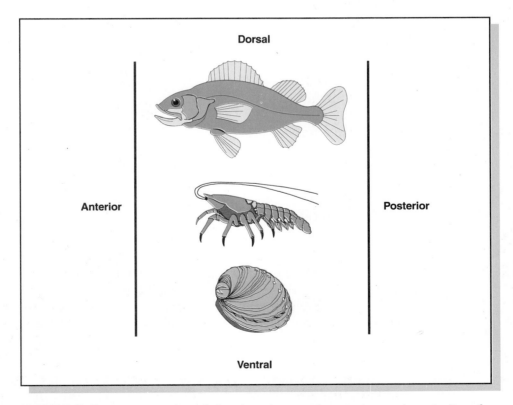

FIGURE 2-5 A perspective of the dorsal, ventral, anterior, and posterior of a fish, crayfish, and clam.

Morphology

Morphology or structure and form of fish can affect feeding and type of culture facility. For example, fish with small, upturned mouths generally are herbivores and/or surface feeders like tilapia. Fish with downturned mouths are generally bottom feeders like catfish.

Single-lobed or homocercal tail fins suggest that fish are slow swimmers and survive well in water free of much movement. Fish with forked or heterocercal tail fins are fast swimmers and prefer flowing water.

The body shape of fish also suggests the type of culture facility. Fish like salmon, with a long body tapered toward the ends (fusiform) are the best swimmers and need water space. Fish that are wide and flat or ventrally compressed tend to stay on the bottom and require lots of bottom space for growth. Laterally compressed fish are rounded and thin from side to side. These fish tend to hover in the water and are not particularly fast swimmers.

Physiology

Physiology is the function of the body systems of aquatic animals. These systems in aquatic species are adapted to the water environment. Nine body systems are found in animals, including aquatic animals. These systems are—

1. Skeletal
2. Muscular
3. Digestive
4. Excretory
5. Respiratory
6. Circulatory
7. Nervous
8. Sensory
9. Reproductive.

Skeletal System. The skeletal system is the rigid framework giving the body shape and protecting the organs. It is composed of bony or hard material and cartilage. Tissues and organs attach to the skeleton. In aquatic animals the skeleton can be internal or external. Fish possess an internal skeleton or endoskeleton. Oysters, shrimp, and crawfish possess an external skeleton or exoskeleton.

Muscular System. The muscular system provides movement internally and externally. Muscles vary in strength and function. Muscles contract and relax to cause movement. Organisms require movement for such functions as obtaining food and oxygen and eliminating wastes.

Digestive System. The digestive system converts feed into a form that can be used by the body for maintenance, growth, and reproduction. It consists of all the parts of an organism involved in taking food into the body and preparing it for assimilation, incorporation into the body. In its simplest form, the digestive system is a tube extending from the mouth to the anus with associated organs. In most species this includes the mouth, esophagus, stomach, intestines, anus, and other associated organs like the liver. Digestive systems vary according to whether the animals are herbivores eating only plants, carnivores eating only animals, or omnivores eating plants and animals.

Excretory System. Life processes produce waste products. The excretory system eliminates wastes from the body. Typically it consists of the kidneys, urinary ducts, urinary bladder, and urinary opening. Kidneys filter the wastes from the blood. The urinary bladder holds the wastes until they are excreted through the urinary opening.

Respiratory System. The respiratory system takes in oxygen from the environment, delivers it to the tissues and cells of the body, and it picks up carbon dioxide from the tissues and cells delivering it to the environment. Gills are the respiratory organs of fish, shellfish, and crustaceans. Water taken in is forced over the gills where oxygen is removed by diffusion into the blood.

Circulatory System. The circulatory system distributes blood throughout the body. Generally this system consists of a heart, veins, and arteries. Pumping action of the heart causes blood to flow through the arteries to the gills, where it picks up oxygen and carries it to the rest of the body. Oxygen is necessary for all cells of the body. As the blood delivers oxygen to the cells of the body, it picks up carbon dioxide, a waste product, which is carried in the blood back through the veins to the heart and gills. The gills release the carbon dioxide to the environment and pick up more oxygen.

Nervous System. The nervous system supplies the body with information about its internal and external environment. This system conveys sensation impulses—electrical-chemical changes—between the brain or spinal cord and other parts of the body. It consists of the brain, spinal cord, many nerve fibers, and sensory receptors. It is a complex system. The sense organs or receptors receive stimuli and convey these by the nerve fibers to the brain or spinal cord where they are interpreted. The brain or spinal cord may send responses to the stimuli back through the nerve fibers.

Sensory System. The sensory system includes the five senses—sight, touch, taste, smell, and hearing. The sensory system relays information through the nervous system. Fish use eyes to find food and identify predators. Ear bones in the skull pick up water vibrations as sound. The sense of taste is important to the aquafarmer when selecting and preparing feed for fish. Some species have an enhanced sense of touch through organs like the barbels on catfish. Lateral lines in fish contain nerves that detect water vibrations and motion. This helps keep fish in schools.

Reproductive System. Sexual reproduction is the process of creating new organisms of the same species through the union of the male and female sex cells—sperm and eggs. Males and females exist in most species. Testes in the males produce sperm. Ovaries in the females produce eggs or ova. Fertilization occurs when the sperm unites with the egg forming a zygote. After a period of incubation the zygote develops into a new organism. An understanding of the reproductive process is important to the success of the culture of a species. Some aquatic species reproduce asexually.

Anatomy

An understanding of the anatomy, the internal and external structure, of aquatic animals is essential for the successful aquaculturalist. External anatomy aids in distinguishing between the sexes and spotting abnormalities caused by disease.

Anatomy of Finfish. Most all fish used in aquaculture are considered bony fish with hard calcium-based endoskeletons. The skeleton gives the fish form and protects the internal organs, such as the digestive system, nervous system, and reproductive system. Figure 2-6 illustrates the external anatomy of a typical finfish.

Exterior coverings of fish vary. Bony plates or scales cover the skin of many fish such as trout and carp. Scales grow as the fish grows. A few species, such as the catfish, have skin without scales.

Figure 2-7 shows the typical internal anatomy of a finfish and the location of major organs in the body system. Depending on the species, organs vary in size and shape.

Digestive systems of fish vary depending on the type of food eaten. Fish consuming algae and detrital matter have a small stomach and long intestines. Carnivorous fish possess large stomachs and short intestines.

The nervous system of fish is well-developed with a brain and spinal column. The lateral line plays an important part as a sensory organ. It helps the fish maintain balance and position in the water.

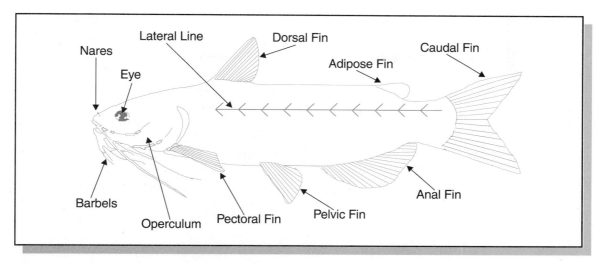

FIGURE 2-6 External anatomy of a finfish.

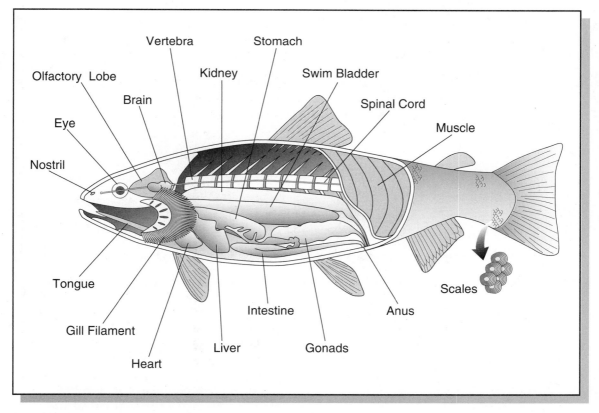

FIGURE 2-7 Basic internal anatomy of a finfish.

Males and females are fairly easy to distinguish in fish. The reproductive organs are located in the body cavity, but reproduction involves the fertilization of eggs laid in the water.

Gills remove oxygen from the water. Transfer of oxygen from the water to blood occurs by diffusion in the cells of the gills. Deoxygenated blood from the body pumped to the gills picks up oxygen and releases carbon dioxide. The higher concentration of oxygen in the water causes the oxygen to enter the blood, moving to an area of lower concentration (diffusion). Carbon dioxide in the blood is higher than in the surrounding water. Thus, it diffuses out of the blood through the gills. Cell membranes of gill cells are very thin and semipermeable, allowing gases to pass through. Figure 2-8 shows two views of fish gills.

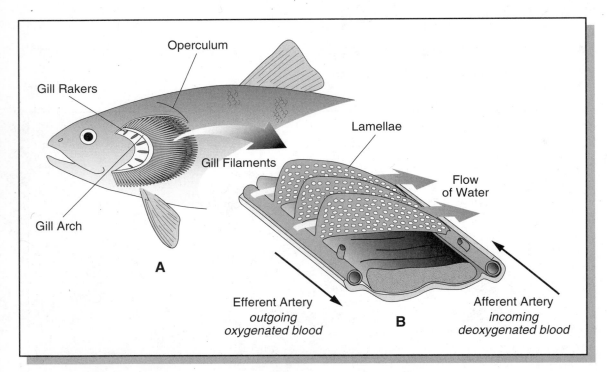

FIGURE 2-8 Gills are the lungs for fish and shellfish. Gills of fish are located on each side of the head (view A). They are covered by a protective movable flap of skin called the operculum. Four gills on both sides of the head each have a double row of slender gill filaments supported by a flexible white gill arch. Each side of the filament has many small cross plates called lamellae (view B). It is across the gill lamellae that the oxygen and carbon dioxide gases are exchanged. The lamellae have spaces through which blood rapidly percolates. Oxygen that is picked up at the gill lamellar surface is carried throughout the body in the blood. Waste carbon dioxide is also carried in the blood for release into the water at the lamellar surface.

Anatomy of Crustaceans. Crustaceans include shrimp, prawns, lobsters, crabs, and crayfish (crawfish). They all possess an exoskeleton made of chitinous material. **Chitin** is a polysaccharide—a carbohydrate—of a hexose (sugar) and also contains some tightly bound noncarbohydrate material, including proteins and **inorganic** salts. Most of the crustaceans considered for aquaculture are known as **decapods** (ten legs). The exoskeleton protects and supports the soft body, since all the muscles are attached to the inside of the exoskeleton. As a crustacean grows, the shell is cast off in a process called **molting**. When crustaceans molt they are known as softshell animals. No more than a day is usually required to regrow the shell. During molting crustaceans are subject to attack by other aquatic animals, including their own species.

Figure 2-9 shows the external anatomy of the crawfish, a representative crustacean. The bodies of crustaceans are divided into three sections—

1. Head
2. **Thorax** (carapace)
3. Abdomen.

Each segment has a pair of **appendages**. The head has two pairs of **antennae**. In crayfish, next to the antennae are the **mandibles** or true jaws and then two pairs of **maxillae** or little jaws that aid in chewing food. The jaws work from side to side, not up and down. The first appendage of the thorax are three pairs of maxillipeds or jaw feet. These hold food during chewing. Next come the large claws, obviously for protection and food getting. The last four

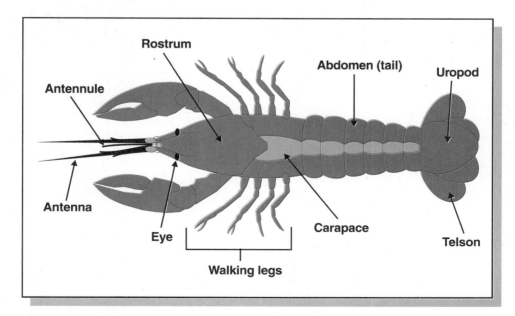

FIGURE 2-9 External anatomy of a crawfish, dorsal view.

pairs of legs on the thorax are two pairs with tiny pincers at the tip and two more pairs with claws. The abdominal appendages of the crayfish are called **swimmerets** and are small on the first five segments. In the female during reproduction, the eggs attach to her swimmerets. The sixth swimmeret develops into a flipper or **uropod** for locomotion.

Through a process known as **regeneration**, crustaceans regrow limbs that have broken off. This process is used to produce crab legs. One claw is removed and the crab is returned to the water to grow another.

Figure 2-10 illustrates the internal anatomy of a crawfish. Crustaceans possess simple circulatory, nervous, and excretory systems. In the crayfish, the colorless blood is pumped by a very simple heart into several large arteries,

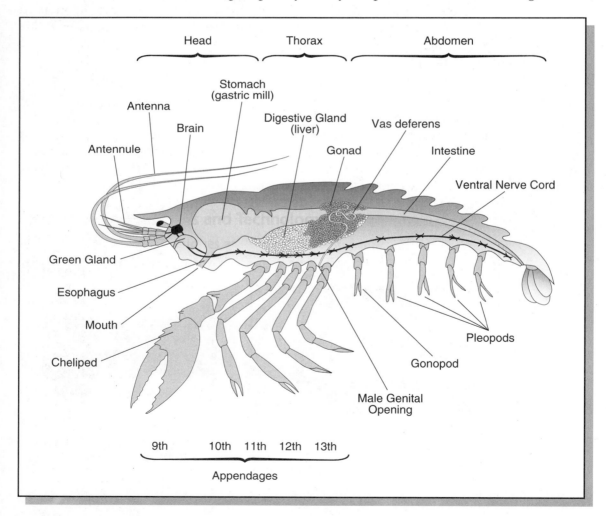

FIGURE 2-10 Basic internal anatomy of a crayfish.

which pour the blood over the major organs. Then the blood collects in spaces called **sinuses** and eventually returns to the heart. The nervous system of the crayfish consists of a brain and ventral nerve cord. The acute senses of smell and touch are located in the antennae, maxillae, and maxillipeds. The compound eyes are on movable stalks and sight is probably not keen. Hearing is poorly developed, but ear sacs located at the base of the antennules probably aid in balance.

Crustaceans use gills to breathe. In the crayfish, gills are located at the base of the legs or maxillipeds and protected by the thorax or carapace. Thus the gills are exposed to water every time the legs or maxillipeds move.

Crustacean life cycle and reproduction is quite complex. Testes of males and ovaries of females are located inside the exoskeleton. A duct from the testes or ovaries leads to the outside for the release of sperm or eggs. Using one of the appendages, the male deposits sperm into a receptacle on the abdomen of the female.

Members of the pandalids group of shrimp all begin life as males. At about two years of age, they change to females.

Anatomy of Mollusks. In the United States, the most commonly cultured mollusks species include oysters and clams. These are **bivalve** mollusks. Two shells completely enclose the animal. These shells are made of a **calcareous** material that is very hard and resembles limestone. An anterior and posterior **adductor** muscles hold the shells together.

A muscular, hatchet-shaped foot can extend from between the shells. In the clam this is used for digging. The **mantle** lays over the internal organs and secretes the hard shell. The digestive system and nervous system are simple. A simple heart pumps the colorless blood to all parts of the body. Figure 2-11 shows the basic internal anatomy of a clam.

The gills not only serve as a respiratory system, but they filter material from the water that is consumed or discharged. Small particles of matter stick to a thin mucous layer on the gills. The surface of the gills have **cilia**, small hair-like structures, which continually beat, carrying trapped material to the mouth. Water enters the mollusk through a **siphon** and passes over the gills. Then water exits the mollusk through another siphon passing the anus where undigested matter is excreted.

Mollusks reproduce with eggs and sperms. Typically, eggs are released into the water and fertilized by water-borne sperm. Some bivalves are **protandrous**, meaning they may change their sex one or more times during their lives. Some mollusks, such as scallops, are **hermaphroditic**, meaning individual organisms have **gonads** (testes and ovaries) for both sexes.

Univalves or gastropods include the snails, conches, and abalones. They have the same general anatomy as the bivalves but only one shell.

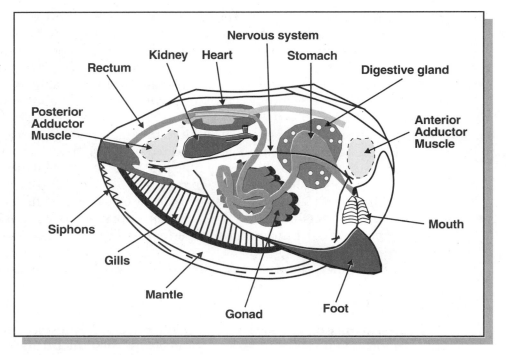

FIGURE 2-11 Basic internal anatomy of a clam.

Aquatic Plants

Aquatic plants share many characteristics of terrestrial (land) plants but are unique in other ways. Like land plants, aquatic plants make their own food through the process of photosynthesis. Photosynthesis requires light and **chlorophyll** to convert carbon dioxide and water to sugar (carbohydrates), oxygen, and water. Chlorophyll allows photosynthesis. For plants to use the sun energy stored in the carbohydrates, they respire. This process uses carbohydrates and oxygen to produce carbon dioxide, water, and energy. Plants use this energy for growth and reproduction. Figure 2-12 shows the complete chemical formulas for photosynthesis and **respiration**.

Microscopic algae such as diatoms, desmoids, blue-green algae, euglena, volvox, and filamentous green algae represent the smallest of the aquatic plants. Giant kelp reaching 200 ft (61.0 m) or more in length represent the largest of the aquatic plants.

Aquatic plants reproduce sexually—the fusion of sex cells—or asexually. Many algae reproduce asexually by forming **spores**. Some algae produce spores that produce **gametes** or sex cells that fuse. Asexual spores called monospores produced by young plants become new plants. All forms of algae reproduce by the simple splitting of a cell into two. Some aquatic plants produce seeds, for

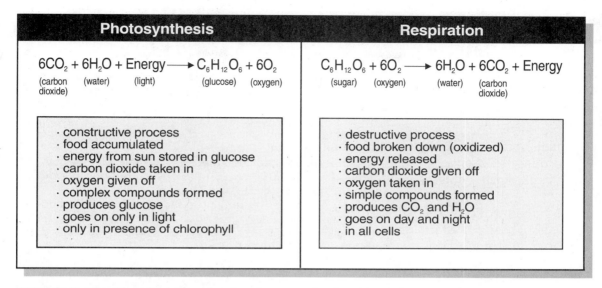

Photosynthesis	Respiration
$6CO_2 + 6H_2O + \text{Energy} \longrightarrow C_6H_{12}O_6 + 6O_2$ (carbon dioxide) (water) (light) (glucose) (oxygen)	$C_6H_{12}O_6 + 6O_2 \longrightarrow 6H_2O + 6CO_2 + \text{Energy}$ (sugar) (oxygen) (water) (carbon dioxide)
· constructive process · food accumulated · energy from sun stored in glucose · carbon dioxide taken in · oxygen given off · complex compounds formed · produces glucose · goes on only in light · only in presence of chlorophyll	· destructive process · food broken down (oxidized) · energy released · carbon dioxide given off · oxygen taken in · simple compounds formed · produces CO_2 and H_2O · goes on day and night · in all cells

FIGURE 2-12 Chemical formulas for photosynthesis and respiration.

example watercress. Like potatoes, the fleshy corms of Chinese waterchestnuts planted below the ground produce more corms. Finally, like terrestrial plants, some aquatic plants can be propagated by cuttings.

Because aquatic plants obtain most of their nutrients from the water, they are very useful for removing ammonia and nitrite wastes from the water. This makes aquatic plants an ideal component for polyculture since the wastes produced by the aquatic animals provide sufficient nutrients to the plants.

Like aquatic animals, aquatic plants differ in structure and occupy different positions in their environment. The two main groups of aquatic plants based

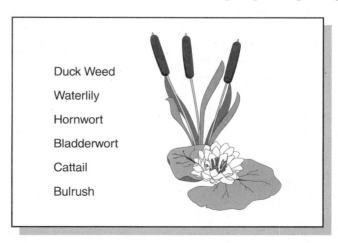

Duck Weed

Waterlily

Hornwort

Bladderwort

Cattail

Bulrush

FIGURE 2-13 Common aquatic plants.

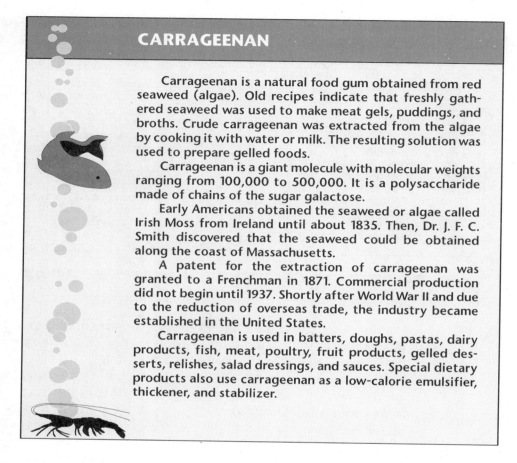

CARRAGEENAN

Carrageenan is a natural food gum obtained from red seaweed (algae). Old recipes indicate that freshly gathered seaweed was used to make meat gels, puddings, and broths. Crude carrageenan was extracted from the algae by cooking it with water or milk. The resulting solution was used to prepare gelled foods.

Carrageenan is a giant molecule with molecular weights ranging from 100,000 to 500,000. It is a polysaccharide made of chains of the sugar galactose.

Early Americans obtained the seaweed or algae called Irish Moss from Ireland until about 1835. Then, Dr. J. F. C. Smith discovered that the seaweed could be obtained along the coast of Massachusetts.

A patent for the extraction of carrageenan was granted to a Frenchman in 1871. Commercial production did not begin until 1937. Shortly after World War II and due to the reduction of overseas trade, the industry became established in the United States.

Carrageenan is used in batters, doughs, pastas, dairy products, fish, meat, poultry, fruit products, gelled desserts, relishes, salad dressings, and sauces. Special dietary products also use carrageenan as a low-calorie emulsifier, thickener, and stabilizer.

on structure are algae and macrophytes. Algae are primitive plants without true roots, stems, or leaves. Macrophytes are vascular plants with true roots, stems, and leaves. Planktonic algae occupy the space between the pond bottom and the surface. Filamentous algae form floating mats or hairlike strands attached to underwater objects, often called moss or pond scum. Macrophytic algae resemble true plants that are large and attached to the pond bottom.

Free-floating macrophytes are tiny green plants floating on the surface that resemble algae, but they have small leaves and roots that hang down into the water, for example duckweed or watermeal. Emergent macrophytes root in the pond bottom but the leaves float or extend above the water surface, for example water lilies or lotus. Submergent macrophytes such as pondweed or hornwort are rooted in the bottom and grow completely underwater. Marginal macrophytes grow in very shallow water or wet soil on the edge of a pond or ditch, for example cattails and bulrushes. Figure 2-13 illustrates some common aquatic plants.

SUMMARY

Other countries rely more heavily on aquaculture for food. Aquaculture is gaining prominence in the United States. Catfish, trout, baitfish, and crawfish dominate U.S. aquaculture. Other aquatic species are being tried and some hold great promise for aquaculture after more research and development of the technology required for successful culture. Potential aquatic species for culture include a long list of finfish, crustaceans, mollusks (shellfish), and some aquatic plants.

For culture, aquatic species possess some advantages over land species. Selecting and successfully culturing an aquatic species requires an understanding of their biology, morphology, anatomy, and physiology. Successfully meeting the demands of a growing human population for food, feed, and energy means finding the best aquatic animals and plants for culture.

STUDY/REVIEW

Success in any career requires knowledge. Test your knowledge of this chapter by answering these questions or solving these problems.

True or False

1. North America ranks first in aquatic plant production worldwide.

2. Carrageenan is an important chemical obtained from aquatic plants.

3. It is against the law to culture alligators commercially.

4. More feed is required to produce one pound of catfish than to produce one pound of beef.

5. Aquatic animals require large amounts of energy to regulate their body temperature.

6. Primary producers grow rapidly and form the first link in the food chain.

Short Answer

7. _____ is a group of minute plants and _____ is a group of minute animals important as food sources.

8. As a group, colorful, small fish used in aquariums are called _____.

9. After spawning, small eels called _____ migrate upstream from the sea.

10. Successful culture of any aquatic animal requires a stable supply of _____ .

11. Name three things that make up a market.

12. Give the scientific and common names for the two freshwater finfish and one crustacean that currently dominate U.S. aquaculture.

13. Name two aquatic plants used for food, two used for waste water treatment, two used for food feed, and two used for phycocolloid production.

14. List five salmonids.

15. List three catfish besides channel catfish.

16. List six saltwater or brackish water aquatic animals that could be or are being cultured.

17. Name two aquatic animals and two aquatic plants that could be used in polyculture.

18. With a pond of 200 female channel catfish weighing an average of 4 lbs (1.8 kg) each, how many total eggs could a producer expect when each female spawned?

19. Name the nine body systems found in aquatic animals.

20. List three different methods of reproduction in aquatic species.

21. For which aquatic species are molting and regeneration important processes?

22. Give two examples of algae and four examples of macrophytes.

23. Name two plants without true roots, stems, and leaves and two plants with true roots, stems, and leaves.

Essay

24. Explain the difference between phytoplankton and zooplankton.

25. Explain the differences between mollusks and crustaceans.

26. Why do eels, bullfrogs, and alligators hold potential as aquaculture species?

27. Using specific plants and animals, describe a polyculture system.

28. In what ways do aquatic animals save energy compared to terrestrial animals?

29. When selecting a species for possible aquaculture, identify six factors to consider.

30. What is the function of the skeleton? Name two types of skeletons.

31. Give an example of how morphology can affect the culture techniques chosen.

32. Briefly describe how gills work.

33. Briefly define the following external anatomical features: adipose fin, lateral line, operculum, uropod, thorax, maxillipeds, and foot.

34. Compare the location of the gills in finfish, crustaceans, and mollusks.

35. Why is a basic understanding of internal and external anatomy important to the future aquaculturalist?

KNOWLEDGE APPLIED

1. Obtain some common pond water and using a microscope and a picture guide such as *Field Book of Ponds and Streams* by Ann Haven Morgan, identify the aquatic life found in a pond. Collect the water in a clean, one quart jar. To increase the number of plankton in the sample, pass a plankton net or nylon stocking through the pond and put the contents into the sample. Prepare microscope slides using a drop of pond water and a drop of methyl cellulose to slow the plankton movement. Add a cover slip and examine the sample at different levels of magnification.

2. As an alternative to using pictures to identify microorganisms in a pond water sample, use the dichotomous key in the Appendix Table A-14. Place the sample of water under a microscope as described in question 1. Once the microorganism to be identified is in view, read the first two descriptions in the key. Decide which one better describes the microorganism. Then, go to the description number indicated at the right of the best description. Continue the process until the name of the microorganism occurs to the right of the description. Draw the microorganisms observed.

3. Choose an aquatic species from Table 2-2 or Table 2-3. Learn the common and scientific names. Develop a more complete profile of the species selected. Include such items as description, natural food, habitat, distribution, behavior, reproduction, larvae, adult size, edible qualities, culture

possibilities, yield in culture, feeding, predators, diseases, harvest, and marketing. If a whole class did this for a number of species, the information could be entered into a computerized database so all species could be compared on a variety of characteristics. For many species, more information can be found in Chapter 4, Aquatic Management Practices.

4. Investigate the external and internal anatomy of a finfish. Obtain some fresh, frozen, or preserved finfish. Identify all the external structures as well as the dorsal, ventral, anterior, and posterior areas. Lift the operculum and identify the gill arch, gill rakers, and gills. Dissect the fish and identify parts of digestive system, reproductive system, and circulatory system. Open the skull and find the brain. Identify any sensory organs located internally or externally. Several books listed in the Learning/Teaching Aids will help with the anatomy of finfish.

5. Obtain fresh, frozen, or preserved crayfish. Identify the following external features: abdomen, carapace, head, swimmerets, maxilliped, and uropod. Dissect the crayfish and identify the internal structures including the gonads, hind gut, gills, abdomen muscle, and any other parts of the body systems.

6. Obtain fresh, frozen, or preserved clams or oysters. Before opening the shells, identify the dorsal, ventral, anterior, and posterior surfaces. Pry open the shell and cut the adductor muscle. Identify the mantle, gonad, stomach, gills, mouth, food, intestine, kidney, heart, and siphons.

7. Conduct a survey of food labels. Find out how frequently and in what foods a phycocolloid, namely carrageenan, is used.

8. Conduct a taste test of some aquatic plants used for food, like brown algae (wakame), water spinach, watercress, and Chinese waterchestnut. Investigate how these plants are used in recipes by different cultures.

LEARNING/TEACHING AIDS

Books

Bouc, K. (1987). *The fish book.* Lincoln, NE: Nebraska Game and Parks Commission.

Chakroff, M. (1982). *Freshwater fish pond culture and management.* Washington, DC: Peace Corps Information Collection and Exchange Manual and Volunteers in Technical Assistance, VITA Publications.

Council for Agricultural Education. (1992). *Aquaculture.* Module II: Discovering plants and animals in aquaculture. Alexandria, VA: Council for Agricultural Education.

Dupree, H. K., & Huner, J. V. (1984). *Third report to the fish farmers: The status of warmwater fish farming and progress in fish farming research.* Washington, DC: U.S. Fish and Wildlife Service.

Klontz, G. W. (1991). *Fish for the future: Concepts and methods of intensive aquaculture.* Moscow, ID: Idaho Forest, Wildlife and Range Experiment Station College of Forestry, Wildlife and Range Sciences, University of Idaho.

Lee, J. S., & Newman, M. E. (1992). *Aquaculture: An introduction.* Danville, IL: Interstate Publishers, Inc.

Piper, R. G., McElwain, I. B., Orme, L. E., McCraren, J. P., Fowler, L. G., & Leonard, J. R. (1982). *Fish hatchery management.* Washington, DC: U.S. Department of the Interior, Fish and Wildlife Service.

Walker, S. S. (1990). *Aquaculture.* Stillwater, OK: Mid-America Vocational Curriculum Consortium, Inc.

Videos

Louisiana Cooperative Extension Service. (N.D.). *Crawfish: A culinary crustacean.* Baton Rouge, LA.

Marketing Aquaculture

M arketing is the process of getting the product from the producer to the consumer. It is the final step in food production but should rate top priority in the mind of an aquaculturalist. While the aquaculturalist may possess the skills and resources to grow a crop, those efforts are in vain without a place to sell the product.

The first step in marketing is to understand the current production and consumption of a product. Next, producers need to understand marketing—its functions and a strategy. Some of the specific details of marketing and processing for each species are covered in Chapter 4, Aquatic Management Practices.

Learning Objectives

After completing this chapter, the student should be able to:

In Aquaculture

- Define marketing
- Describe the process of marketing aquaculture
- Explain the elements in developing a marketing strategy
- Explain the importance of developing a marketing plan
- Identify possible market outlets for aquaculture products

■ Select an appropriate market
■ Explain costs in marketing
■ Describe the process of market promotion in aquaculture
■ Identify terms related to marketing with their correct definitions
■ Discuss quality control

In Science

■ Describe some scientific skills required to maintain the quality fish and fish products
■ Recognize that development of a marketing plan and strategy requires research
■ Describe processing
■ Describe the grading process
■ List factors to consider when exploring marketing alternatives
■ Identify food fish processing cuts and forms with their correct descriptions

Understanding of this chapter will be enhanced if the following terms are known. Many are defined in the text and others are defined in the glossary.

KEY TERMS		
Advertising	Grading	Product push
Assembling	Inputs	Promotion
Branded	Inspections	Quality assurance or
Control	Live-haulers	control
Deheading	Marketing	Shelf life
Demand	Off-flavor	Stunned
Distributors	Offal	Supplier
Eviscerator	Processors	Trademarks
Fillets	Product pull	Value-added

INTERNATIONAL PRODUCTION

Fifteen countries or areas produce about 90 percent of all aquaculture products. As Figure 3-1 shows, the United States rates among the top ten. Production in China, Japan, India, and Korea overshadow all other countries as illustrated by Figure 3-2.

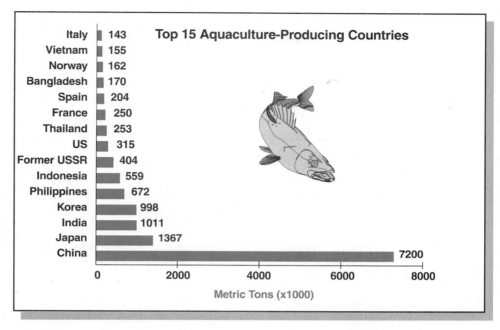

FIGURE 3-1 Top 15 aquaculture-producing countries based on tonnage, including aquatic plants.

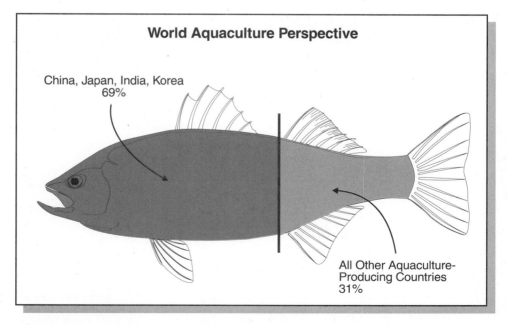

FIGURE 3-2 More than two-thirds of the world's aquaculture is produced in China, Japan, India, and Korea.

Table 3-1 indicates the type of species cultured in these countries.

Table 3-2 shows by continent aquaculture production of major groups. Asia produces more finfish, crustaceans, mollusks, and aquatic plants than any other continent.

This chapter is about marketing aquaculture products produced in the United States in the U.S. market. Trout and catfish represent two success stories in marketing. Other species in the United States represent success stories for niche marketing.

TABLE 3-1 Aquaculture Species Cultured Worldwide[1]

Country or Area	Species or Species Group Cultured
China	Carps, mollusks, shrimp
Japan	Amberjack, mollusks, algae
Taiwan PC	Eel, mollusks, shrimp
Philippines	Milkfish, shrimp, algae
USA	Catfish, pacific salmon, trout, mollusks
USSR[2]	Carps
Norway	Atlantic salmon, rainbow trout
Ecuador	Whiteleg shrimp
Indonesia	Carps, milkfish, shrimp, algae
Korea	Mollusks, algae
France	Rainbow trout, mollusks
Vietnam	Freshwater fish, crustaceans
India	Freshwater fish, crustaceans
Spain	Rainbow trout, mollusks
Thailand	Crustaceans, mollusks
Bangladesh	Freshwater fish
Italy	Rainbow trout, mollusks
Scotland	Atlantic salmon

[1] FAO, (1990) *Aquaculture Minutes*, Number 8, Inland Water Resources and Aquaculture Service.
[2] Represents all the countries that made up the former USSR.

TABLE 3-2 Metric Tons of Aquaculture Products by Continent[1]

Continent (regions)	Finfish	Crustaceans	Mollusks	Misc.	Aquatic Plants	% of Total
Asia	7,019,000	574,700	2,155,300	44,000	3,143,500	84.4
Europe	593,700	3,000	627,200	2	0	8.0
N. America	261,200	47,000	135,000	20	14	2.9
USSR[2]	398,700	0	159	0	5,400	2.6
S. America	49,400	89,200	4,100	0	38,000	1.2
Africa	82,300	375	2,946	80	250	0.6
Oceania	6,400	1,300	40,300	156	550	0.3
Totals	8,410,700	715,600	2,965,000	43,700	3,187,700	100.0

[1] FAO (1990) *Aquaculture Minutes*, Number 8, Inland Water Resources and Aquaculture Service.
[2] Represents all the countries that made up the former USSR.

CONSUMPTION

Marketing requires understanding the competition. Other meats compete with fish. Slowly, gradually, fish and seafood consumption in the United States has increased, as Figure 3-3 illustrates.

Catfish, salmon, trout, and crawfish make up 90 percent of U.S. production. The other 10 percent includes striped bass, eel, alligators, and tilapia.

While the per person consumption of fish and its increase looks small compared to other meats, a small increase in per person consumption translates to big increases in production. For example, the per person consumption of catfish increased from 0.08 lb in 1966 to about 0.7 lb in 1993. Assuming that the U.S. population is about 260 million, this translates to about 364 million lbs of live fish (165,000 metric tons). Figure 3-4 shows the historical trend of the per person consumption of fish only. This illustrates the gradual increase.

MARKETING BASICS

Efficient aquaculture production matters little if the crop cannot be sold for a profit. Where and how an aquacrop will be sold should be the first concern of a producer. This means developing a marketing strategy or plan.

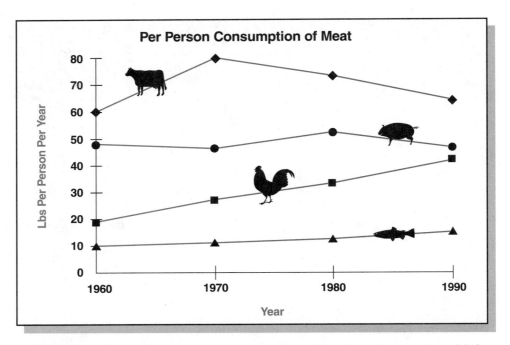

FIGURE 3-3 Trends in per person consumption of beef, poultry, pork, and fish.

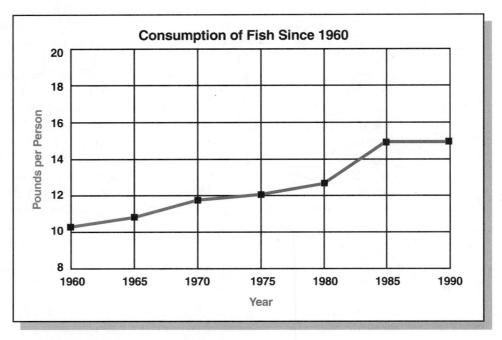

FIGURE 3-4 History of the per person consumption of fish only.

Marketing Plan

Depending on the operation, marketing plans can be long documents or informal plans of several pages. Development of a plan allows the producer to analyze the opportunities and needs. With this done, the producer can focus on production and make better decisions. Market plans or strategies contain three key elements:

1. Determination of the present situation
2. Determination of market goals
3. Developed plans to reach the goals.

Often, separate marketing plans are developed for new products and the continuing or annual plan.

Targeting the Buyer

New product introduction is a risky business. Successful new products can make a company and keep it competitive in its industry, while providing steady outlets for its input **suppliers**, such as farmers. Failure rates are high, so it is very important for suppliers, manufacturers, and **distributors** to understand the forces that affect new product success.

A new product's path from development to market acceptance depends on the type of buyer targeted. There are two basic buyer types: intermediate users, such as **processors** and manufacturers, and final consumers. Marketing channels and the prerequisites of success vary depending on which is targeted.

Manufacturers—intermediate users—use new crops and new products from existing crops and foodstuffs as intermediate **inputs** in producing final goods. This market is made up of professional buyers who base purchasing decisions on strict price/quality specifications and who are highly knowledgeable about the availability of substitute inputs. Convincing them to buy a new product requires being responsive to their price, quality, quantity, and delivery needs.

Where the targeted buyer is the final consumer, the selling environment differs. While consumers make the same type of price/quality comparisons as professional buyers do, they usually have less complete information, and factors such as brand name, advertising, packaging, coupons, convenience, and image play a bigger role. These elements make communication a crucial factor in successful new product introduction.

The consumer market also differs because often the producers are not in direct selling contact with the buyers, as they are in the intermediate goods market. The retail distribution chain links the two so that the producer is faced with a double selling job—to convince the retailer to carry the product and the consumer to buy it. For foodstuffs, this can become a triple selling job.

In the two general areas—intermediate buyers or consumers—aquaculturalist sell or market their product to—

■ Processing plants
■ Live haulers
■ Local stores and restaurants
■ Backyard or pond bank sales
■ Fee-fishing.

A market should be selected based on the potential profits according to the scale of the operation. Each option should be carefully analyzed.

Intermediate Goods. Three key elements that largely shape the environment in which intermediate goods compete are characteristics of—

1. The product
2. The buyer
3. The marketing system.

Product characteristics. Intermediate agricultural goods are sold as inputs for further processing and distribution. Buyers need reliable information on the product's technical and functional characteristics. The producer must be able to demonstrate how the new product performs in its intended application. Buying decisions hinge on whether the new product contributes to the buyer's profit. Price is clearly an important consideration.

If the new product does not offer a price advantage relative to alternative inputs, then it must offer some performance edge in the manufacturing process.

Selling an intermediate good may also require the ability to customize the product to the buyer's specifications. Adapting to a particular buyer's needs may entail physical changes in the product, packaging changes, or changes in delivery methods.

Buyer characteristics. Intermediate buyers differ substantially from household buyers. Buyers of intermediate goods are well informed about prices and product characteristics. While consumers of final goods may be willing to buy a new product on impulse, an intermediate buyer purchasing a vital input needs to know much more about a product before committing to a new supplier.

Compared to most household consumers, intermediate buyers face higher switching costs and risks in trying new products. Switching costs are one-time costs of changing to a new supplier. Such costs and risks for a consumer trying a new food or fiber product generally are small—an improperly cooked meal or an unenjoyable dining experience may be the only result of an unsuccessful experiment. For an intermediate buyer, switching costs and risks may be large.

In many cases, the seller of a new intermediate good needs to work with buyers to develop new product formulas. To provide assurance to buyers, the supplier may have to assume some of the financial risk that accompanies the switch to a new product.

Prices in producer goods markets adjust frequently to changing market conditions. Sellers of a new product must be prepared to negotiate prices with the buyer rather than simply offering a take-it-or-leave-it price.

Marketing system characteristics. Distribution networks play an important role in determining a product's success. For agricultural goods, shipping costs frequently are high relative to the product's value. Many agricultural products are bulky and costly to move long distances. Markets are likely to be local or regional in scope, presenting opportunities for entry by smaller scale operations—a niche market.

In addition to geographic market concentration, both buyer and seller concentration are important strategic considerations for the developer of a new product. Seller concentration helps determine the potential response by competitors to a new product's introduction. If concentration is high, with just a few large sellers, then existing sellers may compete vigorously to minimize sales lost to a new entrant. If the selling industry is competitively structured with many smaller producers, then a new product may be able to enter the market with little response from existing firms.

Product to the Consumer. Winning acceptance for new products is of key importance to food manufacturers because introducing such products

FIGURE 3-5 Many large fish processors have their own trucks for distribution.

ADVERTISING

Advertising is any paid form of nonpersonal presentation of goods. It is a part of marketing and as old as recorded civilization.

Ancient Greeks and Romans used advertising. The walls of ancient Pompeii and Herculaneum contain notices painted in black and red. One painted on a wall in Pompeii tells travelers about a tavern in another town. Another advertisement excavated in Rome offers property for rent. In the Middle Ages merchants hired criers to walk the streets and cry the wares for their clients. Later, town criers became familiar figures on the streets of colonial America.

The first printed advertisement in English appeared in 1648. In America in 1704 *The Boston News Letter* contained the first newspaper advertisement. By the middle of the 20th century advertising appeared in newspapers, magazines, direct mail, on billboards, cars, matchbook covers, radio, and television. Anywhere else an ad agency can think to place a piece of advertising, it will—T-shirts, public transportation, lighters, walls, and in movies. It is big business.

American companies spend about $38 billion a year on advertising. This money is spent for advertisements mainly in magazines, in newspapers, on television, and on radio. Only a little more than 9 percent of this advertising money goes to advertise food. If candy, snacks, soft drinks, beer, and wine are considered part of the food advertising budget, then about 15 percent of the advertising money goes to food and food-like products.

The top five business categories for spending advertising money are—

1. Automotive
2. Retail
3. Business/consumer services
4. Food
5. Entertainment.

Each year the top ten advertisers, in terms of dollars spent, usually include Procter & Gamble, Philip Morris, General Motors, Sears Roebuck, PepsiCo, Grand Metropolitan, Johnson & Johnson, McDonald's, Ford, and Eastman Kodak.

Ad agencies know what they are doing. They employ a research staff, creative layout artists, copywriters, scriptwriters, graphic artists, and sales people. Often they employ a recognizable face or voice to deliver their message. What they do works and advertising is an important part of American business.

is a centerpiece of their marketing strategies. The process has two stages: creating consumer demand (product pull) and encouraging distributors to give the product shelf space (product push).

Creating product pull. The ultimate success of a new product depends on generating strong consumer demand. Manufacturers of branded products seek to develop offerings with the price, quality, and convenience consumers will want, using advertising and coupons to make consumers aware of them. Advertising is a particularly important strategy for gaining new product acceptance because it plays a dual role. It builds consumer demand and signals to retailers and other manufacturers that the company is committed to spending the resources necessary to support the product in the early stages. Large advertising expenditures are routinely involved in introducing new branded products. This type of support builds the demand pull necessary to establish a new grocery store product.

Providing product push. Successful introductions in the retail channel also require the manufacturer to provide push for the new product. Push refers to incentives offered to wholesalers and retailers to carry the product. Some are offered across the board, such as special introductory prices and free goods, while others are negotiated individually. Wholesalers and retailers frequently use buyers and buying committees to evaluate whether a new product is

FIGURE 3-6 Product push helps create space on the supermarket shelf for fish and seafood products.

unique enough and has sufficient manufacturer support to merit shelf space. Recent research indicates that most products do not make it past this stage. This is not surprising, since about 90 percent of new products are extensions, for example, new flavors, of existing lines.

Push is necessary because retailers face restrictions in accommodating new products. Although average store size has increased, product numbers far outpace available shelf space, giving retailers a strategic advantage in choosing products to carry.

Additionally, new product introductions generate costs such as establishing warehouse slots, resetting retail shelves and changing store computer files. Product failures also generate costs. Given these strategic and cost factors, wholesalers and retailers have increasingly demanded more trade support (push) dollars for new products. One form this takes is charging slotting (and sometimes failure) fees to manufacturers.

Figure 3-7 summarizes how aquaculture products can be marketed.

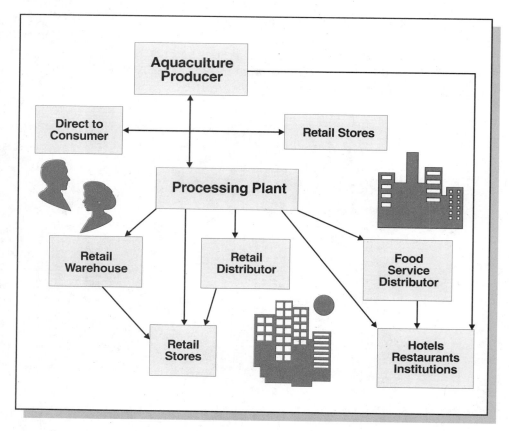

FIGURE 3-7 Marketing flowchart in aquaculture.

Marketing Activities

Marketing of aquaculture products involves some characteristic activities. These include **assembling**, **grading**, transporting, changing ownership, processing, packaging, storing, wholesaling, retailing, and advertising. Not all marketing involves all of these activities.

Assembling. Assembling is collecting aquaculture crops from different production sites at a central location so the volume to be processed will be large enough for efficient use of the processing facilities.

Grading. Grading is ensuring the aquaculture crop batch is of uniform size and species. A grader may be used to screen out animals that are too large or too small. Some of this is done at harvest.

Transporting. Transporting means moving the aquaculture product to a location where it is to be processed. Most animals should be kept alive and in good condition until the time of processing. Specialized haul tanks with aerators and oxygen injection systems may be needed.

Changing Ownership. Most crops are sold several times between the farm and the consumer. Initially, the producer sells fish to the processor based on the weight of the fish at the time of delivery. The change of ownership involves the seller and buyer agreeing on the amount sold and on a price.

Processing. This involves a number of procedures to prepare fish for consumption. With fish, processing typically involves removing the skin and viscera, cutting into portions, preseasoning or cooking, and properly disposing of waste products from the processing.

Packaging. Consumers want to buy products that are packaged attractively and easy to use. Packaging should also keep the food safe and wholesome. Package labels describe the product and how it is to be prepared.

Storing. Aquaculture products are stored several times between the farm and the consumer. Tanks are used at the processing plant to keep the fish alive until processing. Refrigeration and freezing are used with many fish and shellfish to preserve and store them. Canned products may be stored in large warehouses and at supermarkets.

Wholesaling. The processor sells the product to distributors (jobbers) or retail outlets. A price level is established so the processor can make a profit.

Retailing. Selling to the consumer, restaurants, supermarkets, and fish markets may be involved. Attractive merchandising is needed.

FIGURE 3-8 Attractive packaging and brand names assure the consumer of quality.

Advertising. Consumers need to be aware of aquaculture products. Advertising develops awareness and encourages consumers to buy the product. Newspapers, radio and television, signs, and other means of advertising may be used. Grower associations, processors, and local stores may sponsor the advertisements. Advertising is also known as product **promotion**.

FIGURE 3-9 Depending on location and size of market, any promotion is helpful.

Selecting a Market

The producer must select a market. Markets vary according to the species produced, the location, and the amount returned to the producer. Some factors to consider when selecting a market for aquaculture products include profit, equipment, accessibility, species, quantity, size, and quality.

Profitability. Select the market that provides the greatest return on investment to the producer. The highest price per pound may not provide the largest profit if expenses are involved. Producers need to keep records that allow them to calculate the cost per pound to obtain or produce the product.

Specialized Equipment. Some marketing approaches may require expensive equipment for additional processing or packaging. This is especially true when individuals direct market some specialty or value-added products.

Accessibility. Good markets may exist but may not be readily accessible to the aquafarmer. Costs to deliver products to the market may be more than the increased price for the product. Many areas where aquaculture flourishes or can flourish are some distance from large markets for the product.

Species. Market channels tend to vary by species. Some species have a fairly well developed marketing system while others do not. Some species tap niche or ethnic markets depending on the species and location. For example, crawfish are available in the southern United States but hard to buy in the northwest.

Quantity. Large quantities can be marketed through processing plants. Small quantities are more suited to direct marketing. Processing plants may not be interested in small quantities.

Size or Maturity. Aquatic animals that are immature (small) or oversize may require special efforts to market. Processing plants may not accept them and, if they do, may penalize the price paid per pound.

Quality. Most processors insist on an animal that is healthy, free of injury, and has the right flavor. Animals that do not meet these standards should not be marketed. Processors and producers establish in-house quality assurance programs. These programs assure that a standard routine is followed. The required limits of incoming raw materials and finished food products are continuously monitored. Compliance with compositional standards of identity for various products is guaranteed and government regulations are met.

FIGURE 3-10 Cases of catfish stacked on pallets and wrapped in plastic. Large quantities of raw product are necessary to keep large processors operating. (Photo courtesy Chuck Weirich, Delta Research Center, Stoneville, MS)

FIGURE 3-11 One producer looking for a market for trout broodstock decided to process and can old broodstock.

Marketing Costs

Like production, marketing has some costs associated with it. These costs add to the retail price that must be charged to the consumer. These costs can be—

- Transportation
- Grading
- Harvesting
- Packaging
- Storing
- Advertising.

Promotion

Promotion entices the buyer to purchase the product. A major part of promotion is advertising. Various groups in aquaculture promote the products.

- Growers form associations to promote the consumption of the crop produced.
- Check-off systems of fees are assessed to growers or other individuals involved with aquaculture. Almost everyone is familiar with the national radio and television advertisements for beef and pork. Check-off money from the sale of cattle and pigs funded these ad campaigns.
- Processors join together to promote the consumption of their products. The promotion is not brand-specific.
- Federal, state, and local government agencies may develop promotions for certain products or for an industry in general.
- Individuals promote their products or service. For example, individuals promote a fee-fishing operation, a restaurant serving fish, or some value-added product such as smoked trout.

Promotion educates the consumer, and it is achieved on all scales—big and small—in the United States. Anyone living in the United States knows that promotions of all types of products surround us.

Aquaculture producers can promote or sell their product to four general markets—processors, fee-fishing, wholesale and retail, and live-haulers.

Processors

Large processors generally harvest fish for producers within a short radius of the processing plant—50 to 75 mi (80 to 121 km). Some accept fish delivered live by the producer. Fish producers within range of large processing plants arrange harvest or delivery dates before fingerlings are stocked. When producers want

FIGURE 3-12 Signs are a form of individual promotion. This sign leaves no doubt about the product served at the restaurant.

to sell their fish the same time of year, this creates an oversupply of fish for the processors. Fish harvested when supplies are low usually command a higher price. Some producers are able to market their fish more profitably during times of short supplies by manipulating the fingerling size and the stocking date and by partial harvesting.

Small-scale processors in some areas process small quantities of catfish for sale to local businesses and individuals. These processors often produce much of their own fish but, at times, buy from local producers. Some build their own small-scale processing plants.

Fee-Fishing

Fish-out, or fee-fishing, is another market option for many producers. A fish-out business depends on the numbers of fishers in the area and their ability to catch fish. Fishing ponds located near cities are usually more in demand than in remote areas.

Small, densely-stocked ponds are best for fish-out purposes. Fish should be replenished when stocks become low so that the fish will keep biting. Many successful fee-fishing operations buy fish from other producers or produce them in their own ponds to stock fish-out ponds. This results in better fishing success, more customers, and more sales.

Wholesale and Retail

Wholesale and retail sales of live fish are other ways for producers to sell their product. Fish can be captured to order or captured and held live for later sale. Local newspaper ads, road signs, and word-of-mouth establish a good market. Providing a consistent supply of high-quality fish throughout the year maintains consumer demand.

Live-Haulers

Live-haulers, people who buy and haul live fish from producers to retail outlets, are important buyers of farm-raised fish. Usually these haulers want producers to harvest and load the fish into their tank trucks. Live-haulers often transport fish to fish-out ponds or other live markets near large cities such as Chicago or Atlanta. A producer selling to live-haulers exclusively often needs all the necessary equipment for seining and loading, plus all-weather access to the production facility.

PROCESSING

Processing fish through several steps turns it into a salable product. The following steps are typical for catfish processing, but the steps are similar for trout and other finfish.

- Receiving and weighing the live fish at the processing plant
- Holding them alive until needed
- Stunning
- Deheading
- Eviscerating
- Skinning
- Chilling
- Size grading
- Freezing or ice packing
- Packaging
- Warehousing
- Icing
- Shipping the finished product.

Receiving

Before being purchased for processing, catfish are evaluated for flavor quality by experienced tasters. Sample fish are taken from the pond at least three times—

1. Normally two weeks before harvest
2. The day before harvest
3. The day of harvest.

At the pond, fish are loaded into aerated water tanks and transported to the processing facility. The fish are unloaded from the truck into baskets (Figure 3-13) for weighing and then put into an aerated holding vat (Figure 3-14) or directly into the plant. In most cases, fish enter the processing line directly from the trucks and are only held in tanks to keep the plant in operation when fish delivery is delayed.

Deheading

The fish are removed from holding tanks and **stunned** with electrical current, which makes them easier and safer to handle by workers. The fish are moved into the processing plant on a distribution conveyor belt (Figure 3-15). From the distribution conveyor, the stunned catfish drop into a holding bin for each processing line.

FIGURE 3-13 Basket for moving fish from the truck to the aerated holding vat or into the processing plant.

FIGURE 3-14 Aerated holding vat at a catfish processing plant.

FIGURE 3-15 Fish traveling into the processing plant on a conveyor after being stunned. (Photo courtesy Chuck Weirich, Delta Research Center, Stoneville, MS)

The first line operator is the lay-up person. This operator positions each catfish in the proper orientation for the band saw operator to remove the head quickly and efficiently (Figure 3-16). The head is pushed into a chute that routes it to a waste disposal conveyor belt below the band saw, and the carcass proceeds to the evisceration operation. Band saw operators process from 40 to 50 fish per minute.

Automatic **deheading** and eviscerating machines are also being used by most processing plants. Automation requires some size grading of fish for efficient use.

Evisceration

The body cavity is opened by hand with a knife, and viscera are withdrawn by use of a vacuum **eviscerator** (Figure 3-17). Viscera are conveyed to the **offal** collector, and the eviscerated carcass proceeds to the skinning operation.

Skinning

The membrane skinner has been the standard industry machine for skinning channel catfish since its introduction. A rotating roller with sharp "teeth" present the fish to a sharp blade held in place by spring pressure. Very close tolerances between the blade and roller teeth make it possible to remove only

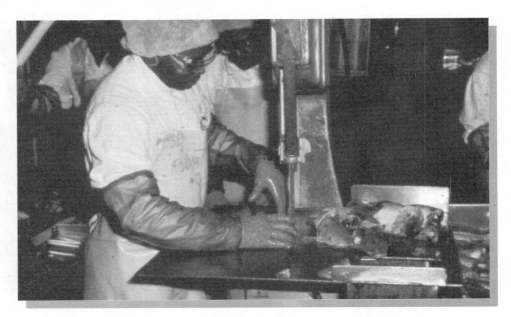

FIGURE 3-16 Deheading channel catfish with a band saw. (Photo courtesy Chuck Weirich, Delta Research Center, Stoneville, MS)

the skin as the fish is passed over the roller. Each membrane skinner processes about 12 to 14 fish per minute per operator.

Chilling

After deheading, eviscerating, and skinning, the whole dressed fish is lightly spray washed and conveyed into the chill tank where it is immersed in a mixture of ice and water. Fish are held in the chill tank from 10 to 30 minutes at a temperature of 38°F (3.3°C) or less. Fish must be cooled rapidly and held below 40°F (4.4°C) to achieve low microorganism numbers, good flavor, and maximum shelf life, and to ensure overall quality. Keeping the microorganism numbers low increases the shelf life of the fish. Some processors add up to 20 ppm chlorine to the chill-tank water or rinse water.

Size Grading

When fish exit the chill tank, they are conveyed to a sizing station where they are sorted by weight. Small and some medium sized fish are usually processed as whole fish. Medium to large fish are typically processed as fillets or steaks. In smaller plants, grading is a hand operation. In larger plants some mechanical or electronic sizing systems are used.

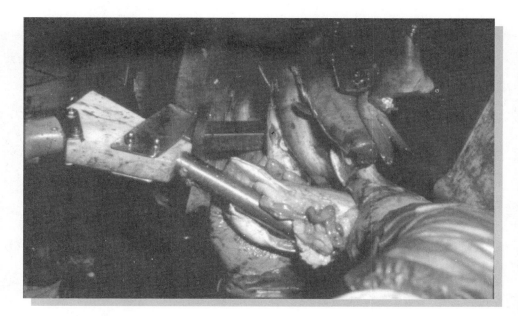

FIGURE 3-17   Removing viscera from the body cavity of the fish with a vacuum eviscerator. (Photo courtesy Chuck Weirich, Delta Research Center, Stoneville, MS)

Product Form

Catfish of the proper size are sold iced, frozen, or prebreaded in the following forms: whole fish, fillets, steaks, strips, and nuggets. Size **control** of fillets to within 1 to 2 oz (28.4 to 56.7 gm) weight increments is essential for marketing of the filleted product. Fish are filleted by hand at filleting tables or by automatic filleting machines. The fillets are trimmed to produce the nugget, then sized and either frozen or packed in ice for shipment (see Figure 3-21).

Channel catfish "steaks" are prepared by cutting size-graded fish into cross-section pieces. The steaks are then individually quick-frozen or packed in ice and sold in 15-lb (6.8-kg) boxes.

Freezing

Before freezing, channel catfish products are injected with or tumbled in a polyphosphate solution, which acts as an antioxidant and prevents excessive water loss during freezing.

The most important consideration in maintaining excellent quality fish in the frozen state is to ensure that they are processed, frozen rapidly, and held at 1° to 10°F (–17.2° to –12.2°C) or below until used. The temperature of the fish must be reduced from 32° to 15°F (0° to –9.4°C) in 30 minutes or less to be considered quick-frozen and to retain the original quality. Tunnel or spiral freezers individually freeze fish to 1°F (–17.2°C) (Figure 3-18). Carbon dioxide,

FIGURE 3-18 Fillets are quick-frozen on a spiral freezer or a tunnel-type freezer. (Photo courtesy Chuck Weirich, Delta Research Center, Stoneville, MS)

liquid nitrogen, or conventional mechanical freezing systems are used in some plants to freeze channel catfish. The choice of freezing media and machinery is mainly a question of economics.

Packaging

Fish are packaged frozen or ice-packed. When whole frozen fish exit the freezer, they are conveyed through a water bath or sprayer. A coating of ice (glaze) is formed over the fish, which is the first step in packaging. The individual quick-frozen glazed whole fish or fillets are sized and packed in cardboard shipping cases that are lined with plastic bags. The whole frozen fish are divided into increments of 2 oz (56.7 gm) each and packed into 15-lb (6.8-kg) boxes. Frozen fillets are packed in 15-lb (6.8-kg) boxes with fillets divided into lots with a 1 to 2 oz (28.4 to 56.7 gm) range.

Whole iced fish are divided into the same size categories as frozen whole fish and packed in ice in 50-lb (22.7-kg) shipping boxes that contain 30 lbs (13.6 kg) of fish and 20 lbs (9 kg) of ice. Steaks are packed in 15-lb (6.8-kg) shipping cartons.

Warehousing

Frozen channel catfish are held at 0°F (–17.2°C) or below if required by state law in a frozen storage warehouse until shipped. The iced product is usually packed and shipped within 48 hours in refrigerated trucks. It is held at the processing plant in refrigerated storage at 30° to 38°F (–1.1° to 3.3°C) until shipped.

INSPECTION

Unlike the red meat and poultry processing industries, fish processing does not fall under the regulations of the United States Department of Agriculture (USDA). Before beginning operation, fish processors must contact local county health officials to comply with county health regulations and to obtain a health permit. Fish processing operations also must adhere to standards set forth by the Good Manufacturing Practice Code of Federal Regulations, Title 21, Part 110, and are subject to announced and unannounced inspections by the Food and Drug Administration (FDA).

Quality

As in other industries, the aquaculture industry considers quality a high priority. Without a quality product, sales of products would quickly decrease. Tables 3-3 and 3-4 list the generic quality control procedures and functions used by any industry concerned with quality.

TABLE 3-3 Quality Control Procedures

Type of Test	Purpose and Example
Chemical	Waste discharge analysis; antibiotic presence analyses; nutrient presence analysis, calcium, vitamins, raw material and finished product analyses—fat, moisture, protein.
Microbiological	Culture testing on fermented products; efficiency of procedures for cleaning, swab testing of equipment. Uncovering microbiological problems in processing lines; running raw material and finished product plate counts, for example, coliforms, yeasts and molds, and tranquilizers.
Sensory	Conducting taste tests between competitor products and own products; grading products by sensory tests; periodic routine testing of own products for quality maintenance; testing for aroma and color.
Other	Finished product shelf life testing by acceleration means; retail store display cabinet testing by random control of temperature in cabinets.

TABLE 3-4 Quality Control Functions

Type of Activity	Example
Records and Reporting	Develop the mechanisms and forms necessary for maintenance of quality control records for use in responding to legal requirements and consumer complaints.
Sampling Schedules	Designate a schedule of sampling that requires the minimum amount of work while maximizing the detection of noncompliance to standards.
Special Problems	Personnel training, short-course work, consumer complaints, bad product lots, and associated production problems.
Compliance with Specifications	Meet compliance standards set by company policy, set by buyer specifications, set by shelf life needs, and set by applicable law.
Test Procedures	Develop criteria and perform tests on raw materials, processes, and end products.
Troubleshooting	Investigate and resolve problems associated with processing supplies such as poor quality materials, erratic supplies and malfunctioning machines as well as nonstandard final products.

In order to maintain a quality product and promote consumer confidence, the major commercial fish processors contracted voluntarily with National Marine Fisheries Service (NMFS) to have their plants inspected. NMFS is an agency service of the National Oceanic and Atmospheric Administration (NOAA), an agency of the U.S. Department of Commerce (USDC). Federal inspectors with the NMFS perform unbiased, official inspections of plants, procedures, and products for firms that pay for these services. The inspectors issue certificates indicating quality and condition of the products.

The NMFS voluntary inspection program provides for the inspection of products and facilities and the grading of products.

Inspection is the examination of fish (seafood) products by a U.S. Department of Commerce inspector or a cross-licensed state or U.S. Department of Agriculture inspector. They determine whether the product is safe, clean, wholesome, and properly labeled. The equipment, facility, and food-handling personnel must also meet established sanitation and hygienic standards. Products that pass inspection can display the federal "Packed Under Federal Inspection" or PUFI mark on the label and/or carton.

Grading

After inspection, grading determines the quality level. Only products that have an established grade standard can be graded. Industry uses the grade standards to buy and sell products. Consumers rely on grading as a guide to purchasing products of high quality. Graded products can bear a U.S. grade mark that shows their quality level. The "U.S. Grade A" mark indicates that the product is of high quality—that it is uniform in size, practically free of blemishes and defects, and has good flavor and odor.

A grading scheme used by trout processors provides an example of how grading works to provide a Grade A mark. In determining the grade of processed trout, each trout is scored for the following factors:

- **Appearance**—The overall appearance of the fish, including consistency of flesh, odor, eyes, gills, and skin. A minor defect is one that is slightly noticeable. A major defect is one that is conspicuously noticeable, but neither seriously affects the appearance, desirability, and eating quality of the fish.

- **Discoloration**—This refers to any color not characteristic to the species. A minor defect is a discoloration of significant intensity involving up to 10 percent of the total area. A major defect is a discoloration of significant intensity involving between 10 and 50 percent of the total area.

- **Surface defects**—These include the presence of unspecified fins in a particular style—ragged, torn or loose fins, bruises and damaged portions of fish muscle, red and opaque in appearance. A minor defect has 3 to 10 percent of the total area affected. A major defect has greater than 10 percent of the total area affected.

- **Cutting and trimming defects**—Four separate categories are scored independently under this factor.

  1. Body cavity cuts are misplaced cuts made during evisceration. Their presence is a minor defect.

  2. Improper washing results in inadequate removal of blood and bits of viscera from the surface or body cavity of the fish. A minor defect is an excessive amount of blood or viscera present.

  3. Improper heading refers to the presence of pieces of gills, gill cover, pectoral fins, or collarbones and ragged cuts after heading. A minor defect is a condition that is scarcely noticeable but does not affect the appearance, desirability, or eating qualities of the trout. A major defect is a condition that is conspicuously noticeable but does not seriously affect the appearance, desirability, or eating qualities of the trout.

  4. Evisceration defects refer to inadequate cleaning of the belly cavity. A minor defect is a condition that is scarcely noticeable but does not affect the appearance, desirability, or eating qualities of the trout. A major defect is a condition that is conspicuously noticeable but does not seriously affect the appearance, desirability, or eating qualities of the trout.

- **Improper boning**—For boned styles only, this refers to the presence of an unspecified bone or piece of bone. Each area of one square inch that contains an unspecified bone or cluster of unspecified bones is counted as one instance.

After inspecting each fish, the number of major and minor defects is totaled. Grade A is given when the maximum number of minor defects is three or less with no major defects. Grade B is given to fish with up to five minor defects and one major defect.

Grade A fish must also possess good flavor and odor for the species, and Grade B must possess reasonably good flavor and odor for the species. In each sample unit of 10 fish, at least 8 of these must meet Grade A standards for the unit to be Grade A.

Detailed information regarding inspection requirements can be found in the *Federal Standard Sanitation Standards for Fish Plants*, FED-STD-369, August 2, 1977. Additional information regarding inspection and standards for products is in the *Code of Federal Regulations*, Title 50, parts 260 and 267.

FIGURE 3-19 The Federal inspection mark (top); the U.S. grade shield. (Photo courtesy Chuck Weirich, Delta Research Center, Stoneville, MS)

Figure 3-19 is an example of the Inspection Mark and the Grade Shield that would be displayed on fish products that meet specific requirements. Products may have one or both of these symbols, depending upon the degree of inspection effort performed and the grade of the product.

In early 1988, The Catfish Institute (TCI), in cooperation with the USDC and the NMFS, began a voluntary inspection program to ensure and promote quality catfish products. Processors who meet the criteria set by this program are able to use TCI's registered **trademarks**, the Mississippi Prime name and logo (Figure 3-20), on their catfish products. To be able to use these trademarks,

FIGURE 3-20 The Catfish Institute's Mississippi Prime logo. (Courtesy Chuck Weirich, Delta Research Center, Stoneville, MS)

a processor must be licensed by TCI and can process only grain-fed channel catfish delivered live to the plant. The plant must be USDC certified as a "Sanitarily Inspected Fish Establishment." Weekly inspections to maintain this certification are required. In addition, weekly unannounced lot inspections by USDC are mandatory.

In addition to federal inspection, major commercial fish processors have in-house quality assurance programs and are often inspected by quality assurance staffs from various customers.

Quality Control

Catfish processing quality control begins at the pond before the fish are harvested for processing. Off-flavor catfish is a major source of concern to catfish producers and processors. This condition is usually generated by minute amounts of chemicals produced from an algae imbalance. Ideally, flavor checks on fish to be processed are done by qualified personnel one week before harvest, one day before harvest, and on the day of harvest.

An overview of specific quality control procedures for catfish processing plants include the following. Processing plants for other species follow similar quality control procedures.

- Fish should be checked for pesticide, herbicide, and heavy metal residue, as well as diseases and off-flavor.

- Holding tanks that are used to store live fish prior to processing should be kept free of algae growth, and proper levels of dissolved oxygen should be maintained. High quality water should be used.

- Proper cleaning procedures, including heading, eviscerating and skinning, should be conducted at all times. Periodic checks should be made at every location during the processing day.

- Proper offal removal procedures should be carefully monitored and maintained.

- A proper chilling procedure, using the latest chilling techniques, should be used to reduce and then maintain the temperature of the catfish at 38°F (3.3°C) throughout processing.

- All surfaces in contact with the fish should be sanitary and not have contact with the floor.

- Fish dropped on the floor should be handled in a proper manner using correct washing methods.

- Temperature of fish products to be frozen should be reduced to 0°F (−17.8°C) as quickly as possible and promptly stored in a freezer at −10° to −20°F (−23.3° to −28.9°C).

- All work-in-process fresh inventory should be promptly iced and stored at approximately 34°F (1.1°C).

- Every effort should be made to keep bacteria counts low. Routine monitoring of product and equipment is encouraged.

- Frozen product should be stored properly in freezer.

- Freezer stock should be rotated regularly.

- Proper clean-up in plant is essential.

- Product should be checked throughout the processing operation with regard to weight, size, visual appearance, proper temperature, and correct packaging.

- Value-added products should be checked on line routinely to ensure proper percentages of breading, glaze, marinade, and other ingredients.

- Product recall procedures, including proper coding of a product, should be used.

With the assistance of U.S. Department of Commerce inspection programs and in-depth quality assurance programs, today's commercial fish processors provide the consumer with quality fish products.

Off-Flavor. Off-flavor in farm-raised catfish is a very important problem to producers. Off-flavor is the presence of objectionable flavors in the fish's flesh. The off-flavor may be so intense that it makes the fish unmarketable. During the fall, more than 50 percent of production ponds may have off-flavor fish. This means that ponds cannot be harvested, and harvest and processing schedules are disrupted. Producers are left feeding and maintaining these fish, which increases production costs, disrupts cash flow and extends risks. Off-flavor is a complicated problem and requires that producers understand the probable causes, possible cures and, most important, how to check the fish before they are marketed.

Off-flavor is caused by chemical compounds that enter the fish from the water. Some of these compounds are produced by certain pond bacteria and algae.

The bacteria belong to a group of filamentous bacteria called the actinomycetes. These bacteria are found in the water column, but they are most abundant in the bottom mud. Actinomycetes thrive in ponds during warm weather, using nutrients from fish wastes and uneaten feed. Algae commonly associated with off-flavors belong to the blue-green group. Blue-green algae, though always present in ponds, are most abundant in the summer and fall. Blue-green algae also thrive in nutrient-rich ponds and can dominate other types of algae. Blue-green algae often float and form paint-like scum or a "soupy" layer near the surface. Off-flavors can be described in many ways.

Possible descriptions include: earthy, musty, rancid, woody, nutty, stale, moldy, metallic, painty, weedy, putrid, sewage, petroleum, and lagoon-like.

Obviously, many compounds and causes are involved. The causes of some off-flavors are still to be identified. Two specific compounds have definitely been identified as producers of off-flavors: geosmin and 2-methylisoborneal (MIB). These compounds are produced by both blue-green algae and actinomycetes. Geosmin causes a musty or woody off-flavor, and MIB causes a musty or weedy off-flavor. Both produce off-flavor in minute concentrations of 2 to 3 parts per billion (ppb) in pond water.

Off-flavor compounds are eliminated from the flesh of the fish in time, if the compounds are no longer in the pond. Depending on temperature and other weather conditions, it can take from a few days to more than a month for the sources of off-flavor and the off-flavor itself to dissipate. A producer can do very little about off-flavor except wait for it to go away. It is nearly impossible to control the bacteria or algae in the pond. The use of herbicides to control the algae is not effective. Stocking catfish ponds with tilapia can reduce the occurrence of off-flavor. Problems of obtaining tilapia fingerlings, controlling reproduction, and finding a market for them are still to be solved. Placing fish in clean water is another option. This method works well, but it is costly in terms of facilities, labor, energy, time, and weight loss of the fish being held.

Processors check fish for off-flavor before scheduling harvests. Producers should check fish for off-flavor also. The first check occurs at least two weeks before the planned harvest, again three days before harvest, and finally the day of harvest. Fish can go off-flavor within a few hours and even during harvest operations. If off-flavor is found, weekly tests are required.

The human nose is the best equipment for determining off-flavor. The following procedure tests catfish for off-flavor—

■ Select one fish from each pond.

■ Head and gut, but do not skin the fish.

■ Cut off the tail section (the last third) with skin intact. Use this part for the test.

■ Cook the tail section until the flesh is flaky, using one of the following methods. Do not season the fish with any spices, not even salt. Wrap the fish in foil and bake at 425°F (218°C) for about 20 minutes. Or, place the fish in a small paper or plastic bag or a covered dish and microwave at high power for 1½ minutes per ounce.

■ After cooking, smell the fish first. Do you notice any foul odors?

■ Next, taste the fish. Do you notice any foul or bad flavors?

The future of the catfish industry depends on a quality product. Catfish producers know that a first-time catfish consumer who eats an off-flavor fish may be a one-time customer.

Fresh

Fish is sold fresh or frozen in a variety of forms. Figures 3-21 and 3-22 indicate the cuts from catfish or tilapia. Other fish are similar.

Processed

Although no national catfish chain rivals poultry product outlets, during the past few years many catfish houses or restaurants have served breaded catfish to consumers.

Breaded, uncooked catfish fillets and whole fish products contain 20 to 30 percent cornmeal breading and usually are sold as a raw product. Sizes

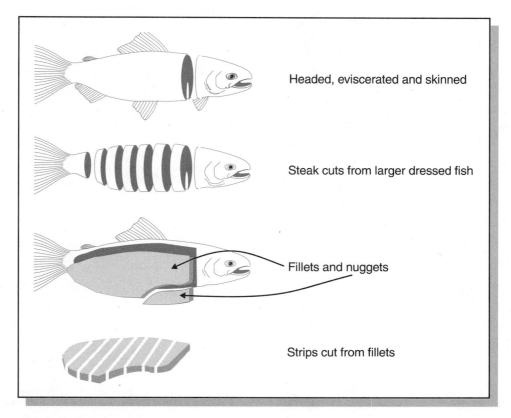

FIGURE 3-21 Catfish cuts available fresh, packed on ice, or frozen.

usually range from 3 to 7 oz (85 to 198 gm) for fillets. One potential product is a formed portion controlled product that is exact in shape, size, and weight. This product is breaded and can be formed into 1- to 5-oz (28.4- to 141-gm) portions. Breaded fillet strips also have been a popular item for fast food outlets or restaurants.

Enrobing is a recent method of further processing catfish products. The enrobing medium usually consists of vegetable oil or oil/water coatings that are applied to fillets, which are then frozen. Some flavors and types include lemon-butter, cajun, and blackened. These types of coatings provide an up-scale catfish product suitable for baking or broiling at restaurants. Combinations of light coatings and bread crumbs are also available.

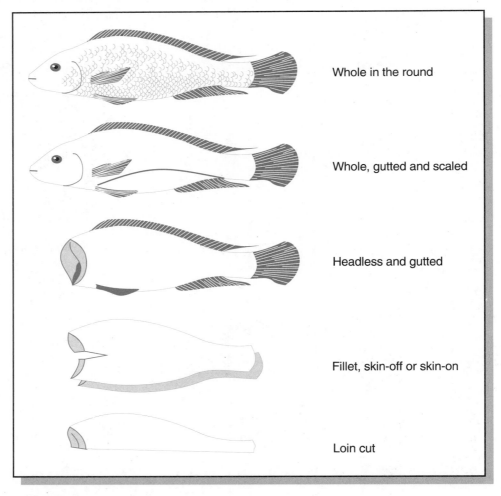

FIGURE 3-22 Tilapia cuts available fresh or frozen.

FIGURE 3-23 Individually packaged crawfish ready for market.
(Photo courtesy Jay Huner, Director, Crawfish Research
Center, Southwestern Louisiana University, Lafayette, LA)

The use of phosphates as a processing aid has provided another area for new processed products. Injectors provide a means of carrying flavors and spices to the core of catfish fillets. Products such as lemon-butter, hot and spicy, and smoked fillets, can be prepared with this technology. Another method of carrying spices and flavors into the catfish is vacuum marination. Marination provides a vehicle to carry spices and flavors of larger particle size and heavier coatings to the surface and interior of fillets.

Many of the further processed catfish products have been packaged with companion products such as hushpuppies. Fully cooked, frozen catfish products have potential in the future.

Minced catfish, deboned from the skeletal frames after filleting, offers several opportunities for further processing. The minced meat is formed into patties and breaded. These patties have been successful for school lunch programs. The catfish mince can be frozen in 16-lb (17.3-kg) blocks for making breaded fish sticks, gumbo, or any product requiring fish in the recipe. Surimi has been successfully made from minced catfish.

FIGURE 3-24 Product on the consumer's plate. That is the goal of marketing. Here sauteed California farmed hybrid striped bass is featured. (Photo courtesy Kent Seafarms, San Diego, CA)

A stuffed catfish fillet is almost certain to make an entrance into the marketplace. Stuffings could include crab and shrimp flavored surimi, shrimp stuffing, cornbread stuffing, and many others.

Another opportunity for further processed products is the use of minced catfish in conjunction with other seafood products such as shrimp. Minced catfish often can be used in lieu of surimi. An example would be a formed, breaded shrimp product containing shrimp pieces and deboned catfish.

Processed trout and other species follow some of the same trends as processed catfish.

SUMMARY

Marketing is the process of getting the product from the producer to the consumer. Production matters little if the producer did not identify markets for the product. Successful marketing also involves developing a marketing strategy and plan. Like production costs, marketing costs add to the final price of the product.

Processing is a part of marketing. It produces a product that the consumer will and can purchase. Some processing is essential, such as killing the fish, eviscerating, and filleting. Further processing, such as breading and adding flavors, attempts to increase consumer demand. As a protection and guarantee to the consumer, processing procedures are inspected, products are graded, and strict quality control procedures are followed.

STUDY/REVIEW

Success in any career requires knowledge. Test your knowledge of this chapter by answering these questions or solving these problems.

True or False

1. The United States ranks first in the production of aquaculture products.

2. The consumption of fish and seafood compared to beef, poultry, and pork is low.

3. Production is the primary concern of an aquatic enterprise.

4. Consumers and processors are both basic buyers of aquatic products.

5. Processing maintains quality control.

6. Off-flavor in catfish is caused by blue-green algae.

Short Answer

7. Which of the following is not a key element of marketing?
   a. Present situation
   b. Harvesting
   c. Goals
   d. Plan

8. Generating consumer demand is a part of _____ _____.

9. Convincing wholesalers and retailers to carry a product is _____ _____.

10. Name five activities that are a part of marketing.

11. Enticing the buyer to purchase a product is called _____.

12. Name the four top aquaculture-producing countries.

13. Name two product characteristics that affect buying decisions.

14. Give examples of three types of selling where aquaculture producers sell direct to the consumer.

15. Name four general quality control procedures.

16. List six quality control functions.

17. Name the agency with which fish processors have contracted voluntarily for plant inspection.

18. List five mistakes during processing that could affect the quality of a fish product.

19. Which species of fish have a problem with off-flavor and what causes this off-flavor?

20. List four aquaculture products that result from further processing.

Essay

21. Define marketing and describe a marketing plan.

22. Worldwide, name four general groups of aquaculture products.

23. Compare the consumption and trends in consumption for fish and shell-fish and other meats in the United States.

24. Name two general marketing channels and give an example of each.

25. List three concerns of an intermediate buyer.

26. Define product pull and give an example.

27. Define product push and give an example.

28. Describe what Grade A indicates to a consumer.

29. What is the role of The Catfish Institute in marketing catfish?

30. When catfish are to be harvested, what schedule should be followed to test for off-flavors?

KNOWLEDGE APPLIED

1. Choose an agricultural or aquacultural product or, preferably, one common to your location. Draw a flowchart tracing this product from the producer to the consumer. At each stage in the flowchart, identify the cost of the product.

2. Visit a processing plant, preferably aquaculture or agriculture. Determine the type of quality control measures taken before the product arrives at the plant and during processing.

3. Invite a retailer to describe his marketing plan. Specifically, have the retailer discuss the value of advertising in creating consumer pull.

4. Visit a supermarket or grocery store. Find out what kinds of fresh and frozen fish and seafood products are sold. Ask the manager where these products are purchased.

5. Obtain menus from several local restaurants. Determine the types of fish on their menus. Ask the managers where they obtain fish for the restaurants.

6. Collect nutritional labels from the packages of fish, seafood, and other meats. Develop a table and compare the nutritional value of fish and seafood to beef, pork, and poultry. For example, compare protein, fat, and calorie content.

LEARNING/TEACHING AIDS

Books

Chaston, I. (1991). *Marketing in fisheries and aquaculture*. Cambridge, MA: Blackwell Scientific Publications.

Chen, L. C. (1990). *Aquaculture in Taiwan*. Cambridge, MA: Blackwell Scientific Publications.

Connell, J. J. (1990). *Control of fish quality* (3rd ed.). Cambridge, MA: Blackwell Scientific Publications.

Council for Agricultural Education. (1992). *Aquaculture.* Module VB: Marketing the product. Alexandria, VA: Council for Agricultural Education.

Dupree, H. K., & Huner, J. V. (1984). *Third report to the fish farmers: The status of warmwater fish farming and progress in fish farming research*. Washington, DC: U.S. Fish and Wildlife Service.

Walker, S. S. (1990). *Aquaculture*. Stillwater, OK: Mid-America Vocational Curriculum Consortium, Inc.

Booklets

Effects of catfish advertising on consumers' attitudes, purchase frequency, and farmers' incomes. (1990). Alabama Agricultural Experiment Station. Auburn, AL: Auburn University.

Retail grocery markets for catfish. (1991). Alabama Agricultural Experiment Station. Auburn, AL: Auburn University.

The U.S. market for farm-raised catfish—an overview of consumer, supermarket, and restaurant surveys. (1990). Arkansas Agricultural Experiment Station. Fayetteville, AR: University of Arkansas.

Aquatic Management Practices

M anagement is the secret ingredient to successful aquaculture. Management involves knowledge of the species being cultured—sources of the species, habitat, seedstock and breeding, accepted culture methods, stocking rates, feeding diseases, processing, and marketing. With knowledge, the successful aquaculturalist uses good judgment.

No attempt is made in this chapter to cover all species that are, can be, or were cultured. Rather, this chapter covers a wide range of species. Management practices are similar for many species. For specific management details of a species, many books and articles listed at the end of the chapter deal with only one species.

Learning Objectives

After completing this chapter, the student should be able to:

In Aquaculture

- Describe the purpose and functions of a hatchery
- Describe the spawning facilities used in aquaculture
- Define harvesting

■ Describe harvesting methods
■ Define terms related to commercial catfish production with their definitions
■ Arrange in order the phases of fingerling production
■ Describe stocking rates for various stages of production and various species
■ List guidelines for obtaining fingerlings for food-fish production
■ Describe trout culture
■ Identify the types of trout farming enterprises
■ List the phases of trout production
■ Explain broodfish management
■ Discuss egg management after fertilization
■ Describe fry and fingerling management
■ List general management guidelines for different species
■ Define terms associated with commercial baitfish production
■ Describe the baitfish industry
■ Distinguish between descriptions and uses of common and Chinese carps
■ Describe hobby fish production and management
■ Explain the methods of pond preparation and fertilization for different species
■ List control techniques for predators
■ List guidelines for transporting fish to long-distance markets
■ Define terms associated with commercial crayfish production
■ Describe the commercial culture of tilapia
■ List methods of managing tilapia to control overpopulation
■ Discuss common production concerns for different species
■ Describe different production systems used by various species
■ Define terms related to commercial trout production
■ Define terms related to harvesting and hauling
■ Compare total and partial harvest

In Science

■ Describe ways seeds are produced for different species
■ Explain how sex is determined in fish

■ Discuss methods of controlling reproduction in fish

■ Describe procedures in reproducing aquatic animals

■ Describe the sexual reproduction processes of aquatic animals

■ List salmonids that could be or are cultured

■ Describe the reproduction and life cycle of crayfish

■ Describe the reproduction and life cycle of a shrimp

■ Distinguish between red swamp and white river crayfishes

■ Describe the commercial production of hybrid striped bass

■ Identify popular baitfish species

■ Describe aquatic species and their current culture or potential for culture

■ Demonstrate a familiarity with the scientific names for different aquatic animals

■ Describe breeding systems and their purposes

Understanding of this chapter will be enhanced if the following terms are known. Many are defined in the text and others are defined in the glossary.

KEY TERMS

Anadromous	Hybridization	Rotational line-crossing
Benthic	Hybrids	Salmon farming
Byssus	Hydrological	Salmon ranching
Catadromous	Inbreeding	Salmonids
Clarification	Inventory	Selective breeding
Clutch	Metabolites	Sex determination
Crossbreeding	Metamorphosis	Sexing
Density Index	Microsporidean	Spat
Detritus	Milt	Spawning
Dredges	Mouthbrooders	Standing crop
Eyed stage	Nursery	Stocking rate
Feeding chart	Phenotypes	Substrate
Genes	Photoperiods	Volumetric
Hormones	Production ponds	
Hybrid vigor	Recruitment	

SPAWNING

Sexual reproduction involves the production of eggs by the ovaries of the female and sperm by the testes of the male. Reproduction is critical to successful aquaculture, as is an understanding of reproduction. Spawning is the act of obtaining eggs from female and sperm or milt from the male.

In nature, most finfish are seasonal breeders. Reproductive cycles are controlled by hormones produced by endocrine glands. Figure 4-1 shows the approximate location of the endocrine glands in a fish. The production and release of the hormones is controlled by environmental stimuli—internal or external. Under natural conditions, climatic changes such as day length and temperature act as stimuli. Environmental stimuli interpreted by areas of the brain influence the release of hormones, as shown in Figure 4-2. Besides controlling the production of eggs and sperm cells, the reproductive hormones control secondary sexual characteristics like coloration and breeding behavior. Hormones in fish and their actions are described in Table 4-1.

Often, reproductive cycles are artificially controlled to ensure the continuous production of seed. Three approaches can be used to control reproduction—

■ Genetic
■ Environmental
■ Hormonal.

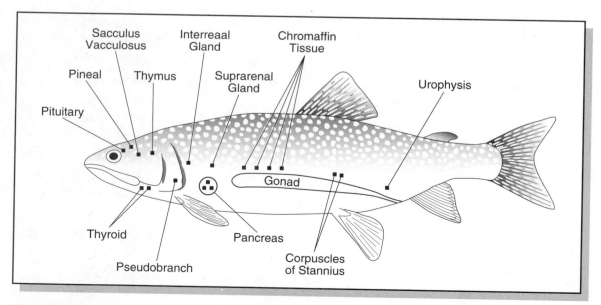

FIGURE 4-1 Location of endocrine glands in fish.

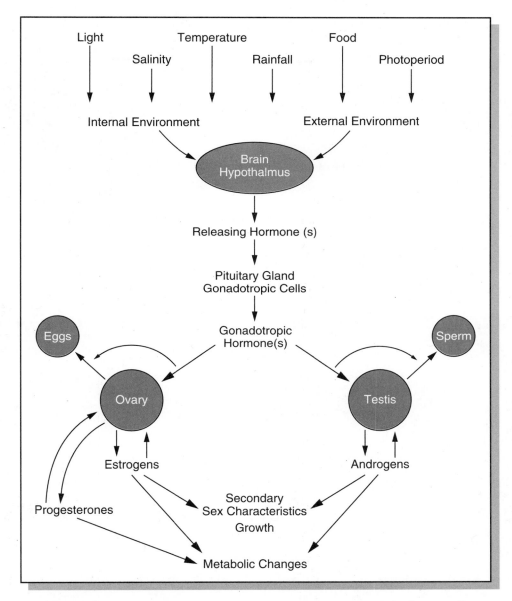

FIGURE 4-2 Environmental stimuli intrepreted by areas of the brain influence the release of hormones.

Genetic

Through genetic selection early-maturing or late-maturing broodstocks can be developed. Management of the broodstock stretches the breeding season. This is a relatively difficult task.

TABLE 4-1 Major Endocrine Glands and Hormones in Fish

Gland	Hormone	Type	Function
Hypothalamus	Releasing hormones	Peptides	Controls the pituitary gland
Pituitary gland	Growth hormone (STH)	Protein	Controls growth
	Prolactin (LTH)	Protein	Controls ion balance
	Adrenocorticotropic hormone (ACTH)	Peptide	Controls interrenal gland
	Melanocyte-stimulating hormone (MSH)	Peptide	Controls color change
	Thyroid-stimulating hormone (TSH)	Protein	Controls thyroid gland
	Gonadotropic hormone (GTH)	Protein	Controls reproduction and gonads
Thyroid gland	Thyroxine (T4) Triiodothyronine (T3)	Amino acids	Controls growth, reproduction, metabolism, and nutrient assimilation
Interrenal (adrenal gland)	Adrenaline	Amino acids	Counteracts stress
	Cortisol	Steroid	Controls ion balance
Testis	Testosterone	Steroids	Metabolic effects. Control secondary sex characteristics. Sperm production.
Ovary	Estrogen Progesterone		Metabolic effects. Yolk and egg production
Pancreas	Insulin Glucagon Somatostatin	Peptides	Protein metabolism and control of endocrine pancreas
Pineal gland	Melatonin	Peptide	Provides information about day/night and seasonal time
Ultimobranchial gland	Calcitonin	Peptide	Controls calcium levels
Stannous corpuscles	Hypocalcin	Protein	Controls calcium balance

Environmental

Controlled light periods—photoperiods—have been used with several species of fish to manipulate spawning time. Salmon exposed to shortened periods of light spawn appreciably earlier. Egg mortalities can be significantly higher. Light, not temperature, is apparently the prime factor in accelerating or retarding sexual maturation in this species. Artificial light has been used successfully to induce early spawning in brook, brown, and rainbow trout. The rearing facilities are enclosed and lightproof, and all light is artificial. Broodstock often have at least one previous spawning season before being used in a light-controlled spawning program.

The light schedule used to induce early spawning in trout follows this scheme. An additional hour of light is provided each week until the fish are exposed to nine hours of artificial light in excess of the normal light period. The light is maintained at this schedule for a period of four weeks and then decreased one hour per week until the fish are receiving four hours less light than is normal for that period. By this schedule, the spawning period can be advanced several months.

Most attempts at modifying the spawning date of fish have been to accelerate rather than retard the maturation process. Artificial light periods longer than normal delays the spawning activity of eastern brook trout and sockeye salmon. Temperature and light control are factors in manipulating spawning

FIGURE 4-3 Bins used to control artificial light, which controls spawning in trout.

time of channel catfish. Reducing the light cycle to eight hours per day and lowering the water temperature by 14°F (8°C) will delay spawning for approximately 60 to 150 days.

Hormonal

Spawning of fish can be induced by hormone injection. Fish must be fairly close to spawning to have any effect, as the hormones generally bring about the early release of mature eggs and sperm rather than the promotion of their development. Both pituitary material extracted from fish, mammals, and human chorionic gonadotropin (HCG) have been used successfully. Recently, synthetic releasing hormone has also been used successfully on some species.

Use of hormones may produce disappointing results if broodfish are not of high quality. Under such conditions, a partial spawn, or no spawn at all, may result. Some strains of fish do not respond to hormone treatment in a predictable way, even when they are in good spawning condition.

Injection of salmon pituitary extract into adult salmon hastens the development of spawning coloration and other secondary sex characteristics, ripens males as early as three days after injection, and advances slightly the spawning period for females, but may lower the fertility of the eggs.

Dried fish pituitaries from common carp, buffalo, flathead catfish, and channel catfish will all induce spawning when injected into channel catfish. The pituitary material is finely ground, suspended in clean water or saline solution, and injected intraperitoneally. One treatment is given each day until the fish spawns. Generally the treatment should be successful by the third or fourth day.

Channel catfish can also be successfully induced to spawn by intraperitoneal injections of HCG. One injection of HCG normally is sufficient.

SEX DETERMINATION

Genes are the basic unit of inheritance. Genes are carried on the chromosomes in the gametes—eggs or sperm. Genes contain the blueprint or code that determines how the animal will look and interact with its environment. The number of chromosomes varies from species to species but is consistent for a species.

Chromosomes also determine the sex of the fish. Although, not as well understood as in mammals, the most common system of sex determination in commonly cultured fish is the XY system like that of mammals. In this system, females carry the XX chromosomes and males carry the XY chromosomes. When females produce eggs, every egg will possess one X chromosome. When males produce sperm, half the sperm will carry the X chromosome and

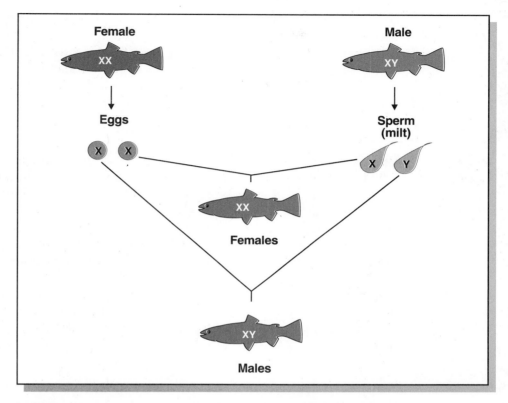

FIGURE 4-4 How sex is determined with the XY system.

half will carry the Y chromosome. When the eggs and sperm unite, half the zygotes will be XX and the other half will be XY. On the average, in a normal population, half of the offspring are males and half are females. Figure 4-4 illustrates how sex is determined with the XY system.

The method for sex determination in fish is not simple. At least eight chromosome systems may control the sex of different species. Also, while sex determination is primarily controlled genetically, environmental factors such as temperature, photoperiod, salinity, and crowding can help determine the sex of fish. Once sex determination is understood, this knowledge can be used to produce monosex cultures of fish.

FINFISH

Catfish and trout represent the major culture of finfish in the United States. Both are freshwater fish, but catfish is a warmwater species and trout a cold-water species. The culture of many other species of finfish is at various stages of development. Some of these are discussed in the sections that follow. For each species the discussion includes sources of species, habitat, seedstock and breeding, culture method, stocking rate, feeding, diseases, harvesting and yields, processing, and marketing. Many of these topics are covered in more detail in other chapters, but not necessarily by individual species.

Channel Catfish

Channel catfish, *Ictalurus punctalus*, is the most important species of aquatic animal commercially cultured in the United States. It belongs to the family Ictaluridae, order Siluriformes. Relatives of the catfish are found in fresh and saltwater worldwide.

In natural waters, channel catfish caught by fishermen are usually less than 3 lbs (1.4 kg), but the world record of 58 lbs (26.3 kg) was caught in Santee Cooper Reservoir, South Carolina, in 1964. The size and age that channel catfish reach in natural waters depends on many factors. Age and growth studies suggest that in many natural waters channel catfish do not reach 1 lb (0.45 kg) in size until they are two to four years old. The maximum age ever recorded for channel catfish is 40 years. Most commercially raised catfish are harvested before they are two years old.

Sources of Species.   At least 39 species of catfish exist in North America, but only seven have been cultured or represent potential for commercial production. These include channel catfish, *Ictalurus punctalus*; the blue catfish, *Ictalurus furcatus*; the white catfish, *Ictalurus catus*; the black bullhead, *Ictalurus melas*; the brown bullhead, *Ictalurus nebulosus*; the yellow bullhead, *Ictalurus natalis*; and the flathead catfish, *Pylodictis olivaris*.

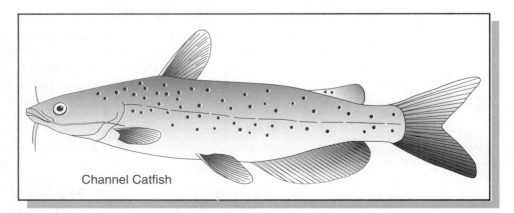

Channel Catfish

FIGURE 4-5a Channel catfish.

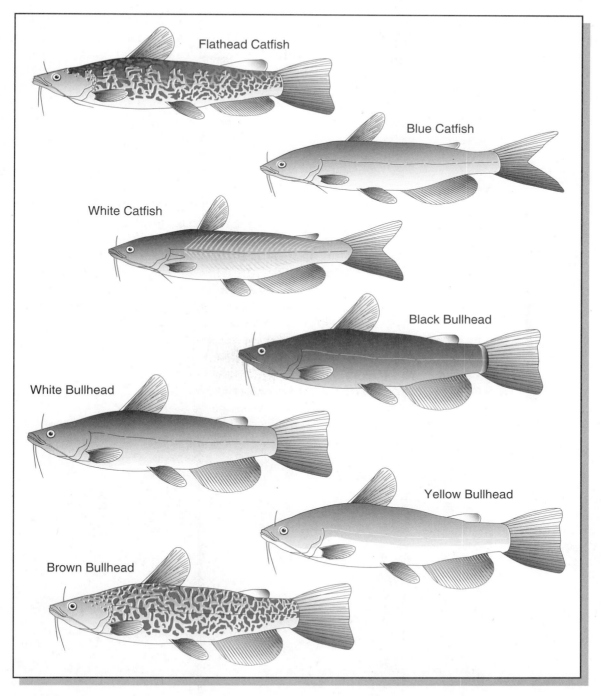

FIGURE 4-5b Other types of catfish are Flathead, Blue, White, Black Bullhead, Yellow Bullhead, and Brown Bullhead.

Habitat. Channel catfish were originally found only in the Gulf States and the Mississippi Valley north to the prairie provinces of Canada, and to Mexico, but were not found in the Atlantic coastal plain or west of the Rocky Mountains. Since then channel catfish have been widely introduced throughout the United States and the world.

In natural waters channel catfish live in moderate to swiftly flowing streams, but they are also abundant in large reservoirs, lakes, ponds, and some sluggish streams. They are usually found where bottoms are sand, gravel or rubble, in preference to mud bottoms. They are seldom found in dense aquatic weeds. Channel catfish are freshwater fish but they can thrive in brackish water.

Channel catfish generally prefer clear water streams, but are common and do well in muddy water. During the day they are usually found in deep holes wherever the protection of logs and rocks can be found. Most movement and feeding activity occurs at night just after sunset and just before sunrise. Young channel catfish frequently feed in shallow river areas while the adults seem to feed in deeper water immediately downstream from sandbars. Adults rarely move much from one area to another and are rather sedentary, while young fish tend to move about much more extensively, particularly at night when feeding.

FIGURE 4-6 Raising catfish in raceways. (Photo courtesy Cooperative Extension Service, Mississippi State University, Stoneville, MS)

Seedstock and Breeding. Channel catfish spawn when the water temperature is between 75° and 85°F (23° to 30°C) with about 80°F (27°C) being best. Wild populations of catfish may spawn as early as late February or as late as August depending on the location. The length and dates of the spawning season vary from year to year depending on the weather and area, but peak spawning time in Mississippi usually occurs in May.

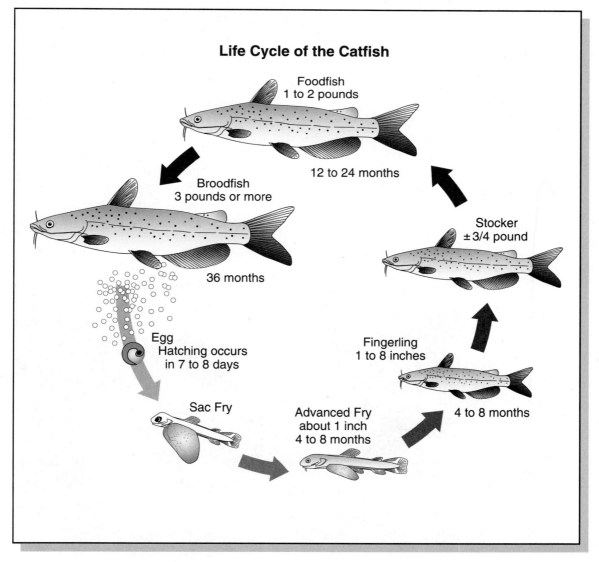

FIGURE 4-7 Stages of the catfish.

Channel catfish are cavity spawners and will spawn only in secluded, semi-dark areas. In natural waters male catfish will build a nest in holes in the banks, undercut banks, hollow logs, log jams, or rocks. This behavior requires the use of spawning containers in order to successfully spawn channel catfish in commercial ponds.

The male selects and prepares the nest by fanning out as much mud and debris as possible. He will then defend this location against any intruder until spawning is completed and the fry leave the nest.

The female is attracted to the nest and spawning occurs within the nest with eggs being laid in a gelatinous mass on the bottom. After the eggs are laid, the male takes over and cares for the eggs by constantly fanning them with his fins to provide aeration and to remove waste products given off by the developing eggs.

Females spawn only once a year, producing about 3,000 to 4,000 eggs per pound of body weight, while the males may spawn more than once. In wild populations, males seldom spawn more than once a year, but in hatcheries where the eggs are removed from the spawning container soon after being laid, males may spawn three or four times.

Channel catfish usually become sexually mature at three years of age, although some may spawn when two years old. In wild populations they may not spawn until after the age of five years. Channel catfish weighing as little as 0.75 lb (0.34 kg) may spawn if old enough. Farm-raised catfish usually weigh in excess of 2 lbs (0.91 kg) when they spawn.

FIGURE 4-8 Old ammunition and milk cans used for catfish spawning.

After the eggs are laid they will usually hatch in five to ten days depending on water temperature. At 78°F (26°C) the eggs will hatch in about eight days. Each 2°F (1°C) rise in temperature above 78°F (26°C) requires one less day, and each 2°F (1°C) fall in temperature below 78°F requires one more day for hatching. Water temperatures below 65°F (18°C) and above 85°F (30°C) will reduce hatching success. Newly hatched fry have a large yolk sac that contains the nourishment they need for the next two to five days until they are fully developed and are ready to start feeding. After the yolk sac is absorbed, the fry take on their typical dark color and will begin to swim-up looking for food. At first swim-up fry will gulp air to fill their swim bladders, which helps them maintain and regulate their buoyancy.

Spawning containers in commercial production are such things as milk cans, nail kegs, earthen crocks, ammunition cans, wooden boxes, and plastic buckets. The spawning container must be large enough to accommodate the brooding pair. The opening should be just large enough for them to enter.

The containers are placed in 1 to 2.5 ft (0.3 to 0.8 m) of water, 1 to 10 yds (0.9 to 9.1 m) apart with the open end toward the pond center. Floats mark container locations. Enough containers are provided for 50 to 90 percent of the males.

Spawning activity sometimes diminishes for no apparent reason. Lowering the water level about a foot and rapidly refilling the pond may encourage additional spawning. Moving the spawning containers may also stimulate spawning.

Considering that not all females spawn and not all eggs, fry, and fingerlings survive, about 1,000 fingerlings will be produced per pound of healthy female brooder with the use of proper broodstock, hatchery, and rearing techniques.

Four methods are used in spawning channel catfish in ponds:

1. Spawning and rearing pond method. This approach requires the least skill, labor, and facilities. It is unreliable and not recommended for commercial operations. Spawning containers are placed in the pond, and the fish are allowed to spawn and hatch the eggs. The fry are left in the pond until ready for harvest.

2. Fry transfer method, open pond spawning. The fry transfer method is more productive than the spawning and rearing pond method but requires more skill and labor. The newly hatched fry are transferred from the spawning containers to previously prepared **nursery** ponds. Spawning containers are checked every three days.

    Males incubate the eggs. One day after the predicted hatching date, the fry are removed. The male catfish can bite hands and bare feet, so he should be chased from the spawning containers.

The fry are transferred to a bucket containing pond water by gently pouring them from the spawning container after counting. Next, fry are released into the nursery pond by slowly submerging the bucket, allowing them to escape into the pond near a shelter.

If the water temperature or chemistry is not the same in both ponds, the fry must be slowly acclimated to the nursery pond water temperature before stocking. When temperature differences are more than 2 to 3 degrees, the water in the bucket is replaced slowly with nursery pond water until the temperature is equalized.

3. Egg transfer method, open pond spawning. Egg transfer is the most productive of the four methods but also requires the most skill, labor, and facilities. The fish are allowed to spawn in the containers as with the other methods, but the eggs are removed and incubated in a hatchery.

   Spawning containers are checked every two to four days. Late afternoon is the best time because most spawning probably occurs at night or early morning. Checking at this time does not interrupt spawning activity and allows for timely removal of eggs. Eggs are removed immediately after finding them. Disturbed broodfish may sometimes eat eggs or dislodge them.

   The egg mass sticks to the container floor and must be gently scraped free. Egg masses are placed into a bucket and carried immersed in water to the hatchery. Eggs can be left in buckets in a shaded area for up to 15 minutes, but no longer without aeration. Eggs must be shielded from sunlight. Egg masses near hatching must be taken to the hatchery immediately because they require more oxygen than young or "green" egg masses.

   Transporting egg masses in a cooler or other container can cause egg death due to suffocation. If a long time in transport is expected from the pond to the hatchery, aeration is required.

4. Pen spawning. Each pen contains a pair of brood fish. The fish should be about equal in size. Daily checks ensure welfare of the brooders and that the females are not being harassed or injured by the male fish. Females should be removed immediately after spawning to keep them from being injured or killed by the male. More than one female or male in the pen at a time can lead to fighting and injury to the females. Eggs can be left with the male or taken to the hatchery to incubate.

In maximum-production systems, eggs are transferred to a hatchery, incubated, and the fry started on food before they are moved into nursery ponds. The hatchery need not be elaborate. The critical ingredient is a water supply of the right quality and quantity.

FIGURE 4-9 Pens for catfish spawning.

Water temperature must be between 75°F (24°C) and 82°F (28°C) for proper hatching. Because eggs and fry have high oxygen requirements, oxygen levels should be maintained at a minimum of 6 parts per million (ppm). Water pH must be between 6.5 and 8.5 for best results. Risk of disease is less if no fish are in the water supply. The best water is clean and free of organic matter such as algae and decaying leaves. A water flow of about 2 gal (7.6 L) per minute is needed for a 100-gal (378.5 L) hatching trough, or about one complete water change every 45 to 60 minutes.

Well water is probably best for the hatchery. It is usually clean and free of disease organisms. Well water is usually too cold for optimum hatching, but it can be warmed in a conventional water heater or stored and warmed in a small pond built specifically for this purpose. Some farmers have two wells, one from a deep aquifer that has warm water, and one from a shallow aquifer with cold water. A mixing valve is used to mix the two in the right proportions to provide uniform 80°F (26°C) water to the hatching troughs. The aeration tank should have a capacity of 25 percent of the hatchery's entire water volume. This will ensure at least a 15-minute retention time with a 60-minute exchange rate.

Total hardness and total alkalinity should exceed 20 ppm, and the pH should range between 6.5 and 8.5. Acidic or soft pond water can usually be corrected by adding agricultural limestone.

Eggs are commonly incubated in flat-bottomed, wooden, fiberglass, or metal troughs about 8 to 10 ft long, 18 to 24 in wide and 10 to 12 in deep, holding about 100 gal (2.4 to 3 m long, 45.7 to 61 cm wide, and 25.4 to 30.5 cm deep, approximately 378.5 L). A series of paddles attached to a shaft are suspended in the trough. Paddles are spaced to allow wire mesh baskets holding the egg masses to fit between them. The paddles should reach about halfway to the bottom of the trough and should extend below the bottoms of the baskets. Baskets are made from ¼-in (0.6 cm) plastic-coated hardware cloth. An electric motor turns the paddles at 30 rpm. This motion gently rocks the egg masses and causes oxygen-rich water to flow through them. An 8-ft (2.4 m) trough can hold six to eight egg baskets. A standpipe fitted into a drain at the other end controls water depth. A window screen over the standpipe prevents fry from escaping.

Bacterial diseases and fungal infections are constant threats to eggs. The best disease control is prevention. A clean water supply of the proper temperature and frequent scrubbing and disinfection of troughs and equipment are essential. Debris and egg shells must be removed regularly with a siphon. Eggs are checked daily for bacterial egg rot or fungus. Bacterial egg rot appears as a milky-white dead patch, usually on the underside and in the center of the mass. Generally, bacterial egg rot occurs when water temperatures are higher than 82°F (28°C). The best preventative measure, besides maintaining good sanitation, is to keep water temperature at 78° to 80°F (25° to 26°C).

FIGURE 4-10 Troughs in catfish hatchery. Paddles simulate fanning by male fish.

Fungus grows on infertile or dead eggs usually when the pond water temperature is below 75°F (24°C) or hatching water is below 78°F (25°C). It appears as a white or brown cotton-like growth made of many small filaments that can invade and kill healthy eggs. Formalin treatment controls fungus.

Bacterial egg rot or fungus seldom causes problems with good aeration and proper temperature, water between 78° and 80°F (25° and 26°C).

As the eggs hatch, sac fry emerge, swim through the screen baskets and school together in a tight cluster on the bottom of the trough. For each pound of egg mass, 10,000 to 11,000 eggs will be present. Approximately 95 percent of these will hatch.

Sac fry do not eat. They receive nourishment from the attached yolk sac. The yolk sac is gradually absorbed by the fry, and after about three days the fry begin swimming up to the water surface searching for food. Their color at this time changes from pink to black. They are called swim-up fry and begin feeding at this stage.

Sac fry can be left in the hatching trough for one to two days and then moved to rearing tanks or troughs. Many types of tanks can be used for holding fry, the most common in Mississippi being an 8 ft × 2 ft × 10 in (2.4 m × 0.6 m × 25 cm) flat bottom trough, which will hold about 100,000 fry. If a large tank is used, a fry holding box is desirable. This is a 2 × 2 × 1-ft (61 × 61 × 0.3-cm) wooden box made from boards or marine plywood and caulked with silicone. The bottom is made of $\frac{1}{16}$-in (1.6 mm) plastic window screen. One box can hold 20,000 to 30,000 fry, the quantity obtained from a large egg mass. A tank that can hold ten fry holding boxes should be supplied with 10 gal (37.8 L) of water per minute.

Sac fry from the hatching trough can be siphoned into a bucket using a 0.5-in (13 mm) hose and transferred to the rearing tanks. Oxygen in the rearing tank should remain about 5 ppm.

An estimation of fry number is crucial so that rearing ponds can be stocked correctly. A convenient time to do this is when they are being transferred to the pond. Two acceptable methods are the **volumetric** and weight comparison methods.

Culture Method. Channel catfish grow best in warm water with optimum growth occurring at temperatures of about 85°F (29.4°C). With each 18°F (10°C) change in temperature the metabolic rate doubles or halves. This means that within limits, their appetite increases with increasing water temperatures or decreases with decreasing water temperatures.

Water quality preferences and limitations for wild channel catfish are not any different from those of farm-raised channel catfish. The lethal oxygen level for both wild and farm-raised catfish is about 1 ppm, and reduced growth occurs at oxygen concentrations of less than 4 ppm.

FIGURE 4-11 Aerators are frequently used in pond production of catfish.

Stocking Rate. Initially, 4- to 6-in (10.2- to 15.2-cm) fingerlings can be stocked at 3,000 to 4,000 per A (7,410 to 9,880 per ha). New producers should not exceed a stocking rate of 3,000 to 3,500 catfish per A (7,410 to 8,645 per ha) for the first growing season. This allows new producers to gain experience in management procedures while reducing potential problems. Exceeding this

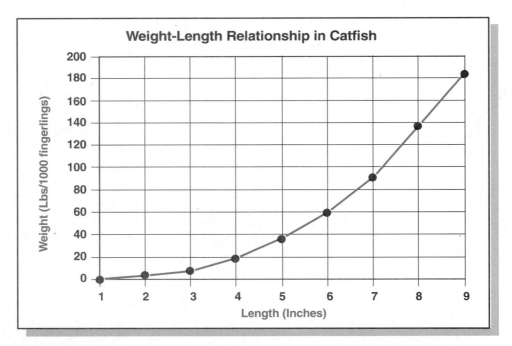

FIGURE 4-12 Weight-length relationship in catfish.

TABLE 4-2 Composite Length-Weight Catfish Fingerling Chart

Length (Inches)	Weight (Lbs/1000 fing.)	Length (Inches)	Weight (Lbs/1000 fing.)
1.0	0.7	2.9	8.1
1.1	0.8	3.0	8.8
1.2	1.0	3.1	9.6
1.3	1.2	3.2	10.4
1.4	1.4	3.3	11.3
1.5	1.6	3.4	12.3
1.6	1.8	3.5	13.3
1.7	2.1	3.6	14.3
1.8	2.4	3.7	15.4
1.9	2.8	3.8	16.6
2.0	3.1	3.9	17.8
2.1	3.5	4.0	19.1
2.2	4.0	4.1	20.4
2.3	4.4	4.2	21.8
2.4	4.9	4.3	23.2
2.5	5.5	4.4	24.8
2.6	6.1	4.5	26.3
2.7	6.7	4.6	28.0
2.8	7.3	4.7	29.7
4.8	31.5	7.0	91.0
4.9	33.3	7.1	94.8
5.0	35.3	7.2	98.6
5.1	37.3	7.3	102.6
5.2	39.3	7.4	106.7
5.3	41.5	7.5	110.8
5.4	43.7	7.6	115.1
5.5	46.0	7.7	119.5
5.6	48.4	7.8	124.0
5.7	50.9	7.9	128.6
5.8	53.4	8.0	133.3
5.9	56.1	8.1	138.2
6.0	58.8	8.2	143.1
6.1	61.6	8.3	148.2
6.2	64.5	8.4	153.4
6.3	67.5	8.5	158.7
6.4	70.6	8.6	164.1
6.5	73.7	8.7	169.7
6.6	77.0	8.8	175.4
6.7	80.4	8.9	181.2
6.8	83.8	9.0	187.1
6.9	87.4		

rate increases the chance of substantial losses caused by water quality problems and diseases. In intensive pond culture systems, the stocking rate varies from 3,000 catfish per A (7,410 per ha) and upward. As the number per acre increases, management problems increase. In ponds with limited or no water available except runoff, stocking rates should not exceed 2,000 catfish per A (4,940 per ha) and a rate of 1,000 to 1,500 per A (2,470 to 3,705 per ha) would be better.

Stockers 6 to 8 in (15.2 to 20.3 cm) long are preferred when available since they will reach a size of 1.5 lbs (0.68 kg) in about 210 feeding days when water temperatures are above 70°F (21°C).

To help determine the number and weight of catfish stocked, the average weight per 1,000 channel catfish and the number of catfish per pound for lengths from 1 to 10 in are given in Table 4-2. Figures given are averages and can vary a great deal depending on the condition of the fish and when they were last fed.

Initial stocking is done as soon as there is water in the pond and catfish of an acceptable size are available. When a pond is clean cropped (all the fish are harvested at one time) restock the pond as soon as it is one-fourth to one-half full and stocker-sized catfish are available.

When a pond is topped (multiple harvested) the pond is restocked as soon as possible after harvest with one 5- to 8-in (12.7 to 20.3 cm) fingerling for each fish harvested.

Feeding. Feeding can occur during day or night, and they will eat a wide variety of both plant and animal material. Channel catfish usually feed near the bottom in natural waters but will take some food from the surface. Based on stomach analysis, young catfish feed primarily on aquatic insects. The adults have a much more varied diet that includes insects, snails, crawfish, green algae, aquatic plants, seeds, and small fish. When available, they will feed avidly on terrestrial insects, and there are even records of birds being eaten. Fish become an important part of the diet for channel catfish larger than 18 in (45.7 cm) long. In natural waters, fish may constitute as much as 75 percent of their diet. Commercial channel catfish are fed complete diets of pellets sprayed over the surface of the pond.

Channel catfish primarily detect food with their sense of taste. Taste buds are found over the entire external surface of catfish as well as inside the mouth, pharynx, and gill arches. They are most numerous on the barbels and gill arches. In clear water, eyesight can be an important means of finding food. In turbid water, taste is the primary way catfish locate food. The organ of smell (olfactory organs) may play some role. The olfactory organs are found in the nostrils (nares), which are located on top of the head just in front of the eyes.

Chapter 5, Fundamentals of Nutrition in Aquaculture, provides more details on feeding catfish.

Diseases. Intensive catfish culture can set a producer up for problems with many diseases. The key to disease prevention is good management that does not stress fish or introduce disease-causing conditions. Chapter 6, Health of Aquatic Animals, discusses diseases affecting catfish.

Harvesting and Yields. In production ponds the growth rate of channel catfish is determined by water temperature, length of time held at different water temperatures, quantity and quality of food fed, palatability or taste of food, frequency of feeding, and water quality. Most farm-raised catfish are harvested at a weight of 1.25 lbs (0.57 kg) at an age of about 18 months.

In a topping or multiple harvest production system, a pond is stocked initially and fed until about one-fourth to one-third of the fish are larger than 0.75 lb (0.34 kg). Then the pond is seined with a seine having a mesh size of 1⅜ to 1⅝ in (3.5 to 4.1 cm). The seine captures those fish that weigh 0.75 lb (0.34 kg) or more and will allow smaller fish to escape. After partial harvesting, catfish fingerlings are restocked at a rate of one for each one harvested.

Processing and Marketing. Where and how the catfish will be sold should be the first concern of anyone thinking about raising catfish. Catfish farmers traditionally sell or market their catfish to—

- Processing plants
- Live haulers
- Local stores and restaurants

FIGURE 4-13 Catfish being harvested from a pond in Mississippi and loaded into a live haul truck.

■ Backyard or pond bank sales to local residents

■ Use their catfish in a fee-fishing operation.

Obviously, some variations of these marketing schemes are used, but these are the main outlets.

In Mississippi, processing plants will not send a harvesting crew more than 50 miles (80 km) from the plant, and they charge about three cents a pound for harvesting. In addition, they charge from one to three cents per pound for transportation. The minimum load that the processing plants will take is 8,000 to 10,000 lbs (3,629 to 4,536 kg). Arrangements for selling fish to a processing plant usually must be made 7 to 60 days before harvest.

Like the processing plants, most live haulers will not take less than 8,000 lbs (3,629 kg) a load. Also, they do not provide harvesting crews. This means the farmers must harvest the fish. Live haulers want catfish only during a four- to five-month period, mid-April to mid-September. The farmer must set production and harvesting schedules to the live hauler's schedule.

Local stores and restaurants usually want fish all year on a weekly basis. This means a farmer must be able to harvest fish weekly either by seining or trapping. One main problem is that many stores and restaurants will take only dressed fish, so the small catfish farmer must be willing to hand-process fish.

Depending on location, area population, size of the catfish operation, the number and size of other catfish operations in the area, and other factors, the backyard sales method can be excellent or poor. Fish are available year-round and are sold live or dressed. Another method used is to harvest once a year and advertise by local radio and newspapers that fish will be available live at the pond bank on a certain date.

The fee-fishing method of marketing catfish, allows the farmer to grow the fish in one or more ponds and permits fishing in any or all the ponds for a fee, usually so much per day or rod, and so much per pound. The pond may be open for fishing all year or just on certain days or weeks. In addition to the usual management problems, this system means that someone must be at the pond when it is open for fishing.

Trout

Salmonids include the members of the trout group and the salmon. Trout live in freshwater. Salmon hatch in freshwater then swim to saltwater where they grow to maturity and return to freshwater to spawn. Salmon are anadromous.

Sources of Species. Table 4-3 lists cultured trout and chars. The most commonly cultured trout is the rainbow trout.

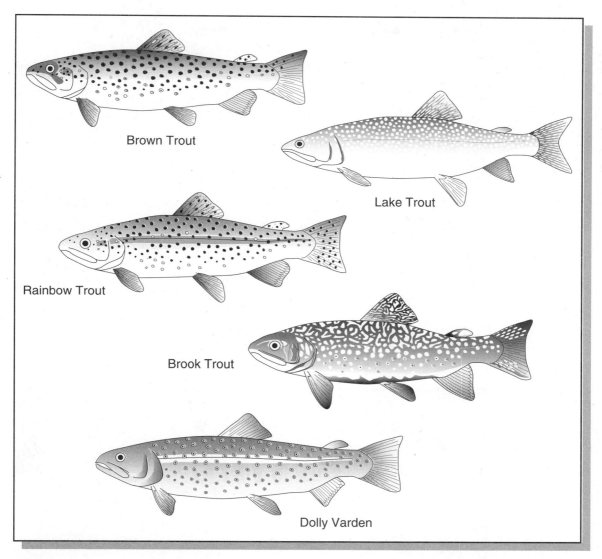

FIGURE 4-14 Types of trout.

Habitat. Trout grow naturally in the streams and lakes of the northern half of the United States. They are cold-water fish, preferring water of 50° to 68°F (10° to 20°C). Growth slows below 50°F (10°C) and above 68°F (20°C). Water temperatures above 75°F are lethal.

Seedstock and Breeding. In commercial production of trout and other salmonids in the United States, eggs are typically produced on broodfish farms, which are separate from farms used for the production of fish for food or for

TABLE 4-3 Current Status of Cultured Trouts and Chars of the Genus *Salmo* and *Salvelinus*.

Scientific Name	Common Name	Current culture status[1]	
		Public	Private
Salmo aguabonita	Golden trout	LIM	LIM
Salmo clarki	Cutthroat trout	MC	LIM
Oncorhynchus mykiss	Rainbow trout	EC	EC
Salmo trutta	Brown trout	MC	LIM
Salvelinus aureolus	Sunapee trout	LIM	NA
Salvelinus fontinalis	Brook trout	MC	LIM
Salvelinus malma	Dolly Varden	LIM	NC
Salvelinus namaycush	Lake trout	MC	NC

[1] Status: EC—Extensively cultured; MC—Moderately cultured; LIM—Cultured on a limited basis; NC—Not cultured; NA—Data not available.

FIGURE 4-15 Trout eggs in the eyed stage. (Photo courtesy Terry Patterson, College of Southern Idaho, Twin Falls, ID)

stocking. The production of good quality, disease-free eggs is a specialized activity requiring a high degree of skill and management.

Most of the eggs used in the commercial production of trout in the southeastern United States are produced in the Pacific Northwest region. Trout eggs are usually shipped when they reach the eyed stage, which is over halfway through the incubation period.

Incubation time is temperature dependent. At 55°F (12.8°C) rainbow trout eggs will hatch approximately three weeks after fertilization, or within four to seven days after received as eyed eggs.

Trout eggs arrive "on ice." The first step is tempering, or gradually bringing them up to the incubation temperature of the hatchery. Any water loss in the eggs from shipping is replaced.

Tempering of the eggs should be done in a clean bucket or other hatchery container by adding the eggs to water that is the same temperature. Temperature of the eggs can be increased to the hatchery water temperature over a 30 to 60 minute time period by adding small amounts of clean water. The eggs need to be gently stirred once or twice during the tempering process to ensure adequate water circulation to all eggs.

Three types of incubator systems are commonly used: California trays, vertical tray or Heath incubators, and upwelling incubators. California trays are screened, flat-bottomed trays that fit inside rearing troughs in series hori-

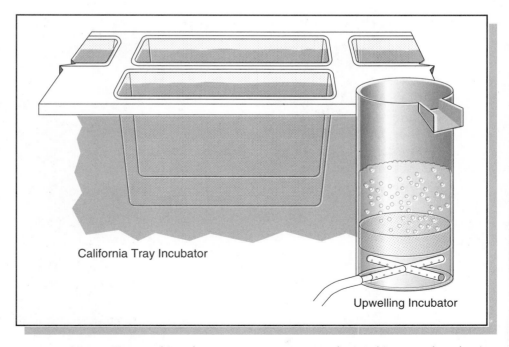

California Tray Incubator

Upwelling Incubator

FIGURE 4-16 Types of incubator systems commonly used in trout hatcheries.

zontally. Between each tray, a partition extending to the bottom of the trough forces water through the eggs from below. Vertical tray incubators are essentially California trays arranged in stacks, having the advantage of requiring relatively little floor space to incubate large numbers of eggs. The water also is aerated as it flows down through the stack. Upwelling incubators are commercially available in several different models, or can be easily constructed from PVC or other materials.

To prevent smothering, trout eggs are placed no more than two layers deep in either California or vertical incubator trays. The recommended water flow in tray incubators is from 4 to 6 gal (15.1 to 22.7 L) per minute (gpm). Upwelling incubators maintain adequate circulation by using the water flow to partially suspend the eggs, but should contain no more than two-thirds of the total volume in eggs. The flow rate in upwelling units should be adjusted so that eggs are lifted approximately 50 percent of their static depth. If eggs are 6 in (15.2 cm) deep with water off, they should be approximately 9 in (22.9 cm) deep with water on. All types of egg incubating containers should be covered to protect developing embryos from direct light.

If the eggs are more than three days from hatching, then dead eggs should be removed regularly to limit fungal infections. Siphoning off dead eggs is more effective than chemical treatment at controlling fungus but can be very time consuming. Formalin added to the inflowing water controls fungus.

FIGURE 4-17 Trout eggs being incubated in upwelling incubators.

Trout eggs should not be treated with formalin within 24 hours of hatching. The eggs will concentrate the chemical inside the shell and die. Once hatching begins, the eggs and sac fry should not be treated with any chemicals.

Hatching rate depends on water temperature, but will usually be completed within two to four days after starting. Empty shells should not be allowed to accumulate in the incubating units. If the eggs are incubated separately from the rearing troughs, the sac fry are transferred into troughs shortly after hatching is complete. Up to 30,000 fry can be stocked into a standard fry trough 10 ft (3 m) long and 18 in (45.7 cm) wide. The water level in the trough is kept fairly shallow (3 to 4 in, 7.6 to 10.2 cm) until fry swim up, approximately two weeks after hatching at 55°F (12.8°C). Any mortalities or deformed fish are removed regularly.

When about 50 percent of fry swim up, feeding begins with small amounts of starter mash on the surface three to four times daily, until most of the fish begin active feeding. Then, if possible, fry are fed every 15 minutes (but not less than every 60 minutes) at this stage. Automatic feeders usually are better and certainly are more convenient than feeding by hand.

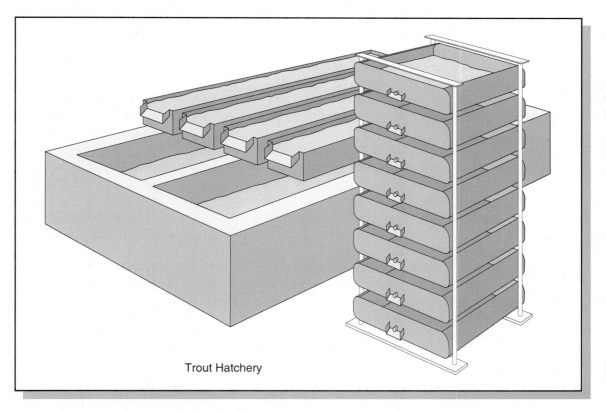

Trout Hatchery

FIGURE 4-18 Typical small trout hatchery with hatching trays, fry troughs, and fingerling tanks.

After the fry have been actively feeding for two weeks, they are counted every week and the feeding rate and feed size are adjusted accordingly. Monitoring of dissolved oxygen levels is a good way to help determine when to reduce the density. Ideally, the oxygen level should not be allowed to be lower than 6 ppm. The fry will be ready to move into larger grow-out tanks when they grow to 200 to 250 per lb (90.1 to 113.4 per kg). In areas where *Yersinia Ruckeri*, the causative agent of enteric redmouth disease (ERM) is present, the fish should be vaccinated seven to ten days before moving into a production facility.

Hatcheries avoid the possibility of disease being inadvertently introduced by restricting traffic and using a footbath.

All equipment used in the hatchery is reserved for hatchery use only. The hatchery and equipment is cleaned and disinfected regularly with a hypochlorite solution, or an approved quaternary ammonium disinfectant. Troughs and floors are also disinfected between groups of fish. Additional ventilation prevents condensation from forming on walls or the ceiling and spreading disease.

Culture Method. Raceways are generally constructed in a ratio of 5 to 1 (or greater) length to width, and with a depth of 3 to 5 ft (0.9 to 1.5 m). Water should flow evenly through the system to eliminate areas of poor water circulation where waste materials or sediment may accumulate. Raceways may be constructed above ground or in the ground from cement or fiberglass, and even wood has been used. Fish cultured in raceways require a large quantity of good-quality water, preferably supplied by gravity flow from artesian wells or higher elevations. If pumping is required, operating cost may be high and risks increased due to possible failure of pumps or power supply.

On the average, 1 to 3 gal (3.8 to 11 L) per minute of flow should be available for each cubic foot of raceway volume at densities of 3 lbs of fish per cubic foot. If supplemental aeration is used, the water requirement may be somewhat reduced. Water flow should be sufficient to keep solid waste material from accumulating in the raceway and to dilute liquid waste (primarily ammonia) excreted by fish.

To achieve good production and minimize problems of stress and disease, water quality should be sustained within desirable ranges at all times. Oxygen should be maintained above 60 percent of saturation. Ammonia levels should remain below 0.1 mg/L in the discharge. Water quality should be monitored frequently, especially oxygen and ammonia, to ensure that conditions remain suitable. This enables the producer to learn more about the production system and its operating characteristics.

Traditionally, raceways are considered to be single pass, flow-through systems. Some fish farmers have developed raceways that are joined with ponds and use the ponds to clean the water prior to reuse. If such a system is designed, the pond(s) should have a volume of at least seven times the total daily discharge volume of the raceway. This allows sufficient time for water quality improvement.

Recirculating systems are often proposed as a type of closed or semi-closed raceway. The water is reconditioned by **clarification**, biological filtration, and reaeration so that most of the water is re-used and only a fraction of the total daily flow is made up of new water. The productive capacity of this system is dependent on the ability of the filtration system to remove wastes and on the volume of replacement water used to improve water quality. Fish production in systems of this sort may reach levels similar to that achieved in raceways. Water quality should be monitored frequently in such a system since without high rates of water exchange, toxic **metabolites** may accumulate rapidly if the biological filtration system is not sufficient to handle the wastes. Chapter 8, Aquatic Structures and Equipment, describes recirculating systems.

Trout may also be raised in earth ponds or cages.

Stocking Rate. The quantity of fish that can be grown intensively in a raceway is more dependent on the quantity and quality of the water than on the size of the facility. Small fish consume proportionally more oxygen per unit of body weight than larger fish, and are normally stocked at lower densities. Densities of fish stocked in raceways may range from 1 to 10 lbs per cubic foot of water, depending on the capacity of the system to support the population. In practice, stocking densities can be calculated based on expected harvest weight of fish to be produced, or based on the carrying capacity of the system. With the latter method, the number of fish is reduced as their size increases.

The carrying capacity of a trout rearing unit is dependent upon fish size, and on several water quality factors, principally oxygen content, temperature, water flow, and volume. Carrying capacity is usually expressed in terms of pounds of fish per cubic foot of rearing space or water flow. A number of different formulas have been devised to calculate carrying capacities, taking into account oxygen consumption, rate of increase in fish length, water volume and temperature, feeding rates, and other factors. As long as the appropriate limiting factors are monitored by the operator, the choice of a particular estimator is a matter of preference.

The easiest method for estimating maximum fish density for a rearing unit is to keep tank loadings within a level of 0.5 to 1 times the length of the fish (in inches) in lbs per cubic ft (ft^3), for example, 2-in fish at 1 to 2 lbs per ft^3, 4-in fish at 2 to 4 lbs per ft^3. The multiplying factor is referred to as a **density index**. Many trout farmers simply stock all sizes of fish at 4.5 lbs per ft^3 as an upper limit for fish density. With proper management, the density can be much higher.

The density index estimates only the appropriate density of fish without regard to water flow in the system. Water flow rate will determine how quickly other water quality characteristics become limiting in each unit. An estimate of the appropriate capacities of trout relative to water flow is to keep loadings within a range of 0.5 to 1 times the fish's length in pounds per gallon per minute (gpm) of water flow, for example, 2-in fish at 1 to 2 lbs per gpm, 4-in fish at 2 to 4 lbs per gpm. This factor is referred to as a flow index, and works on the assumption that inflowing water is at or near saturation of dissolved oxygen. In a properly designed facility, the estimate of carrying capacity obtained from the flow index and the density index will be nearly equal.

Grading. During the production cycle, fish should be graded periodically to maintain size uniformity. Trout usually are graded four times during the period from fingerling stocking, about 3 in (7.6 cm) until they reach a marketable size of 12 to 16 in (30.5 to 40.6 cm). Frequency of grading will vary according to individual circumstances, but should routinely be done whenever loadings need to be decreased.

The simplest graders are made of wooden frames that are as long as the tank is wide and slightly higher than the water is deep. Pieces of aluminum tubing, PVC pipe, or smooth wood are spaced at regular intervals across the frame to perform the grading (Figure 4-19). The grader is put in the top of the tank and fish are crowded down toward the tailscreen. Fish too large to pass through the bars remain at the bottom of the tank, where they can be moved to another tank with fish of a similar size. The smaller fish swim through the bars and remain in the same tank, although 10 percent or more usually remain behind the grader. This method works best with fish larger than 2 to 3 in (5 to 7.6 cm)

FIGURE 4-19 Bar grader for trout production.

long. Grading fish smaller than this is usually not necessary and will be stressful for the fish. Mechanized graders are available, and function by pumping the fish onto a series of grading bars. These systems are very effective when properly sized for the fish to be graded, but are difficult to justify economically for most smaller trout farms.

Inventory. Fish in each tank should be sample counted at least monthly to assure that the fish are growing as expected and to keep track of loading rates. Feeding according to a feeding rate chart allows a check of daily ration amounts and adjustment as necessary. When sample counting, the fish should be crowded starting from two-thirds of the way down the length of the raceway moving toward the head of the tank. The smallest, weakest fish, which will linger toward the tail of the tank, are not representative of the general fish population and will be left behind. With the fish loosely crowded at the head end of the tank, a sample of fish is netted into a bucket of water suspended from a spring tension scale. The weight is recorded and the number of fish is determined as they are poured back into the tank. If fish are graded rather uniformly, three or four samples from different areas are sufficient. Fish size (expressed as number per pound) is calculated by dividing the number of fish in each sample by the total sample weight. The average for each tank is then used to estimate the weight of fish in the entire raceway.

Removing mortalities from each tank on a daily basis and recording their numbers is an important management detail. Dead fish left in tanks are a potential source of disease and indicate poor farm hygiene. Analyzing mortality rates in each tank may indicate developing fish health problems before they become severe. Also, mortalities (morts) should be subtracted each month from estimated population totals to maintain an accurate inventory.

Feeding. Research on trout nutrition has been conducted for more than 40 years. Aside from final sale price of the fish, the amount and suitability of feed used for trout farming will be the primary factor determining the profitability of production. Digestive systems of trout and other salmonids are naturally equipped to process foods consisting primarily of protein (mostly from fish), and can obtain a limited amount of energy from fat and carbohydrates. Diets for fry and fingerling trout require a higher protein and energy content than diets for larger fish. Fry and fingerling feed should contain approximately 50 percent protein and 15 percent fat; feed for larger fish should contain about 40 percent protein and 10 percent fat. The switch to lower protein formulations usually occurs at transition from a crumble feed to a pelleted ration, called a growout or production diet. Several brands of high-quality commercial trout diets are available. Although a farm could produce its own fish food, it is usually uneconomical to do so.

The primary goals in feeding trout are to grow the fish as fast and effi-ciently as possible, maintaining uniformity of growth with the least degrada-tion of water quality. The amount of feed required by trout is dependent on water temperature and fish size. During normal production trout should be fed seven days per week with a high-quality commercially prepared diet formu-lated for trout. Due to higher metabolic rates, smaller fish need more feed relative to their body weight than do larger fish, and fish in warmer water need more feed than fish in cooler water. Because fish are poikilothermic (cold-blooded) their body temperatures and metabolic rates vary with environ-mental temperatures.

In trout, the minimum temperature for growth is approximately 38°F (3°C). At this temperature and below, appetites may be suppressed and their digestive systems operate very slowly. Trout will require only a maintenance diet—0.5 percent to 1.8 percent body weight per day, depending upon fish size—at these temperatures. More than this will result in poor food conversion and wasted feed. Above 38°F (3°C), the metabolism and growth rate of trout will increase with temperature until approximately 65°F (18°C), depending upon the genetic strain being cultured. Optimum temperatures for efficient growth are from 55° to 65°F (12.8° to 18°C). Feeding rates should be at maximum levels—1.5 to 6.0 percent body weight per day. Above 65°F (18°C), the metabolic rate will con-tinue to increase until the temperature approaches lethal levels, but the oxy-gen-carrying capacity of the water and respiratory requirements of the fish will limit the amount of food to be processed efficiently.

In water above 68°F (20°C), a trout's digestive system does not use nutrients well, and more of the consumed feed is only partially digested before being eliminated. This nutrient loading of the water, coupled with generally lower oxygen levels in warm water, can easily lead to respiratory distress. Under these conditions, feeding rates should be reduced enough to maintain good water quality and avoid wasting feed.

The best way to determine the correct amount and sizes of feed needed for trout production is to use a published **feeding chart**, usually provided by the feed manufacturer. The chart should be used as a guide but may need adjust-ment to fit conditions on individual farms. Overfeeding will cause the fish to use the feed less efficiently and will not increase growth rates significantly.

Chapter 5, Fundamentals of Nutrition in Aquaculture, provides more com-plete details on feeding limit.

Diseases. Trout are susceptible to a myriad of bacterial, viral, protozoan, metazoan, and mycotic pathogens, and to environmental alterations such as nitrogen supersaturation, free ammonia, low dissolved oxygen, and a host of environmental contaminants from industrial and agricultural sources. During the course of hatchery rearing, an estimated 50 percent of the fish die. Losses are greatest in the yolk-sac and swim-up stages.

Most of the infectious and noninfectious diseases are easily diagnosed and the causes identified. Currently, oxytetracycline, sulfamerazine, salt, and more recently formalin, are approved for use with fish for human consumption.

Chapter 6, Health of Aquatic Animals, discusses diseases affecting trout.

Harvesting and Yields. Trout can be harvested by seining, trapping, netting, or draining the raceway or pond. Unless the processing plant is on-site, fish are transported in a live-haul truck.

Trout are harvested when they are 7 to 14 in (17.8 to 43.2 cm) long and weigh 0.5 to 1 lb (227 to 454 gm). Depending on the culture conditions, food size fish can be produced in 7 to 14 months.

Processing and Marketing. Trout are marketed at several stages in their production process. Culturalists specialize depending on which of the following markets they intend to target—

- Broodfish marketed to hatcheries
- Eyed eggs
- Fingerlings
- Food fish for processing
- Fee for fishing or other recreational businesses
- Live haulers.

Food fish are transported to a processor where they are killed, graded, dressed, boned, and packaged. Some are sold whole and others are sold as fillets. Processed trout are sold fresh or frozen. Some trout are further processed into specialty products like smoked trout.

Salmon

Salmonids include the members of the trout group and the salmon. Salmon hatch in freshwater then swim to saltwater where they grow to maturity and return to freshwater to spawn. Salmon are anadromous.

The world supply of Pacific salmon, determined by commercial catches, declined from 7.65 million metric tons (16.9 billion lbs) annually from 1935 to 1939 to 405,000 metric tons (893 million lbs) in the 1970s.

Source of Species. Table 4-4 briefly describes the commercially important species of salmon.

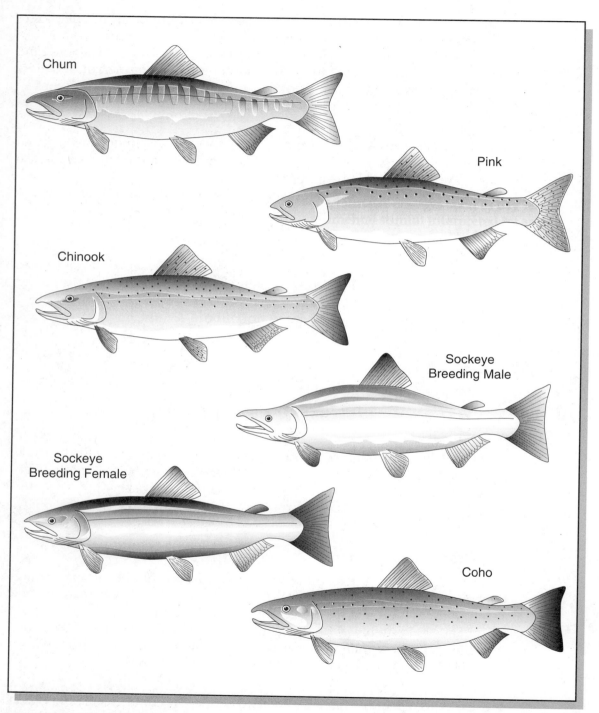

FIGURE 4-20 Types of salmon.

TABLE 4-4 Characteristics of the Major Commercially Important Salmon

Common Name	Scientific Name	Range	Time in Sea Water	Mature Weight
Atlantic Salmon	*Salmo salar*	North Atlantic from New England-Ungava Bay on the west, Iceland, Greenland, from northern Portugal to the Kara Sea on the east	1 to 5 years	2 to 66 lbs (1 to 30 kg)
Sea trout, many local names for the "jack" form, which spends only a few months at sea (finnock, whitling, sewin)	*Salmo trutta*	Sea-going forms found in countries bordering the northeast Atlantic where sea temperatures are <70°F (<21°C) maximum	Months to 3 years, then 1 to 2 years between spawning	11 oz to 22 lbs (300 g to 10 kg)
Steelhead	*Oncorhynchus mykiss*	Western North America from Mexico to the Bering Sea	<1 year to 4 years	7 oz to 42 lbs (200 g to 19 kg)
Pink salmon, humpback, humphy, karafuto-maru, gorbuscha	*Oncorhynchus gorbuscha*	East and west Pacific	Always 2 years	2 to 11 lbs (1 to 5 kg)
Chum salmon, dog, sake, keta	*Oncorhynchus keta*	East and west Pacific. Widest distribution of all the Pacific salmon species	3 to 4 years, 5 for Yukon chum	8 to 9 lbs 44 lbs for Yukon chum (3.5 to 4 kg, 20 kg)
Sockeye salmon, red, blueback, beni-masu, nerka	*Oncorhynchus nerka*	East and west Pacific in rivers with lakes in the system. Greatest numbers from Bristol Bay to the Columbia River on the east, Kamchatka on the west	3 years or more, males may mature as jacks	2 to 11 lbs (1 to 5 kg)
Coho salmon, silver, gin-maru, kizhuch	*Oncorhynchus kisutch*	East and west Pacific from coastal California north to Norton Sound, Alaska, Hokkaido (rare) to the Anadyr River	2 years except for jacks	9 to 11 lbs (4 to 5 kg)
Chinook salmon, king, spring, masunosuka, chavycha	*Oncorhynchus tshawytscha*	East and west Pacific, Ventura River California to Point Hope, Alaska, Hokkaido to the Anadyr River	1 to 5 years, 6 to 7 for Yukon females	Avg 22 lbs (10 kg) max 121 lbs (55 kg)
Masu, cherry, yamama, sima	*Oncorhynchus masou*	West Pacific, over the southern part of the range of the other Pacific salmons	1 to 2 years	Avg 9 lbs (4 kg) jacks <1 year

Habitat. Table 4-4 indicates the habitat for the different species of commercially important salmon.

Pacific salmon introduced into the Great Lakes now support a sport fishery. Hatchery-reared stocks must be used because natural reproduction contributes less than 10 percent of the **recruitment** to the fishery. Survival in the Great Lakes is substantially greater than in marine environments. Salmon in the Great Lakes form a self-perpetuating population in fresh water. Salmon help control the alewife (*Alosa pseudoharengus*).

Seedstock and Breeding. Since salmon that are ready to spawn make their way back to their place of hatching, they are easy to catch. Once caught, they are moved to some hatcheries with tanks or some artificial spawning channels in natural streams or rivers.

The salmon and steelhead trout hatcheries of the Pacific states are among the largest and most technically sophisticated aquatic culture systems in the world. These hatcheries provide fish for a significant portion of the U.S. salmon fishery.

Culture Method. Pacific salmon are cultured by two methods—ranching and farming. California, Oregon, and Alaska allow private ocean **salmon ranching**. Companies rear salmon to migratory size in freshwater hatcheries, then release them into a river or estuary. Fish swim to the ocean, where they graze on natural food. These fish are available to common-property fisheries in the ocean and during their return migration. After their return to the point of release, the ocean rancher processes them for spawning or marketing. Depending on the species, the return time can be two to five years.

Salmon farmers use net pens or sublittoral enclosures. When the smolts or fingerlings are about 6 in (15.2 cm) long, the farmer moves them from the hatchery to the enclosure or pen. **Salmon farming** provides a more reliable, year-round source of fish. Net pen farming requires the farmer to be concerned with the management of the dissolved oxygen, nutrition, wastes, temperature, and salinity.

Stocking Rate. Large variations occur in numbers of returning adult salmon. Various climatic and oceanographic factors that control primary productivity most likely affect the size and abundance of returnees.

A major problem involves gradual loss of wild populations as a result of heavy exploitation of hatchery stocks. Heavy fishing of hatchery stock in the ocean results in heavy fishing of wild stocks where they occur together.

Feeding. Nutritional requirements for growth of young chinook salmon and coho salmon in fresh water are established. The food requirements of salmon are similar to trout. As a carnivorous fish, salmon require a high-protein

FIGURE 4-21 At one time salmon could be caught when they returned to spawn in the Middle Fork of the Salmon River in Idaho. This picture was taken in 1961. (Photo courtesy Dick and Louise Parker)

diet. Growers often feed the same diets to different species, although differences exist in their growth rates and length of time spent in fresh water.

Hatchery rearing may increase requirements of young salmon for certain nutrients that become depleted under stress conditions. Commercial diets for salmon are available.

Diseases. Prevention remains the most effective means of control. Conditions under which diseases occur vary among hatcheries, species, and stocks of fish. With salmon, the difficulty is to define conditions that increase or decrease the occurrence of disease.

Requirements for drug certification are expensive and difficult to meet, and the small potential market does not motivate the pharmaceutical industry to develop new drugs.

Commercial vaccines exist for vibriosis and enteric redmouth disease, but the cost and time involved limits use of these.

Harvesting and Yields. Growth of salmon in hatcheries can be enhanced by such means as varying water temperature, better diets, accelerated feeding, and genetic selection. Size at release seems to be a determinant of survival.

Farmed salmon are harvested between 0.5 and 1.5 lbs (227 to 680 gm). At this size the salmon compete and compare with the trout market. To harvest farmed salmon, nets are opened and fish removed with dip nets, or the net pen may be emptied into a boat. Seines harvest salmon from streams or sublittoral enclosures.

Processing and Marketing. Salmon products are highly valued and are in heavy demand throughout the world. Well-established markets and distribution channels exist for these salmon products. The United States consumes 28 percent of the products; Japan, 25 percent; countries that made up the former USSR, 17 percent; Canada, 7 percent; and others, primarily European countries, 23 percent.

Commercial products include fresh and frozen steaks and fillets, canned, salted, pickled, smoked, kippered, and specialty items, and salted roe (caviar) for human consumption, and meal, oil, animal feed, and bait for industrial uses. The United States probably is the largest consumer of canned, fresh, frozen, and smoked salmon. Japan is the largest producer and consumer of salted salmon.

Demand for sport fishing has increased steadily, but outlook for the future remains unknown. Demand is a function of income, leisure time, cost, and availability of fish. Supply, not demand, represents the major problem confronting salmon fisheries worldwide. Limited supply aggravates conflicts over allocation and increases prices paid by U.S. consumers.

Tilapia

In the United States, the most appropriate species of tilapia for culture are the mouthbrooders: *Tilapia nilotica, T. aurea, T. mossambica, T. hornorum,* and the substrate spawners: *T. rendalli and T zillii.* Various hybrids between mouthbrooding species may also be important. For example, most of the reddish-orange tilapias are hybrids.

Sources of Species. Tilapia belong to the cichlid family. They are native to Africa and the Middle East, but have been distributed widely in tropical, subtropical, and temperate areas. Tilapia were introduced in the United States in the early 1950s. Tilapia established wild populations in parts of Texas.

Potential tilapia culturalists in the United States first determine which species, if any, can be legally cultured in their state. Assuming no restrictions, the selection of a species depends mostly on growth rate and cold tolerance. Rankings for growth rate in ponds are—

1.   *T. nilotica* (Nile tilapia)
2.   *T. aurea* (Blue tilapia)
3.   *T. rendalli*
4.   *T. mossambica* or *T. hornorum.*

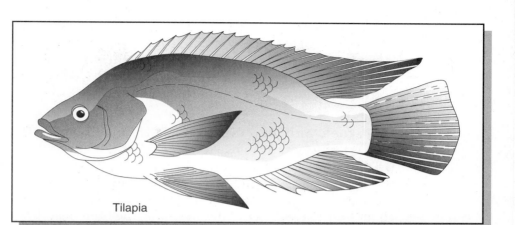

Tilapia

FIGURE 4-22 Tilapia.

Most of the hybrids tested grow as fast as their parent species. Cold tolerance may become an increasingly important criterion for selecting a species in more northerly latitudes. *Tilapia aurea* is generally recognized as being the most cold tolerant.

Habitat. In general, tilapia are extremely hardy animals and can tolerate relatively poor water quality conditions. One of the major constraints to development of the industry in the United States is the inability of tilapia to tolerate low temperatures. Tilapia will not survive at temperatures below 50°F (10°C). Activity is usually reduced at temperatures below 68°F (20°C). Tilapia exposed to low sublethal 50° to 60°F (10° to 16°C) temperatures may develop fungal infections. Ideal temperature for growth is between 79° and 90°F (26° and 32°C).

Tilapia can withstand high water temperatures up to 95°F (35°C). Extremely high mortality rates can result from handling and transporting fish at temperatures above 86°F (30°C), especially smaller fish.

Tilapia probably evolved from marine ancestors and most species can tolerate a wide range of salinities. Changes in salinity greater than 10 to 15 ppt should be gradual. Most important species grow well at salinities up to 30 ppt. Spawning may be inhibited at salinities above 20 ppt, but this may be an advantage in production of food fish in brackish water.

Tilapia tolerate extremely low levels of dissolved oxygen in ponds. Tilapia can use atmospheric oxygen when dissolved oxygen levels drop in ponds. Oxygen levels reach near 0 ppm in heavily manured tilapia ponds without mortality. Although mortality is low, growth rates are reduced during these periods of low oxygen. Tilapia must have access to the water surface. If ponds are covered with duck weed or if fish are crowded in tanks, tilapia cannot survive low dissolved oxygen.

Ammonia can be toxic to tilapia. As with most fish, ammonia toxicity in tilapia is related to pH. The unionized form of ammonia prevalent at higher pH is the toxic form. Chapter 7, Water Requirements for Aquaculture, describes water quality. Tilapia can acclimate gradually to increasing ammonia levels and can tolerate higher levels than most fishes.

Seedstock and Breeding. Tilapia are mouthbrooders. The male establishes a territory and builds a round nest in the pond bottom. The diameter of the nest correlates to the size of the male. The female enters the nest and lays the eggs. The eggs are fertilized by the male. The female then collects and incubates the eggs in her mouth. The eggs are yellow in color. Eggs hatch in about four to eight days. After hatching, the fry remain in the mouth of the female for another three to five days. The fry begin to swim freely in schools, but may return to the mouth of the female when threatened.

Females may spawn every four to six weeks but may spawn sooner if eggs or fry are removed during mouthbrooding. The number of eggs produced per spawn is related to the size of the female. A female of 100 gm can produce approximately 100 to 150 eggs, while a female weighing 1 kg (2.2 lbs) can produce between 1,000 and 1,500 or more eggs per spawning. Males in general will mate with more than one female.

Tilapia can spawn year round if maintained at a temperature between 77° and 86°F (25° and 30°C). Spawning activity is usually inhibited at salinities above 15 to 20 ppt except for *T. mossambica*, which will spawn in seawater.

Tilapia can reach sexual maturity between 50 and 100 gm. If placed in ponds, tilapia will readily spawn and within a short period of time, a pond can become overloaded with fingerlings. At high fingerling densities, growth of food fish is inhibited. This overpopulation of fingerlings is a major constraint to production of larger fish in ponds.

Culture Method. Pond culture is the most popular method of growing tilapia. One advantage is that the fish are able to use natural foods. Management of tilapia ponds ranges from extensive systems, using only organic or inorganic fertilizers, to intensive systems, using high-protein feed, aeration, and water exchange. The major draw-back of pond culture is the high level of uncontrolled reproduction that may occur in growout ponds. Tilapia recruitment, the production of fry and fingerlings, may be so great that offspring compete for food with the adults. The original stock becomes stunted, yielding only a small percentage of marketable fish weighing 1 lb (454 gm) or more. In mixed-sex populations, the weight of recruits may constitute up to 70 percent of the total harvest weight. Two major strategies for producing tilapia in ponds, mixed-sex culture and male monosex culture, revolve around controlling spawning and recruitment.

1. Male and female preparing nest.

2. Female lays eggs in the nest.

3. Male fertilizes the eggs.

4. Eggs picked up in mouth by female.

5. Fertilized eggs held in mouth until they hatch and yolk sac is absorbed – 4 to 5 days.

FIGURE 4-23 Diagram of reproduction in tilapia (mouthbrooders).

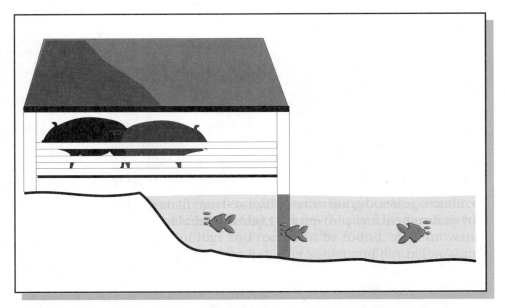

FIGURE 4-24 In other countries, fish and other animals are sometimes managed together. Stables and/or pens are built over fish ponds. Pig manure falls directly into the pond, providing food and fertilizer.

Ponds can be any size, but for ease of management and economical operation, shallow (3 to 6 ft, 0.9 to 1.8 m), small (1 to 10 A, 0.4 to 4.0 ha) ponds with drains are recommended.

Draining is necessary to harvest all of the fish. A harvesting sump is needed to concentrate the fish in the final stage of drainage. Drying the pond eradicates any fry or fingerlings that may interfere with the next production cycle. Geographic range for culturing tilapia in ponds is dependent upon temperature. The preferred temperature range for optimum tilapia growth is 82° to 86°F (27.8° to 30°C). Growth diminishes significantly at temperatures below 68°F (20°C) and death occurs below 50°F (10°C).

In temperate regions, tilapia must be overwintered in heated water. In the continental United States, the southernmost parts of Texas and Florida are the only areas where tilapia survive outdoors year-round with the exception of geothermally-heated waters, most notably in Idaho. In the southern region, tilapia can be held in ponds for 5 to 12 months a year depending on location.

Mixed-sex culture. Mixed-sex populations of fry are cultured together and harvested before or soon after they reach sexual maturity. This eliminates or minimizes recruitment and over-crowding. A restricted culture period limits the size of fish that can be harvested.

In mixed-sex culture, tilapia are usually stocked at lower rates to reduce competition for food and promote rapid growth. One-month-old, 1-gm fry are stocked at 2,000 to 6,000 per A (5,000 to 15,000 per ha) into growout ponds for a four to five month culture period.

Newly-hatched fry should be used because older, stunted fish, such as those held over winter, will reach sexual maturity at a smaller, unmarketable size. Supplemental feeds with 25 to 32 percent protein are generally used. At harvest, average weight is approximately 8 ozs (220 gm), and total production is near 1,400 lbs per A (1,568 kg per ha) for a stocking rate of 4,000 per A (4,481 kg per ha). Expected survival is roughly 70 percent.

Species such as *Tilapia zillii, T. hornorum,* or *T. mossambica* are not suitable for mixed-sex culture because they reproduce at an age of two to three months and at an unmarketable size of 30 gm or less. Tilapia suitable for mixed-sex culture are *T. aurea, T. nilotica,* and their hybrids, all of which reproduce at an age of five to six months.

Male fingerling rearing. With male monosex culture, fry are usually reared to fingerling size in a nursery phase, and then male fingerlings are separated from females for final growout. All-male fingerlings can be obtained by three methods: **hybridization**, sex-reversal, and manual **sexing**. None of these methods is consistently 100 percent effective. A combination of methods works better. Hybridization may be used to produce a high percentage of male fish. The hybrids may then be manually sexed or subjected to a sex-reversal treatment. All three methods are sometimes used.

Hybridization and sex-reversal reduce the number of female fingerlings that must be discarded during manual sexing. This saves time, space, and feed. Problems still exist with hybridization and sex-reversal. Producing sufficient numbers of hybrid fry may be difficult because of spawning incompatibilities between the parent species. Sex-reversal is more technically complicated and requires obtaining recently hatched fry and rearing them in tanks with high-quality water.

Manual sexing is commonly used by producers. Manual sexing (hand sexing) is the process of separating males from females by visual inspection of the external urogenital pores, often with the aid of dye applied to the papillae. Secondary sex characteristics may also be used to help distinguish sex. Reliability of sexing depends on the skill of the workers, the species to be sorted, and its size. Experienced workers can reliably sex 15 to 50 gm fingerlings.

In temperate regions, fingerlings are produced during summer and stored in overwintering facilities for the next growing season. If manual sexing is used, it is done prior to overwintering. The best fingerling size for overwintering depends on the number of fingerlings that will be needed and the available storage capacity. Fingerlings that weigh less than 20 gm should not be overwintered because their survival rate will be low.

Overwintering facilities consist of geothermal springs, greenhouses, and heated buildings. Fingerlings can be held in cages located in geothermal springs or in small ponds or tanks through which warm spring water is diverted. In greenhouses and heated buildings, recirculating systems are used to hold large quantities of fingerlings. Fingerlings can be overwintered in long, narrow ponds that are covered with clear plastic if the winter is mild.

Male monosex culture. Males are used for monosex culture because male tilapia grow faster than females. Females use considerable energy in egg production and do not eat when they are incubating eggs. Male monosex culture permits the use of longer culture periods, higher stocking rates and fingerlings of any age. High stocking densities reduce individual growth rates, but yields per unit area are greater. With a longer growing season, fish weighing 1 lb (454 gm) or more can be produced. Expected survival for all-male culture is 90 percent or greater. A disadvantage of male monosex culture is that female fingerlings are discarded.

The percentage of females mistakenly included in a population of mostly male tilapia affects the maximum attainable size of the original stock in growout. For example, manually sexed *T. nilotica* fingerlings (90 percent males) stocked at 3,848 per A (9,505 per ha) will cease growing after five months when they average about 0.8 lb (365 gm) because of competition from recruits. If larger fish are desired, females should make up 4 percent or less of the original stock and predator fish should be included.

Polyculture. Tilapia are frequently cultured with other species to take advantage of many natural foods available in ponds and to produce a secondary crop, or to control tilapia recruitment. Polyculture uses a combination of species with different feeding niches to increase overall production without a corresponding increase in the quantity of supplemental feed. Polyculture can improve water quality by creating a better balance among the microbial communities of the pond, resulting in enhanced production. The disadvantage of polyculture is the need for special equipment like sorting devices and conveyors. Extra labor is needed to sort the different species at harvest. The role of natural pond foods is less important in the intensive culture of all male populations and may not justify the expense of sorting the various species at harvest.

Tilapia can be cultured with channel catfish with only a minor reduction in catfish yields. Also, catfish production does not decline when cultured in combination with tilapia, silver carp, and grass carp. With no additional feed, total net production increases over that for catfish cultured alone. The incidence of off-flavor catfish may be less in catfish/tilapia polyculture than catfish monoculture. Another promising polyculture system consists of tilapia and prawns. In polyculture, survival and growth of tilapia and prawns are independent. Feed is given to meet the requirements of the fish. Prawns, which are unable to compete for the feed, use waste feed and natural foods that result from the breakdown of fish waste.

Another type of polyculture involves the use of a predatory fish such as largemouth bass to reduce tilapia recruitment. Stocking predators with mixed-sex tilapia populations controls recruitment and allows the original stock to reach a larger market size. Predators must be stocked at a small size to prevent them from eating the original stock. Predators may be stocked when tilapia begin breeding.

The number of predators required to control tilapia recruitment in culture ponds depends primarily on the maximum attainable size of the predator species, the ability of the predator to reproduce, and the number of mature female tilapia. In general, as predators grow they eat larger sized tilapia recruits. More predators are required to control recruitment when there are larger numbers of mature female tilapia.

Stocking Rate. The stocking rate for male monosex culture varies from 4,000 to 20,000 per A (10,000 to 50,000 per ha) or more. At proper feeding rates, densities around 4,000 per A (10,000 per ha) allow the fish to grow rapidly without the need for supplemental aeration. About six months are required to produce 18-oz fish (500 gm) from 1.8-oz (50 gm) fingerlings. Total production approaches 2.2 tons per A (4.9 metric tons per ha). A stocking rate of 8,000 per A (20,000 per ha) is frequently used to achieve yields as high as 4.4 tons per A (9.9 metric tons per ha). At this stocking rate the daily weight gain will range from 1.5 to 2.0 gm. Culture periods of 200 days or more are needed to produce large fish that weigh close to 18 oz (500 gm). To produce an 18-oz fish in temperate regions, overwintered fingerlings should weigh roughly 70 to 100 gm and be started as early as possible in the growing season. A stocking rate of 8,000 per A (20,000 per ha) does require nighttime emergency aeration when the standing crop is high.

Stocking rates of 12,000 to 20,000 per A (30,000 to 50,000 per ha) have been used in 1.2 to 2.5 A (0.5 to 1 ha) ponds, but this requires the continuous use of two to four, 1-hp paddlewheel aerators per pond. Yields for a single crop range from 6 to 10 tons per A (5.4 to 9.1 metric tons per ha).

Feeding. With optimal temperatures, feeding rates depend on fish size and density. Commercial catfish diets can be used but producers usually take advantage of the tilapia's ability to use the natural productivity of the pond. If densities are high, suboptimal feeding rates may have to be used to maintain suitable water quality, increasing culture duration.

The most appropriate mouthbrooding tilapia for culture can feed low on the food chain, on a diet of plankton and detritus. If the natural productivity of a pond is increased through fertilization or manuring, significant production of tilapia can be obtained without supplemental feeds. Although yields are not as high as those obtained with feed and fertilizers, animal manures can be used to reduce the quantity and expense of supplemental feeds.

Inorganic fertilizers are used less often because of their expense, but a single large application of an inorganic fertilizer high in phosphorus is frequently made prior to stocking fish to create an algal bloom. Tilapia productivity is stimulated mainly by an increase in phosphorus and to a lesser extent by an increase in nitrogen. Phosphorus is effectively increased through the application of liquid polyphosphate (13-38-0) at a rate of 20 lbs per A (22.5 kg per ha), or 2.4 gal per A (22.5 L per ha).

Manuring, which is widely used for food fish production overseas, has not been practiced in the United States because of public perception. Manuring may have application in the production of tilapia as a source of fish meal for animal feeds. The quality of manure as a fertilizer depends on several factors. Pig, chicken, and duck manures increase fish production more than cow and sheep manure. Animals fed high quality feeds (grains) produce manure that is better as a fertilizer than those fed diets high in crude fiber. Fresh manure is better than dry manure. Finely-divided manures provide more surface area for the growth of microorganisms and produce better results than large clumps of manure.

Integrated systems. Collection, transport, storage, and distribution of manure involve considerable expense and are major obstacles to manured systems. These problems can be overcome by locating the animal production unit adjacent to or over the fish pond so that fresh manure can easily be delivered to the pond on a continuous basis. Effective and safe manure loading rates are maintained by having the correct number of animals per unit of pond surface area. Animal production units located adjacent to or over fish ponds in some areas include chicken, pigs, and ducks.

1. *Chicken/fish farming.* Maximum tilapia yields are obtained from the manure output of 2,000 to 2,200 chickens per A (5,000 to 5,500 chickens per ha), which deliver 90 to 100 lbs (dry weight) of manure per A per day (101 to 111 kg of manure per ha per day). Several crops of chickens can be produced during a fish production cycle.

2. *Pig/fish farming.* Approximately 24 to 28 pigs per A (60 to 70 pigs per ha) are required to produce a suitable quantity of manure, 90 to 100 lbs of dry matter per A per day, 101 to 111 kg per ha per day for tilapia production. The pigs are usually grown from 44 to 220 lbs (19.9 to 99.7 kg) over a six-month period.

3. *Duck/fish farming.* Ducks are grown on ponds at a density of 300 to 600 per A (750 to 1,500 per ha). The ducks are generally raised in

confinement, fed intensively, and allowed access to only a portion of the pond where they forage for natural foods and deposit their manure. Ducks that are raised on ponds remain healthier than land-raised ducks. Also by raising ducks on ponds, feed wasted by the ducks is consumed directly by the fish. Since ducks reach marketable size in 10 to 11 weeks, staggered production cycles are needed to stabilize manure output.

Although the growing season is adequate to produce marketable size tilapia, economical overwintering systems are essential for broodfish and fingerlings. In order to make the most of the limited growing season, it is essential to have fingerlings stocked early in the season. To do this, fingerlings must be produced in controlled systems during the winter or must be overwintered. This can be done using flow-through systems with warmer deep well water or in closed recirculating systems. Research and commercial applications use both systems.

Diseases. As a group, tilapia are very hardy and thrive under conditions that kill other fish. They have few diseases and parasites. At temperatures below 54°F (12.2°C) tilapia lose their resistance to disease and are subject to infections by bacteria, fungi, and parasites.

Harvesting and Yields. Tilapia are best harvested by seining and draining the pond. A complete harvest is not possible by seining alone. Tilapia are adept at escaping a seine by jumping over or burrowing under it. Only 25 to 40 percent of a *T. nilotica* population can be captured by seine haul in small ponds. Other tilapia species, such as *T. aurea*, are even more difficult to capture. A 1-in (2.54-cm) mesh seine of proper length and width is suitable for harvest.

Yields of male monosex populations in manured ponds have been modest, but production costs are very low if the manure is free. For example, all-male hybrids weighing 29 gm (*T. nilotica x T. hornorum*) stocked at 4,000 per A (10,000 per ha) will produce a net yield of 1,470 lbs per A (1,647 kg per ha) of 200-gm fish in 103 days when given fresh cattle manure at an average rate (dry weight) of 46 lbs per A per day (52 kg per ha per day). In comparison, fish receiving a commercial high-protein feed will give a net yield of 2,370 lbs per A (2,655 kg per ha). Feeding costs per pound of production are 2 to 20 times higher for fish fed the commercial diet compared to fish receiving manure.

Processing and Marketing. Marketing constraints are always a problem when a new product is introduced. Fish retailers indicate a demand for tilapia, but seasonal production is a problem. Ethnic markets hold a lot of potential. Introduction of red varieties definitely increased marketability in parts of the United States.

Hybrid Striped Bass

The hybrid striped bass has become a highly desirable substitute for the declining striped bass seafood industry. As a foodfish, the hybrid exhibits a mild taste and firm texture. Aquaculturalists find these hybrids well-suited to pond culture, and current research is helping to improve culture techniques. Hybrid fry are raised in rearing ponds to 35- to 45-day-old fingerlings. Young fingerlings, when harvested, are graded by size and trained to feed on pelleted feed. Producing market-size fish requires 15 to 18 months.

Sources of Species. Hybrid striped bass generally refers to a cross between striped bass (*Morone saxatilis*) and white bass (*M. chrysops*). This cross, sometimes called the "original cross," was first produced in South Carolina in the mid-1960s using eggs from striped bass and sperm from white bass. The accepted common name of this cross is the Palmetto Bass. The opposite cross using white bass females and striped bass males produces the Sunshine Bass.

Habitat. Hybrid striped bass are stocked into a variety of water types for recreational purposes. They do well in slow-moving streams, large reservoirs, lakes, and ponds. They are seldom found in extremely shallow areas or areas that contain dense growth of aquatic weeds. Because they are pelagic in nature, they are generally found in open water areas. They are generally most active during periods of low light such as dawn and dusk. Beginning in late winter, they tend to concentrate in deep areas near inflowing streams and in the spring may undergo spawning migrations into upstream areas. Hybrids are fertile and there are reports of successful reproduction in a few reservoirs.

FIGURE 4-25 Hybrid striped bass.

Seedstock and Breeding. Broodstock to produce fingerling striped bass and hybrid striped bass are collected from the wild during their spring spawning run. Spawning runs for striped bass species occur from late March to late May depending on location. Spawning grounds for striped bass are usually found near deep, swift, and turbulent sections of a river, well upstream from lakes, reservoirs, and sounds. Males begin their spawning run one to three weeks before the females when water temperature is less than 59°F (15°C). Female striped bass begin their spawning migration when water temperature is around 59°F (15°C). For any given population of striped bass, several periods of spawning may occur during a four to five week period when water temperatures are 61° to 68°F (16° to 20°C). Periods of spawning activity during this time frequently follow sudden temperature increases of 2° to 4°F (1° to 2°C).

White bass, although restricted to freshwater, also make spawning migrations from lakes and reservoirs to inflowing streams. They generally spawn in rocky areas where water flow is turbulent. Their peak spawning season usually occurs from late March to late May depending on location. As with striped bass, male white bass usually arrive at the spawning grounds before the females. Peak spawning activity occurs when water temperatures are 64° to 66°F (18° to 19°C). Usually more than one period of activity occurs for a specific population.

The production of hybrid striped bass must be accomplished by manually stripping the eggs and sperm from the ripe fish into a container. Sperm from two or more white bass or striped bass males is used to ensure fertilization of the eggs.

Fertilization of striped bass eggs is accomplished by using either a wet or a dry method. Little difference exists in the percent fertilization between the two methods.

The most common method of incubating striped bass and white bass eggs is in a modified McDonald hatching jar.

The incubation period varies inversely with water temperature. At 61° to 64°F (16° to 18°C) the incubation period is between 40 and 48 hours. Two hours after fertilization, percent fertilization should be determined by counting the number of eggs with dividing cells. At four hours, an estimate of total number of eggs should be determined volumetrically by letting the eggs settle to the bottom of the jar. The number of eggs per milliliter may be determined by counting the number of eggs in a known volume.

A hatch rate of 50 percent is acceptable and 60 to 80 percent is considered good. The fry are held in 30- to 75-gal (113.6 to 284 L) containers before pond stocking. Water exchange in these containers should be continuous. Newly hatched hybrids have no mouth opening, an enlarged yolk sac, and a large oil globule projecting beyond the head. At four to eight days post-hatch, the yolk sac and oil globule are assimilated, the mouthparts develop, and fry begin to

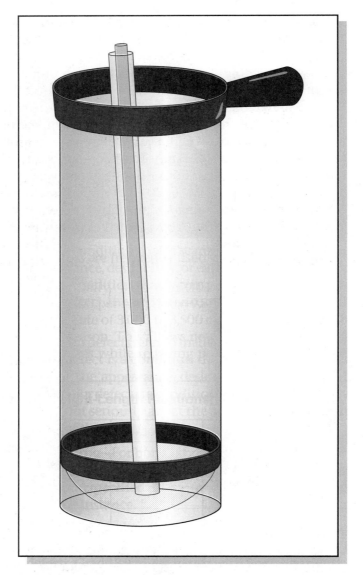

FIGURE 4-26 A modified McDonald hatching jar.

feed. Fry are stocked into fertilized ponds at two to ten days post-hatch. Fry held more than five days must be provided with live food such as brine shrimp nauplii or wild-caught copopod nauplii and cladocerans. Fry should be fed frequently (at least every three hours) during the early rearing period.

Culture Method. Hybrid striped bass survive and grow well in a wide range of water quality variables. Maintaining good water quality is a major part of all phases of production. Temperature and dissolved oxygen levels should be monitored daily, morning and evening, and aerators used to keep dissolved oxygen levels above 4 mg per L. Maximum growth occurs around 77° to 81°F (25° to 27°C). Hybrids can survive a temperature range of 39° to 90°F (4° to 32°C) in culture systems. Below 59°F (15°C), feed consumption is reduced and growth slowed.

Dissolved oxygen is important in any culture operation, and especially for hybrid striped bass. Hybrids may survive dissolved oxygen levels as low as 1 mg per L for a short time, but these levels are very stressful. Dissolved oxygen levels below 4 mg per L reduce feed consumption and growth, while increasing amount of energy needed for respiration, and increasing mortality.

Alkalinity, hardness, and pH levels are usually related, and hybrid striped bass grow well over a wide range of values. Alkalinity of 100 mg per L or above is desirable in culture situations. Mortality can be significant during transfer from water of high alkalinity per hardness to water with low alkalinity per hardness. Hybrids survive in a pH range of 6.0 to 10.0, although 7.0 to 8.5 is optimum for growth. Ammonia, the principal excretory product of fish, should also be monitored regularly in ponds. Concentrations should not exceed 1 mg per L.

Nursery ponds should be filled approximately two weeks prior to stocking fry. Ponds filled too early will develop large populations of predaceous insects that eat hybrid fry. Most hatcheries use freshwater, but brackish water is used in some areas. Generally, hatcheries that use brackish water or hard freshwater (more than 100 ppm Ca hardness) are more successful than those that rely on soft freshwater.

Ponds should be dried and disked prior to filling to promote the breakdown of nutrients in the pond bottom. Agricultural limestone may also be applied to the bottom at this time if necessary. Success in rearing hybrid striped bass depends on the presence of adequate populations of zooplankton. Nursery ponds are usually fertilized with a combination of organic and inorganic fertilizers to enhance the natural production of zooplankters. New ponds or ponds filled with well water may be inoculated with phytoplankton and zooplankton to foster development of the desired zooplankton populations.

Stocking Rate. Fry are generally stocked at a rate of 100,000 to 200,000 fry per A (250,000 to 500,000 fry per ha) at two to ten days of age. Food supply, dissolved oxygen, and other water quality factors are especially important to fish survival. Aeration and circulation of pond water help moderate water quality shifts, improve dissolved oxygen levels, and increase plankton production. Rapid changes in temperature, pH or hardness, and insufficient dissolved oxygen levels all affect survival of larval fish. Constant monitoring of water quality and food supply and remedying problems quickly improves fish survival.

Fingerlings are generally available from producers in the southeastern United States from May to July depending on location. They are stocked at a rate of 8,000 to 12,000 fish per A (20,000 to 300,000 fish per ha) to complete their first year of growth. Ponds of 2 to 4 A (0.8 to 1.6 ha) are recommended for commercial production. Large ponds are more difficult to manage whereas small ponds are expensive to build.

Fingerlings (110 to 225 gm) are stocked into grow-out ponds at a rate of 3,000 to 4,000 fish per A (7,500 to 10,000 fish per ha) depending on the experience of the culturalist. With proper management these fish will reach marketable size by October or November. Survival rates for the second growing season are generally 90 percent or better.

Feeding. Hybrid striped bass is predaceous throughout its life. Survival and production of fingerlings depend upon the culturalist's ability to supply the young fish with live food of good quality and quantity. Striped bass female fry crossed with white bass male fry prefer large crustacean zooplankton as their first food. White bass female fry crossed with striped bass male fry must have an adequate supply of small zooplankters, such as rotifers, because the fry are smaller than fry hatched from striped bass eggs.

FIGURE 4-27 Hybrid striped bass being grown in tanks.
(Courtesy Kent Seafarms Corp., San Diego, CA)

Pond management techniques are used to increase zooplankton populations in the nursery ponds. Pond fertilization requires culturalists to work out the techniques that work best for their situations.

At a size of 1 in (25 mm), fish are introduced to prepared food. The transition to pelleted feed is begun at around 14 to 21 days old when fish are presented prepared food. Particle size of prepared food is critical to successful transition. By 28 days old, fish should be sustained on prepared feed and fed increasing amounts according to growth. Food particle size is increased as fish grow. Food should be offered daily with frequency depending on the amount of natural pond zooplankton.

Advanced fingerlings should be graded before they are stocked for grow-out to reduce the size variation in each pond. Feeding problems will be reduced and all the fish in one pond will reach market size at about the same time.

Fingerlings are fed commercial feed at a rate of 1 to 3 percent of body weight per day. While temperatures are low and dissolved oxygen levels are high, fish can be fed at a rate of 3 percent of body weight per day. As temperature and biomass increase, dissolved oxygen levels become more difficult to manage. The feeding rate should then be around 1 percent of body weight per day. Food conversion ratios of 2:1 or less are expected.

Diseases. Infectious diseases pose few problems in the culture of striped bass even though many disease-causing agents are known. The few disease

FIGURE 4-28 Seine strung across a pond.

FIGURE 4-29 An auger type grading device called a pescalator used successfully in pond culture of hybrid striped bass.

problems that do occur vary greatly from one cultural facility to another. Most of the variation in disease among hatcheries is accounted for by differences in intensity of culture, phase of culture, water quality, and management.

Harvesting and Yields. At the end of the 30- to 45-day nursery period, fingerlings are harvested by seining and draining the ponds. Survival rates are extremely variable, and 0 to 80 percent of the larvae stocked may be harvested as fingerlings. Fish that are to be sold for aquaculture are held in tanks or raceways to be graded and trained to feed on pelleted food.

Grading, or sorting by size, is very important to prevent cannibalism. Losses of 50 percent or more can occur in one to two weeks if fingerlings are not graded every day or two. Cannibalism is prevalent because hybrid fingerlings would normally be switching to a fish diet when they are 2 to 3 in (50 to 75 mm) long. Fast-growing fish should be graded out before they learn to cannibalize. Once trained to take pelleted food, fish are ready to be stocked into ponds.

Mortalities decrease when fish are transported in slightly saline water. Before transferring fish from one system to another, they should be gradually acclimated to the temperature and hardness of the new system.

FIGURE 4-30 Shipping containers and tables for icing hybrid striped bass.

By the end of the first growing season, fish may weigh an average of 0.5 lb (227 gm). Any fish from 3.8 oz (108 gm) should reach a marketable size of 1.25 lbs (0.6 kg) in the second year. Survival rates of 85 percent are common at the end of the first growing season. Fish are harvested after the growing season ends, usually beginning in December, when pond temperatures drop below 52°F (12°C) and continuing through March. Handling fish at 52°F (12°C) or above increases the likelihood of fungus and disease problems. The pond is seined, and the fish are herded through an opening in the seine into a holding net (live-car). The number and weight of fish is estimated by weighing several samples of a known number of fish and taking a total weight of fish. The fish should be weighed in water to reduce stress.

Processing and Marketing. Hybrid striped bass are processed or marketed at 1 to 3 lbs (0.5 to 1.4 kg). They are sold whole in the round, dressed with the head on, and filleted with the skin on or off. Another marketing strategy involves the producer shipping the whole fish—no processing—on ice to distributors in large cities. After harvesting fish, they are chilled on ice and shipped in 50 to 25 lb (22.7 to 11.3 kg) containers shown in Figure 4-30.

Carp

The Chinese cultured grass carp, silver carp, and bighead carp for food for several thousand years. Other Asians and some Europeans cultured carp for lesser periods.

Considerable information exists for the production and management of these fish for food throughout the world, and interest in production for food has increased in the United States. Interest in the grass carp centers on its use as a vegetation control animal. The silver carp has been investigated principally as a nutrient removal species in sewage and animal waste lagoons. Controversy surrounded grass, silver, and bighead carp importation because people fear that they will become established in natural waters, displacing native species and possibly disrupting natural ecosystems, as the common carp is believed to have done. Most states outlawed these exotics. Anyone contemplating their culture must first become acquainted with local restrictions.

Sources of Species. The grass carp was introduced into the United States in 1963, and the silver carp and the bighead carp somewhat later. Carp are native to Asia and were introduced into Europe during the Middle Ages. They were introduced into the United States after the Civil War.

Habitat. Success of carp culture is due to the hardiness of the species at all stages of life. They thrive in pond culture. Carp adapt to acid or alkaline waters. They tolerate varying salinities, a wide temperature range, and turbid waters.

Seedstock and Breeding. Grass carp, silver carp, and bighead carp for potential broodstock should be selected at one to two years of age and stocked in broodfish ponds. Ponds suitable for channel catfish broodfish are adequate. Generally, these ponds are from 1 to 10 A (0.4 to 4.0 ha) in area and have adequate water supplies. Female grass carp and bighead carp mature at about four years, and the silver carp (both sexes), male grass carp, and male bighead carp at about three years. The different species of broodfish can be stocked alone, or some species can be stocked together. Males and females are generally stocked at ratios of 1:1 or 3:2.

In mid-April to mid-May in the southern United States, the broodfish should be seined and inspected. Since common carp spawn at about the same time as the Chinese carps, they can be observed as an indicator of readiness. The female fish should have a distended, flaccid abdomen as compared with that of the males. A pinkish area around the egg vent indicates that a fish is ready to ovulate. The male grass carp should have protuberances on the head and operculum when ripe. Milt usually can be easily expressed from the vent of all mature male carp during the breeding season.

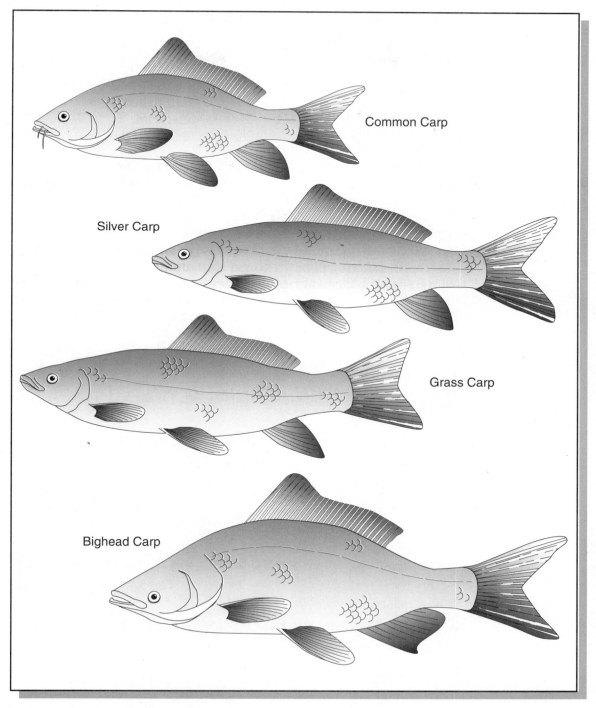

Common Carp

Silver Carp

Grass Carp

Bighead Carp

FIGURE 4-31 Types of carp.

When the seined broodfish appear ready to spawn, fish that will be spawned immediately should be taken from the net and each put into a muslin or porous cloth bag, placed in water in a transport unit, and carried to the spawning shed. Water in the transport unit must be aerated and the fish must be protected from rapid changes in water temperature. Anesthetizing the fish during transport prevents them from injuring themselves.

Broodfish should be stocked in separate tanks in water with sufficient dissolved oxygen and a water temperature near 77°F (25°C). The tanks must be covered to keep the fish from jumping out.

Good quality water is essential for maintaining broodfish, hatching eggs, and culturing larvae and fry. Water obtained from wells is often preferred because it is usually free of pesticides and disease organisms.

Like many fish, hormones can be used to artificially spawn carp. A number of hormone injection schedules are satisfactory.

About 10 hours after the 24-hour hormone injection, the females should be netted and inspected. When slight pressure is applied to the abdomen of a fish that is ready to ovulate, the eggs flow rapidly. If few or no eggs are obtained, the fish should be returned to the vat and examined again after another hour.

When the eggs flow rapidly, the carp is ready to spawn. The following steps are typical for hand-spawning carp and some other fish.

1. Wrap the fish in a towel or place its head in a sack.

2. Dry the fish around the vent.

3. Holding the tail slightly lower than the head, allow the eggs to extrude into a wash pan.

4. Apply additional pressure to eliminate the remaining eggs.

5. Keep all water out of the egg-collecting pan.

6. Remove two males and press the abdomen to ensure that milt is flowing.

7. Dry the fish and press the abdomen to obtain enough milt to cover the eggs.

8. Stir the mixture of eggs and milt with a feather, paint brush, or dry finger.

9. Slowly add water and mix further.

10. Repeat until the water volume is twice the egg volume.

11. Continue stirring for about three minutes.

12. Decant and replace the water.

13. Repeat this process three or four times.

14. After about ten minutes, the eggs are water-hardened.

At this stage a 1-qt (1-L) volume contains 225,000 to 250,000 eggs. Eggs of Chinese carp can be hatched in a variety of devices. For small quantities of eggs, the glass or plastic McDonald hatching jars used by the trout industry are adequate. Larger quantities can be hatched in fiberglass containers. Essentially, all that is necessary is to provide sufficient water exchange to maintain adequate dissolved oxygen and flush out waste metabolites, and sufficient agitation to move the eggs. Movement ensures no dead-water areas in the hatching container where the eggs might suffocate. The outlet must be screened to prevent the loss of the eggs and fry through the hatching tank overflow. The fry can be held in the hatching tanks or removed to other tanks and kept for about five days before they are stocked.

Culture Method. Ponds suitable for the culture of Chinese carp are similar to those used to culture channel catfish fry, bait minnows, and striped bass. Drain and dry the pond and disinfect any wet spots and puddles with hydrated or burnt lime. From one to two weeks before the fry are stocked, the pond should be filled with well water or filtered surface water. The surface water must be free of pesticides, and filtered to remove all fish and predaceous organisms. Any mixture of organic materials and chemical fertilizers that enhances the growth of zooplankton can be applied at the time of the filling.

The fry can be poured or drained from the hatching device into tubs and immediately transported to the ponds. Netting of the delicate fry should be avoided if possible. When transporting fry a distance, oxygenated plastic bags may be used.

At first the fish will feed primarily on zooplankton and other animal organisms. Supplemental feed can be offered. The growth of the fish must be observed. When growth rates diminish, either more supplemental feed must be offered or density must be reduced by transferring some of the fish to other ponds. Water quality must be observed when the fish are fed.

Stocking Rate. Chinese carp are stocked initially in monoculture. The fry can be stocked at the rate of 100,000 to 500,000 per A (250,000 to 1,250,000 per ha). Stocking them in the early morning reduces temperature shock. The vat water temperature can be gradually raised or lowered to that of the pond.

Fingerling grass carp, bighead carp, and silver carp can be produced in monoculture. Grass carp and bighead carp, or grass carp and silver carp, can be produced in polyculture. The date of sale and the desired size at sale dictate the specifics of the production program. For example, relatively large numbers of fish can be stocked when only small fish are needed late in the growing season, but fewer fish must be stocked when the demand is for large fish early in the production period.

Feeding. Feeding and management techniques used depend on whether the fish are being produced alone or in polyculture. Polyculture species could also include some native U.S. fish such as catfish or black bass. Grass carp fingerlings and food-sized fish feed exclusively on aquatic plants in nature.

The fish in managed systems feed on floating pelleted fish feeds and terrestrial grasses such as millet and Bermuda grass. These plants can be fed green (freshly cut) at rates up to 200 lbs per A (224 kg per ha) daily. Some producers feed smaller amounts of green forage to supplement a diet of pelleted feed. Most grass carp producers feed pelleted feed. When small fish first begin feeding, floating meals formulated for minnows may be necessary. The fish appear to feed best near sundown. During winter the fish should be offered feed only on the warmer days.

Bighead carp feed on detritus and zooplankton in the natural environment. In managed environments, they feed on baitfish feeds and sinking channel catfish feeds. Fertilizers, plant materials, and animal wastes that result in the collection of pond bottom detritus can be used to indirectly support the growth of bighead carp. Silver carp feed on plankton in the natural and managed environment.

Chemical fertilizers and organic and animal wastes that stimulate plankton production probably provide the best conditions for growth. Plankton production varies with nutrient source and natural pond fertility. Production systems should be managed for maximum plankton production but without significant reduction in water quality. In polyculture systems consisting of grass carp stocked in combination with silver carp or bighead carp, the production system should be managed for grass carp. The uneaten feed, and other organic materials, provide detritus for bighead carp or nutrients for the phytoplankton that is eaten by silver carp.

Diseases.   Carp are susceptible to the usual array of diseases and parasites, and they respond to the usual treatments and prevention.

Harvesting and Yields.   Seining, handling, and transporting are more difficult for the Chinese carps—especially the grass carp—than for some other warmwater fish. Because grass carp are prone to jump over a seine, the seine must be held up manually, or some other method must be devised to keep the fish inside the net. Some workers use floats to hold the cork line 1 to 2 ft (0.3 to 0.6 m) above the water. Others use a floating seine inside the main seine. The simplest method is for workers to physically hold the net above the water.

Handling can cause self-induced injuries, some of which are fatal. The fish can be placed individually in porous bags (muslin or netting) to restrain them, and can be anesthetized to reduce their activity in the transport tubs and tanks.

All species benefit from the addition of chemical and organic fertilizers that enhance the growth of food organisms. Annual yields of 2,000 to 5,000 lbs per A (2,265 to 5,662 kg per ha) are common.

Processing and Marketing. People in many countries rely on carp as a source of food. So far, they have not been widely accepted in the United States. Some ethnic markets exist. One U.S. market is the sale of grass carp to control vegetation.

Baitfish

Commercial development of baitfish began as early as 1915, but little progress was made until after World War II. During the 1950s and 1960s, rapid expansion occurred. Arkansas traditionally leads as the baitfish producer.

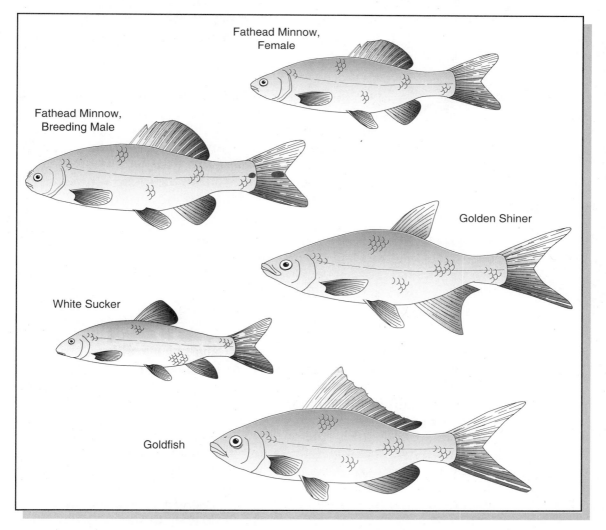

FIGURE 4-32 Some common types of baitfish.

Sources of Species. Of about 100 species of fish used as bait in the United States, only 4 are raised in significant quantities. They include, in order of importance, the golden shiner (*Notemigonus crysoleucas*), the fathead minnow (*Pimephales promelas*), the white sucker (*Catostomus commersoni*), and the goldfish (*Carassius auratus*).

Many farms raise goldfish both for aquarium bait and the trade. Often the most colorful go to aquarium dealers and the others serve as baitfish. Some also serve as feeder fish for aquarium predators.

Other species that may prove profitable to baitfish farmers include killifish (*Fundulus spp.*), chub sucker (*Erimyzon spp.*), stoneroller (*Campostoma spp.*), tilapia (*Tilapia spp.*), Hawaiian topminnow (*Poecilia spp.*), and various other shiners (*Notropis spp.*). Hawaiian topminnows possess many behavioral and biological characteristics necessary as bait for the skipjack tuna (*Euthynnus pelamis*).

Habitat. The golden shiner, produced principally on fish farms in the Midsouth, reaches a size of 10 in (25 cm) when mature. Most are marketed in smaller sizes. The fathead minnow is primarily reared in shallow lakes in Minnesota and the Dakotas, but is also increasing in Arkansas. Fatheads also are produced on Midsouth and Midwest fish farms that use intensive culture systems. White suckers are produced principally in the Upper Midwest as bait for bass, crappie, muskellunge, yellow perch, and northern pike.

Seedstock and Breeding. Almost no research has been done in this field. Selection of broodstock represents a major problem because of the large numbers of fish involved. Baitfish farmers often select broodstock from yearling baitfish. Baitfish begin spawning in the spring. The free spawning, egg transfer, or fry transfer methods are used to produce baitfish.

Culture Method. Most baitfish are cultured in enclosed ponds supplied with water from underground sources. Some small farms use water from nearby creeks. A combination of land, water, and climate favors concentration of the baitfish industry in the Midsouth. Close proximity to markets is also a factor.

A successful baitfish enterprise requires relatively flat land with good water retention. Rocky, gravelly, or sandy soils and rolling or steep terrain usually are undesirable. Adequate quantities of water that is neither too acid nor too alkaline must be available. Surface waters often contain excessive amounts of silt and pesticides, or undesirable fish species with the usual problems of disease and parasites. Absence of adequate underground water generally restricts development of a baitfish farm.

A host of animals prey on baitfish, including otters, diving ducks, egrets, herons, mergansers, bullfrogs, alligators, snakes, snapping turtles, back swimmers, dragonfly nymphs, and other aquatic insects. Baitfish farmers wage a constant battle against these predators.

Stocking Rate. Since baitfish are small and their size varies, so does the stocking rate. Even at market size of, often, no more than 2 in (5.1 cm), a 1-A pond contains about 286,000 fish. The number of fish per pound ranges from 250 to 300. If the young are to be left in the pond to grow, flathead minnows can be stocked at a rate of 2,000 brood females to 400 males per A. If the fry are removed from the pond, 25,000 brood females and 5,000 males may be stocked per A.

Feeding. Under certain water quality and algae conditions, golden shiners produce only 300 lbs per A (336 kg per ha) per year. Farmers need to know what kinds of feed to use before, during, and immediately after spawning. Much of the knowledge about nutrition and diet comes from trout research. Artificial feeding increases production, and commercial diets are available.

Diseases. The nematode *Capillaria* can become a major problem now that more water is being reused. Also, the anchor parasite presents another problem.

A microsporidean, *Pleistophora ovariae*, can progressively damage the ovaries of female golden shiners. One-year-old fish lack significant parasites, and farmers can use them as spawners because golden shiners reach sexual maturity the same year as their birth.

Blooms of large plankton, usually crustaceans or rotifers, compete for food and increase fry mortality in some ponds during the first month.

FIGURE 4-33 Covered concrete holding vats for baitfish.

Harvesting and Yields. Baitfish harvesting techniques have remained constant over the years. Harvesting and grading fish requires considerable hand labor. Baitfish are seined using nets with $\frac{3}{16}$ in mesh. Next, they are dipped from the pond. Tanks or vats hold baitfish until they are sold.

Goldfish farmers produce yields as high as 4,000 lbs per A (4,482 kg per ha) and golden shiner farmers produce yields as high as 1,400 lbs per A (1,568 kg per ha). The average production industrywide varies.

Processing and Marketing. When fishermen buy baitfish they expect healthy, hardy animals. To produce fish that meet these requirements requires quality control throughout the growing and harvesting process. Marketing requires healthy fish graded into sizes suitable for different kinds of sport fishing. Growers use tanks or vats to hold baitfish until they are sold.

Red Drum

Major conflicts have occurred over red drum (*Sciaenops ocellatus Linnaeus*) for 100 years. Recreational fishermen blamed inadequate catches on commercial overfishing, and commercial fishers accused sport fishers of having inadequate skills. Management of the conflict usually attempted to reduce the conflict by addressing recreational concerns without appreciably affecting commercial harvests. Size limits enacted in the 1920s to protect juvenile and adult fish, and seasonal and area net-closures were expanded to protect spawning adults.

In Texas, additional regulations resulted until the sale of native red drum was prohibited in September 1981 because of overfishing.

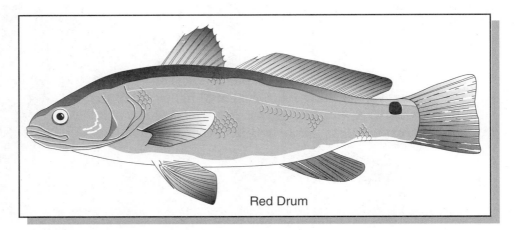

Red Drum

FIGURE 4-34 Red drum.

Sources of Species. Broodstock are captured from natural habitats. Captured red drum can be spawned by controlling the day length and water temperature. Hormone injections can also be used to spawn red drum.

Habitat. The red drum is a quasi-catadromous fish that ranges from Tuxpan, Mexico, on the Gulf of Mexico to Massachusetts on the Atlantic Ocean. Fisheries for red drum have existed since the 1700s. In Texas, red drum are harvested primarily in estuaries where they once comprised as much as 35 percent of the commercial fish landings.

Juvenile red drum occur in all gulf and Atlantic estuaries from Chesapeake Bay to the Laguna Madre of Mexico. Their numbers vary, depending mainly on surface area. Salinity may also affect distribution, probably by affecting survival in early life stages.

Adult red drum permanently emigrate from estuaries to the gulf when sexually mature. Sexual maturity is probably reached in three and one-half years. Adults rarely occur in the estuaries, but commonly occur in the Gulf of Mexico. Adult red drum in the gulf are distributed to at least 70 miles (113 km) from shore. Most occur within 10 miles (16 km) off the shore of Texas.

Seedstock and Breeding. Red drum males court females before spawning. Drumming and nudging intensifies prior to spawning and may be a major stimulus for spawning.

Spawning occurs from August through January and peaks in September or October. Only simulated fall daylight and temperatures induced spawning in the laboratory. Females probably produce over 500,000 eggs annually.

Hormone injections provide seeds at a time out of synchronization with natural spawning and can be conducted in small, inexpensive tanks.

Culture Method. Environmental requirements may change over the life span of the red drum. Red drum begin life in seawater, then passively drift or actively migrate into the brackish waters of bays and estuaries where they pass their larval and juvenile stages. As the fish grow toward adulthood, they migrate seaward and, in the process, encounter progressively increased salinity. In long-lived species such as red drum, the general seaward migration may involve a series of annual migrations, with fish moving into deeper, more saline water in winter and perhaps also during the warmest weather of summer and returning to fresher parts of the estuary during spring and fall.

The more serious obstacle to culture of red drum is their relative intolerance to cold, which limits outdoor overwintering in most of Texas and the other Gulf states. Experiments with juvenile (1 to 4 cm standard length) red drum suggest that cold tolerance is greatest in hard water with a salinity of 5 to 10 ppt.

Stocking Rate. As fry red drum can be stocked in ponds at 296,000 to 492,000 per A (740,000 to 1,230,000 per ha). After 30 to 40 days the fish are harvested as fingerlings and show a survival rate of about 30 to 50 percent. Fish weighing about 0.5 lb (225 gm) have been stocked at 200 per A (500 per ha).

Feeding. Red drum feed throughout the water column, but mainly at the bottom. Prey are sucked up from the bottom by a rapid expansion of the branchial region or captured by biting the bottom.

Red drum feeding in shallow water often exhibit "tailing" behavior. In this activity their caudal and dorsal fins are out of the water as they feed on the bottom. Schools of tailing fish can be easily sighted, which increases their vulnerability to bay fishers.

Occasionally, red drum feed at the surface. Schools have been sighted close to gulf beaches feeding on other fish at the surface. Fish demonstrating this behavior are also easily sighted, which probably increases their susceptibility to fishers.

The most common component of their diet includes crustaceans and fishes. Red drum are omnivorous feeders. Larvae begin feeding within four days of spawning and eat mainly zooplankton. The yolk sac provides nourishment for the first four days post-spawn. Juveniles eat mainly small bottom invertebrates and young fish. Fish 3 to 4 in (76 to 102 mm) rely heavily on amphipods. Shrimp, crabs, and fish (menhaden, gobies, mullet, killifish, and eels) predominate the diet in fish larger than 4 in (102 mm). Adults eat mainly fish.

Diseases. Numerous parasitic organisms have been found on or in red drum, but copepods and cestodes are most frequent.

Harvesting and Yields. Like other fish in ponds or raceways, seines are used to harvest. The yield depends on survival, feeding, growth rate, and feed conversion. Feed conversion should be around 2:1.

Processing and Marketing. Estimated market for red drum is about 14 million lbs. Because the Gulf is closed to commercial capture of red drum, aquaculture production of red drum should have a market. Processing plants on the Gulf processed red drum and provided a market for farm-raised fish. Most of these processors sell to wholesalers who transport the fresh fish directly to retailers. Inland production of red drum will need to find a different market.

Other possible outlets for red drum include fee-fishing operations.

Crappie

In some areas, the favorite sport fish of anglers includes crappie (*Promoxis spp.*). The eating qualities of crappie account for much of their popularity. Because of these qualities, many pond and lake owners request crappie fingerlings for stocking.

Sources of Species. The common names, black and white crappie, are not always a good key to distinguish between the species. The dorsal spines provide positive identification. Black crappie have seven or eight while the white crappie have five or six.

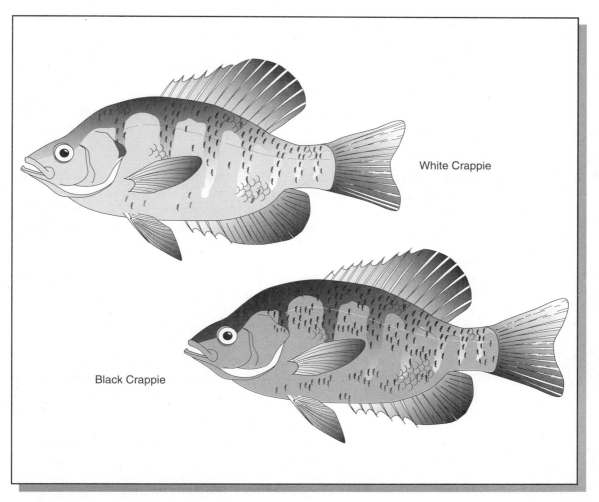

FIGURE 4-35 Some common crappies.

Habitat. Black crappie occur in clearer, slightly acidic waters. The white crappie is native to the more turbid waters of the South and West. Widespread stocking programs caused considerable overlapping of the species range.

Seedstock and Breeding. Despite spawning only once annually, the crappies are among the most prolific species of the sunfish family. White crappie are capable of producing 1,000 to 200,000 eggs per female as compared to 20,000 to 140,000 eggs per female for black crappie. The number of eggs produced depends on the size, condition, and age of females.

Most culturalists prefer broodfish ranging in size from 1 to 1.25 lbs (0.5 to 0.6 kg). Males become darker than females prior to the spawning season, making them easily distinguishable. Females appear swollen with eggs. Broodfish are stocked at the rate of three to eight pairs per surface A.

Black crappie begin spawning activity at 60° to 64°F (15.6° to 17.8°C). The white crappie spawns at slightly higher temperatures (65° to 70°F, 18.3° to 21.1°C). This difference probably reduces the incidence of natural hybridization. Males guard the nests and fry hatch in about five days. Once the yolk sac is absorbed, a ready supply of zooplankton is necessary to maintain growth.

Culture Method. A variety of culture techniques can be used to produce crappie. The extensive method (spawn and rear in the same pond) and intensive method (fry transfer) can be used. Crappie culture techniques are very similar to those used for largemouth bass and bluegill. Culture ponds need a bottom drainpipe, preferably, a concrete drain basin, and a reliable source of high-quality water.

If fingerlings are to be reared using the extensive method, the pond can be dried and planted with a cover crop such as rye grass (10 to 15 lbs per surface A, 11.1 to 16.8 kg per ha) overwinter. The pond is filled two weeks before the onset of spawning. If a water source containing fish is used to fill the pond, water should be carefully filtered. Some ponds may require predacious insect control. To establish a bloom, many culturalists use both inorganic and organic fertilizers (cow manure or cotton seed meal, for example). Small but frequent applications of organic fertilizer maintain the bloom throughout the summer months. Use of inorganic fertilizers throughout the hot months often causes filamentous algae growth, which hampers harvest efforts.

When using the fry transfer method, intensive culture, the spawning pond is not planted with a cover crop or fertilized. Success depends on water clarity in the spawning pond for fry transfer. One to two weeks prior to transferring fry, the grow-out pond is filled, treated, and fertilized as described for extensive production techniques.

Because of their high reproductive potential and forage requirements, crappie often stunt and become undesirable in many ponds. Crappie should not be stocked in ponds that are (1) less than 5 surface A (2.0 ha) in size, (2) turbid (visibility less than 89 in [226.1 cm] throughout most of the year), and/or (3) forage deficient.

Stocking Rate

If the extensive method is utilized, broodfish should be removed from the pond once spawning is completed. A 1 to 2 in (2.5 to 5.0 cm) mesh seine works well for broodfish removal. Broodfish are stocked at the rate of three to eight pairs per surface acre for extensive or intensive techniques.

If the intensive method is used, fry averaging 0.75 to 1 in (1.9 to 2.5 cm) should be transferred to grow-out ponds. Fry can be concentrated by lowering the pond, then running fresh water to the drain basin or one end of the pond. Once the fry are in the drain basin, a fine mesh soft dip net is used for collection.

Crowding the fry should be avoided. A tub containing pond water should be used to transfer the fry. Stocking rates vary between 25,000 to 50,000 per surface A (62,500 to 125,000 per ha). Determination of fry numbers is normally based on volume displacement, weight of a known count, or estimation.

Feeding. A frequently fertilized pond and even a cover crop provide feed for crappies.

Diseases. The fingerlings are extremely delicate and must be handled carefully to avoid stress and bacterial disease.

Harvesting and Yields. Crappie fingerlings should produce 2 to 3 in (5.1 to 7.6 cm) fingerlings by the start of cool weather. The water level in the grow-out pond is lowered. Then fingerlings are dipped from the drain basin or seined with a soft, small-mesh seine. Fingerlings should be loaded into hauling boxes from tubs.

Many culturalists reduce mortality by (1) harvesting and hauling in cool weather only, (2) using additives in transport water, and (3) hauling with oxygen instead of agitators.

Processing and Marketing. Crappie are used to stock farm ponds and for some sport fishing. The individual catching the fish gets to process it—clean it.

Hobby Fish

Most species of hobby fish, also called tropical fish or aquarium fish, occur naturally in tropical or semitropical fresh water, brackish water, or salt water.

Sources of Species. Hobby fish includes representatives of several families, and over 100 species of small, colorful and unique fishes. Table 4-5 describes some common hobby fish.

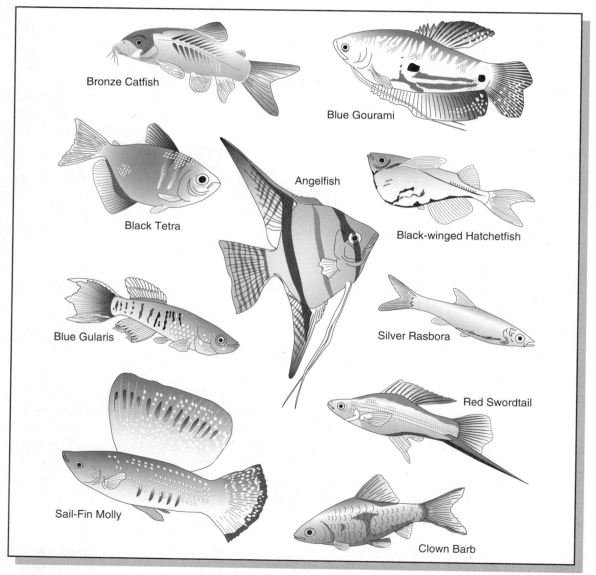

FIGURE 4-36 Some common hobby fish or tropical fish.

TABLE 4-5 Representative Popular Hobby or Tropical Fish[1]

Family	*Scientific* and Common Name	Characteristics
Poeciliidae (livebearers)	*Poecilia latipinna* sailfin molly *P. reticulata* guppy *Xiphophoroug helleri* swordtail *X. maculatus* platy	Often eat the living young they bear and need to be separated from them by using spawning cages or by providing cover for the young
Cyprinidae (carps and minnows)	*Brachydanio rerio* zebra danio *Puntus everetti* clown barb *Rasbora* rasboras	Some of these egg layers scatter eggs in gravel; others scatter them among fine-leaved or broad-leaved aquatic plants, depending on species. Parents do not care for the young and should be removed after spawning has ended.
Cyprinodontidae (killifishes)	*Aphyosemion coeruleum* blue gularis	Eggs laid by these fishes adhere to plants or gravel. Parents do not care for young and should be removed from tank after spawning. The fish are very aggressive.
Characidae (dharacins)	*Carnegiella marthae* black-winged hatchetfish *Copeina gukttata* red-spotted copeina *Gymnocorymbus ternetzi* black tetra *Hemigrammus erythrozonus* glow-light tetra	Egg layers with varied breeding habits. Typically, adhesive eggs are scattered among thick plants. Parents eat the eggs if they are not removed after they are laid. Some species have very exacting water quality requirements and will not spawn if the requirements are not met.
Cichlidae (cichlids)	*Symphysodon discus* discus fish *Cichlasoma meeki* firemouth cichlid *Pterophyllum scalare* angelfish	Includes the mouthbrooding tilapias. Some species build nests on the bottom or in crevices or natural cavities. Still others lay adhesive eggs that adhere to flat surfaces on rocks or vegetation. Parents guard eggs and young. The discus fish "feeds" newly hatched young with secretions in surface mucus.
Anabantidae (gouramies)	*Betta splendens* Siamese fighting fish *Trichogaster trichopterus* blue gourami	Labyrinth fishes with air chambers in their heads. They require atmospheric oxygen and drown if denied access to the surface. The males of the more popular species build bubble nests at the surface for egg incubation and guard eggs and fry. Males of some species must be removed from the tanks because they eat the fry.
Callichthyidae (Corydoras catfishes)	*Corydoras aneneus* bronze catfish	Breeding pairs of species from this family "join" belly to belly, as eggs are extruded into a pocket created by the female's ventral fins, where they are fertilized. The female then deposits them on plants and the process is repeated until all eggs are laid. Parents normally ignore the offspring.

[1] Source: *Third Report to the Fish Farmers.*

Habitat. Habitat varies with the species selected. Table 4-5 provides some information.

Seedstock and Breeding. The livebearers are fairly easy to propagate, but they hybridize so readily that great care must be taken to keep the individual species and strains separate. As a group, they are not as prolific as the egg layers. Livebearers can be propagated in earthen ponds. A producer may stock as many as 400 or as few as 25 broodfish in a pond. It is sometimes desirable to place all the brood fish in one pond, and net or trap the offspring and remove them to other ponds. Specific hybrids of livebearers are usually bred in tanks or vats, and the fry are later transferred to earthen ponds for further growth.

The egg layers are generally difficult to propagate because their breeding requirements can be very exacting. Some species deposit adhesive eggs on vegetation, rocks, tiles, and other surfaces. Other species lay their eggs in small nests that look like depressions in the pond or container bottom. Some species build floating bubble nests in which the embryos develop and hatch. A few popular species are mouthbrooders that gather the fertilized eggs in their mouths and carry them until they hatch.

A few species of egg layers can be propagated in earthen ponds. Mats of Spanish moss may be used as a substrate for species that lay adhesive eggs on plants or rocks. The preferred method for propagating egg layers is inducing them to spawn in tanks or vats. Breeding pairs are stocked into 10- to 30-gal (37.9 to 113.6 L) tanks. Some fish use flat surfaces such as the aquarium walls or

FIGURE 4-37 Hobby fish are popular in offices and homes.

vertical glass plates. Species that eat their eggs must be separated from them almost immediately. For free-spawning fish, breeding traps or boxes with wire bottoms are commonly used to separate the eggs from the parents.

Culture Method. Hobby fish are either domestically produced or imported. Most of the domestically bred fish are cultured at the more than 300 farms located near Tampa in central Florida. Hobby fish are also raised in most of the other states. Most culturalists specialize in the production of the varieties that are most colorful and easy to propagate. All operations, regardless of location, are limited by available space, cost of labor, and the cost of heating water during winter. Even in Florida's warm, semitropical climate, winter temperatures reduce production for a two- to four-month period in a warm-winter year, and virtually stop it in a cold-winter year.

Tanks, vats, and earthen ponds are used to culture hobby fish. A typical farm may propagate 20 to 30 or more species or varieties, each of which requires separate accommodations. Ponds are usually 20×80 ft or 80×100 ft (6.1×24.4 m or 24.4×30.5 m).

Stocking Rate. Stocking rates differ depending on production scheme and species.

Feeding. Feeding of hobby fish is a major consideration to the producer. Sometimes brine shrimp *nauplii* are fed to the fry of some species before they are stocked into ponds. Most of the varieties of fish must be fed in the ponds, and feeding is one of the major costs in production. Often, different formulations and feeding techniques are required for the various varieties of hobby fish. Expensive carotenoids and other pigments are commonly included in the feed to promote bright coloration in the fish.

Diseases. Hobby fish are susceptible to disease. Temperature and low dissolved oxygen levels are common stressors of hobby fish.

Harvesting and Yields. Harvesting is a major consideration because fish that die in retail stores, or soon after being purchased by the aquarium owner, ultimately must be replaced by the producer. Livebearers can be caught with traps and seines. Because trapped fish suffer little trauma, they can be shipped soon after capture. Fish caught with seines must be held for several days before they can be shipped.

Processing and Marketing. Hobby fish are normally shipped by air express to their market, like pet shops and aquaria. The fish are placed in plastic bags partly filled with water, and the rest of the volume of the bag is filled with pure oxygen. These bags are then placed in styrofoam or cardboard shipping containers.

Other Commercial Finfish

A number of marine finfishes, including black drum, dolphinfish, flounder, groupers, rabbitfish, red snapper, sablefish, sea bream, sea trout, threadfish, tunas, and yellowtail may have important possibilities for aquaculture.

In other countries, many marine fish are reared in cages, pens, ponds, or other enclosures. These include sea bream and yellowtail in Japan; groupers, milkfish, and snappers in Southeast Asia; sea bream in Israel; and mullet in Asian and Mediterranean nations. Cage culture of high-priced fish such as groupers, sea bass, sea bream, and snappers appears to be a technique applicable to U.S. legal, social, and economic conditions.

Other fish considered for aquaculture in the United States include the mullet, pike, muskellunge, pompany, sturgeon, sunfishes, walleye, white fish, and yellow perch.

Mullet. The striped mullet (*Mugil cephalus*) inhabits coastal waters and estuaries throughout the tropics and subtropics. The commercial fishery plays an important role in the rural subsistence economics of many developing countries in Asia and the Pacific Basin. In the United States, mullet harvests constitute one of the largest fisheries in Florida.

Commercial mullet culture, as practiced today, is a low-intensity operation dependent on unpredictable natural supplies of fry. Desirable characteristics such as high-quality flesh, extreme tolerance to salinity and temperature, and low position on the food chain give mullet some aquaculture potential in the United States. For this to become a reality in most parts of the country, market development would have to accompany research on food technology, hatchery technology, and food-system dynamics.

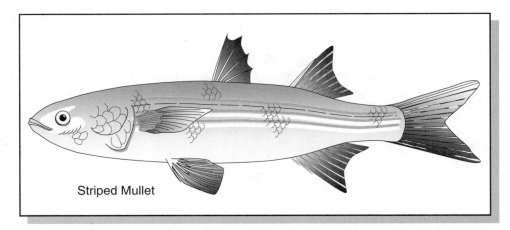

Striped Mullet

FIGURE 4-38 Striped mullet.

Northern Pike and Muskellunge. The most commonly reared pike, the northern pike (*Esox lucius*), may weigh more than 44 lbs (20 kg). It ranges as far south as Iran in the Eastern Hemisphere and as far south as Missouri in the Western Hemisphere. North Americans cultivate it as a game fish and Europeans stock it in carp ponds to control excess reproduction.

The muskellunge (*E. masquinongy*), a larger member of the pike family, also is cultured as a sport fish in the United States. Raising pike species for stocking programs will probably continue in the United States. Research needs include development of artificial feeds and studies on behavior and genetics.

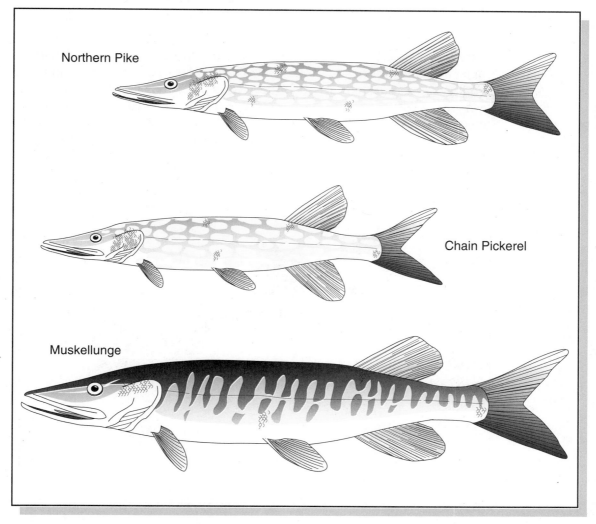

FIGURE 4-39 The pike, pickerel, and muskellunge.

Pompano. The Atlantic pompano (*T. carolinus*), found in Atlantic coastal waters from Massachusetts to Brazil, is the only species cultured experimentally. Several commercial ventures were developed during the 1970s based upon the assumption that these fish could be reared profitably. All failed because of an inadequate technological base. Poorly understood factors such as nutritional and environmental requirements caused extensive mortalities. Pompano consume up to 20 percent of their body weight per day in feed-dry weight of feed to wet weight of fish, and fish may double their weight every two weeks. Because of the metabolic rate and the high rate of oxygen consumption, rapid rates of water exchange in ponds or cages are essential. At temperatures above 86°F (30°C), it becomes virtually impossible to supply enough oxygen to high concentrations of fish.

Sturgeon. The U.S. sturgeon fishery began with production of American caviar in 1855. Smoked and fresh meat, oil, and isinglass also became important products, and a substantial fishery developed on both coasts by 1880. By 1915, overfishing reduced harvests to 10 percent of those in the 1880s. Attempts to replenish stocks through culture failed for both lake sturgeon (*Acipenser fulvescens*) and Atlantic sturgeon (*A. oxyrhynchus*). No commercial culture exists in the United States at present.

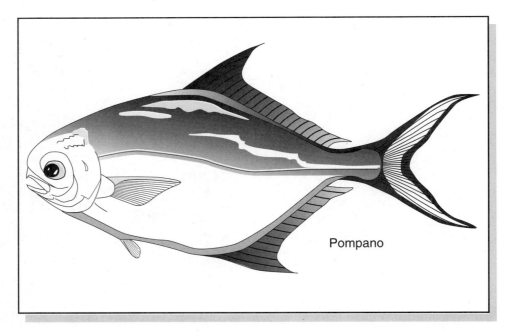

Pompano

FIGURE 4-40 Pompano.

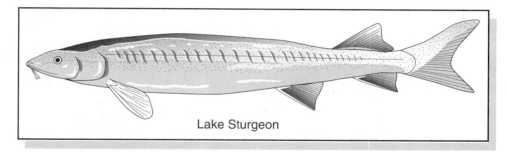

FIGURE 4-41 Lake sturgeon.

Soviet scientists, stimulated by the importance of sturgeon to that nation's economy, developed hormonally-induced spawning techniques. Both Japan and Russia have established aquaculture centers for sturgeon, and the basic hatchery methods are considered routine. Extremely rapid growth rates and hardiness make the fish well suited for culture. In Japan, growth to a size of 7 to 9 lbs (3 to 4 kg) has been recorded in 15 months.

U.S. researchers have induced spawning of the sturgeon and reared wild juveniles in captivity on a variety of feeds. Sturgeon do not spawn until five to six years of age.

FIGURE 4-42 Young Snake River sturgeon at the College of Southern Idaho hatchery.

Sunfishes. Sunfishes have little value in commercial fisheries, but they are popular sport fish. Some states forbid their sale as food. Natural reproduction maintains sport fisheries in most waters. Major U.S. programs of sunfish culture occur in conjunction with farm pond stocking activities. A U.S. farm pond program, which sought to conserve water and wildlife and to provide food and recreation for residents of rural areas, reached a peak in the 1950s. Popular species used in these pond stockings include bluegill (*Lepomis macrochirus*), and redear sunfish (*L. microlophus*). Private hatcheries produce sunfishes for sale to pond owners in the United States.

Bluegill

Redear Sunfish

FIGURE 4-43 Bluegill and Redear Sunfish.

Experimental fish culturalists in this country have not neglected sunfishes, but results are unspectacular. The principal obstacle to commercial culture involves their high rate of reproduction, which often leads to overcrowding and stunting. Sterile hybrids with exceptional growth potential can be produced, but intensive culture has not developed to any great extent. Hatchery propagation for stocking programs probably will continue in private and state hatcheries at the present rate. New aquaculture ventures probably will involve hybrids now cultured to a limited extent in some areas. Additional work is needed on genetics and hybridization.

Walleye. Walleye (*Stizostedion vitreum*) have been artificially propagated since the late 1800s, but spawning and rearing remained inefficient until the 1960s. Cultured fingerlings are used to stock lakes and streams and in a few waters, to enhance commercial fisheries. State hatcheries in northern areas rear large numbers annually.

One culture problem involved maintaining dense populations of copepods and benthic organisms required by large fingerlings. Another problem was that early fertilization methods did not work, and culturalists faced either a high percentage of cannibalism or premature stocking of small fingerlings, which are susceptible to predation. Intensive culture of walleye fingerlings is difficult as the percentage of fry that adapt to formulated feeds is low. Strong interest exists for development of intensive culture of fingerlings to market size. Strong demand, coupled with a drop in landings from Lake Erie, stimulated this interest.

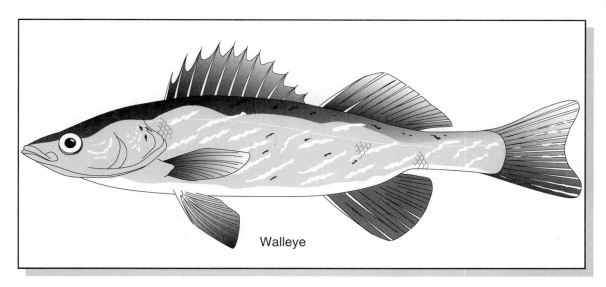

Walleye

FIGURE 4-44 Walleye.

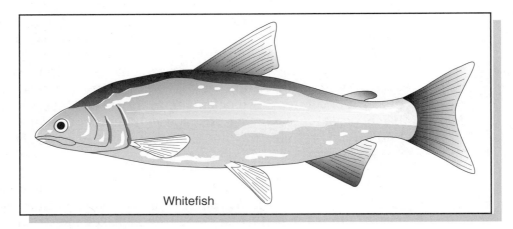

FIGURE 4-45 Whitefish.

Whitefish. Decreasing supplies from natural stocks caused by contamination of the Great Lakes created shortages. Although little work has been done on cultivation of whitefish, an aquacultural potential exists, especially for pen culture in large lakes. Research is needed on nutrition, physiology, and culture systems engineering.

Yellow Perch. Traditionally, the Lake Erie fishery supplied 80 to 88 percent of the Midwestern yellow perch (*Perca flavescens*) market. In 1969, this amounted to 33 million lbs (15,000 metric tons) annually. Threats to the fishery indicate that aquaculture may be needed to supply the market for yellow perch.

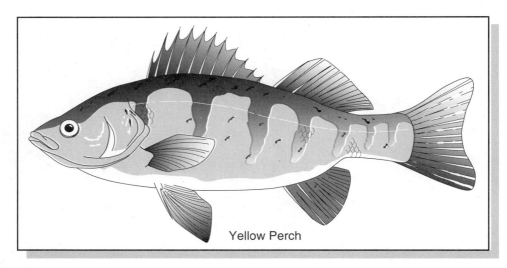

FIGURE 4-46 Yellow perch.

Culture techniques for yellow perch are being developed. Several commercial perch farms started operation. Still, nutritional requirements, disease control methods, and hatchery systems need improvement.

CRUSTACEANS AND MOLLUSKS

The culture of crustaceans and mollusks is not at the level of finfish. In the United States, most crustacean culture is that of crawfish. Shrimp culture is increasing. Mollusk culture is developing.

Crawfish (Crayfish)

The pond culture of crawfish is based primarily on the simulation of the natural hydrological (water) cycle to which the crawfish life cycle is adjusted in natural habitats. The crawfish farmer should establish, maintain, and manage a self-sustaining population of crawfish in the pond. This requires initial stocking, proper control of water and vegetation, and reasonable harvesting that will ensure adequate stock for the next season.

FIGURE 4-47 Crawfish or crayfish. Mature (left) and immature (right) male red swamp crawfish showing variation in body proportion. (Photo courtesy J.V. Huner, Director, Crawfish Research Center, University of Southwestern Louisiana, Lafayette, LA)

MOLTING

For crustaceans to grow they must shed their hard outer shell—the exoskeleton. This process is called molting or ecdysis. The exoskeleton is made primarily of inorganic calcium carbonate (chalk) mixed with chitin and modified proteins. The crawfish provides a good example of the molting process.

The shell or exoskeleton prevents growth of the crawfish between molts. Crawfish grow in a series of abrupt steps rather than in the uninterrupted progression of animals with skeletons. The animal eats for a period and builds tissues internally. Just before the crawfish molts, its shell steadily loses calcium carbonate. This is stored in the stomach in two small stones called gastroliths. With the loss of calcium, the old shell weakens. It splits at a point between the back surfaces of the carapace and the abdomen. The crawfish crawls out, shedding all of its protective shell. The actual shedding lasts only a few seconds.

The body is soft at this time. It rapidly absorbs water and swells, often doubling in weight. Its new exoskeleton remains soft for about 12 hours. Then it gradually hardens as it absorbs calcium carbonate from the supply stored in the gastroliths and from food and water. A crawfish may even eat its old shell for calcium and carbonate.

The molting processes are a complex physiological cycle controlled chiefly by temperature, light, and hormones. The molt itself is hormonally controlled and occurs every 6 to 18 days depending upon the maturity of the crawfish and the water temperature. Great differences in growth rate are usually found among crawfish in a habitat or even in the same brood. These differences are probably due to varying activities and amounts of food consumed.

In the wild, a crawfish may grow about 0.25 in for each molt, which means a maximum size of 2.25 to 3 in. In well-managed culture ponds and wild areas where there is an abundance of food, crawfish often grow 0.5 in or more per molt and reach a maximum size of 3.5 to 5 inches. Under conditions in the southern United States, fast growth occurs in spring. Crawfish require about eleven molts to reach full maturity.

Molting is also a marketing tool in crustacean culture. Soft-shell blue crabs have been produced throughout the coastal areas of the south Atlantic and the Gulf of Mexico for decades. Now crawfish farmers in Louisiana produce soft-shell crawfish.

For a few hours after molting, crawfish shells are butter soft and they are good to eat with very little preparation. Immature, pond-raised crawfish are captured in ponds and transferred to indoor, recirculating aquaculture systems. Here the crawfish are fed and cared for in shallow troughs until they shed their shells. Then the live, soft crawfish are quickly frozen to stop the hardening process. Soft-shell crawfish may be fried or boiled using many different recipes to satisfy the consumer.

A soft-shelled crawfish operation. The technician is removing premolt crawfish and transferring them to a molting tank where the hard crawfish will not kill them. (Photo courtesy J.V. Huner, Director, Crawfish Research Center, University of Southwestern Louisiana, Lafayette, LA)

Sources of Species. Two species, the red swamp crawfish (*Procambarus clarki*) and the white river crawfish (*Procambarus zonangulus*) are cultured in the United States. The red swamp is native to the south-central United States.

Habitat. Crawfish are found in temperate fresh water throughout the world. Their natural habitat includes swamps and marshes. Crawfish are low on the food chain, recycling decaying plant material.

Seedstock and Breeding. Stocking of the "seed" crawfish is necessary the first year, after which the population, if properly managed, should be self-sustaining. Ponds are usually stocked with adult crawfish from late April to mid-May. The ponds should be flooded for at least two weeks before stocking and have some vegetation or other form of cover for the crawfish. Freshly caught crawfish should be used for stocking purposes. Pond-spawned crawfish are preferable since they are more or less used to pond systems and are less likely to migrate from the pond after stocking. The crawfish should be placed in the pond as soon as possible after being caught or harvested and kept cool and damp until that time. To reduce predation on newly stocked crawfish, the stock should be released in densely vegetated areas or in the deepest water far from the pond edge if little cover is present.

Culture Method. An important management procedure in crawfish culture is the manipulation of water level and quality. This involves the draining and flooding of the ponds at the right time to ensure reproduction by the mature crawfish and the production of young crawfish, respectively.

Every year ponds should be drained in late spring or early summer to simulate the summer drought of the hydrological cycle in the natural habitats of the crawfish, during which time the crawfish burrow and reproduce. Although crawfish may be able to reproduce even when there is water year round, draining accomplishes three functions. First, it forces the crawfish to burrow more or less during the same period, ensuring simultaneous reproduction and producing heavy recruitment of young crawfish during the flooding time. Second, draining and the subsequent drying out of the pond allow annual grasses and semiaquatic plants such as alligatorweed, smartweed, and water primrose to grow and become established in the pond. This ensures enough vegetation for food and cover for the young crawfish after the pond is flooded. Finally, draining helps control unwanted vegetation and predators and allows work on pond or dike repair, if needed.

Water should be drained gradually in late June or early July. A quick method to determine when to start draining is to look for burrows of early burrowing females. If burrows can be seen, usually along the banks or under logs or heavy debris in the pond, draining begins. Slow draining allows young crawfish to seek hiding places for the summer and let the adults have time to find suitable burrowing areas. Fast draining will strand some crawfish that are not ready to burrow and expose them to predators. Draining rate should proceed no faster than 2 to 3 in (5.1 to 7.6 cm) per day.

After a period of drying, during which time the crawfish reproduce in the burrows, ponds are flooded. This ensures the newly hatched crawfish will have ample water. Flooding softens the burrow plugs and allows the female and the young crawfish to escape from the burrow and start feeding and growing. Flooding should take place when the water is still warm enough to promote rapid growth and cool enough to hold more oxygen and slow vegetative decay.

Generally, flooding takes place in late September and early October. Digging up a few burrows and observing the condition of the eggs or hatched young of the burrowed females help determine when to flood. If the females are in berry with dark eggs, flooding begins in two or three weeks. If the young crawfish have hatched out but still cling to the underside of the females, flooding begins within a week. If the young crawfish have left the females and are freely swimming in the burrow water, immediate flooding is needed. Checking the burrows should be done at least once a week starting in early September.

Water quality in the pond is a key factor to good crawfish production. Under certain conditions, oxygen depletion may occur. This usually happens during the warm fall and spring months when vegetation is decomposing rapidly. Good water circulation prevents this problem. The cheapest and easiest way to improve circulation is to exchange the pond water with good oxygenated water. This flushes out the deoxygenated water. Another way is to recirculate the water by pumping. In large ponds where water exchange may be a problem, mechanical aerators may be used.

Vegetation in ponds serves as food and cover for the crawfish as well as access to the water surface when dissolved oxygen levels are low. The crawfish will eat a variety of plants. The more tender plants are generally the most desirable, especially for the young crawfish. Any animal matter that pond crawfish get is the result of natural production and their own foraging. The crawfish farmer encourages the growth of suitable food plants and discourages undesirable vegetation. Plants used as crawfish food and cover should be capable of surviving and growing during the dry period and when the pond is reflooded.

Plants generally considered desirable in a crawfish pond include—

1. Alligatorweed (*Alternanthera spp.*)—An excellent food and cover, this plant can grow luxuriantly in many areas.

2. Water primrose (*Ludwigia spp.*)—Considered as good or even better than alligatorweed, this plant does not grow as thick as alligatorweed and is more tolerant to extremely cold weather.

3. Smartweed (*Polygonun spp.*)—A fair food and cover plant, it grows naturally and easily in most ponds and looks like water primrose.

4. Pondweed (*Potamogeton spp.*)—Another fair food and cover plant, it also grows naturally in ponds and is generally submerged.

A number of other plants occurring naturally in ponds, such as duckweed and Elodea, are also fair food and cover plants. Fertilization of the water is not necessary for crawfish culture, unless it is needed to get the food plants growing.

Too much vegetation may hamper harvesting and cause severe oxygen depletion when it decomposes. Too little vegetation may not provide enough food and cover. Generally, cover in the form of rooted, semiaquatic plants should be present over at least 25 percent of the pond bottom during drawdown to provide protection for burrowing and young crawfish. At least a similar amount of vegetation, especially along the edges and shallower areas of the pond, is also needed during flooding to provide food and refuge for young crawfish.

Stocking Rate. Stocking rates vary with existing conditions of the pond. In ponds where crawfish are already present, 20 to 25 lbs per A (23 to 28 kg per ha) may be stocked. In a densely vegetated pond with no existing crawfish population, a stocking rate of 40 to 50 lbs per A (45 to 57 kg per ha) is recommended. Densely wooded ponds and open ponds with sparse vegetation should be stocked at 45 to 60 lbs per A (51 to 68 kg per ha). Ponds with no or very little natural vegetative cover require a higher stocking rate because of higher predation losses. Up to 100 lbs per A (112 kg per ha) may be stocked in these ponds, depending on price and availability.

Stocking rates are based on medium to large crawfish. Sexually mature crawfish in a ratio of at least one female to one male should be used for stocking. A ratio of one male to three or four females will ensure heavier reproduction. In properly managed ponds, further stocking is generally not necessary. Enough adult crawfish present after harvest ensure adequate stock for the next season.

Feeding. A primary advantage of crawfish culture over traditional fish culture is that the crawfish derive their nutrition from natural production of plants and organisms associated with decaying matter. If a good cover crop of desirable vegetation is present in the pond, no supplemental feeding is required. Studies indicate that the addition of pelleted and extruded commercial fish feeds in experimental ponds can produce significant increase in crawfish production. High cost and low feed conversions of the artificial feeds make them uneconomical as yet in commercial crawfish operations. Agricultural forages and byproducts such as hay, sweet potato vines and trimmings, rice bran and stubble, and cottonseed cake, among others, serve as excellent sources of supplemental food and may be added to ponds. Experiments show that pelleted hay sold as rabbit or goat food can support molting and body maintenance requirements of crawfish. These products may be used as supplemental food in ponds with poor vegetation or as starter diets during the first few weeks of crawfish production in newly-built ponds without properly established vegetation. Pelleted hay may also be used as maintenance ration for bait-sized crawfish until needed or sold.

Supplemental feed in crawfish ponds, particularly during warm periods, may cause water quality problems. The farmer should watch for signs of low oxygen levels and be ready to circulate or exchange water and/or remove some of the feed.

The addition of lime in aquaculture ponds is practiced by many farmers to bring up certain soil and water parameters to optimal levels. Liming increases the pH of bottom mud, increasing availability of nutrients, especially phosphorus, for use by plants. It raises the alkalinity of the water, increasing the availability of carbon dioxide needed by plants as well as buffering against drastic daily changes in pH. Liming increases total hardness by addition of minerals such as calcium required by crawfish for their molting and other metabolic processes. It also causes the precipitation of excessive organic matter in suspension in water.

Before adding lime, the farmer should have the soil and water in the pond analyzed. The exact amount to be added depends on factors such as initial soil and water conditions and the chemical relationships among them.

Diseases. Few diseases cause problems in crawfish. Management of oxygen and temperature is critical for growing healthy crawfish. Cold water stresses crawfish, predisposing them to disease. Currently, diseases and parasites are not a major concern to crawfish production. But as culture becomes more intensive, diseases and parasites could become a serious consideration. Intensive culture stretches environment conditions to near extreme.

When extreme environmental conditions are prolonged or intolerable, culture crawfish can more easily become diseased. A bacterial infection known as shell disease could become a problem in intensive culture. Knowledge of crawfish disease lags behind that of finfish.

Crawfish grow well between 72° and 80°F (22° and 27°C) but begin to die when temperatures exceed 91° to 95°F (33° to 35°C). Consumption of oxygen by crawfish increases with temperature. If low oxygen and high temperatures occur at the same time, this enhances the problems. The presence of some bacteria increase when temperatures exceed 83°F (28°C). The most reliable way to manage water is water circulation.

Nutritional diseases also occur when crawfish do not receive the proper diet or enough feed.

Harvesting and Yields. For maximum yield, a crawfish pond must be harvested intensively throughout the production season. Enough reproductively active crawfish are left to serve as broodstock for the next season. Harvesting ponds is quite tedious since it is generally done manually with the use of baited wire traps that should be run daily.

Crawfish traps vary in size, shape, type, and construction, depending on the preference of the crawfish culturalist. The traps are basically cylinders of wire fencing or similar material that have funnel-shaped openings through which the crawfish can enter. The wire mesh depends on the size of crawfish needed to be caught. Traps made of 0.75-in (19 mm) chicken wire or hardware cloth will trap only market-size—3 in (7.6 cm) and larger—crawfish. Traps made of 0.5-in (12.7 mm) mesh will retain bait-size—1.25-in and larger—crawfish. Most crawfish farmers construct traps that are 24 to 36 in (61 to 91 cm) long. Among the preferred types of traps are—

FIGURE 4-48 Harvesting crawfish. (Photo courtesy Chuck Weirich, Delta Research Center, Stoneville, MS)

1. Modified minnow traps with wider funnels and mesh sizes.

2. Stand-up traps with two or three bottom entrance funnels and open top; the traps are propped up on stakes to leave the open end above the water surface.

3. Pillow-shaped traps that are similar in conformation to the stand-up type, except that the nonfunnel end is pinned shut. These traps may be set flat on the bottom in deeper water or propped up partly out of water.

Daily trapping requires 15 to 25 traps per A (37 to 62 per ha). For bait, cut fish and fish heads are usually used, with gizzard shad being the most preferred type. Other good baits are carp, alewifes, and other oily trash fish. Since crawfish prefer fresh animal matter, the bait used should be fresh or fresh frozen. When changing baits, the harvester should not throw the old bait or the remains of it into the pond. Added decayed matter will compound water quality problems.

Crawfish may also be caught with minnow seines in ponds where absence of dense vegetation permits it. Seines are especially useful for collecting soft

FIGURE 4-49 Crawfish are harvested with traps. Pyramid traps, on the floor, are recommended in favor of other designs. There are three entrance funnels around the base of each trap.
(Photo courtesy J.V. Huner, Director, Crawfish Research Center, University of Southwestern Louisiana, Lafayette, LA)

crawfish and those about to molt since these crawfish hardly move at all from their hiding places.

The crawfish crop may be trapped from as early as late November through approximately mid-June or just before water drawdown. The crawfish hatched in late September and early October could attain market size by mid-December, providing there is adequate food and still mild water temperature, above 50°F (10°C). While the young crawfish are growing, many adult crawfish, especially the spawned-out females, are still present. These holdover crawfish should be harvested in late November and early December to provide more space and food for the young of the year. Many of the adults will reach the end of their life span during winter and thus will be lost if not harvested during this period. This one-and-a-half to two months after reflooding gives these adults in the burrows ample time to feed and fill out their tails. Harvesting one or two weeks after flooding will usually yield only hollow-tailed crawfish.

Harvesting may be spotty in January and February when water temperature is well below 50°F (10°C) since crawfish are generally inactive during periods of low temperature. Intensive harvesting should commence when water temperature rises above 50°F (10°C), usually in mid-March or early April, and continue until just before draining. If during the few weeks before drawdown, the average catch is less than 0.5 lb (0.2 kg) per trap per day, it is better to stop trapping. This ensures that enough crawfish will be left as broodstock for the next season.

Processing and Marketing. New product development with whole, cooked, and frozen crawfish and prepared frozen dishes has increased the distribution of crawfish sales.

Crawfish marketed as food are about 3 in (7.6 cm) and larger. Sometimes, incidental catches of small crawfish, less than 3 in (7.6 cm) long may also be sold as bait, which is another lucrative market. Bait-size crawfish are generally available in well-managed ponds from about six weeks after fall flooding into February and March. Since bait crawfish are usually in demand during spring and summer when most of the crawfish have grown to larger food-size ranges, or when ponds are dry, some modifications to the general production schedule for raising food crawfish have to be made.

Prawns and Shrimp

Worldwide, more than 300 species of prawns and shrimp exist. Only about 80 species are commercially important. Shrimp represent the most valuable United States fishery and the most widely cultured saltwater species. Culture of shrimp and prawns is attractive because of the increasing demand. Shrimp culture in the United States is relatively new, while shrimp are commonly cultured in Asia.

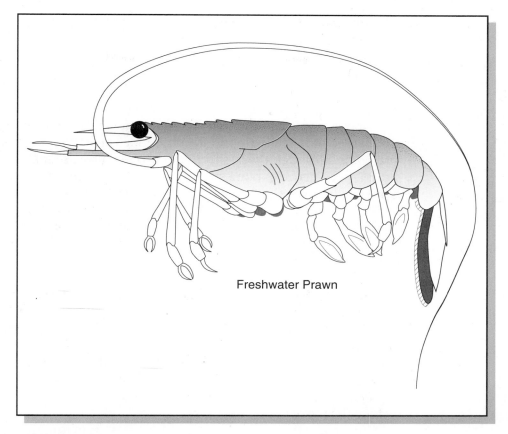

Freshwater Prawn

FIGURE 4-50 Freshwater prawn.

Sources of Species. Throughout the world, most shrimp aquaculture involves the use of penaeid species, such as *Penaeus japonicus,* native to the North Pacific; *Metapenaeus ensis, P. monodon,* and *P. merguiensis,* native to the West-Central and South Pacific; *P. stylirostris, P. vannamei, P. californiensis,* and *P. occidentalis,* native to the eastern Pacific; and *P. setiferus, P. duorarum, P. aztecus,* and *P. schmitti,* native to the western Atlantic.

In the United States, commercial aquaculture of the freshwater prawn (*Macrobrachium rosenbergii*) began in Hawaii and developed most rapidly there.

Habitat. Freshwater prawns, *Macrobrachium rosenbergii,* require warm temperatures of about 82°F (28°C) and saline waters (10 to 20 ppt) for larval stages and fresh or slightly saline waters for grow out. Land and water resources suitable for pond-based aquaculture exist in Hawaii, Puerto Rico, the Southeast and Gulf states, California and perhaps other areas of the United States, plus many foreign locations.

Shrimp live in the oceans from the Atlantic to the Pacific and the Gulf of Mexico.

Seedstock and Breeding. Routine reproduction of *M. rosenbergii* in captivity poses no problem. Domestication is a selection process that favorably modifies economically important physiological, developmental, and behavioral traits. Although the freshwater prawn remains essentially a wild animal, it displays evidence of domestication. Because *M. rosenbergii* breeds readily in captivity, its domestication can be accelerated through selective breeding.

Shrimp are obtained by catching larval shrimp or catching gravid females and spawning them in hatcheries. Eyestalk ablation is an effective method to induce spawners to early maturation. This method may help ensure a constant supply of larvae.

Culture Method. Facilities and equipment for shrimp and prawn aquaculture evolved mainly through trial and error, and through adoption of technologies from other types of operations. Some needs exist for improvement in mechanization and automation.

Prawn culture technology includes three phases: hatchery, nursery, and grow-out. Only the larval stage (hatchery phase) requires saltwater. Nursery production and grow-out are effective in brackish water. Production facilities include the use of ponds and tanks.

Shrimp aquaculture occurs in three phases: (1) maturation and reproduction for production of seedstock (larvae); (2) the hatchery for production of postlarvae; and (3) grow-out to the adult stage in raceways or ponds. Seedstock produced in the maturation and reproduction phase supplies the hatchery phase, and the postlarvae produced in the hatchery supply the grow-out phase.

Figure 4-51 shows the life cycle of a shrimp.

Stocking Rate. Stocking rate varies with the species cultured, the method of culture, and the stage of production.

Feeding. Commercial feeds are available for the various stages of production. What and how much to feed relies on experience and continuous observation of the larvae, sediments, feces, and water conditions. Shrimp are fed on a variety of cultured plankton. Shrimp also eat small fish, other shrimp, soybean cake, peanut cake, and rice bran. Commercial, artificial, complete diets are the main food source for cultured shrimp.

A general lack of information exists on the nutritional requirements of *M. rosenbergii*. A variety of commercial feeds like chicken-broiler starter and

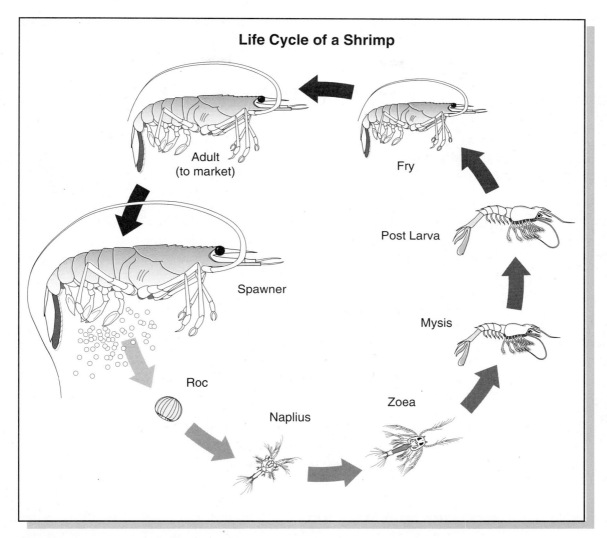

FIGURE 4-51 Life cycle of a shrimp.

catfish and trout feed aids the start of experimental and commercial culture of prawns in ponds. Prawn obtain substantial nutrients from the natural life of the ponds. Since feed accounts for an estimated 20 to 40 percent of operating costs, it can affect the profitability of pond culture operations. Careful selection of an inexpensive, nutritious diet improves profitability.

The principal expenditure in larval culture is for food, mainly brine shrimp. Dependence on expensive and occasionally unobtainable cysts ("eggs") of the brine shrimp troubles hatchery operators.

Diseases. Disease is an important factor in reducing shrimp numbers in natural populations. Natural mortality or death from old age is the potential fate of all shrimp, but the toll taken by predation, starvation, infestation, infection, and adverse environmental conditions is significant. A variety of viral, bacterial, fungal, and protozoan agents affect shrimp. Flukes, tapeworms, and nematodes also occur in shrimp populations.

Microsporidians parasitize crustaceans. In shrimp, microsporidian infections cause a condition known as "milk" or "cotton" shrimp. Microsporidians become abundant in the infected shrimp and cause the white appearance of tissues.

Cramped shrimp is a condition of shrimp kept in a variety of culture situations. The tail is drawn under the body and becomes rigid to the point that it cannot be straightened. The cause of cramping is unknown.

With the development of prawn culture, disease problems can be expected in the list of obstacles to successful production. High-density, confined rearing is unnatural and produces stress.

Harvesting and Yields. Shrimp and prawn are harvested by hand with seines or dip nets. In ponds, the water is drained and shrimp are caught at the drain or lower levels in the pond.

Shrimp and prawn growth depends on the stocking period, stocking density, feed, temperature, salinity, and pond management. Shrimp harvested at 1 to 2 oz (30 to 60 gm) require 100 or more days of culturing from the postlarval period.

Perhaps the most prominent characteristic of a prawn population is its mixed growth rate. Highly variable growth affects the entire industry because it influences the production of large prawns, which command a higher price in the current market.

Processing and Marketing. As soon as the shrimp come out of water they should be placed on ice. Shrimp are beheaded, graded, and frozen. Depending on the location and the market, some may be sold fresh as well as frozen.

As the volume of production increases, so does the need for processing control and improved quality. Currently, most prawns produced in Hawaii are sold fresh, on ice, or live. As this market fluctuates, producers lower prices or process their products to move them. On the mainland, nearly all shrimp are sold as frozen tails or in some processed form.

Oysters

The reasons for the decline in domestic production of oysters includes overfishing, natural disasters, loss of habitat, diseases, economic factors, and pollution.

Overfishing contributed to major declines after the turn of the century. Disease, pollution, and habitat losses have caused major declines since the late 1940s.

Sources of Species. Four species of oysters are harvested commercially in the United States: the American oyster (*Crassostrea virginica*); the Pacific oyster (*C. gigas*); the European oyster (*Ostrea edulis*); and the Olympia oyster (*O. lurida*). The first two species make up the majority.

Habitat. Like other aquaculture species, oyster culture occurs in water free of toxic substances associated with industrial and domestic pollution. Bays and estuaries used for oyster culture require protection. Experiments are underway to culture oysters in controlled environments. These studies should expand, not only to support controlled systems on a commercial scale but to help define water quality in the natural environment.

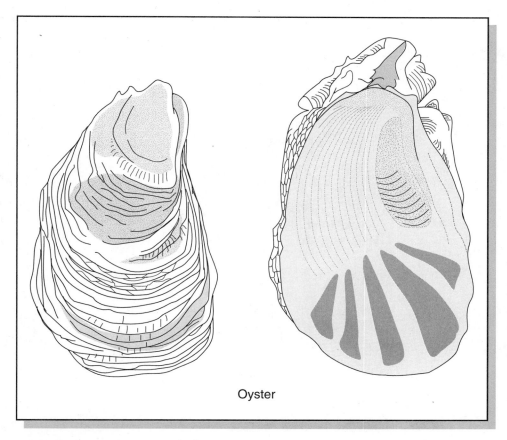

Oyster

FIGURE 4-52 Oyster in shell.

Seedstock and Breeding.   Figure 4-53 illustrates the production of oyster seed. In many areas of the United States, natural seed setting has become sporadic, and in some years a total failure. Advantage should be taken of natural broodstock and setting areas in order to enhance recruitment. New or improved spat (young oysters) capture systems, including development of an

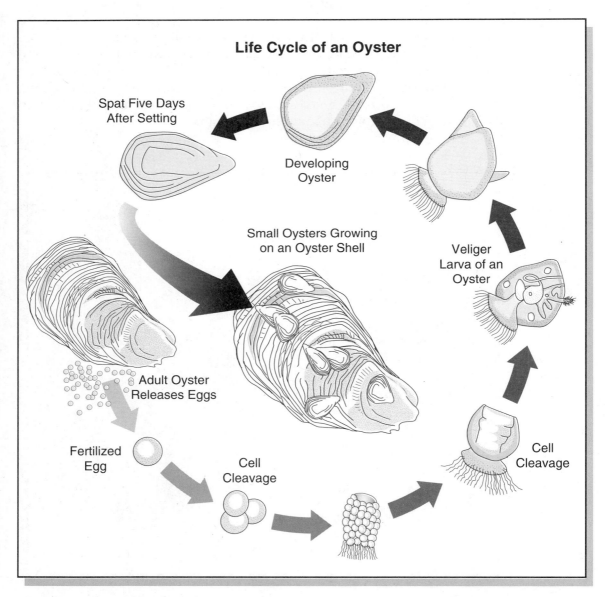

FIGURE 4-53 Production of oyster seed.

economical artificial clutch, are helpful to culture. To increase natural sets, commercial hatcheries were constructed, especially on Long Island Sound and the West Coast.

On the West Coast, the availability of economically competitive seed is a major problem. Three sources of seed include Japan and commercial hatcheries and natural settings in Washington and Canada. No seed was available from Japan in 1978 and 1979 because of high local demand and poor setting.

Culture Method. The oyster industry is seriously constrained by pollution and continued loss of grow-out areas due to water-use conflicts. In many areas it is almost impossible for a new grower to find high-quality grow-out areas. This difficulty may force the use of more efficient culture techniques, such as suspension and off-bottom relaying systems. Cost of these techniques may be higher because of materials and labor, and special permits must be obtained to place such systems in the water. Oyster culture requires space in bays and estuaries that often conflict with other uses.

Oyster culture requires water free from toxicants that inhibit oyster growth and from contaminants that reduce meat quality. Present culture techniques cause high mortalities. For example, the movement of seed from setting to grow-out areas can result in more than 50 percent mortality. On the West Coast, 75 percent of the oysters die during the first year. By increasing survival 25 percent, production would increase substantially and in turn lower culture costs.

Stocking Rate. Stocking rate depends on the culture system used.

Feeding. Oysters are filter feeders. A variety of algae species are needed to rear oyster larvae to metamorphosis. But to grow oysters from postmetamorphosis to seed size or to market size in intensive systems is too expensive because of the kinds and quantities of algae required. Growers need to know the nutritional requirements of oysters at every stage and how algae can be mass cultured to fulfill these requirements.

Diseases. A number of diseases and parasites plague oysters. The most serious ones that affect adult oysters are *Minchinia nelsoni* (MSX), *M. costalis* (SSO), and *Dermocystidium marinum* (Dermo). The first two organisms cause heavy mortalities in the mid-Atlantic area—Delaware, Chesapeake, and Chincoteague Bays. Dermo causes mortalities in the Gulf Coast states. Mortalities from Dermo also have been recorded as far north as the Chesapeake Bay. Stress conditions implicate viral and other microbial diseases in shellfish mortalities.

Major predators include oyster drills, starfish, flatworms, crabs, and fish. Mortality varies with geographic area, salinity, and other factors.

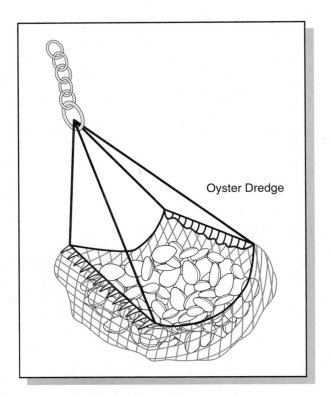

Oyster Dredge

FIGURE 4-54 Oyster dredge for harvesting.

Researchers and growers regard oyster drills as the most damaging predator. Successful research efforts to reduce or limit drill predation could save the industry millions of dollars.

Harvesting and Yields. With intensively managed bottom culture such as in Long Island Sound, it is possible to produce 4,450 lbs per A (5,000 kg per ha) per year. Public grounds, with little or no management, produce only 9 to 90 lbs per A per year (10 to 101 kg per ha). Studies of off-bottom culture indicate that 0.25 A (0.1 ha) covered by rack cultures could yield 2.9 tons (2.6 metric tons) of meats per year. Such yields depend on factors like productivity of the waters and total flow.

The size of an oyster culture operation varies from 2 to 3 A (1 to 1.2 ha) in Maryland to 10 to 15 A (4 to 6 ha) in Virginia to 200,000 acres (80,940 ha) in Long Island Sound. Many states limit the number of acres a person can lease.

Harvesting methods range from primitive tools, such as hand tongs, to modern hydraulic dredges. In some cases, state laws dictate the type of gear that can be used. Also, the size of the operation limits the gear that can be used economically.

Processing and Marketing. Oyster processing often begins with hand shucking—unpleasant work. Oyster shucking machines that produce a fresh product need development. Oysters are sold whole, fresh, or canned.

Crabs

While crab meat is high priced, ocean fisheries provide most of the crab for the markets. Generally, crab are not considered a major aquaculture species. The blue crab (*Callinectes sapidus*) fills a special market niche, the demand for soft-shelled crabs or soft crabs, which sell for five to six times more than hard-shell crabs.

To produce soft-shelled crabs, blue crab, just about to molt, are harvested. These crabs are placed in holding boxes, fenced pens, or indoor recirculating systems. Crabs are observed for signs of molting. After a split occurs along the back of the shell, molting takes place in a few hours. Within four or five hours after molting the crabs are removed and sold. If left in the water their shells begin to harden.

Lobsters

Analysis of catch statistics shows that lobster fisheries are harvested at or near maximum sustainable yields. This is an especially severe problem for the American lobster (*Homarus americanus*), whose populations are on the verge of a major decline. Since 1955, annual U.S. landings have been fairly constant at

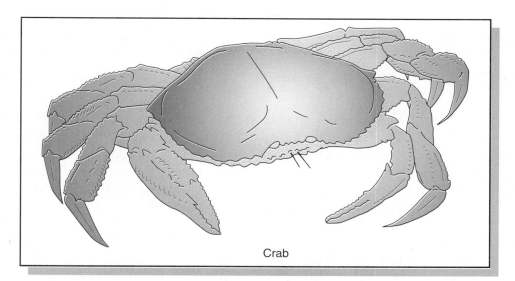

Crab

FIGURE 4-55 Crab.

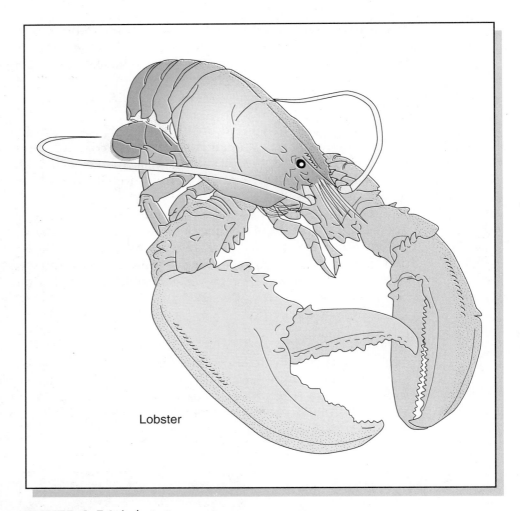

FIGURE 4-56 Lobster.

approximately 30 million lbs (13,620 metric tons). Fishing efforts quadrupled. Awareness of the limits of the fishery stimulated strong commercial interest in lobster aquaculture.

Hatcheries are used to supplement the number of lobsters available for commercial catch. The farming of lobsters from the egg to the market-sized adult is not profitable yet. One reason is the five to six years required to grow a lobster to market size.

Individualized containers will probably remain a requirement of lobster aquaculture and is a major part of the large initial capital investment. Container size affects growth of the American lobster, but the reason is not entirely known.

Shell erosion, a chronic problem, becomes more critical as desirable brood-stock is held for longer periods. Shell erosion also influences marketability and probably predisposes the animal to other infections.

The basic technology for lobster aquaculture is available, but a number of problems hinder rapid commercialization.

Clams

Hard clam harvesting traditionally provided supplementary employment for those in other seafood industries. For example, during closed seasons for oysters and crabs, oystermen and crabbers often go clamming. Because harvesting equipment is relatively simple and inexpensive, clamming also attracts many part-time and recreational clammers. Recently, clamming became more of a full-time occupation. Several trends encourage aquaculture of hard clams—

■ A chronic shortage of smaller size clams—littlenecks and cherry-stones—and a strong demand

■ The increased price of small clams

■ Declining harvests in New York, New Jersey, and Virginia—major producers in the past.

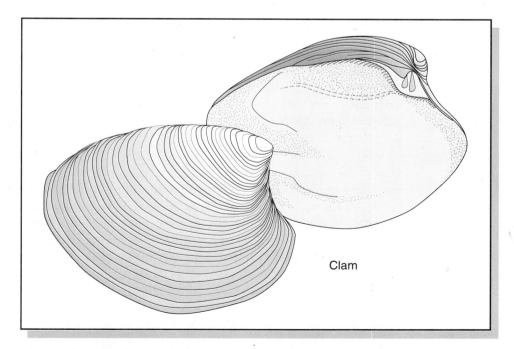

Clam

FIGURE 4-57 Clam in shell.

Sources of Species. The hard clam (*Mercenaria mercenaria*) is harvested commercially in 16 states.

Seedstock and Breeding. Gonadal development begins at 46° to 50°F (8° to 10°C). Spawning occurs between 72° and 82°F (22° and 28°C). Growth takes place in water between 46° and 82°F (8° and 28°C).

Culture Method. Clams require relatively clean seawater with a salinity range between 20 and 38 ppt. Salinities may go above or below that range for short periods without causing losses of larger seed or adult clams, but larvae and small clams are vulnerable beyond this range.

Clams grow intertidally and subtidally in virtually any bottom where they can burrow. They do well in soft mud, sand, shell, and rubble. Growth is maximum in coarse sand and inhibited in areas with high silt clay content. Shell or rubble offers some predator protection, and clams often are more abundant in these and other coarse substrates. Where currents run one knot or less, clams grow very well. Because they do not require a firm bottom, clams do not compete for oyster growing grounds.

Stocking Rate. They can be grown at relatively high densities of 1 to 2 million per A (2.47 to 4.94 million per ha). Clams reach littleneck size throughout most of their range.

Feeding. Clams are filter feeders, feeding on the phytoplankton in their environment.

Diseases. Bacteria are considered responsible for culture failures. The clam leech (*Malacobdella grossa*) and other parasites, and commensals are found in clams.

Harvesting and Yields. Harvesting methods in use include rakes, hoes, patent tongs, modified oyster dredges, modified surf clam dredges—hydraulic dredges attached to cages—and hydraulic dredges attached to conveyors. Legal constraints usually dictate the type of harvesting used in a given area. Although some justification exists for restricting methods of harvesting wild stocks, this is counterproductive when applied to aquaculture. No unreasonable restrictions should be placed on gathering cultured clams, provided the method used does not adversely affect the environment.

Processing and Marketing. Clams are usually sold in the shell, eliminating the need for packaging or processing. If the harvest increases because of aquaculture of larger numbers of individuals, some clams probably will be processed. Hard clams usually are sold in the shell, either by the bushel or individual count. Prices vary—smaller sizes or grades (littlenecks) have a higher value than larger clams. Most littleneck clams are consumed raw, steamed, or as specialty items, such as clam casino or deviled clams.

Mussels

In the United States, the blue mussel (*Mytilus edulis*) is the only indigenous mussel species cultured commercially. Mussels are commonly cultured in other parts of the world. Most of the mussels cultured in the United States are grown in suspended culture systems such as longlines or rafts. Bottom culture is also profitable.

Mussels are easy to raise, requiring little attention after being placed in the growing area. The market for mussels is increasing in the United States. Most mussels are being cultured in the Northeast Atlantic states.

Seedstock for culture is obtained from wild stock. Mussels attach to almost any structure by means of a thread-like byssus, which they secrete. Unlike oysters, mussels can discard their byssus and move to a new location.

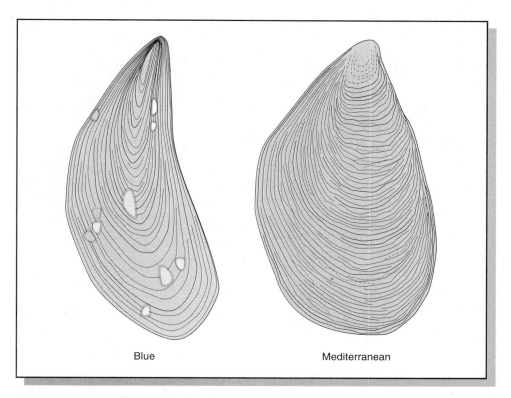

Blue   Mediterranean

FIGURE 4-58 Mussels.

With good management and a good location mussels will grow to market size of 2 to 3 in (5.1 to 7.6 cm) in one or two years depending on the size of the seed stock and the climate.

Abalone

The largest and most commercially important species of abalone, the red abalone (*Haliotis refescens*), occurs mostly in California waters. Fishermen harvest less than 1 million lbs (450 metric tons) annually from the wild, but demand and prices are high and are increasing.

Japanese scientists developed hatchery culture procedures for abalone and Japan has large stocking programs. Private firms tried abalone culture in the United States in 1965, 1968, and 1972. These attempts, hampered by technical problems, resulted in an orientation toward research rather than production. Both tank culture and stocking juveniles in natural habitats have been tested. A few California firms are currently showing a good potential for abalone culture.

Problems include slow growth rate, high post-larval mortality, design of tank culture systems, protection from predation in open coastal waters, cost-effective feeds and feeding systems, and legal aspects of obtaining adequate space for production facilities. The time required to reach market size may be three or more years, depending upon species, water temperatures, and feeding rates. Research is needed on the physiology, nutrition, and genetics of abalone.

OTHER COMMERCIAL SPECIES

Depending on the market, location, and the entrepreneurship of an individual, many other species are or could be cultured. Some species need more research. Some species need markets developed. For some of the other species that hold potential refer to Chapter 2, Aquatic Plants and Animals, and Appendix Table A-10. Some of these include—

■ Alligators

■ Eels

■ Scallops

■ Bullfrogs

■ Octopuses

■ Squids

■ Conch

■ Snails.

AQUATIC PLANTS

A number of marine plants are used extensively by humans. Commonly referred to as seaweeds, the worldwide harvest of these attached algal forms was an estimated 5.3 billion lbs (2.4 million metric tons), valued at about $800 million. Aquaculture produced a significant percentage of the plants. The principal use of marine algae in some countries, especially in Asia, is for food. In the United States, seaweeds are a source of chemical extracts used in the food-processing and chemical industries.

The United States harvests 254,000 lbs (115 metric tons) of marine algae with a value of $2 million. At present, the only commercial-scale culture in the United States involves "seeding" and transplanting the giant brown alga Macrocystis (known as kelp) along the California coast. Other seaweeds are being reared in experimental or pilot-scale operations.

Demand for marine algae grew steadily during the past 20 years. The worldwide harvest more than doubled between 1960 and 1973. With a finite natural supply, commercial culture will become a significant industry. If present efforts to use kelp as a biomass source of energy succeed, vast areas of the oceans may be devoted to algae farms. Problems of culture include predation and undesirable growth of competitive seaweeds with low commercial value.

Freshwater plants, such as waterchestnut and watercress, are grown widely for food. Water hyacinth, duckweed, and other species serve as feed and fuel. Use of freshwater plants in recycling wastes holds great potential, and investigations into this topic have just begun. Field testing of and applied research on plants in polyculture and waste recycling systems are needed.

Culturalists raise unicellular algae on both small and large scales as food for larval fish, oysters, and other invertebrates. Some species have value as human food. Spirulina is grown for this purpose in Mexico and Chad. Pigments, glycerol, and other hydrocarbons can be produced from unicellular algae, and pilot-scale production of several species is underway in several countries. Blue-green algae have nitrogen-fixing abilities similar to leguminous plants, and researchers are experimentally putting these algae on rice field soils.

The broad potential of aquatic plants to produce protein, energy, chemicals, pigments, hydrocarbons, and other special products makes them deserving of further research. Areas of study should include basic physiology and reproduction, nutrition, environmental requirements, food chain dynamics, and germ-plasm maintenance and transfer.

CONTROLS

Proper management of aquatic production requires control of water, disease, predators, and vegetation. Table 4-6 shows some the items controlled in a cycle of catfish production.

Water Supply and Levels

Water quality and quantity are critical to a successful aquaculture venture. Requirements of many species are given in this chapter. Chapter 7, Water Requirements for Aquaculture, details the water requirements for aquaculture.

Disease

Disease represents a substantial source of lost income to aquaculturalists. Production costs are increased by fish disease outbreaks because of the investment lost in dead animals, cost of treatment, and decreased growth during convalescence. In nature we are less aware of fish disease problems because sick animals are quickly removed from the population by predators. In addition, aquatic

TABLE 4-6 Disease, Parasite, and Weed Control Frequency of Occurrence, Acres Requiring Treatment, Delta of Mississippi

Item	Expected Frequency of Occurrence	Acres Requiring Treatment (%)	Treatment
Parasite incidence	Once annually	26.6	8 ppm potassium permanganate per acre ft of water
Bacterial incidence	Four times annually	17.5	Maintain feeding schedule with medicated feed (Romet-30 or TM-50) for 5 or 10 days
Weed control	Once annually	75.0	1 ppm copper sulphate per acre foot of water
Fungus	Once annually	17.5	4 ppm potassium permanganate per acre foot of water
Nitrite	Once annually	100.0	30 ppm sodium chloride per acre foot of water
Ammonia	Once annually	100.0	Flush one-half of pond with fresh water and add 20 lbs of triple superphosphate per surface acre

animals are much less crowded in natural systems than in captivity. Parasites and bacteria may be of minimal significance under natural conditions but can contribute to substantial problems when animals are crowded and stressed under culture conditions.

Disease is rarely a simple association between a pathogen and a host. Usually other circumstances must be present for active diseases to develop in a population. These circumstances are generally grouped under the umbrella term "stress." Management practices directed at limiting stress are likely to be most effective in preventing disease outbreaks. Complete details on diseases are contained in Chapter 6, Health of Aquatic Animals.

Bird Predation. Bird predation is a serious problem for producers. Most birds that cause problems are migratory and are protected under the federal Migratory Bird Treaty Act (MBTA). The birds' migratory nature complicates the problem, because predation varies greatly depending on migration patterns, time of the year, migratory concentrations, and the location of catfish ponds. Proximity to nesting or rookery sites can also compound the problem.

Besides eating the fish, these birds can damage property. They are known to transmit fish diseases. Predatory birds consume the individual fish that are easiest to catch. Fish that are easily caught are often those that are diseased. So, the birds pick up diseases and transmit them to other ponds through their excrement and through simple body contact.

Catfish are preyed upon by cormorants, egrets, and herons throughout the year and by kingfishers and anhingas (water turkeys) during the warmer months. Ospreys and pelicans can sometimes cause problems, too. Frequent

FIGURE 4-59 Egrets and herons are attracted to crawfish ponds in huge numbers, especially when ponds are drained at the end of the production cycle. (Photo courtesy J.V. Huner, Director, Crawfish Research Center, University of Southwestern Louisiana, Lafayette, LA)

visits by flocks of anhingas, herons, egrets, pelicans, and cormorants can be devastating. The problem is generally most pronounced in fingerling ponds.

The MBTA is often confused with the endangered species laws. Under the MBTA, migratory birds may not be killed or trapped without permits. But the species mentioned above can be harassed or frightened away from ponds, and habitat alteration and physical barriers are possible methods of control.

Physical barriers can include hanging netting or wires over ponds and erecting fences around the edge of ponds. These measures are expensive and may cause physical problems to the producer during harvest.

The most common control measures are harassment techniques to frighten birds away from ponds. Birds can be frightened away by—

■ Gunfire
■ Fireworks
■ Gas-powered noise cannons
■ Electronic noisemakers
■ Flashing lights
■ Reflecting material
■ Repellents
■ Bird distress calls
■ Water fountains or cannons
■ Scarecrows
■ Electronic shocking devices.

FIGURE 4-60 Wire netting used to keep birds out of trout raceways.

FIGURE 4-61 Propane gun to scare birds away from baitfish ponds.

These measures have had mixed success. Most methods appear to be effective at first, but they become ineffective as the birds get used to them. The best approach is to use a combination of techniques and to frequently move the devices randomly around the ponds.

Producers may contact the U.S. Fish and Wildlife Service and the USDA's Animal and Plant Health Inspection Service (APHIS) for assistance. These agencies will recommend control measures. Permits can be issued to kill birds if producers keep good records of control measures and estimated losses of fish. These permits are only issued after other methods have proven ineffective as certified by the APHIS Animal Damage Control Office.

More information about predators is given with the details of some of the species in this chapter.

Human Predators

Unfortunately, having a pond or raceway full of ready-to-harvest fish is a temptation for some individuals. Depending on the location, this can be a big problem. High numbers in ponds and raceways guarantee a potential thief a catch.

Control of Unwanted or Excess Vegetation

Some naturally occurring plants in ponds interfere with harvesting, shade out the more desirable ones, and cause water quality problems when they decay. Measures must be taken to control them. Three types of control measures may be used to rid ponds of nuisance vegetation: mechanical, biological, and chemical.

In mechanical vegetation control, the farmer either physically removes the undesirable plants or alters the environment to create conditions discouraging the growth of nuisance plants. Hand removal of undesirable plants by pulling, raking, cutting, or digging may be accomplished in small ponds during the drying period. This may not be practical in large ponds. Some vegetation may also be mowed and disked under. Water drawdown is also an effective tool for controlling some aquatic weeds since it results in the drying out of underwater weeds and compaction of bottom mud. Prevention of the build-up of nutrients in the pond water by periodic exchange of the water from a nutrient-free source will reduce some aquatic plant growth. Rooted aquatic plants such as alligatorweed, smartweed, and primrose use nutrients in bottom sediments and will remain unaffected.

Biological vegetation control relies on the use of animals to consume the unwanted vegetation. Plant-eating fish such as grass carp and mouth-brooders (tilapia) are usually used to consume weeds and algae. Their introduction into ponds may cause some legal problems. The farmer should check state regulations.

Chemical control of nuisance weed and algae growths is generally an effective means but it also involves certain risks. In many cases, an application of weed-killing chemicals kills a lot of plant material that can cause oxygen depletion. Some algaecides and herbicides used to control unwanted plants also kill desirable vegetation. Use is limited in crawfish ponds. If used, assistance from proper authorities must be sought, and care taken in handling and applying them by following the recommendations or directions on the manufacturers' labels.

Whatever method is used to control unwanted or excess vegetation, weeds and algae are a continuing problem from year to year, and procedures must be taken regularly to control them.

GENETICS

Improving fish through genetic research is a relatively new activity. Much of the improvement in the last 40 years in all phases of agricultural production, both plant and animal, resulted from genetic selection and hybridization. Faster growth, higher yield, better feed conversion, and increased resistance to disease can all be improved through genetic manipulation.

Several universities in the Southeast are involved in catfish genetic research. Scientists are doing work in selection, strain identification and evaluation, cross-breeding, hybridization, polyploidy, sex reversing, and gene splicing. The results of this research are encouraging, and research is expanding. Some of the more traditional genetic systems applied to aquaculture include—

- Selective breeding—the choosing of individuals of a single strain and species
- Hybridization—the crossing of different species
- Crossbreeding—the mating of unrelated strains of the same species to avoid inbreeding.

Selective Breeding

Selective breeding is artificial selection, as opposed to natural selection. It involves selected mating of fish with a resulting reduction in genetic variability in the population. Criteria that often influence broodfish selection for selective breeding include—

- Size
- Color
- Shape
- Growth
- Feed conversion
- Time of spawning
- Age at maturity
- Reproductive capacity
- Past survival rates.

These may vary with conditions at different hatcheries. No matter what type of selection program is chosen, an elaborate recordkeeping system is necessary in order to evaluate progress of the program.

Inbreeding occurs whenever mates selected from a population of hatchery broodfish are more closely related than they would be if they had been chosen at random from the population. The extent to which a particular fish has been inbred is determined by the proportion of genes that its parents had in common. Inbreeding leads to an increased incidence of phenotypes—visible characteristics—that are recessive and that seldom occur in wild stocks. An albino fish is an example of a fish with a recessive phenotype. Such fish typically are less fit to survive in nature. Animals with recessive phenotypes occur less frequently in populations where mating is random.

Problems after only one generation of brother-sister mating include reduced growth rate, lower survival, poor feed conversion, and increased numbers of deformed fry. Broodstock managers must be aware of the problems that can result from inbreeding and minimize potential breeding problems. To avoid inbreeding, managers should select their broodstocks from large, randomly mated populations.

If inbreeding is avoided, selective breeding is an effective way to improve a strain of fish.

A system for maintaining trout broodstocks for long periods with lower levels of inbreeding requires the maintenance of three or more distinct breeding lines in a rotational line-crossing system. The lines can be formed by—

1. An existing broodstock arbitrarily subdivided into three groups

2. Eggs taken on three different spawning dates and the fry reared separately to adulthood

3. Three different strains or strain hybrids.

Rotational line-crossing does nothing to reduce the level of inbreeding in the base broodstock but serves only to reduce the rate at which further inbreeding occurs. Consequently, a relatively high level of genetic diversity must be present in the starting broodstock. The use of three different strains or the subdivision of a first generation strain hybrid is the preferred method for line formation. Either of these tends to maximize the initial genetic diversity within the base population. After the three lines have been formed, the rotational line-crossing system can be implemented. At maturity, matings are made between lines. Figure 4-62 shows how a rotational line-crossing system works.

The rotational line-crossing system is flexible enough to fit into most broodstock operations. At least 300 fish—50 males and 50 females from each of the three lines—are needed for maintenance of the population, but this number could be set at any level necessary to meet the egg production needs of a particular hatchery operation. One potential problem with the system is the amount of separate holding facilities required for maintaining up to 15 groups if each line and year class are held separately. This problem sometimes can be overcome by using marks such as fin clips, brands, or tags to identify the three lines and then combining all broodfish of each year class in a single rearing unit. The total number of broodfish to be retained in each year class would be determined by the production goals, but equal numbers of fish should come from each line. This method will not only slow down inbreeding, but will also make a selection program more effective.

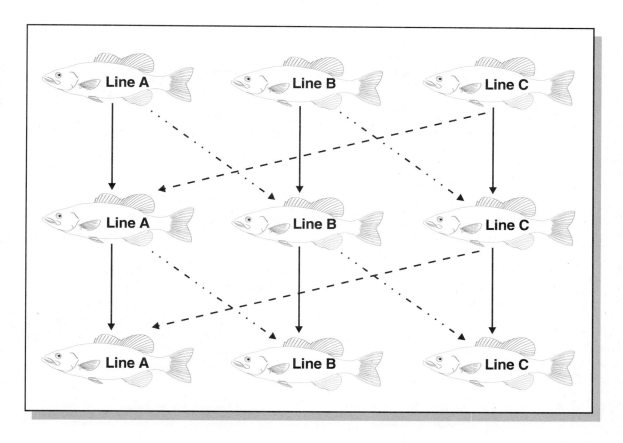

FIGURE 4-62 A line crossing system. Rotational line crossing based on three lines. Each fish represents a pool of fish belonging to a specific line. Each row of fish represents a generation of fish. The solid arrow lines show the females used to produce the next generation. The broken arrow lines shown represent the males used in the mating system.

Hybridization and Crossbreeding

Hybridization between species of fish and crossbreeding between strains of the same species have resulted in growth increases as great as 100 percent, improved feed conversions, increased disease resistance, and tolerance to environmental stresses. These improvements are the result of hybrid vigor, the ability of hybrids or strain crosses to exceed the parents in performance. Most interspecific hybrids are sterile. Those that are fertile often produce highly variable offspring and are not useful as broodstock themselves. Hybrids can be released from the hatchery if they cause no ecological problems in the wild. Several species of trout have been successfully crossed, for example, the splake, a cross between brook and lake trout. A cross between northern pike males and muskellunge females produced tiger muskie. The hybrid striped bass was developed by fertilizing striped bass eggs with sperm from white bass. The hybrids had faster growth and better survival than striped bass. The chief advantage of the reciprocal hybrid, from white bass eggs and striped bass sperm, is that female white bass are usually more available than striped bass females and are easier to spawn.

Both hybridization and crossbreeding of various species of catfish have been successfully accomplished at the Fish Farming Experimental Station, Stuttgart, Arkansas. Hybrid catfishes have been tested in the laboratory for improved growth rate and food conversion. Two hybrids, the white catfish crossed with channel catfish and the channel catfish crossed with blue catfish, demonstrated a greater growth rate than the parents.

Various hybrids of sunfish species also have been successful. The most commonly produced hybrid sunfish are crosses of male bluegill crossed with female green sunfish and male redear sunfish crossed with female green sunfish. They are popular for farm pond stocking because they do not reproduce as readily as the purebred parental stocks and grow much larger than their parents.

SUMMARY

Successful management requires a knowledge of a potential species that could be cultured. Managing the culture of a species involves knowledge of the species, where to obtain seedstock, the life cycle and biology, how to culture, stocking rates, feeding, potential diseases, harvesting, and processing.

Production systems and cultural methods may vary depending on the aquatic animal and the stage of production. Many management concerns remain constant, such as water quality and prevention of stress and disease. Also, successful management demands a thorough understanding of the reproductive process, feeding behavior, and nutritional needs. Once managers understand the reproductive process, they try to enhance it. Once nutritional needs are identified, managers try to completely and economically meet these needs.

Some species may never be cultured from egg to adult without drastic changes in current cultural practices. Other species that are successfully cultured may be improved through genetic selection similar to the improvement of livestock and poultry.

STUDY/REVIEW

Success in any career requires knowledge. Test your knowledge of this chapter by answering these questions or solving these problems.

True or False

1. Sexual reproduction involves the production of eggs from the testes and sperm from the ovaries.
2. With a permit, predatory birds can be killed around fish ponds.
3. Humans stealing fish from ponds can become a problem.
4. Management is the secret ingredient to successful aquaculture.
5. The color of the water controls spawning time in catfish.
6. Channel catfish spawn on spawning mats.
7. Oxygen levels of 5 ppm are lethal for fish.
8. Rainbow trout hatch in freshwater, swim to saltwater, mature, and return to freshwater.
9. Water temperatures above 75°F are lethal for trout.
10. Hatching rate depends on water temperature.

Short Answer

11. Give the scientific name for the majority of cultured catfish.
12. When do catfish spawn?
13. Besides fertilizing the eggs, what role does the male catfish play?
14. What is the optimum temperature for growing catfish?
15. At what weight are catfish harvested?
16. Name three types of incubators used for trout.
17. In a trout hatchery, how many fry can be stocked in a trough 10 ft long and 18 in wide with 3 to 4 in of water?
18. What two things should be considered when selecting a species of tilapia?
19. Tilapia are native to what parts of the world?
20. What is the major problem associated with the pond culture of tilapia?

21. At what size can tilapia reach sexual maturity?

22. Trout grow best in water temperatures of:
    a. 35° to 45°F
    b. 30° to 40°C
    c. 50° to 68°F
    d. 50° to 68°C

23. Name three types of incubator systems used in fish culture.

24. Trout fry are fed when 50 percent _____ _____.

25. Most trout are commercially raised in _____ while most catfish are commercially raised in _____.

26. The process of adding oxygen to a pond or raceway is called _____.

27. Which of the following do not affect the carrying capacity of a trout production facility?
    a. water quality
    b. concrete thickness
    c. temperature
    d. water flow

28. How much time is required to raise market-size hybrid striped bass?

29. Grass carp, silver carp, and bighead carp are native to what continent?

30. Name three hormones involved in the natural spawning process.

31. Name two ways to increase spawning frequency.

32. What effect does temperature have on the development of fertilized eggs.

33. Name four baitfish.

34. What is the most common use for crappies?

35. Name four common hobby fish.

36. Develop a list of fish that would be stressed if the water turned cold.

37. Name the most commonly cultured crustacean in the United States.

38. Name two markets for crawfish.

39. What type of a climate is necessary for a successful shrimp or prawn culture?

40. List three breeding techniques for improving any aquatic species.

Essay

41. Why would a producer consider raising tilapia in a male monosex culture?

42. Describe two polyculture systems using tilapia.

43. Describe three methods of fertilizing a tilapia pond for increased production.

44. How are hybrid striped bass produced?

45. Describe why an understanding of sex determination is important.

46. What does it mean to harvest fish by topping?

47. Give three reasons why culturing fish or other aquatic animals increases the concern for disease.

48. Define flow and density indexes.

49. Compare salmon ranching to salmon farming.

50. Give three reasons why tilapia are not widely available in the United States.

51. Explain why grading or sorting is necessary for hybrid striped bass culture.

52. Suggest two reasons why carp culture is dominant in countries other than the United States, especially in Asia.

53. Explain why lobster culture from egg to adult does not occur.

54. Give some indication of the worldwide distribution of crawfish and shrimp.

55. Indicate the main reason for growing aquatic plants in the United States.

56. Define a hybrid and give two examples.

KNOWLEDGE APPLIED

1. Visit or call a local pet shop or department store that sells hobby or tropical fish. Make an inventory of the fish they sell and the price of each. Ask the manager who supplies the fish and how are they shipped. Compare your list with Table 4-5.

2. Using the information in this chapter and information from the Learning/Teaching Aids identified at the end of the chapter, construct a specie profile sheet for five aquatic animals. The specie profile sheet should include the following information: common name, scientific name, brief

description, habitat, distribution, reproduction, special features, problems, potential yield, feeds, and marketing potential.

3. Some hobby fish are easily spawned in an aquarium. Obtain hobby fish and research the requirements for spawning these fish. Compare how these fish spawn with others described in this chapter.

4. Using an aquarium, monitor some of the water quality parameters that would be important to pond or raceway production. For example, monitor pH, oxygen, and temperature. If possible, monitor the pH and oxygen with test kits and electronically. Compare the results.

5. Survey your community and determine if any aquaculture products have small niche markets or cultural markets.

6. Evaluate local water resources. Determine if any of the water resources lend themselves to aquaculture. Determine which species could be raised in the water resources available. Perhaps the best way to do this is invite a guest speaker who can discuss local water resources and possibly bring maps.

7. Since catfish, trout, and crawfish are most frequently cultured in the United States, call local grocery stores and restaurants. Ask if they sell trout, catfish, or crawfish. If they do, find out how it is sold and where they purchase it.

8. Obtain any of the videos listed at the end of this chapter. Show these and develop reports based on the videos.

LEARNING/TEACHING AIDS

Books

Axelrod, H. R., & Schultz, L. P. (1990). *Handbook of tropical aquarium fishes.* Neptune City, NJ: TFH Publications, Inc.

Bardach, J. E. (1972). *Aquaculture: The farming and husbandry of freshwater and marine organisms.* New York: John Wiley and Sons.

Barnabe, G. (1990). *Aquaculture* (Vols. 1 and 2). New York: L. Horwood.

Bouc, K. (1987). *The fish book.* Lincoln, NE: Nebraska Game and Parks Commission.

Chakroff, M. *Freshwater fish pond culture and management.* Peace Corps Information Collection and Exchange, Volunteers in Technical Assistance, VITA Publications, April 1977, 1978, March 1981, September, 1982.

Chen, K. J. (1989). *Prawn farming: Hatchery and grow-out operations* (Distributed by Evergreen Agri-Aqua, Inc., Nurich Vitameal Corporation, Dagupan City, Philippines). Manila, Philippines: West Point Aquaculture Corporation.

Chew, K. K., & Toba, D. (1991). *Western region aquaculture industry situation and outlook report*. Seattle, WA: Western Regional Aquaculture Consortium, University of Washington.

Conte F., Doroshov, S., & Lutes, P. (1988). *Hatchery manual for the white sturgeon*. Oakland, CA: ANR Publications.

Council for Agricultural Education. (1992). *Aquaculture*. Module IV: Farming in water. Alexandria, VA: Council for Agricultural Education.

Dore, I., & Frimodt, C. (1987). *An illustrated guide to shrimp of the world*. Huntington, NY: Osprey Books.

Dupree, H. K., & Huner, J. V. (1984). *Third report to the fish farmers: The status of warmwater fish farming and progress in fish farming research*. Washington, DC: U.S. Fish and Wildlife Service.

Fallu, R. (1991). *Abalone farming*. Cambridge, MA: Blackwell Scientific.

Fast, A. W., & Lester, L. J. (1992). *Marine shrimp culture: Principles and practices*. New York: Elsevier Publishing.

Gall, G. A. (1992). *The rainbow trout*. New York: Elsevier Publishing.

Groot, C., & Margolis, L. (1991). *Pacific salmon life histories*. Vancouver, BC: University of British Columbia Press.

Groves, R. E. (1985). *The crayfish: Its nature and nurture*. Farnham, Surrey, England: Fishing News Books.

Gulland, J. A., & Rothschild, B. J. (1984). *Penaeid shrimps: Their biology and management*. Farnham, Surrey, England: Fishing News Books.

Horvath, L., Tamas, G., & Seagrave C. (1992). *Carp and pond fish culture: Including Chinese herbivorous species, pike, tench, zander, wels catfish and goldfish*. New York: Halstead Press.

Huet, M. (1986). *Textbook of fish culture* (2nd ed.). Farnham, Surrey, England: Fishing News Books.

Innes, W. T. (1979). *Exotic aquarium fish* (2nd ed.). Neptune City, NJ: TFH Publications Inc.

Iversen, E., Allen, D., & Higman, J. (1993). *Shrimp capture and culture fisheries of the United States*. New York: Halstead Press.

Jung, C. (1988). *Prawn culture: Scientific and practical approach*. Dagupen City, Philippines: West Point Aquaculture Corporation.

Klontz, G. W. (1991). *Fish for the future: Concepts and methods of intensive aquaculture* (Text Number 5). Moscow, ID: University of Idaho, College of Forestry, Wildlife and Range Sciences.

Landau, M. (1992). *Introduction to aquaculture*. New York: Wiley Press.

Lee, J. S., & Newman, M. E. (1992). *Aquaculture: An introduction*. Danville, IL: Interstate Publishers, Inc.

Lee, E., & Wickens, J. (1992). *Crustacean farming.* New York: Halstead Press.

Manzi, J., & Castagna, M. (1989). *Clam mariculture in North America.* New York: Elsevier Publishing.

McLarney, W. (1987). *The freshwater aquaculture book.* Point Roberts, WA: Hartley & Marks, Inc.

Muir, J. F., & Roberts, R. J. *Recent advances in aquaculture* (Vol. 4). Boulder, CO: Westview Press.

Nelson, J. S. (1994). *Fishes of the world* (3rd ed.). New York: John Wiley & Sons.

Nielsen, L. A., & Johnson, D. L. (1983). *Fisheries techniques.* Bethesda, MD: American Fisheries Society.

Pillay, T. V. R. (1990). *Aquaculture: Principles and practices.* Cambridge, MA: Blackwell Scientific Publications.

Piper, R. G., McElwain, I. B., Orme, L. E., McCraren, J. P., Fowler, L. G., & Leonard, J. R. (1982). *Fish hatchery management.* Washington, DC: U.S. Department of the Interior, Fish and Wildlife Service.

Seagrave, C. (1991). *Aquatic weed control.* Cambridge, MA: Blackwell Scientific.

Shearer, W. M. (1992). *The Atlantic salmon: Natural history, exploration and future management.* New York: Halstead Press.

Shepherd, C. (1988). *Intensive fish farming.* Boston, MA: BSP Professional Books.

Shumway, S. E. (1991). *Scallops: Biology, ecology and aquaculture.* New York: Elsevier Publishing.

Tave, D. (1993). *Genetics for fish hatchery managers.* New York: Van Nostrand, Reinhold.

U.S. Department of Agriculture, Office of Aquaculture. (1988). *Aquacultural genetics and breeding* (Vols. 1 & 2). Washington, DC.

Usui, A. (1991). *Eel culture* (2nd ed.). Cambridge, MA: Blackwell Scientific.

Walker, S. S. (1990). *Aquaculture.* Stillwater, OK: Mid-America Vocational Curriculum Consortium, Inc.

Williams, K., Schwartz, D. P., Gebhart, G. E., & Maughan, O. E. (1987). *Small-scale caged fish culture in Oklahoma farm ponds* (5th ed.). Stillwater, OK: Oklahoma State University.

Extension Pamphlets/Fact Sheets

Beem, M. D. (N.D.) Caged catfish culture problems. *Extension Facts.* Langston, OK: Langston University.

Beem, M., & Anderson, S. (1989). *Catfish farming.* Stillwater, OK: Oklahoma State Cooperative Extension, Oklahoma State University.

Davis, J. T. (1990, June). Red drum: Biology and life history (Publ. No. 320). Stoneville, MS: Southern Regional Aquaculture Center.

Davis, J. T. (1990, June). Red drum: Production of food fish (Publ. No. 322). Stoneville, MS: Southern Regional Aquaculture Center.

Davis, J. T. (N.D.). Red drum: Production of fingerlings and stockers. Stoneville, MS: Southern Regional Aquaculture Center.

de la Bretonne, Jr., L. W., & Romaire, R.P. (N.D.) Crawfish culture: Site selection, pond construction and water quality. Stoneville, MS: Southern Regional Aquaculture Center.

de la Bretonne, Jr., L. W., & Romaire, R. P. (N.D.) Crawfish production: Harvesting, marketing and economics. Stoneville, MS: Southern Regional Aquaculture Center.

de la Bretonne, Jr., L. W., & Romaire, R. P. (N.D.) Crawfish production systems. Stoneville, MS: Southern Regional Aquaculture Center.

Gray, D. L. (N.D.) Baitfish. Stoneville, MS: Southern Regional Aquaculture Center.

Higginbotham, B. (N.D.). Forage species: Production techniques. Stoneville, MS: Southern Regional Aquaculture Center.

Higginbotham, B. (N.D.). Forage species: Range, description and life history. Stoneville, MS: Southern Regional Aquaculture Center.

Hinshaw, J. M. (1990, January). Trout farming: A guide to production and inventory management (Publ. No. 222). Stoneville, MS: Southern Regional Aquaculture Center.

Hodson, R. G., & Hayes, M. (N.D.). Hybrid striped bass: Hatchery phase. Stoneville, MS: Southern Regional Aquaculture Center.

Hodson, R. G. (1989, July). Hybrid striped bass: Biology and life history (Publ. No. 300). Stoneville, MS: Southern Regional Aquaculture Center.

Jensen, G. L. (1990, June). Sorting and grading warmwater fish (Publ. No. 391). Stoneville, MS: Southern Regional Aquaculture Center.

Kleinholz, C. (N.D.) Water quality management for fish farmers. *Extension Facts.* Langston, OK: Langston University.

Littauer, G. (1990, November). Avian predators: Frightening techniques for reducing bird damage at aquaculture facilities (Publ. No. 401). Stoneville, MS: Southern Regional Aquaculture Center.

Littauer, G. A. (1990, November). Control of bird predation at aquaculture facilities: Strategies and cost estimates (Publ. No. 402). Stoneville, MS: Southern Regional Aquaculture Center.

Louisiana Cooperative Extension Service (N.D.). Aquafacts: Nitrite management in commercial fish ponds. Baton Rouge, LA: Louisiana State University Agricultural Center.

Rottmann, R. W., Shireman, J. W., & Chapman, F.A. (1991, November). Introduction to hormone-induced spawning of fish (Publ. No. 421). Stoneville, MS: Southern Regional Aquaculture Center.

Shelton, J. L., & Murphy, T. R. (N.D.). Aquatic weed management: Control methods. Stoneville, MS: Southern Regional Aquaculture Center.

Wellborn, T. L. (N.D.). Channel catfish: Life history and biology. Stoneville, MS: Southern Regional Aquaculture Center.

(*Note:* These publications and others are available from any of the five Regional Aquaculture Centers listed in Appendix Table A-11, or by contacting your state aquaculture specialist.)

Videos

University of Washington. (1991). *Alligator aquaculture in the south.* Instructional Media Services.

University of Washington. (1991). *Catfish farming in the south.* Instructional Media Services.

University of Washington. (1991). *Crawfish aquaculture in the south.* Instructional Media Services.

University of Washington. (1991). *Induced spawning of fish.* Instructional Media Services.

University of Washington. (1991). *Red drum aquaculture.* Instructional Media Services.

University of Washington. (1991). *Southern hybrid striped bass production in ponds.* Instructional Media Services.

University of Washington. (1991). *Trout production in the southwest.* Instructional Media Services.

University of Washington. (1993). *Avian depredation of southern aquaculture.* Instructional Media Services.

(*Note:* These videos are available from any of the Regional Aquaculture Centers listed in Appendix Table A-11.)

Fundamentals of Nutrition in Aquaculture

T he primary purpose of aquaculture is the efficient conversion of feed into meat for consumption. Fish convert feed to food for human consumption very efficiently, especially compared to other meat-producing animals. An understanding of the digestive system and the nutrition of fish is essential for successful aquaculture.

Learning Objectives

After completing this chapter, the student should be able to:

In Aquaculture

- ■ Find the protein, energy, vitamin, and mineral requirements for fish
- ■ Understand the role of nonnutritive factors in feed
- ■ Know what toxic substances to watch for in fish feed
- ■ Identify methods for preparing and feeding fish
- ■ Identify potential ingredients for fish diets
- ■ List the information required by a least-cost ration program

■ Explain the importance of winter feeding catfish

■ Explain different feeding practices for different species

■ Calculate the amount of feed required

■ Calculate the feed conversion ratio (FCR)

In Science

■ Identify the parts of the digestive system

■ Explain the role of the digestive system in absorption

■ Explain how anatomy and behavior affect feeding

■ List factors that influence energy requirements

■ List three sources of energy

■ Identify factors that affect the digestibility of fat

■ Explain the role of essential fatty acids and essential amino acids

■ Name ten essential amino acids

■ Name two essential fatty acids

■ List the fat-soluble and water-soluble vitamins

■ Describe ten effects of vitamin deficient diets

■ Name the macrominerals and the microminerals

■ List ten functions of minerals

Understanding of this chapter will be enhanced if the following terms are known. Many are defined in the text and others are defined in the glossary.

KEY TERMS		
Absorption	Carnivores	Enzyme
Amino acid	Carotenoids	Essential amino acids
Anemia	Coenzymes	Essential fatty acids
Antinutrients	Connective tissue	Extrusion
Antioxidants	Coolwater	Feed conversion ratio
Automatic feeder	Demand feeder	Fines
Binders	Diet	Gall bladder
Breakdown	Digestion	Gastrointestinal
Carbohydrates	Dry feed	Grazers

KEY TERMS		
Heme	Natural foods	Premix
Herbivores	Nutrition	Protein
Humectants	Omnivores	Proximate
Intestine	Osmoregulation	composition
Least-cost	Oxidation	Pylorus
Lipids	Parasites	Saturated
Liver	Peptide bond	Sedentary
Macrominerals	Peroxide	Strainers
Metabolic rate	Pharynx	Suckers
Microencapsulation	Phytin	Synthesize
Microminerals	Planktonic	Vitamins
Mycotoxins	Predators	Yolk sac

NUTRITION OF FISH

The science of nutrition draws heavily on findings of chemistry, biochemistry, physics, microbiology, physiology, medicine, genetics, mathematics, endocrinology, cellular biology, and animal behavior. To the individual involved in aquaculture, nutrition represents more than just feeding. Nutrition becomes the science of the interaction of a nutrient with some part of a living organism, including the composition of the feed, ingestion, liberation of energy, elimination of wastes, and synthesis for maintenance, growth, and reproduction.

FIGURE 5-1 A truck driver delivering catfish feed to farmers on the Mississippi delta stops for human food.

Feeds and feedstuffs contain the energy and nutrients essential for the growth, reproduction, and health of aquatic animals. Deficiencies or excesses can reduce growth or lead to disease. Dietary requirements set the necessary levels for energy, protein, amino acids, lipids (fat), minerals, and vitamins.

The subcommittee on Fish Nutrition of the Committee on Animal Nutrition of the National Research Council (NRC) examines the literature and current practices in aquaculture. The NRC publishes recommendations on fish nutrition. The latest NRC publication on fish nutrition was issued late in 1993. Many of the recommendations in this chapter are based on this NRC publication.

To make money in aquaculture, transforming feed to food must be done efficiently and economically. Fish can do this, as Table 5-1 shows, but the principles of nutrition must be applied. Successful nutritional practices also depend on breeding, health, and management.

Digestion and Absorption

Figure 5-2 represents the digestive system of fish. The digestive or gastrointestinal tract is described as a continuous hollow tube extending from the mouth to the anus with the body built around it.

The digestive system of fish includes the mouth, pharynx, esophagus, stomach, pylorus, intestine, liver, and gall bladder. It acts like an assembly line in reverse, taking the feedstuffs apart to their basic chemical components so the fish can absorb them and rearrange them into its own characteristic body composition.

TABLE 5-1 Feed Conversion Comparisons

Species	Unit of Production	Average Lb of Feed Per Lb of Product
Broiler	1 lb chicken	2.4
Dairy cow	1 lb milk	1.1
Turkey	1 lb turkey	5.2
Layer	1 lb eggs (8 eggs)	4.6
Rabbit	1 lb fryer	3.0
Hog	1 lb pork	4.9
Beef Steer	1 lb beef	9.0
Lamb	1 lb lamb	8.0
Fish	1 lb fish	1.6

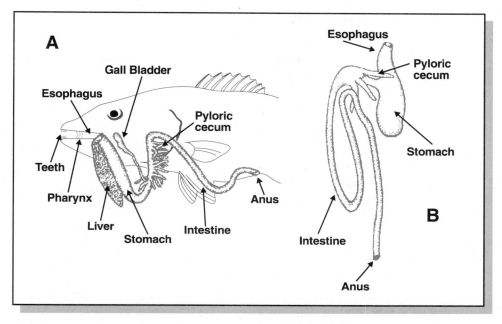

FIGURE 5-2 The digestive system of fish.

Table 5-2 summarizes all structures of the digestive system in digestion and absorption.

TABLE 5-2 Digestive System Structures and Functions

Structure	Functions
Teeth	Grasping, holding, crushing, depending on species
Pharynx	Opening to the gills
Esophagus	Short, simple passage to stomach, lined with mucus-secreting cells
Stomach	Walls lined with cells secreting hydrochloric acid and pepsinogen for initial stages of protein digestion; holding compartment for feed
Pyloric ceca	Secretes enzymes for digestion; increased surface area for absorption of nutrients
Intestine	Secretes enzymes for digestion; increased surface area for absorption of nutrients
Gall bladder	Stores and releases bile for digestion and absorption of fats
Liver	Synthesis or storage from absorbed nutrients, production of bile, removal of some waste products from blood

Other considerations of digestion and absorption in fish include the type of eaters, the anatomy of the mouth, and feeding behavior.

Feeding Type and Anatomy. Fish can be divided into three types of eaters—

- **Carnivores** consume primarily animal material. Foods consumed by this type of fish may be as small as a microscopic crustacean or insect or as large as an amphibian or a small mammal.
- **Herbivores** subsist primarily on vegetation and decayed organic material in the environment.
- **Omnivores** consume almost any food source, either plant or animal in origin.

Certain anatomic changes in the mouth of fish occurred through evolutionary development. Fish can be classified according to their feeding habits into the following categories—

- **Predators.** Trout are an example of fish that feed on animals that are generally large enough to be seen with the naked eye. Teeth are well developed and act as a means of grasping and holding the prey. Some predators rely primarily on sight to hunt while others rely on the senses of taste and touch or on lateral-line sense organs.
- **Grazers.** The mullet is an example of fish that graze in the same sense as mammalian grazers. Generally, they graze continuously on the bottom of the water habitat for either plants or small animal organisms. Food is taken by well-defined bites.
- **Strainers.** Menhaden are an example of fish that select food primarily by size rather than by type. An adult menhaden can strain in excess of 6 gal (22.7 L) of water per minute through its gill rakers. Through this process of rapid straining, the menhaden is able to concentrate a relatively large mass of plankton and other organisms.
- **Suckers.** Buffalofish are an example of fish that feed primarily on the bottom of their habitat, sucking in mud and filtering and extracting digestible material.
- **Parasites.** Some fish, like the lamprey, attach themselves to other animals and exist on the body fluids of the host.

Behavior. Fish also developed behavioral feeding patterns that are sensitive to environmental stimuli. By knowing the behavioral patterns of the particular species of fish, the producer can adapt a system of feeding that will best use labor and feed. Some environmental influences on feeding behavior include:

- **Sensory use.** Some fish depend largely on their sense of sight in hunting food, while others rely primarily on taste, touch, and smell.

- **Season of the year.** Some fish cease feeding activity during their spawning season. Most fish start increasing feed intake in the spring in the temperate climatic regions when the water temperature starts to rise. The peak growth period for most fish occurs in the spring and summer.

- **Time of day.** Some fish show peak feed activity in the dawn and dusk hours of the day.

- **Physical contact with the food.** Quite often the texture of a potential food source is felt before the fish will consume it.

Even though the digestive anatomy, physiology, and feeding habits differ in fish compared to warm-blooded domestic animals, the nutritional requirements remain expressed in the same terms—energy, protein, vitamins, and minerals. Most of the research on these nutritional requirements centers around catfish, salmon, and trout.

Energy Requirements

Energy is not a nutrient but it is released during the breakdown (metabolic oxidation) of carbohydrates, amino acids, and fats. Since fish are cold-blooded, they expend no energy to maintain body temperature. This allows more energy for growth, activity, and reproduction. Several other factors affect energy use by fish:

- **Age.** As age increases, the metabolic rate of fish generally decreases.

- **Composition of the diet.** If a diet has a high protein or mineral content, metabolism increases in order that the fish can eliminate waste products that could possibly build up and become toxic.

- **Light exposure.** Darkness decreases the energy requirement in some species. Fish grown in constant light do not grow as well as those of the same species having a "rest period" of darkness.

- **Physiological activity.** Salmon have high metabolic rates during the spawning season. Conversely, during winter rest fish have extremely low metabolic rates.

- **Size.** In general, smaller fish have higher metabolic rates than larger fish.

- **Species.** Metabolic rates vary according to the characteristic behavioral patterns of the species. For example, sedentary fish have lower metabolic rates than do pelagic (open sea) fish.

■ **Temperature of the water**. As the temperature of water increases, its ability to carry oxygen decreases. In response to the reduced oxygen-carrying capacity of the water, the respiration rates of fish increase, resulting in higher metabolic rates. For every 18°F (10°C) increase in water temperature, the metabolic rate doubles. Figure 5-3 shows how increasing temperature increases the amount of feed recommended for trout.

Additionally, the metabolic rate decreases 5 percent for each degree F decrease from Standard Environmental Temperature (SET). Thus, when the water temperature is 20°F (6.7°C) below Standard Environmental Temperature, growth for all intents and purposes ceases. However, the fish still eat to accommodate their metabolic needs for swimming, osmoregulation, and respiration.

Other environmental factors such as water flow rates, water composition and pollution put certain stresses on fish and result in their metabolic rates being altered in keeping with the severity of the stress.

Energy Losses. Feeds and feedstuffs contain energy but not all the energy goes toward growth and reproduction. Energy losses occur as feed is digested

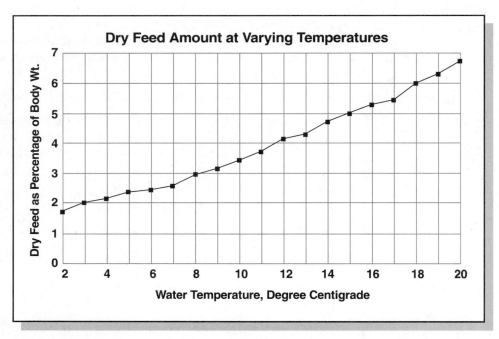

FIGURE 5-3 Increasing water temperature increases the recommended amount of feed for trout, 2 to 3 in (5 to 7.5 cm).

and metabolized. As feed moves through the digestive processes, energy is lost in the feces, urine, and gill excretions. Energy is also lost as heat. Figure 5-4 illustrates the loss of energy as intake energy (gross) loses fecal energy, becoming digestible energy. Then digestible energy loses energy to urine or gill excretions to become metabolizable energy. Metabolizable energy loses heat energy, becoming net energy. This is the energy available for maintenance, growth, and reproduction. Digestible energy (DE) and metabolizable energy (ME) are more exact measures of the energy required by fish.

Like traditional livestock, fish derive their energy from three sources—carbohydrates, fats, and proteins. Since carbohydrates are used rather inefficiently in most fish systems, the primary energy sources are fats and proteins.

Carbohydrates. Fish digest simple sugars efficiently. As the sugar becomes larger and more complex, digestibility decreases rapidly. Warmwater fish digest dietary carbohydrates better than coldwater or marine fish. The ability to use carbohydrates as an energy source varies among the fish species. No specific NRC requirement is established for carbohydrates in the diet of fish. Some

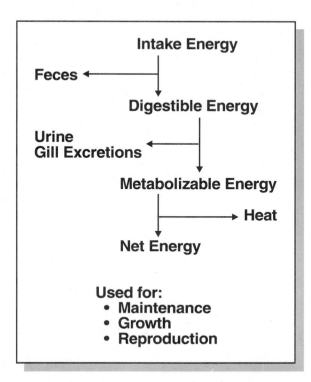

FIGURE 5-4 Energy contained in a feed is lost to digestive processes and heat, leaving net energy to be used for maintenance, growth, and reproduction.

form of digestible carbohydrate should be included in the diet. Carbohydrates improve growth and provide precursors for some amino acids and nucleic acids. Also, carbohydrate is the least expensive source of dietary energy. In warmwater fish, cereal grains provide inexpensive sources of carbohydrates, but their use is limited in coldwater fish. Digestible carbohydrates in trout feed are generally lower than the levels in catfish feed. In nutrition, carbohydrates spare protein since less protein will be used for energy. An excess of dietary carbohydrates can cause livers to enlarge and glycogen to accumulate in the liver. A general recommendation is a diet of no more than 12 percent digestible carbohydrates. Fats and proteins supply most of the energy in fish diets.

Fat. Each gram of fat contains 2.5 times the energy in a gram of carbohydrates or proteins. The digestibility of fat varies depending on—

- Amount in the diet
- Type of fat
- Water temperature
- Degree of unsaturation
- Length of carbon chain.

Animal fats and fats that are highly saturated have a lower digestibility. On the other hand, in highly unsaturated fats—fats that fish can readily digest—there is danger of oxidation of the fats, resulting in feed spoilage. Antioxidants are routinely added to most fish diets to prevent fats from becoming rancid in storage.

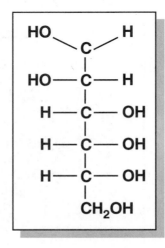

FIGURE 5-5 Starch is a long chain of glucose molecules. Digestion requires the starch to be broken down into simple sugars-glucose.

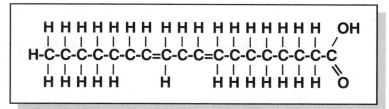

FIGURE 5-6 Diagram of linoleic acid, an essential fatty acid.

Besides being an important source of energy for fish, dietary fats provide essential fatty acids (EFA) needed for normal growth and development. Fish cannot synthesize these fatty acids. Also, dietary fats assist in the absorption of fat-soluble vitamins. Freshwater fish require a dietary source of linoleic acid and/or linolenic acid (Figure 5-6). These are both 18-carbon fatty acids. Marine fish, like the yellowtail or red sea bream require a dietary source of eicosapentaenoic acid (EPA) and/or docosahexaenoic acid (DHA). These are 20- and 22-carbon fatty acids respectively.

Channel catfish, coho salmon and rainbow trout require linolenic acid or EPA and/or DHA. Table 5-3 indicates the essential fatty acid requirements for several species of fish.

TABLE 5-3 Essential Fatty Acid Requirement of Fish[1]

Species	Requirement
Channel catfish	1.0 to 2.0% linolenic acid or 0.5 to 0.75% EPA[2] and DHA[2]
Chum salmon	1.0% linoleic acid and 1.0% linolenic acid
Coho salmon	1.0 to 2.5% linolenic acid
Common carp	1.0% linoleic acid and 1.0% linolenic acid
Rainbow trout	0.8 to 1.0% linolenic acid 20% of fat as linolenic acid or 10% of fat as EPA and DHA
Tilapia	0.5 to 1.0% linoleic acid
Red sea bream	0.5% EPA and DHA
Yellowtail	2.0% EPA and DHA

[1] National Research Council. (1993). *Nutrient requirements of fish.* Washington, DC: National Academy Press.
[2] EPA is eicosapentaenoic acid and DHA is docosahexaenoic acid.

Essential fatty acid deficiency signs include skin lesions, shock syndrome, heart problems, reduced growth rate, reduced feed efficiency, reduced reproductive performance, and increased mortality. In the body essential fatty acids function as a part of cell membranes and the precursor for biochemicals that perform a variety of metabolic functions.

Fish diets are formulated to meet the optimum ratio of energy to protein for each species. Fats serve as an important source of energy, but not a definite percentage of dietary fat can be given without considering the type of fat, as well as the protein and energy content of the diet. Table 5-4 lists some general guidelines for dietary fat in different species of fish for different situations. Too much dietary fat can result in an imbalance of the digestible energy to crude protein ratio and excessive deposition of fat in the body cavity and tissues.

Protein Requirements. Proteins are long chains of amino acids linked by bonds called **peptide bonds**. Figure 5-7 shows diagrams of several amino acids. All amino acids contain nitrogen, so all proteins contain nitrogen. In fact, measuring the nitrogen content is a method of calculating protein content. Metabolism of protein for energy produces nitrogen end products. Fish eliminate these through the gills, feces, and urine. These nitrogen end products can cause problems in fish ponds.

Protein serves three purposes in the nutrition of fish—

1. Provide energy
2. Supply amino acids
3. Meet requirements for functional proteins—enzymes and hormones—and structural proteins.

The requirement for protein in fish diets is essentially a requirement for the amino acids in the dietary proteins. Some amino acids the fish cannot

TABLE 5-4 Guidelines for Dietary Fat in Fish

Species	Situation	Percentage of Fat
Trout	Starter diet	12 to 16
Trout	Grower diet	8 to 10
Trout	Production diet	6 to 8
Catfish	82°F (28°C)	12
Catfish	73°F (23°C)	5
Carp	82° to 73°F (28° to 23°C)	10 to 15
Carp	<68°F (20°C)	10

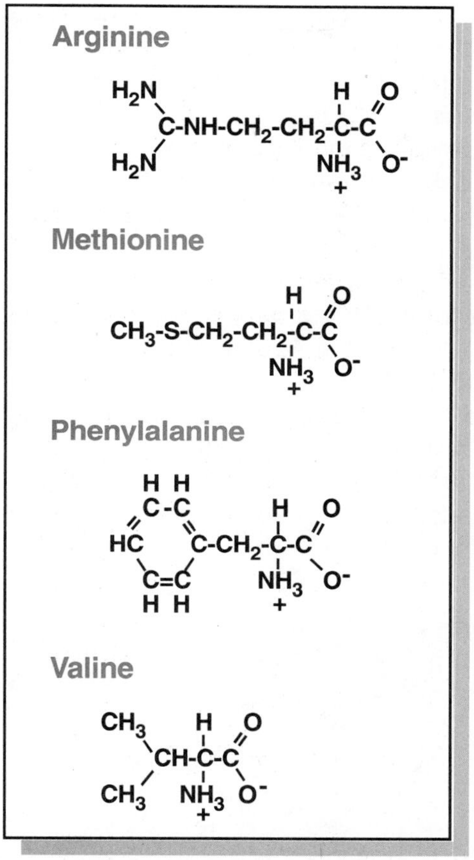

FIGURE 5-7 Diagram of some essential amino acids.

synthesize are called indispensable or essential amino acids. These include ten amino acids—

- Arginine
- Histidine
- Isoleucine
- Leucine
- Lysine
- Methionine
- Phenylalanine
- Threonine
- Tryptophan
- Valine.

Some of the dietary requirements for methionine and phenylalanine can be met by the amino acids cystine and tyrosine, respectively.

The amino acid requirement given by the NRC is shown in Table 5-5 for catfish, trout, salmon, carp, and tilapia.

Research evidence suggests that large differences exist among fish species in their requirements for amino acids. Some of these differences are probably caused by differences in growth rate, feed intake, and source of the amino acid in the diet.

When proteins in most feedstuffs are properly processed they are highly digestible. For a variety of protein-rich feedstuffs the digestibility ranges from 75 to 95 percent. As dietary carbohydrate increases the digestibility of protein tends to decline. Also, overheating during drying or processing reduces the nutritive value of protein. But, insufficient heating of soybean meal decreases the availability of protein.

Protein requirements for fish are considerably higher than those for warm-blooded land animals. Protein requirements of fish decline with age.

TABLE 5-5 Amino Acid Requirement for Finfish[1]

Energy Base[b] (kcal DE/Kg diet) →	Channel Catfish 3,000	Rainbow Trout 3,600	Pacific Salmon 3,600	Common Carp 3,200	Tilapia 3,000
Protein, crude (digestible), %	32(28)	38(34)	38(34)	35(30.5)	32(28)
Amino acids					
Arginine, %	1.20	1.5	2.04	1.31	1.18
Histidine, %	0.42	0.7	0.61	0.64	0.48
Isoleucine, %	0.73	0.9	0.75	0.76	0.87
Leucine, %	0.98	1.4	1.33	1.00	0.95
Lysine, %	1.43	1.8	1.70	1.74	1.43
Methionine + cystine, %	0.64	1.0	1.36	0.94	0.90
Phenylalanine + tyrosine, %	1.40	1.8	1.73	1.98	1.55
Threonine, %	0.56	0.8	0.75	1.19	1.05
Tryptophan, %	0.14	0.2	0.17	0.24	0.28
Valine, %	0.84	1.2	1.09	1.10	0.78

[1] National Research Council. (1993). *Nutrient requirements of fish.* Washington, DC: National Academy Press.

Animal protein sources are generally considered to be of higher quality than plant sources, but animal protein costs more. In diets, a combination of protein sources yields better conversion rates than any single source.

Fish do not have the ability to use nonprotein nitrogen sources. Such nonprotein nitrogen sources as urea and diammonium citrate, which even nonruminant animals can use to a limited extent, have no value as a feed source for fish. In fact, nonprotein nitrogen can be toxic at high levels.

Factors such as age, species, stocking density, water temperature, and production stage should govern the protein level incorporated in the diet. Table 5-6 indicates some recommended protein levels in practical fish diets.

A protein deficiency or indispensable amino acid deficiency is observed as a reduction in weight gain. But some specific amino acid deficiencies manifest as disease conditions. Cataracts form in salmonids, including rainbow trout, when given diets are deficient in methionine or tryptophan. A tryptophan deficiency also causes a lateral curvature of the spinal column or scoliosis in some salmonids. In trout, a tryptophan deficiency disrupts the metabolism of the minerals calcium, magnesium, sodium, and potassium.

In fish diets protein and energy should be kept in balance. A deficiency or excess of energy in the diet reduces growth rates. When dietary energy is deficient, protein is used for energy. When dietary energy is in excess, feed consumption drops and this lowers the intake of the necessary amounts of protein for growth.

Vitamin Requirements. Vitamins are organic compounds required in the diet for normal growth, reproduction, and health. They function in a variety of chemical reactions in the body. The simple digestive system of the fish establishes a definite need for the supplementation of vitamins in fish diets. Vitamin requirements for fish resemble those of nonruminant animals, like pigs and chickens. Fish and humans are among the few higher animals that require a dietary source of vitamin C.

TABLE 5-6 Protein Levels in Practical Fish Diets

Species	Percentage As-Fed Basis		
	Starter	Grower	Production
Trout	40 to 55	35 to 40	30 to 40
Chinook Salmon	40	—	—
Catfish	35 to 40	25 to 36	28 to 32
Carp	43 to 47	37 to 42	—

The vitamins are divided into two categories, water-soluble and fat-soluble. The water-soluble vitamins include—

■ Thiamin
■ Riboflavin
■ Pyridoxine (Vitamin B_6)
■ Pantothenic
■ Niacin
■ Biotin
■ Folate
■ Vitamin B_{12}
■ Choline
■ Myoinositol
■ Vitamin C.

FIGURE 5-8 Vitamin C deficiency in channel catfish causes a spinal deformity. The top catfish is normal and the bottom catfish was deprived of dietary vitamin C for eight weeks. (Photo courtesy Department of Fisheries and Aquaculture, Auburn University, Auburn, AL)

Choline, myocinositol, and vitamin C serve a variety of functions. Choline functions as a—

■ Component of membranes

■ Precursor of acetylcholine, a chemical for nerve transmission

■ Provider of methyl (CH_3) groups for chemical reactions.

Myoinositol is also a component of membranes and involved in sending signals in several body processes. Vitamin C is involved in the formation of connective tissue, bone matrix, and wound repairs. It also facilitates the absorption of iron from the intestine and helps prevent the peroxidation of fats (lipids) in tissues.

STABLE VITAMIN C

Although essential for growing fish, the instability of vitamin C provided difficulty for manufacturers of aquaculture feeds. Heat and water from the milling process tended to accelerate the oxidation and eventual destruction of vitamin C. In 1988, a cooperative effort by Kansas State University, Manhattan, Kansas; Rangen, Inc., Buhl, Idaho; and Zeigler Brothers, Inc., Gardner, Pennsylvania, resulted in the development of a relatively inexpensive, stable vitamin C product to use in aquaculture feeds and possibly other animal and human health and nutrition industries.

The compound, L-ascorbyl-2-polyphosphate, or ASPP for short, allows aquaculture nutritionists to formulate aquatic feeds economically with a high degree of assurance that adequate levels of the essential nutrient vitamin C will be present in the feed. Phosphate attached to the vitamin C protects it from oxygen during feed processing and storage. Upon digestion by fish, the digestive processes split ASPP back into two nutrients—vitamin C and phosphate.

In 1989, Vitamin Technologies International, Buhl, Idaho, received approval from the Food and Drug Administration to manufacture and market ASPP for use in aquaculture feeds.

Since fish feeds usually contain high levels of oil, oxidation may inactivate many of the vitamins. Amounts of vitamins in excess of the requirements ensure that fish receive adequate levels.

Most water-soluble vitamins serve as coenzymes in the biochemical reactions of the body. Enzymes are biological catalysts. Most enzymes are proteins, and they are unique for each biochemical reaction. Coenzymes then work with or become a part of an enzyme.

The fat-soluble vitamins are—

- Vitamin A
- Vitamin D
- Vitamin E
- Vitamin K.

Fat-soluble vitamins are absorbed in the intestine along with fats in the diet. Unlike water-soluble vitamins, fat-soluble vitamins can be stored in body tissues. Excessive amounts in the diet can cause a toxic condition called hypervitaminosis. Functions of the fat-soluble vitamins are quite specific.

Vitamin A is necessary for sight, proper growth, reproduction, resistance to infection, and maintenance of body coverings. As in many land animals, fish can use beta-carotene as a vitamin A precursor.

Vitamin D helps the body mobilize, transport, absorb, and use calcium and phosphorus. It works with two hormones from an endocrine gland, the parathyroid.

Vitamin E is a name that is given to all substances that act like alpha-tocopherol. Vitamin E, working with selenium, protects cells against adverse effects of oxidation.

Vitamin K is required for the normal blood clotting process. Many animals can synthesize Vitamin K in their intestines. Fish do not possess the bacteria to do this.

Table 5-7 summarizes the NRC recommendations and the effects of a vitamin deficiency for each of the fat-soluble and water-soluble vitamins.

Mineral Requirements. Fish can absorb a number of minerals directly from the water: calcium (Ca), magnesium (Mg), sodium (Na), potassium (K), iron (Fe), zinc (Zn), copper (Cu), and selenium (Se). This reduces the mineral requirement in the diet. But this also makes research on dietary mineral requirements difficult and inconclusive. Most researchers agree that fish require all of the minerals required by other animals.

Based on their requirement or use by an animal, minerals are divided into two groups: macrominerals and microminerals. Macrominerals are present in the body in relatively large quantities. The macrominerals include—

- Calcium (Ca)
- Chlorine (Cl)
- Magnesium (Mg)

TABLE 5-7 Vitamin Recommendations and Deficiencies in Fish

Vitamin	NRC[1] Recommended Requirements (mg/kg as fed)	Vitamin Deficiency Symptoms
Fat soluble:		
A (IU/kg)	1,000 to 2,500	Hemorrhages; ascites, edema, visual dysfunctions; poor growth; high mortality; anemia; twisted gill opercula
D (IU/kg)	500 to 2,400	Poor growth; muscular disorders; increased fat content of liver
E	50	Anemia; red blood cell fragility; clubbed gills; low hematocrit; poor growth; edema; ascites; high mortality; ceroid in spleen, kidney, liver
K	R[2]	Anemia; prolonged blood coagulation time
Water soluble:		
B_{12}	0.01	Anorexia; anemia; depressed growth
Biotin	0.15	Lesions of colon; anemia; blue slime disease; spasms; muscle degeneration; anorexia; red blood cell fragility
Choline	400 to 1,000	Depressed growth and feed conversion; internal hemorrhaging; anemia; yellow-colored livers
Folate	1.0 to 1.5	Dark coloration; anemia; fragility of caudal fin; anorexia; lethargy; depressed growth; erythropenia; ascites
Myoinositol	300	Depressed growth; anemia; fin erosion; dark coloration; anorexia; stomach distension
Niacin	10 to 14	Lesions of colon; lethargy; edema of stomach and intestines; loss of coordination; photophobia; swollen gills; tetany; lesions of skin; anorexia; sunburn
Pantothenic acid	15 to 20	Clubbed gills; mumpy appearance; anorexia; lethargy; prostration; poor weight gain; flared opercles (nutritional gill disease)
Pyridoxine B_6	3	Irritability and fits; gasping; anorexia; ataxia; weight loss; ascites; anemia; rapid onset of rigor mortis; blue-green coloration on back
Riboflavin	4 to 9	Anorexia and poor growth; cloudy lens; hemorrhaging in eyes; photophobia; loss of coordination; xerophthalmia
Thiamin	1	Muscle and kidney degeneration; ataxia; anorexia; fatty liver; loss of coordination; depressed growth; internal hemorrhaging; anemia
Vitamin C (Ascorbic acid)	25 to 30	Impaired collagen production; lordosis and scoliosis; poor growth; anorexia; impairment of wound healing

[1] Values in the table are ranges for catfish, trout, salmon, carp, and tilapia. The National Research Council (NRC), *Nutrient Requirements of Fish* (1993), National Academy Press, Washington, DC lists individual requirements.
[2] R—required in diet but quantity not determined.

■ Phosphorus (P)

■ Potassium (K)

■ Sodium (Na).

Calcium and phosphorus are most directly involved in the development and growth of the skeleton, and they act in several other biochemical reactions. Fish absorb calcium directly from the water by the gills and skin. The requirement for calcium is determined by the water chemistry.

Dietary phosphorus is more critical. Phosphorus is derived from dietary phosphate. Phosphorus deficiency signs include poor growth, reduced feed efficiency, and bone deformities. The availability of phosphorus in feedstuffs varies widely. Feedstuffs from seeds contain phosphorus in a form known as phytin. The availability of phosphorus in phytin is low. Simple stomach animals lack the enzyme to release the phosphorus.

Magnesium functions with many enzymes as a cofactor. The dietary requirement can be met from the water or the feed. Deficiencies of magnesium cause anorexia, reduced growth, lethargy, vertebrae deformity, cell degeneration, and convulsions.

Sodium, potassium, and chlorine are electrolytes. Sodium and chlorine reside in the fluid outside the cells. Potassium resides inside the cells—an intracellular cation. Because of the abundance of these elements in the environment, deficiency signs are difficult to produce.

Table 5-8 lists the NRC requirements for the macrominerals.

FIGURE 5-9 Feed mills in Mississippi prepare complete catfish diets to truck out to farmers.

TABLE 5-8 Macromineral Requirements for Finfish[1]

Energy Base[b] (kcal DE/Kg diet) →	Channel Catfish 3,000	Rainbow Trout 3,600	Pacific Salmon 3,600	Common Carp 3,200	Tilapia 3,000
Protein, crude (digestible), %	32(28)	38(34)	38(34)	35(30.5)	32(28)
Macrominerals					
Calcium, %	R[2]	1E[3]	NT[4]	NT	R
Chlorine, %	R	0.9E	NT	NT	NT
Magnesium, %	0.04	0.05	NT	0.05	0.06
Phosphorus, %	0.45	0.6	0.6	0.6	0.5
Potassium, %	R	0.7	0.8	NT	NT
Sodium, %	R	0.6E	NT	NT	NT

[1] National Research Council. (1993). *Nutrient requirements of fish.* Washington, DC: National Academy Press.
[2] R—required in diet but quantity not determined.
[3] E—estimated.
[4] NT—Not tested.

Microminerals are present in very small (micro) amounts in the bodies of fish, but they are still important to the health of fish.

The microminerals include—

- Copper (Cu)
- Iodine (I)
- Iron (Fe)
- Manganese (Mn)
- Selenium (Se)
- Zinc (Zn).

Copper is a part of many enzymes and required for their activity. While it is required by fish, copper can be toxic at concentrations of 0.8 to 1.0 m per L in the water. Fish are more tolerant of copper in feed than in water.

Iodine is necessary for the formation of hormones from the thyroid gland. Fish can obtain iodine from the water or feed. Similar to land animals, a deficiency causes the thyroid gland to grow, a condition like goiter.

Iron is necessary for the formation of heme compounds. These compounds carry oxygen. Since natural waters are low in iron, feed is considered the major source of iron. Iron deficiency causes a form of **anemia**. At high levels iron can be toxic and cause reduced growth, diarrhea, liver damage, and death.

Manganese functions as a part of enzymes or as a cofactor. While it can be absorbed from the water, it is more efficiently absorbed from the feed. A deficiency causes reduced growth and skeletal abnormalities.

Selenium protects cells and membranes against **peroxide** danger. Deficiencies of selenium cause reduced growth. Both selenium and vitamin E are required to prevent muscular dystrophy in some species. When dietary selenium exceeds 13 to 15 mg per kg of **dry feed** it becomes toxic, resulting in reduced growth, poor feed efficiency, and death.

Zinc is also a part of numerous enzymes. Dietary zinc is more efficiently absorbed than that dissolved in water. Dietary calcium and phosphorus, phytic acid protein type, a form of zinc, all affect zinc absorption and use. A zinc deficiency causes suppressed growth, cataracts, fin and skin erosion, dwarfism, and death.

Other trace minerals such as fluoride and chromium may be important but evidence is limited.

Other Dietary Components

Many diets contain other ingredients that can affect the fish. Some of these ingredients are natural, others are added. These ingredients include substances

FIGURE 5-10 In small ponds, fish can be hand-fed.

such as water, fiber, hormones, antibiotics, antioxidants, pigments, binders, and feeding stimulants.

Water. Diets contain water. The water may be a part of the feedstuffs, come from the air, or be added. The less water in a diet, the easier the storage and handling. When moisture in a diet exceeds 12 percent, the feed is more susceptible to spoilage. Some commercial diets contain high moisture levels because fish seem to prefer the moist feed.

Fiber. Fiber refers to plant material such as cellulose, hemicellulose, lignin, pentosums, and other complex carbohydrates. These are indigestible and do not play an important role in nutrition. Fiber adds bulk to a feed but increases the amount of fecal material produced. The goal in commercial aquaculture is to limit fiber content of the diet and use highly digestible feeds.

Hormones. Researchers have evaluated the use of various natural and synthetic hormones on fish. These hormones include growth hormone, thyroid hormones, gonadotropin (GnH), prolactin, insulin, and various steroids—androgens and estrogens. Hormones are used for two purposes: (1) induced or synchronized spawning and (2) sex reversal. Induced or synchronized spawning increases the availability and dependability of seed. Sex steroids reverse the sex of salmonids, carps, and tilapia, producing a monosex culture of sterile fish. This improves growth rate, prevents sexual maturation, and reduces flesh quality.

Antibiotics. With the arsenal of antibiotics available for humans and other livestock, only two have received FDA approval for use in fish—sulfadimethoxine/ormetoprim and oxytetracycline. When these antibiotics are used in feed, the quantity fed, the feeding rate, and the withdrawal time must be strictly controlled. Only licensed manufacturers can add antibiotics to feeds in the United States. Unlike land livestock, fish do not demonstrate any benefit from subtherapeutic levels of antibiotics in their feed.

Antioxidants. Fish feeds containing high levels of fats often use antioxidants. Oxidation of the fats affects the nutritional values of the fat and some vitamins. Synthetic vitamin E in diets usually has little antioxidant activity, so synthetic antioxidants like ethoxyquin, BHT, BHA, and propyl gallate are used.

Pigments. Pigmentation of the skin and flesh in fish comes from carotenoids. Fish cannot make these carotenoids so they must be present in the diet. In salmonids the carotenoids astaxanthin and canthaxanthin are responsible for the red to orange color of their flesh. In the wild these carotenoids come mainly from zooplankton. Some of the natural materials used to pigment the flesh of salmonids include crab, brill, shrimp, and yeast.

Yellow pigmentation of the flesh of catfish is undesirable. It is caused by the carotenoids lutein and zeaxanthin from plant material in the diet.

Pellet Binders. Binders improve stability in the water, firmness, and reduce fines during processing and handling. Widely used binders are sodium and calcium bentonites, lignosulfates, carboxymethylcellulose, hemicellulose, guar gum alginate, and some new inert polymers.

Feeding Stimulants. Fish use sight and smell to find food, but the taste of the food determines its acceptance. Researchers and manufacturers continue

```
                        RANGEN INC.
                         EXTR 450
                    EXTRUDED TROUT DIET

                     GUARANTEED ANALYSIS

      Crude Protein, Not Less Than                    44.0%
      Crude Fat, Not Less Than                        15.0%
      Crude Fiber, Not More Than                       5.0%
      Ash, Not More Than                              12.0%
      Added Mineral Ingredients, Not More Than         2.0%

      INGREDIENTS: Fish Meal, Wheat Flour, Soybean Meal, Blood Meal,
      Feather Meal, Fish Oil, Soy Lecithin, L-ascorbyl-2-polyphosphate
      (Vitamin C), Biotin, Choline Chloride, Folic Acid, Niacin,
      d-Calcium Pantothenate (Pantothenic Acid) Pyridoxine
      Hydrochloride (Pyridoxine), Riboflavin, Thiamine Mononitrate
      (Thiamine), Vitamin B-12 Supplement, Vitamin A Acetate (Vitamin
      A), D-activated Animal Sterol (Vitamin D-3), dl-alpha tocopheryl
      Acetate (Vitamin E), Menadione Sodium Bisulfate Complex (Vitamin
      K3), Copper Sulfate, Manganese Sulfate, Sodium Selenite,
      Potassium Iodate, Salt, Zinc Sulfate, Ethoxyquin (preservative).

                         Manufactured By
                           RANGEN INC.
           115 13th Ave. So., Buhl, Idaho 83316 U.S.A.

      Net Weight 50 lbs.                  Net Weight 22.7 Kg.
```

| 1/8 | 3/32 | 5/32 | 3/16 | 1/4 | 3/8 |

FIGURE 5-11 Feed tag from trout feed.

to attempt to increase the palatability and acceptance of feed. This is especially important in starter and larval feeds. In general, carnivorous fish respond to alkaline and neutral substances. Herbivorous fish respond to acid substance. Besides increasing feed consumption some compounds act as deterrents. This is the last thing an aquaculturalist wants.

Antinutrients and Toxins

Some naturally occurring substances can contaminate fish feeds and affect the performance of fish. Some antinutrients occur in plant and animal feedstuffs. These are described in Table 5-9.

TABLE 5-9 Antinutrients in Feedstuffs

Name	Source	Effect	Prevention
Trypsin inhibitors	Raw soybeans	Inhibits digestive enzyme trypsin	Heat processing
Phytic acid (Phytate)	Soybean meal and other feedstuff from plants	Reduce availability of protein and minerals like Zn, Mn, Cu, Ca, Fe	Limit plant feedstuff; increase nutrient level
Gossypol	Glanded cottonseed meal	Depress growth; damage organs and tissues; acts as a carcinogen	Limit use of glanded cottonseed meal
Cyclopropenoic fatty acids (CFAs)	Cottonseed meal	Lesions, glycogen deposition; elevated fatty acids; act as a carcinogen	Limit use of cottonseed meal
Glucosinolates	Rapeseed	Act as antithyroid agents upon enzymatic hydrolyses	Use low glucosinolate variety like canola; limit use of rapeseed or canola
Erucic acid	Rapeseed oil	Death; problem with skin, gills, kidneys, and heart	Avoid rapeseed oil
Alkaloids	Contamination of cottonseed or soybean meal	Growth depression and death	Quality cottonseed and soybean meal
Thiaminase	Some raw fish preparation	Destroys the vitamin thiamin	Heating or ensiling; feed thiamin in separate diet

Other feed contaminats occur by natural processes or find their way into the feed. These are listed and briefly described in Table 5-10.

TABLE 5-10 Feed Contaminats from Natural Processes and Environmental Contamination

Name	Source	Effect	Prevention
Mycotoxins	Aflatoxins produced by the mold *Aspergillus Flavus* or fungus *Fusarium tricintum*	Carcinogenic; death; reduced growth; reduced feed consumption	Check for contamination; dry feed
Algal and marine toxins	Toxins of algae such as *Ganyaulax spp.* *Gyrodiniun spp.*	Death	Identify toxic algae and eliminate
Oxidative rancidity	Autoxidation of unsaturated fats	Production of free radicals, peroxides, aldehydes, and ketones reduces nutritional value	Add synthetic or natural antioxidants to feed
Mercury	Industrial contamination	Gill problems; but may accumulate in muscle tissues and be a human health hazard when fish consumed	Selenium reduces toxicity and rate of accumulation
Cadmium	Industrial water	Liver necrosis; death	EDTA to chelate
Arsenic	Marine fishmeal	Potential toxicity in organic complex not known	
Polychlorinated biphenyls (PCBs)	Industrial wastes; fish oil and fishmeal	Accumulate in fat; widespread; can cause liver enlargement, liver dysfunction, and decreased thyroid activity	Check feedstuffs and water quality
Pesticides	Accidental	Varies depending on pesticide; most accumulate in tissues; may affect human health or marketability of product	Careful application

DIET FORMULATION AND PROCESSING

The goal of feeding fish is to provide a nutritionally balanced diet that supports maintenance, growth, reproduction, and health at a reasonable cost. Ingredients of the diet should facilitate manufacturing processes and produce a palatable feed with the desired physical properties. Finally, the diet should have minimal effect on water quality.

Nutrient and energy recommendations made by the NRC or anyone else are just that—recommendations. Fish size, management, and environment can profoundly influence dietary requirements. Recommendations are adjusted up or down depending on the best judgment of the aquaculturalist or nutritionist.

Formulating Diets

When formulating diets, protein is the first consideration with the energy being adjusted to the optimal ration. Protein is also considered for the essential amino acids it provides. The amount of carbohydrate used depends on fish species and the type of processing. Fat added to the diet provides energy and needs to meet the essential fatty acid requirement. Mineral supplements are considered after determining the mineral content of the major feedstuffs. Vitamin premixes supply vitamins in diets because of the variation in vitamin content and availability of feedstuffs. Processing destroys some nutrients. To overcome this these nutrients are supplied at higher rates.

Feeding represents a major share of the costs associated with production. Least-cost formulation is used to develop nutritionally complete diets at the minimum cost. Least-cost computer programs require the following information—

- Nutrient requirements of the fish
- Availability of the nutrients to the species of fish
- Energy content of the ingredients
- Minimum and maximum restrictions supplied in ingredients
- Costs of the ingredients.

Table 5-11 shows a typical catfish feed.

Ingredients

Feedstuffs used to prepare fish diets can be classified as sources of protein, energy, essential fatty acids (EFS), minerals, and vitamins. Special ingredients include those to enhance palatability, to preserve, to increase growth, or to pigment the flesh.

TABLE 5-11 Formula for a Nutritionally Complete 32 Percent Protein Catfish Feed Suitable for Pelleting or Extruding

Ingredient	Lbs/Ton	Percent
Menhaden fish meal	160.0	8.00
Soybean meal, 48% protein	965.0	48.25
Corn	582.0	29.10
Rice bran or wheat shorts	100.0	10.00
Dicalcium phosphate	20.0	1.00
Pellet binder	40.0	2.00
Fat (sprayed on finished feed)	30.0	1.50
Trace mineral mix:	1.0	0.05
Manganese	23.00 gms/ton	
Iodine	5.00 gms/ton	
Copper	3.00 gms/ton	
Zinc	136.00 gms/ton	
Iron	40.00 gms/ton	
Cobalt	0.05 gm/ton	
Vitamin mix:	2.5	0.125
Thiamine	10.00 gms/ton	
Riboflavin	12.00 gms/ton	
Pyridoxine	10.00 gms/ton	
Pantothenic acid	32.00 gms/ton	
Nicotinic acid	80.00 gms/ton	
Folic acid	2.00 gms/ton	
Vitamin B_{12}	0.008 gm/ton	
Choline chloride (70%)	500.00 gms/ton	
Ascorbic acid	340.50 gms/ton	
Vitamin A (4,000,000 IU)		
Vitamin D3 (2,000,000 IU)		
Vitamin E	50.00 gms/ton	
Vitamin K	10.00 gms/ton	
Coated ascorbic acid	0.75	0.0375

Feed composition tables list the common feedstuffs used in fish diets and indicate the energy content, protein composition, amino acid content, vitamin content, and mineral content of the feedstuff. These tables are used by aquaculturalists and fish nutritionists to prepare diets that economically meet the nutritional needs of fish.

The best source of protein for fish diets is generally an animal source. Oddly, the best source of a high-quality protein is fishmeal prepared from good-quality whole fish. It is highly digestible and palatable. Fishmeal from processing waste has less high-quality protein and a high mineral (ash) content. Meat and bone meal from livestock and poultry by-products are fair but not as good as fishmeal. Dried blood meal is rich in protein but low in some essential amino acids.

Of the plant protein sources for fish diets, soybean meal is widely available and its amino acid content is the best of the plant proteins. Other plant protein sources such as cottonseed and peanut meal have been used in fish diets, but they lack some essential amino acids. Table 5-12 compares the composition of some protein sources.

Grains contribute the carbohydrates to fish diets. Grains such as corn, wheat, and barley contain 60 to 70 percent starch. Besides supplying energy to a diet, starch acts as a binding agent in steam-pelleted and extruded fish feed.

Fats and oils used in fish diets supply energy and coat pellets to reduce abrasiveness and dustiness. Fats and oils also provide the essential fatty acids (EFAs). Marine fish oils represent one of the best sources of EFAs.

Feed composition tables provide useful guides to selecting and planning ingredients for fish diets. However, major ingredients of a fish diet should be

FIGURE 5-12 Large feed mills in the Mississippi delta produce complete catfish feeds and buy large quantities of the feed ingredients.

TABLE 5-12 Protein Sources for Fish Diets

Feedstuff	Dry Matter (%)	Average Digestible Energy (Kcal/kg)	Crude Protein (%)	Fat (%)	Ash (%)	Problems
Blood meal	93	3188	89.2	0.7	2.3	Low in methionine and unbalanced
Canola meal	93	2549	38.0	3.8	11.1	Higher in fiber and tanins
Cottonseed meal	92	2792	41.7	1.8	6.4	Limiting in lysine and methionine
Fishmeal	92	3778	62.0	7.1	20.7	None
Fishmeal, by-product	92	—	50.8	9.6	10.4	Less high-quality protein
Meat and bone meal	94	3058	50.9	9.7	29.2	High mineral (ash) content; use high quality ration
Peanut meal	92	3370	49.0	1.3	5.9	Limiting in lysine and methionine
Poultry by-product	93	3546	59.7	13.6	14.5	Less high-quality protein
Poultry feather meal	93	3325	83.3	5.4	2.9	Digestibility low unless hydrolyzed
Soybean meal	90	3010	44.0	0.9	5.8	Heating destroys most antinutritional factors

subjected to a complete analysis to assure the producers of its quality. This analysis should include—

■ **Proximate composition** (protein, fat, ash)

■ Limiting amino acids

■ Essential fatty acids

■ Digestibility

■ Test for **mycotoxins**

■ Screening for pesticides and other contaminants.

Some feedstuffs vary considerably in quality and composition. Established standards help producers evaluate an ingredient's value.

Processing

To maintain water quality and allow efficient consumption, feeds are processed into water stable granules or pellets. Five processing methods prepare fish feeds—

1. Steam pelleting
2. Extrusion
3. Moist or semi-moist
4. Crumbling
5. Microencapsulation.

In steam pelleting, moisture, heat, and pressure force ingredients into compact, large particles that sink in water. Steam helps gelatinize any starch, which helps bind the ingredients. All ingredients are finely ground before pelleting. Fiber and fat content can prevent firm bonding during the process. Sometimes special binding agents are used. After pelleting, the pellets are cooled and dried immediately.

The **extrusion** process uses the feed mixture in the form of a dough. To produce the feed, this dough is forced through a small opening at high pressure and temperatures. This process produces a feed that will float because of water vapor trapped in the feed. After extrusion the feed is dried immediately. Overall, the extrusion process requires more elaborate equipment and more moisture, heat, and pressure than pelleting. Aquaculturalists prefer extruded feeds in large ponds because they can see the fish feed.

FIGURE 5-13 Large trucks deliver processed feeds to bulk storage tanks on catfish farms.

Moist or semi-moist feeds are prepared by adding moisture and a binding agent such as carboxymethyl-cellulose or ground wet animal tissue to the dry feed ingredients. These soft pellets are more palatable but they are more susceptible to spoilage. Moist or semi-moist feeds are frozen, or they contain humectants like propylene glycol that prevent bacterial growth. Also, they contain fungistats like propionic or sorbic acid. For best quality, these diets are stored in hermetically sealed containers, under nitrogen, and kept at a low temperature until used.

Crumbling prepares diets for small fish. In this process the diet is first pelleted. The size of the pellets are reduced by crumbling. Then screening separates the crumbles by particle size. Crumbling increases the surface area of the feed exposed to water so it requires some compaction to prevent leaching of water-soluble vitamins.

Microencapsulation is the coating of small feed particles with a substance that is insoluble in water but digestible by the enzymes in the digestive tract of the fish. This reduces disintegration and leaching in the water. Some compounds used include nylon cross-linked proteins, calcium alginate, and oils.

FEEDING AQUATIC ANIMALS

Feeding crustaceans and mollusks is important to commercial ventures. Chapter 4, Aquatic Management Practices, contains limited but sufficient information about feeding these species. What follows are common feeding practices for prominent finfish: channel catfish, rainbow trout, striped bass, tilapia, and larval fish.

Catfish

Most catfish feed manufacturers now use the least-cost instead of fixed-feed method of feed formulation where the formula varies, within limits, as ingredient prices change. The kind or amount of ingredients needed to provide essential nutrients for catfish is not secret. Feed manufacturers should be willing to reveal the type and amount of ingredients in their feed.

Form and Size. Not only must the feed contain all of the essential nutrients, it must also be palatable to the catfish and of a size that can be eaten. If catfish do not eat it or cannot eat it, maximum growth is not achieved, costing the producer money. The feed must be offered in a way and at a time that promotes total consumption. Form and size of feed available include four types—

- Meal
- Crumbles
- Floating (expanded or extruded) pellets
- Sinking (hard or compacted) pellets.

Feed size and form used depends on fish size, water temperature, and type of management. Meal and crumbles are used for fry and small fingerlings. Although more expensive, extruded or floating feed is generally preferred when water temperatures are above 65°F (18°C) because feeding behavior is much easier to monitor. Most producers feel that seeing the fish when they are feeding is well worth the extra cost.

Sinking feed is used when the water temperature falls below 65°F (18°C) since catfish reduce their feeding activity at colder temperatures and seldom come to the surface.

Topping or multiple harvest is the most common production scheme used in intensive pond production of catfish. The size of catfish in a pond at any given time may vary from 0.02 to 2 lbs (9 to 900 g) or larger. Since it is not practical to feed the catfish two or more sizes of feed every day, most farmers compromise by feeding one size pellet.

Feeders. Catfish may be fed by hand from the bank or a boat, or using some type of mechanical feeder. Hand feeding more than 10 A (4 ha) of intensively cultured catfish ponds is too time consuming and laborious. Larger farms use some type of mechanical feeder.

The blower-type feeder with a 1- to 3-ton hopper, mounted on a truck bed or pulled by a tractor, is best. The blower-type feeder can be calibrated to blow a known amount of feed per minute or equipped with a scale that allows the operator to know the amount fed. The hopper of the blower is filled from a bulk storage tank.

Two other types of mechanical feeders are the demand feeder and the automatic feeder. Neither has any place in intensive pond production of catfish because frequent observation during feeding is not possible. The demand feeder is activated by the catfish, thus allowing a few large, aggressive "hogs" to consume most of the feed. When demand feeders are used in intensive production ponds, there is a large difference in the size of the fish produced. Waterfowl, particularly coots, quickly learn to use demand feeders and may consume more feed than the catfish. The automatic feeder is programmed to release specific amounts of feed at predetermined times during the day. It must be carefully monitored to avoid over- or underfeeding the catfish. Unless large numbers of automatic feeders are used, many catfish will

FIGURE 5-14 Blower type feeder mounted on a truck bed, receiving feed from a storage bin located between catfish ponds.

not get enough feed while a small number of the more aggressive fish will eat most of the feed.

Feeding Rates. Several factors affect the amount of feed a catfish will eat—

- Water temperature
- Water quality
- Size of the feed
- Palatability or taste of the feed
- Frequency of feeding
- Feeding method
- Location of feeding sites
- Type of pellet used
- Health of the fish.

Table 5-13 gives the amount to feed daily, based on average expected gains, at stocking rates of 1,000 5-in fingerlings per A (2,470 12.7-cm fingerlings per ha). To get the amount to feed per A at higher stocking rates, divide the number of catfish stocked per A by 1,000 and multiply the answer by the daily amount to feed per A in Table 5-13.

TABLE 5-13 Feeding Guide Based on Average Expected Gains with a Feed Conversion of 1,000 5-inch Fingerlings per A

Week	Water temp °F	Column 1 Wt. of 1,000 fish at beginning	Column 2 % of body wt. fed daily	Column 3 Wt. of food fed/acre/day 1,000 fish	Column 4 Feed conversion
1	55–60	34.0	1.0	0.3	1.75
3	60–65	37.4	1.5	0.6	1.75
5	65–70	41.9	2.0	0.8	1.75
7	70–75	49.4	2.5	1.2	1.75
9	75–80	59.9	3.0	1.8	1.75
11	80–85	75.9	3.0	2.3	1.75
13	85–90	95.4	3.0	2.9	1.75
15	90–95	120.9	3.0	3.6	1.75
17	90–95	151.8	3.0	4.6	1.75
19	90–100	193.4	3.0	5.8	1.75
21	90–95	242.9	3.0	7.3	1.75
23	85–90	310.1	3.0	9.3	1.75
25	75–85	389.6	3.0	11.7	1.75
27	65–75	490.1	2.5	12.3	1.75
29	60–65	595.1	2.0	11.9	1.75

Estimated feeding rates given in Table 5-14 calculate the amount to feed different size catfish at water temperatures above 70°F (21°C). A good estimate of the total weight of catfish in the pond uses the formula: Amount to feed daily = percentage body weight fed × total weight of fish in pond.

EXAMPLE About 40,000 lbs of catfish in ponds are being fed at a rate of 2.5 percent of their body weight daily. Amount to feed daily = 0.025 × 40,000 lbs = 1,000 lbs of feed.

The amount fed daily must be adjusted at least every two weeks or the catfish soon will be underfed, causing a reduction in both growth and profits. This is best done by taking a sample of fish from the pond, usually by seine, and counting and weighing them. Then use the formula given to calculate the total weight of catfish in the pond at that time.

Total weight in pond = wt. fish in sample × no. fish in pond/no. fish in sample

EXAMPLE There are 45,000 catfish being fed at 3 percent of their body weight daily, stocked in a 10-A pond. To adjust the amount to be fed daily for the next two weeks, a sample of fish is seined, counted, and weighed. The sample contains 200 fish weighing 80 lbs. The new amount to feed daily is calculated:

$$\text{Total weight in pond} = \text{wt. of fish in sample} \times \frac{\text{no. fish in pond}}{\text{no. fish in sample}}$$

$$= 80 \text{ lbs} \times \frac{45,000 \text{ fish}}{200 \text{ fish}} = 18,000 \text{ lbs of catfish in the pond}$$

Lbs to feed daily
for next 2 weeks $= 3\% \times 18,000 \text{ lbs} = 0.03 \times 18,000 \text{ lbs} = 540 \text{ lbs of feed}$

Total expected weight of fish $= 795.4 \text{ lbs}$ Total weight of food fed $= 1,331.2 \text{ lbs}$

Rather than seining a sample of fish and counting and weighing them every two weeks, growth of the fish can be estimated. Some assumptions can be based on a pond's historical data or on industry averages about the percent of body weight fed daily and feed conversion factors. Table 5-13 is a feeding guide that shows how to make these calculations. It can be used with good results, but it is much better to use feed conversion ratios and percent of body weight to feed daily, which are valid for ponds. It is a good practice to remove a sample of fish occasionally for counting and weighing to calculate the weight of fish present and see how close growth estimates have been.

TABLE 5-14 Estimated Percent of Body Weight Consumed Daily by Different Size Channel Catfish at Temperatures of 70°F (21°C) and Above

Average Weight (Pounds)	Pounds per 1,000 Fish	Estimated % Body Weight Consumed Daily
0.02	20	4.0
0.06	60	3.0
0.25	250	2.7
0.50	500	2.5
0.70	750	2.2
1.00	1,000	1.6
1.50	1,500	1.3

Another method of estimating the amount of feed to use daily when the water temperature is above 65°F (18°C) is to feed the fish what they will eat in 10 to 15 minutes. If feed is still floating on the surface at the end of 15 minutes, the fish are being overfed and increasing costs.

Feeding Practices. Manner and time of feeding, as well as the amount and type of feed, can have a profound effect on the growth and size variation and the quality of the catfish produced. A large variation in the size of catfish produced usually is the result of underfeeding or feeding in a small area of the pond. In underfeeding, the larger, more aggressive catfish eat a larger share of the feed and become bigger at the expense of the smaller catfish. This also happens when feed is offered in only a small area of the pond since the larger, more aggressive catfish quickly learn where the feed will be put in the pond and are there waiting for it. To produce catfish uniform in size, and to maximize profits, it is equally important that catfish be fed the proper amount of feed daily and the food be distributed as evenly over the pond as possible.

Feeding twice daily, if possible, will usually improve feed consumption and feed conversion. This means that one-half of the daily allowance is fed in the early morning, and the other half in the late morning. If the catfish are fed only once a day, morning is the preferred time since feeding in the late afternoon increases the amount of fat deposited, and this can affect the quality of the processed fish.

Feed should not be offered until the oxygen level of the pond water is at least 4 parts per million (ppm) or milligrams per liter (mg/L) or higher since feed consumption goes down dramatically at lower oxygen concentrations.

FIGURE 5-15 Tractor-pulled feeder and feed storage bins near catfish ponds in Mississippi.

Oxygen requirements for catfish increase greatly during feeding, so it is best not to feed in the late afternoon when oxygen concentrations in the water are decreasing.

Feeding seven days a week maximizes growth. Production time can be decreased by four weeks when compared to feeding only six days a week.

Winter Feeding. Winter feeding is an important management practice. It means more profit for the farmer, and the catfish will be in better condition during the winter and spring to withstand stresses that can cause disease outbreaks.

Two basic winter feeding programs are: (1) feed sinking feed at 0.5 to 1 percent of body weight on alternate days when water temperature is above 49°F (9°C); or (2) feed sinking feed at 0.5 to 1 percent of the body weight whenever the water temperature at a depth of 3 ft (0.9 m) is 54° F (12° C) or higher.

Trout (Salmonids)

Aside from the final sale price of the fish, the amount and suitability of feed used for trout farming will be the primary factor determining the profitability of production. Digestive systems of trout and other salmonids are naturally equipped to process foods consisting primarily of protein (mostly from fish), and can obtain a limited amount of energy from fat and carbohydrates. Although a farm could produce its own fish food, it is usually uneconomical to do so.

Form and Size. Diets for fry and fingerling trout require a higher protein and energy content than diets for larger fish. Fry and fingerling feed should contain approximately 50 percent protein and 15 percent fat. Feed for larger fish should contain about 40 percent protein and 10 percent fat. The switch to lower protein formulations usually occurs at transition from a crumble feed to a pelleted ration, called a grow-out or production diet.

Feeders. After selecting a quality feed and setting the amount of feed, the next consideration is how to feed the fish. Specific methods for feeding trout are somewhat dependent upon the size of the fish. First-feeding fry should be fed a small amount by hand at least ten times per day until all the fish are actively feeding. After this period, an automatic feeder is most practical, with two or three hand feedings daily to observe the fish.

As the fry grow, frequency of feeding can be gradually decreased to about five times per day. Trout can hold roughly 1 percent of their body weight in dry feed at each feeding, so frequency should be adjusted accordingly. Fry gain weight rapidly so they should be sample counted weekly for the first four to six weeks on feed and the daily feed ration adjusted according to their weight. Feed should be distributed over at least two-thirds of the water surface when fry are less than 2 in (5 cm). This assures easy access to the feed and will help to achieve size uniformity within the population.

Though the use of a published feeding chart is strongly recommended, charts are only guides and individual judgment should be exercised based on observations. Overfeeding allows feed to settle to the bottom of the tank, where small trout will ignore it. Excess feed leads to deterioration of water quality and promotes disease. Remove excess feed promptly.

After fingerlings are moved out to tanks or earthen ponds, a variety of feeding alternatives are available. Hand-feeding is generally not practical on a large commercial farm except in certain situations. Several types of automatic and mechanical feeders are available for trout farming, including electric, water-powered, and solar-powered feeders with variable timers. Some automatic feeders use compressed air to blow feed out over the water surface at set intervals, and truck or trailer mounted units that have hydraulically operated blower feeders.

One type of feeder commonly used on commercial trout farms is the demand feeder (Figure 5-15). This consists of a hopper for holding the feed pellets and a movable disc below the hopper opening that is attached to a pendulum extending to the water. Trout greater than 5 in (12.7 cm) can be

FIGURE 5-16 Demand feeder for trout.

readily trained to feed themselves, and with careful adjustment of the feeders, rapid weight gain and efficient feed use can be reached. The use of demand feeders can eliminate the sharp oxygen decline that occurs when fish are fed by hand or machine a few times each day. Demand feeders also reduce labor costs associated with daily hand feeding. Enough feed for several days can be loaded. Disadvantages include the tendency to allow overfeeding due to improper adjustment of the feeders, and food release only in a small section of the pond or tank. Overfeeding with demand feeders can be a problem with larger trout.

Even if demand feeders are used, feeding according to a feed chart is recommended for best performance. When feeding by hand or with a mechanical distribution system, feed should be distributed throughout the pond and should not accumulate on the bottom. In concrete tanks, trout will feed on some pellets that fall to the bottom, but trout will rarely pick up pellets from the bottom of earthen ponds.

Feeding Rates. The primary goals in feeding trout are to grow the fish as fast and efficiently as possible, maintaining uniformity of growth with the least degradation of water quality.

The amount of feed required by trout is dependent on water temperature and fish size; during normal production, trout should be fed seven days per week with a high-quality commercially prepared diet formulated for trout. Due to higher metabolic rates, smaller fish need more feed relative to their body weight than do larger fish, and fish in warmer water need more feed than fish in cooler water. Because fish are poikilothermic (cold-blooded) their body temperatures and metabolic rates vary with environmental temperatures.

In trout, the minimum temperature for growth is approximately 38°F (3.3°C). At this temperature and below, appetites may be suppressed and their digestive systems operate very slowly. Trout will require only a maintenance diet of 0.5 percent to 1.8 percent body weight per day, depending upon fish size at these temperatures; more than this will result in poor food conversion and wasted feed. Above 38°F (3.3°C), the metabolism and growth rate of trout will increase with temperature until approximately 65°F (18.3°C), depending upon the genetic strain being cultured. Optimum temperatures for efficient growth are from 55° to 65°F (12.8° to 18.3°C). At these temperatures, feeding rates should be at maximum levels (1.5 percent to 6.0 percent body weight per day). Above 65°F (18.3°C), the metabolic rate will continue to increase until the temperature approaches lethal levels. Oxygen-carrying capacity of the water and respiratory requirements of the fish will limit the amount of food to be processed efficiently.

In very warm water (above 68°F, 20°C), a trout's digestive system does not use nutrients well, and more of the consumed feed is only partially digested before being eliminated. This nutrient loading of the water, coupled with generally lower oxygen levels in warm water, can easily lead to respiratory distress and should be avoided. Under these conditions, feeding rates should be reduced enough to maintain good water quality and avoid wasting feed.

The best way to determine the correct amount and sizes of feed needed for trout production is to use a published feeding chart, usually provided by the feed manufacturer. The chart should be used as a guide, and may need adjustment to fit specific conditions on individual farms. Overfeeding will cause the fish to use the feed less efficiently, and will not increase growth rates significantly. Knowing the number and size of the fish in a facility helps provide an appropriate amount of feed. At water temperatures above 55°F (12.8°C), a sample count the fish every week can be used to adjust feeding percentages. In cooler waters, a sample count every two weeks usually is adequate. Good growth records help to predict the seasonal growth rate.

At times, feeding should be restricted or stopped altogether, such as when water temperatures drop much below 40°F (4.4°C) or rise much above 68°F (20°C). Feeding rates should also be reduced when fish are sick, as appetite will be depressed. Fish should always be kept off feed for a period before handling or transporting. For routine handling, such as grading or vaccinating, 24 hours

FIGURE 5-17 An automatic feeder used at a large trout facility in Idaho.

FIGURE 5-18 Feed mill in Idaho producing trout feed. (Photo courtesy Rangen, Inc., Buhl, ID)

without food is sufficient. If fish are to be transported a long distance or are to be processed, they should be kept off feed for a minimum of three to four days, longer if temperatures are low.

Special Purpose Feeding. Commonly used specialty feeds for trout include those containing antibiotics (tetracycline hydrochloride or potentiated sufadimethoxine) or carotenoid pigments (canthaxanthin). Antibiotic treated feeds should be used only with diagnosis of a bacterial condition susceptible to treatment. Carotenoid pigmented feeds impart a pink or red coloration to the flesh of fish and do not affect their health or growth rate. Successful pigmentation can be achieved in approximately three months when fish are actively growing and in approximately six months during colder water conditions. Other specialty diets include an enriched diet for broodfish and a high-fat diet (16 to 24 percent fat) for producing an oilier fish used for smoking or for specialty markets.

Other Coolwater Fish

For many years, fish culture was classified into two major groups. Coldwater hatcheries cultured trout and salmon, and warmwater hatcheries cultured any fish not a salmonid. Muskellunge, northern pike, walleye, and yellow perch prefer temperatures warmer than those suited for trout, but colder than those water temperatures most favorable for bass and catfish. The term coolwater species gained general acceptance in referring to this intermediate group.

Extensive pond culture of coolwater fish involves providing sufficient quantities of microorganisms and plankton as natural foods through pond fertilization programs. If larger fingerlings are to be reared the fry are transferred, when they reach approximately 1.5 in (3.8 cm) in length, to growing ponds where minnows are provided for food. A major problem in extensive pond culture is that the fish culturalist is unable to control the food supply, diseases, or other factors. Many times it is extremely difficult to determine the health and growth of fish in a pond.

Coldwater fishes can be successfully reared on dry feed. Starter feed is distributed in the trough by automatic feeders set to feed at five-minute intervals from dawn to dusk.

Coolwater fish will not pick food pellets off the bottom of the tank, so it is necessary to continually present small amounts of feed with an automatic feeder. In some situations, coolwater fry are started on brine shrimp and then converted to dry feed.

Tiger muskie fry aggressively feed on dry feeds. Fry often follow a food particle through the entire water column before striking it. Hand-feeding or human presence at the trough does not disrupt feeding activity. When fish reach a length of 5 to 6 in (12.7 to 15.2 cm), human presence next to a trough or tank can disrupt feeding activity completely. Cannibalism generally is a problem only during the first 10 to 12 days after initial feeding, when the fish are less than 2 to 3 in (5 to 7.6 cm) in length. Removing weak and dying fry reduces cannibalism.

The methods developed for estimating feeding rates for salmonids can be adapted for use with coolwater species.

Tilapia

Pond culture is the most popular method of growing tilapia. One advantage is that the fish are able to use natural foods. Management of tilapia ponds range from extensive systems, using only organic or inorganic fertilizers, to intensive systems, using high-protein feed, aeration, and water exchange. The major draw-back of pond culture is the high level of uncontrolled reproduction that may occur in grow-out ponds. Tilapia recruitment, the production of fry and fingerlings, may be so great that offspring compete for food with the adults. The original stock becomes stunted, yielding only a small percentage of marketable fish weighing 1 lb (454 gm) or more.

With optimal temperatures, feeding rates depend on fish size and density. Optimal daily feeding rates for tilapia are—

Weight (g)	Percent of Body Weight
30	3.5
50	3.0
100	2.5
175	2.0
450	1.5

If densities are high, suboptimal feeding rates may have to be used to maintain suitable water quality, increasing culture duration.

Feeding in Cage Culture. After proper stocking, the most important aspect of cage culture is providing good quality feed in the correct amounts to the caged fish. The diet should be nutritionally complete, containing vitamins and minerals. Commercial pellet diets for tilapia, catfish, or trout are best. Protein content should be 32 to 36 percent for 1- to 25-gm tilapia and 28 to 32 percent for larger fish.

Floating feeds allow observation of the feeding response and are effectively retained by a feeding ring on the cage. Since it takes about 24 hours for high-quality floating pellets to disintegrate, fish may be fed once daily in the proper amount, but twice-daily feedings are better.

Sinking pellets can be used, but extra care must be taken to ensure they are not wasted. Sinking pellets disintegrate quickly in water and have a greater tendency to be swept through the cage sides. More than one feeding is needed each day. Tilapia cannot consume their daily requirement of feed for maximum growth in a single meal of short duration. Fish of less than 25 gm should be fed at least three times day.

Sinking pellets may be—

■ Slowly fed by hand, allowing time for the fish to eat the feed before it sinks through or is swept out of the cage

■ Placed in shallow, submerged trays

■ Placed in demand feeders.

Feeding slowly by hand is inefficient. Use of a tray allows quick placement of feed onto the tray, but multiple daily feedings are still required. The correct amount of feed must be weighed daily. Feeding rate tables or programs are required to make periodic increments in the daily ration. Feeding adjustments can be made daily, weekly, or every two weeks. The fish should be sampled every four to six weeks to determine their average weight and the correct feeding rate for calculating adjustments in the daily ration. Adjustments can be made between sampling periods by estimating fish growth based on an assumed feed conversion ratio (feed weight divided by fish weight gain).

EXAMPLE With a feed conversion ratio of 1.5, the fish would gain 10 gm for every 15 gm of feed. The correct feeding rate, expressed as percent of body weight, is multiplied by the estimated weight to determine the daily ration. Recommended feeding rates are listed in Tables 5-15 and 5-16.

TABLE 5-15 Expected Average Final Weights for Different Culture Periods and Initial Weights of Tilapia

Length of Growing Season (Wks)	Starting Weight 30 g	Starting Weight 60 g	Starting Weight 100 g
	Expected average final weight (g)[1]		
12	200	270	350
16	250	340	440
20	310	410	520
24	370	480	600
28	420	550	690

[1] Values are for male populations.

TABLE 5-16 Recommended Daily Feeding Rates, Expressed as Percentage of Body Weight, for Tilapia of Different Sizes

Fish Weight (grams)	Feeding Rate (percent)	Fish Weight (grams)	Feeding Rate (percent)
1	11.0	30	3.6
2	9.0	60	3.0
5	6.5	100	2.5
10	5.2	175	2.5
15	4.6	300	2.1
20	4.2	400+	1.5

Feeding rate tables serve as guides for estimating the optimum daily ration but are not always accurate under a wide range of conditions, such as fluctuating temperatures or dissolved oxygen. Demand feeders can be used to eliminate the work (feed weighing, fish sampling, calculations) and uncertainty of feeding rate schedules by letting the fish feed themselves. Fish quickly learn that feed is released when they hit a rod that extends from the funnel into the water. Demand feeders and feeding rate schedules produce comparable growth and feed conversion, but demand feeders reduce labor by nearly 90 percent. Feeding rate schedules may still be used with demand feeders by adding a computed amount of feed daily instead of refilling the feeder whenever it is nearly empty.

With high quality feeds, good growing conditions, and effective feeding practices, feed conversion ratios as low as 1:3 have been obtained. Generally, feed conversion ratios will range from 1.5 to 1.8 lbs of feed per lb of fish.

Tank Culture of Tilapia. Fry are given a complete diet of powdered feed (40 percent protein) that is fed continuously throughout the day with automatic feeders. The initial feeding rate can be as high as 20 percent of body weight per day under ideal conditions—good water quality and temperature (86°F, 30°C). It is gradually lowered to 15 percent by day 30. During this period, fry grow rapidly and will gain close to 50 percent in body weight every three days. Therefore, the daily feed ration is adjusted every three days by weighing a small sample of fish in water on a sensitive balance. If feeding vigor diminishes, the feeding rate is cut back immediately and water quality is checked.

Feed size can be increased to various grades of crumbles for fingerlings (1 to 50 gm), which also require continuous feeding for fast growth.

During the grow-out stages, the feed is changed to floating pellets to allow visual observation of the feeding response. Recommended protein levels are 32 to 36 percent in fingerling feed and 28 to 32 percent in feed for larger fish. Adjustments in the daily ration can be made less often (for example, weekly) because relative growth, expressed as a percentage of body weight, gradually decreases to 1 percent per day as tilapia reach 1 lb in weight. The daily ration for adult fish is divided into three to six feedings that are evenly spaced throughout the day.

If feed is not consumed rapidly (within 15 minutes), feeding levels are reduced. Dissolved oxygen (DO) concentrations decline suddenly in response to feeding activity. Although DO levels generally decline during the day in tanks, feeding intervals provide time for DO concentrations to increase somewhat before the next feeding. Continuous feeding of adult fish favors the more aggressive fish, which guard the feeding area, and causes the fish to be less uniform in size. With high quality feeds and proper feeding techniques, the feed conversion ratio should average 1:5 for a 1 lb (454 gm) fish.

OTHER WARMWATER FISH

As long ago as 1924, fish culturalists attempted to increase yield and survival of smallmouth bass by providing a supplemental feed of zooplankton.

Ground fresh-fish flesh also was successfully used, but costs were prohibitive. These early attempts were discouraging but culturalists have continued to rear bass fry to fingerling size on naturally occurring foods in fertilized earthen ponds. This method generally results in low yields and is unpredictable.

Interest in supplemental feeding of bass has been renewed in recent years due to successful experimental use of formulated pelleted feeds with large mouth bass fingerlings. The best success in supplemental feeding has been obtained by rearing bass fry on natural feed to an average length of 2 in (5 cm) in earthen ponds before they are put on an intensive training program to accept formulated feed.

Striped Bass

Striped bass fingerlings often are fed supplemental diets in earthen ponds when zooplankton blooms have deteriorated or larger fish are desired. The fingerlings are fed a high-protein (40 to 50 percent) salmonid type of formulated feed at the rate of 5 lbs per A (5.7 kg per ha) per day. This is increased gradually to a maximum of 20.0 lbs per A (22.2 kg per ha) per day by the time of harvest. The fish are fed two to six times daily.

When striped bass fingerlings reach a length of approximately 1.5 in (3.8 cm) they will accept salmonid-type feeds readily. Good success can be anticipated when a training program is followed. Striped bass fingerlings can be grown to advanced sizes in ponds, cages, or raceways.

Young striped bass and hybrid striped bass require the essential fatty acids eicosapentaenoic or docosahenaenic acid in the diet for normal growth.

TIME OF FIRST FEEDING

The time at first feeding depends on the type of digestive system in the larval form and the changes that occur during metamorphosis. Larval fish can be divided into three groups.

The first group includes salmonids and catfish. These fish have a functional stomach before changing from the food supplied by the yolk sac to external feed.

The second group includes fish such as the striped bass and many marine fish. Larval stages of these fish possess rudimentary digestive tracts with no

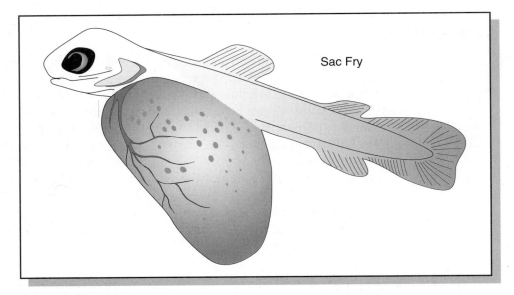

FIGURE 5-19 Sketch of fry with yolk sac.

functional stomach or digestive glands. The digestive systems of these fish go through a complex metamorphosis.

Larvae of fish such as carp represent the third group. These larval fish develop a functional digestive system but remain stomachless throughout life.

Obviously, those larval fish with immature digestive tracts represent greater difficulty feeding.

The most common practice is to offer food when the fry swim up. Swim-up occurs when the fry have absorbed enough of their yolk sac to enable them to rise from the bottom of the trough and maintain a position in the water column. A considerable amount of work has been conducted to determine when various salmonid fry first take food. Brown trout begin feeding on food approximately 31 days after hatching in 52°F (11°C) water, while food was first found in the stomach of rainbow trout fry 21 days after hatching in 50°F (10°C) water. The upper alimentary tract of rainbow trout fry remains closed by a tissue plug until several days before swim-up. Thus, feeding of rainbow trout fry before swim-up is useless. Yolk absorption is a useful visual guide to determine the initial feeding of most species of fish. Most studies indicate that early feeding of fry during swim-up does not provide them with any advantage over fry that are fed later, after the yolk sac has been absorbed. Many culturalists start feeding when 50 percent of the fry are swimming up because if fry are denied food much beyond yolk-sac absorption, some will refuse to feed. No doubt, starvation from a lack of food will lead to a weakened fry that cannot feed even when food is abundant.

The initial feeding time for warmwater fishes is much more critical than for coldwater species because metabolic rates are much higher at warmer water temperatures. This leads to more rapid yolk absorption and a need for fish to be introduced to feed at an earlier date.

FEED CALCULATIONS

Nutrition of fish requires that the aquaculturalist make some calculations. Specifically, aquaculturalists need to calculate the amount of feed required, the amount of storage required and, finally, the feed conversion ratio. These calculations help determine the profitability of an operation.

Feed Requirements

The feed requirements for fish change with age, size, health, and water conditions. Feeding charts are available for catfish and rainbow trout of different sizes and at different water temperatures. The following examples illustrate problems and solutions related to feed requirements for fish.

EXAMPLE 1. A pond is stocked with 45,000 fish that weigh 50 lbs per 1,000 fish. The desired feeding rate is 3 percent of their weight daily. How much feed is needed for one day and for one week?

STEP 1. $\dfrac{45,000 \text{ fish}}{1,000 \text{ fish}} \times 50 \text{ lbs} = 2,250 \text{ lbs of fish stocked}$

STEP 2. 2,250 lbs × 0.03/day = 67.5 lbs of feed daily

STEP 3. For 1 week:
67.5 lbs/day × 7 days/week = 472.5 lbs

EXAMPLE 2. A 12-A pond contains 2,000 lbs of fish per A. A bacterial disease is diagnosed and double-strength (2×) Terramycin medicated feed is needed for disease treatment. The daily recommended feeding rate is 1.5 percent body weight per day for a total of ten days. How much medicated feed should be ordered and fed to the sick fish?

STEP 1. Total pounds of fish in pond:
2,000 lbs/A × 12 A = 24,000 lbs of fish

STEP 2. Pounds of medicated feed required per day:
24,000 lbs × 0.015/day = 360 lbs/day

STEP 3. Total pounds of medicated feed required for total treatment time of ten days:

5 360 lbs/day × 10 days = 3,600 lbs

STEP 4. If the medicated feed comes in 50-lb bags, how many are needed?

$$\frac{3,600 \text{ lbs}}{50 \text{ lbs/bag}} = 72 \text{ bags}$$

Storage Requirement

A producer needs to determine whether buying bulk feed is feasible based on farm size, storage time, and availability of feed sizes and loads. Table 5-17 illustrates how farm size and feeding ratios affect storage.

The following formula can be used to calculate feed storage time.

$$\text{Storage time in days} = \frac{\text{Feed bin capacity in pounds}}{\text{Maximum or average feeding rate in lbs/A/day}}$$

EXAMPLE 3. The storage time or capacity in days of feed for a 5-A pond fed at a maximum of 75 lbs/A/day can be calculated as follows:

STEP 1. Determine the maximum feed requirement for 1 day:

5 A × 75 lbs/A/day = 375 lbs/day

STEP 2. If a person is considering bulk purchase and storage in a 10-ton bin, then the capacity in days for feed is:

10 tons × 2,000 lbs/ton = 20,000 lbs feed capacity of 10-ton bin

TABLE 5-17 Capacity in Days of Feed for Two Sizes of Bulk Storage Feed Bins for Five Farm Sizes and Three Feeding Rates

Farm Size in Water Acres	Maximum Feeding Rates					
	50 lbs/acre/day Bin Size		75 lbs/acre/day Bin Size		100 lbs/acre/day Bin Size	
	10 Ton	23 Ton	10 Ton	23 Ton	10 Ton	23 Ton
15	26.7	61.3	17.8	40.9	13.3	30.7
40	10.0	23.0	6.7	15.3	5.0	11.5
70	5.7	13.1	3.8	8.8	2.9	6.6
100	4.0	9.2	2.7	6.1	2.0	4.6
140	2.9	5.8	1.9	4.4	1.4	3.3

$$\frac{20{,}000\,\text{lbs}}{375\,\text{lbs/day}} = 53.3 \text{ days storage time or capacity}$$

This is a long storage time in the bin before new feed is purchased. Storage time should not exceed 30 to 45 days in the summer. For this small farm, bagged feed is recommended to reduce storage time and assure quality and freshness of feed.

Feed Conversion Ratio

Feed conversion ratios (FCR) are calculated to determine the cost and efficiency of feeding the fish. They are affected by the quality of feed, size and condition of fish, number of good feeding days related to temperature and water quality, and feeding practices. Feed conversion ratios for catfish can vary from less than 1.5 to as high as 4 or more. Feed conversion ratios much higher than 2 should be reduced. FCR is a ratio of the pounds of weight gained by fish after consuming a known amount of feed. In commercial fish ponds, the fish get little nutrition from natural food organisms.

To determine the FCR requires records of the amount of feed fed to fish in each pond, fish losses, and pounds of fish harvested. The feed conversion ratio can be calculated monthly when fish are sampled and when fish are harvested.

Using this formula, FCR for any period can be determined—

$$\text{FCR} = \frac{\text{Total pounds feed fed}}{\substack{\text{Final fish weight—Initial fish weight or weight gain} \\ \text{between sampling periods}}}$$

EXAMPLE 1. About 67,500 fingerlings weighing 50 lbs per 1,000 fish were stocked in a pond. Later, the fish were sampled and the average weight of fish was 0.25 lb, or 250 lbs per 1,000 fish. During this time, 10 tons plus 1,600 lbs of feed were fed. No fish losses were observed. What is the feed conversion ratio?

$$\text{Feed Conversion Ratio (FCR)} = \frac{\text{Amount feed fed}}{\text{Weight gain of fish}}$$

STEP 1. Convert all feed weight to one unit (pounds) instead of having two units (tons and pounds).

One ton = 2,000 lbs

10 tons × 2,000 lbs/ton + 1,600 lbs

= 20,000 lbs + 1,600 lbs

= 21,600 lbs total feed

STEP 2.   Determine the weight gain of fish for this period. The final weight can be calculated using two methods:

Average fish weight × Number fish = Total weight

$$\frac{0.25 \text{ lb}}{\text{fish}} \times 67{,}500 \text{ fish} = 16{,}875 \text{ lbs}$$

$$\frac{67{,}500 \text{ fish}}{1{,}000 \text{ fish}} \times 250 \text{ lbs} = 16{,}875 \text{ lbs}$$

STEP 3.   The initial weight of fish was:

$$\frac{67{,}500 \text{ fish}}{1{,}000 \text{ fish}} \times 50 \text{ lbs} = 3{,}375 \text{ lbs}$$

STEP 4.   Determine the feed conversion ratio using the formula below:

$$\text{Feed Conversion Ratio} = \frac{\text{Amount feed fed}}{\text{Final weight} - \text{Initial weight}}$$

$$\text{FCR} = \frac{21{,}600 \text{ lbs feed}}{16{,}875 \text{ lbs} - 3{,}375 \text{ lbs}}$$

$$= \frac{21{,}600 \text{ lbs feed}}{13{,}500 \text{ lbs weight gained}} = 1.6$$

The FCR means that during the grow-out time, the fish consumed an average of 1.6 lbs of feed to gain 1 lb in weight.

EXAMPLE 2.   A pond had an estimated standing crop of 22,500 lbs of fish at the last sampling. A new sample estimated the total fish weight at 33,000 lbs. Between these two samplings, about 2,500 pounds of fish were lost. During this time, 11 tons plus 1,400 lbs of feed were fed. What is the feed conversion ratio?

STEP 1.   Convert all feed weights to pounds.

11 tons × 2,000 lbs/ton + 1,400 lbs

22,000 lbs + 1,400 lbs = 23,400 lbs feed

STEP 2.   Final weight – Last weight = Weight gain

33,000 lbs – 22,500 lbs = 10,500 lbs gained

During this period, 2,500 lbs of fish were lost. These fish were part of the last weight sampling when fish weighed 22,500 lbs. Also, these fish did consume feed before they were lost. From the standpoint of the feed conversion ratio, these lost fish should be included. From the standpoint of economic return, both the weight gain and the feed that the fish consumed were lost. The bottom line is how many pounds of feed were fed, and how many pounds of fish were produced and marketed? Examine the feed conversion ratio from both standpoints. First, determine the FCR including the fish that were lost.

$$\text{FCR (Including lost fish)} = \frac{23{,}400 \text{ lbs feed}}{10{,}500 \text{ lbs} + 2{,}500 \text{ lbs}}$$

$$= \frac{23{,}400 \text{ lbs feed}}{13{,}000 \text{ lbs gain}} = 1.8$$

Next, determine the feed conversion ratio that does not include the lost fish that consumed feed but which you will not market.

$$\text{FCR} = \frac{23{,}400 \text{ lbs feed}}{10{,}500 \text{ lbs (alive)}} = 2.23$$

Fish losses increase the conversion ratio. Even though fish were converting well at 1.8 lbs, the cost of production is 2.23 lbs because of the fish losses.

When fish are harvested, their weight should always be recorded and included in the total weight of fish produced to determine the feed conversion ratio. Recordkeeping forms should be used to record and use the information that is needed to determine the feed conversion ratio.

EXAMPLE 3. A fish farm has 45 A of water. The expected annual average production per A is 3,500 lbs of fish. Approximately how much feed will need to be purchased for the year, and what will be the total feed cost if feed is expected to cost $240/ton? From past experience, the producer expects an FCR of 1.8.

STEP 1. Determine the total pounds of fish expected to be produced on the farm for the year.

3,500 lbs per A × 45 A = 157,500 lbs fish

STEP 2. Determine the amount of feed required to produce 157,500 lbs of fish assuming that this weight represents weight gained and not the total weight of fish produced. Remember that the fish weighed something when they were stocked, but this initial weight is not taken into account in this example.

$$\text{FCR} = \frac{\text{Pounds feed fed}}{\text{Weight gain of fish}}$$

$$= \frac{\text{Pounds of feed}}{157{,}500 \text{ lbs weight gain}} = 1.8$$

Pounds of feed = 1.8 × 157,500 lbs = 283,500 lbs

$$\frac{283{,}500 \text{ lbs}}{2{,}000 \text{ lbs}} \times 1 \text{ ton} = 141.75 \text{ tons of feed}$$

STEP 3. Determine the approximate cost of feeding the fish for the year.

$$141.75 \text{ tons} \times \$240/\text{ton} = \$34{,}020$$

Feed Costs

The production of food fish in the U.S. involves the use of high-protein feeds. A major production cost is the feed bill. The cost of feeding fish is determined by the feed conversion efficiency of fish and cost of feed. Fish farmers should know how to estimate their feed requirements over time for planning purposes and know their feed costs.

Table 5-18 illustrates how the feed conversion ratio (FCR) and price of feed affect the cost of producing fish. With this information, the per acre feed cost can be estimated.

EXAMPLE If 4,000 lbs of fish were produced per A, the feed was $250/ton, and the FCR was 1.8, then what is the total cost of feeding these fish per A?

From Table 5-18, the cost of feed in cents to produce a 1-lb fish with feed at $250/ton and FCR at 1.8 is 22.5 cents or $.225.

$$4{,}000 \text{ lbs}/A \times \$.225/\text{lb} = \$900/A$$

If the same fish had an FCR of 2.0 instead of 1.8 with feed at $250/ton, then:

$$4{,}000 \text{ lbs}/A \times \$.25/\text{lb} = \$1{,}000/A \text{ or } \$100/A \text{ higher.}$$

For a 15-A pond, that means $100/A difference × 15 A = $1,500 either saved or spent.

TABLE 5-18 Cost of Feed in Cents to Produce a 1-lb Fish at Different Feed Conversion Rates and Feed Prices

FCR	Feed Cost Per Ton				
	$200	$225	$250	$275	$300
1.5	15.0	16.9	18.8	20.6	22.5
1.6	16.0	18.0	20.0	22.0	24.0
1.7	17.0	19.1	21.3	23.4	25.5
1.8	18.0	20.3	22.5	24.8	27.0
1.9	19.0	21.4	23.4	26.1	28.5
2.0	20.0	22.4	25.0	27.5	30.0

AQUATIC PLANTS

In the presence of sunlight and chlorophyll plants convert carbon dioxide and water to sugar and then to starch. This process is photosynthesis.

To release the energy stored in starches during photosynthesis plants use the process of respiration. The energy released is used for growth and reproduction.

Aquatic plants obtain most of the nutrients they need from the water. Depending on the plants' water temperature, salinity and soil (substrate) varies. Water temperature varies from 50°F (10°C) to 85°F (29°C) depending on the plant. Salinity varies from 35 to 0 ppt. Some plants actually attach themselves to the ground or bottom of the water facility. These are called benthic. Other plants float free. These are called planktonic.

SUMMARY

While fish convert feed to human food very efficiently, the feeding cost of production needs to be controlled. Feeding fish requires an understanding of the process of digestion, the digestive system, and fish nutrition. Some anatomical and behavioral traits of fish species influence how they are fed. Finally, digestion and feeding are affected by factors such as age, diet, light, activity, size, species, and water temperature.

Fish consume feed for energy. This energy is used for growth, activity, and reproduction. Feeds containing protein, fats, and carbohydrates supply energy in the fish diet. These feeds enter the digestive system where enzymes break down the protein, fats, and carbohydrates to simpler compounds that the fish uses for energy and to form tissue, enzymes, and bone. Protein in the diet also supplies ten essential amino acids and fat in the diet supplies essential fatty acids. Fat-soluble and water-soluble vitamins are supplied by the diet. Minerals are supplied by the diet and by the water.

Feeding practices vary depending on the species of fish being fed and the facility in which they are being raised. For all producers, keeping records and calculating the feed conversion ratio (FCR) is essential. The FCR indicates how efficiently feed is being converted to human food.

STUDY/REVIEW

Success in any career requires knowledge. Test your knowledge of this chapter by answering these questions or solving these problems.

True or False

1. Carbohydrates are made up of amino acids.

2. Antinutrients are necessary components in a fish diet.

3. Microminerals are just as important as macrominerals.

4. Protein contains nitrogen.

5. Vitamins A, D, E, and K are carried in the water portion of the diet.

Short Answer

6. Name two land animals that come close to being as efficient as fish in the conversion of feed to food.

7. Name five structures in the digestive tract of fish.

8. List five categories of feeding habits.

9. Give two examples of feeding behavior.

10. List four factors that increase the energy requirement for fish.

11. List three factors that decrease the energy requirement for fish.

12. As fish digest feeds, in what forms is energy lost?

13. List three factors that affect the digestibility of fat.

14. What chemical element is contained in all amino acids and becomes a waste product that can present problems?

15. Name ten water-soluble vitamins.

16. Name all the fat-soluble vitamins.

17. Vitamin E acts with which micromineral?

18. Deficiency of which vitamin causes nutritional gill disease?

19. Name the six macrominerals.

20. Name six microminerals.

21. Name two reasons for using hormones.

22. What two antibiotics have FDA approval for use in fish?

23. Name six naturally occurring substances that could contaminate fish feed and adversely affect performance.

24. What is the first consideration when formulating a fish diet?

25. What is the best source of high-quality protein for fish diets?

26. Name five methods of processing fish feed.

27. What is the advantage of feeding fish sinking pellets?

28. What is the common method of feeding intensively cultured catfish?

29. Give three disadvantages of using a demand feeder.

30. As catfish grow, what happens to daily feed consumption?

31. What is the most common method of feeding trout?

32. Some fish are able to use natural foods in the pond. Name two general categories of natural foods.

33. Name two factors that can effect the time of first feeding.

34. A pond is stocked with 50,000 fish that weigh 75 lbs per 1,000 fish. The desired feeding rate is 2.2 percent of their daily weight. How much feed is needed for one day and for one week?

35. Fifty thousand fish weighing 50 lbs per thousand were stocked in a pond. Later, the fish were sampled and the average weight of the fish was 250 lbs per thousand fish. During this time, 14 tons of feed were fed. No fish losses were observed. What is the feed conversion ratio?

36. A pond has an estimated standing crop of 30,000 lbs of fish. A new sampling estimated the total fish weight at 40,000 lbs. Between the two samplings, about 3,000 lbs of fish were lost. During this time, 16 tons of feed were fed. What is the feed conversion ratio?

37. A fish farm has 50 A of water. The expected annual average production per A is 2,800 lbs of fish. Approximately how much feed will need to be purchased for the year, and what will be the total cost of the feed if the feed costs $275 per ton? The producer expects a feed conversion ratio of 1.6.

38. If 4,500 lbs of fish were produced per A, the feed cost was $300 per ton and the feed conversion ratio was 1.9, what is the total cost of feeding these fish per A?

Essay

39. What is the function of the digestive tract?

40. Define nutrition.

41. Name two essential fatty acids and describe the effect of a deficiency.

42. Give two reasons for including dietary fat in fish diets.

43. Give three reasons for including protein in the diet.

44. Name eight essential amino acids.

45. Describe a protein deficiency.

46. Describe the effects of a vitamin C deficiency.

47. Why are mineral requirements difficult to determine in fish diets?

48. Why should fiber be avoided in fish diets?

49. Why are antioxidants used in fish diets?

50. Describe why pigments are important in fish diets?

51. Computer programs that compute least-cost rations require what type of information?

52. Why do producers feed floating feed?

53. What effect does warm water have on the trout digestive system?

54. Why is it important that fish rapidly consume all of the feed?

55. Commonly, fish are offered food when the fry swim-up. Describe the swim-up.

KNOWLEDGE APPLIED

1. Obtain some freshly killed fish for dissection. Identify the external structures. Then identify the major internal structures, especially those associated with the digestive system.

2. Obtain five samples of high-protein feeds common to your area. Take these samples to a laboratory for protein analysis. Compare their values to those obtained from feed composition tables.

3. Fecal material is a reality of any livestock production operation. Research the differences and similarities in the composition of trout (salmonid), sheep, beef, and dairy fecal waste.

4. Compare the nutrition information contained on the side panel of a cereal box for humans to the information contained on a feed tag for fish feed. What did you learn about human nutrition?

5. Obtain samples of protein feeds, for example, soybean meal, cottonseed meal, meat meal, and fish meal. Use published composition tables and compare each feed. Observe differences in the smell, texture, and other characteristics. Compare cost.

6. Use a computer program to balance a catfish or trout diet. Figure the cost of the diet.

7. Obtain feed labels from trout, catfish, pig, and chicken feed. Compare the contents of each and compare the price.

8. Contact suppliers of aquaculture equipment and obtain information on demand feeders and compare the different types.

9. Obtain samples of floating feed and sinking feed. Place these in water and observe their behavior.

10. Collect samples of fish feed. Develop a display in small bottles. Label the type, size, protein, and energy content.

LEARNING/TEACHING AIDS

Books

Committee on Animal Nutrition, National Research Council. (1993). *Nutrient requirements of fish.* Washington, DC: National Academy Press.

Food and Agriculture Organization of the United Nations and the United Nations Development Program. (1978). *Fish feed technology.* Rome: Food and Agriculture Organization of the United Nations.

Groot, C., & Margolis, L. (Eds.). (1991). *Pacific salmon life histories.* Vancouver, BC: University of British Columbia.

Halver, J. E. (Ed.). (1989). *Fish nutrition.* New York: Academic Press.

Hepher, B. (1988). *Nutrition of pond fishes.* Cambridge: Cambridge University Press.

Hodson, R., McVey, J., Hawell, R., & Davis, N. (1987). *Hybrid striped bass culture: Status and perspective.* (University of North Carolina Sea Grant College Publ. UNC-SG-87-03.) Chapel Hill, NC: University of North Carolina Press.

Lovell, R. T. (1989). *Nutrition and feeding of fish.* New York: Van Nostrand Reinhold.

Stickney, R. R. (1991). *Culture of salmonid fishes.* Boca Raton, FL: CRC Press.

Tucker, C., & Robinson, E. H. (1991). *Channel catfish farming handbook.* New York: Van Nostrand Reinhold.

Wilson, R. P. (1991). *Handbook of nutrient requirements of finfish.* Boca Raton, FL: CRC Press.

Articles

Alanara, A. (1992). Demand-feeding as a self-regulating feeding system for rainbow trout in net-pens. *Aquaculture, 100* (1–3), 167.

Arnason, A. M., et al. A growth model fitting procedure that simultaneously provides an accurate feeding table. *Aquaculture, 100* (1–3), 168.

Bryant, P. L., & Matty, A. J. (1981). Adaptation of carp (*Cyprinus carpio L.*) larvae to artificial diets: Optimum feeding rate and adaptation age for commercial diet. *Aquaculture, 23*, 275–286.

Burgner, R. L. (1991). Life history of sockeye salmon *Oncorhynchus-nerka.* In C. Groot & L. Margolis (Eds.), *Pacific salmon life histories* (pp. 3–118). Vancouver, BC: University of British Columbia.

Cho, C. Y. (1990). Fish nutrition, feeds, and feeding: With special emphasis on salmonid aquaculture. *Food Review International 6* (3), 333–357.

El-Sayed, A. F. M., & Teshima, S. I. (1991). Tilapia nutrition in aquaculture. *Reviews in Aquatic Sciences, 5* (3–4), 247–266.

Green, B. W. (1992). Substitution of organic manure for pelleted feed in tilapia production. *Aquaculture, 101* (3–4), 213–222.

House, C. (1991, July). Commercial feed may spur crawfish industry. *Feedstuffs,* p. 12.

Johnsen, P. B. (1990, December). Least cost ingredients work well for catfish feeds. *Feedstuffs,* p. 15–17.

Johnsen, P. B., & Dupree, H. K. (1991). Influence of feed ingredients on the flavor quality of farm-raised catfish. *Aquaculture, 96* (2). 139–150.

Jorgensen, E. H., & Jobling, M. (1992). Feeding behavior and effect of feeding regime on growth of Atlantic salmon *Salmo-salar. Aquaculture Engineering, 101* (1–2), 135–146.

Kadri, S., et al. (1991). Daily feeding rhythms in Atlantic salmon in sea cages. *Aquaculture, 92* (2–3), 219–224.

Killian, S. (1991, July). Winter feeding adds to catfish profits. *Feedstuffs, 63* (31), 10.

Lim, C. (1989). Practical feeds—tilapias. In R.T. Lovell (Ed.), *Nutrition and Feeding of Fish* (pp. 163–167). New York: Van Nostrand Reinhold.

Lovell, R. T. (1991). Nutrition of aquaculture species. *Journal of Animal Science, 69* (10), 4193–4200.

McCarthy, I. D., Carter, C. G., & Houlihan, D. F. (1992). Correlating food intake and specific growth rate of rainbow trout of different sizes. *Aquaculture, 100* (1-3), 326.

Meyer, F. P. (1991). Aquaculture disease and health management. *Journal of Animal Science, 69* (10), 4201-4208.

No, H. K., & Storebakken, T. (1992). Pigmentation of rainbow trout with astaxanthin and canthaxanthin in freshwater and saltwater. *Aquaculture, 101* (1-2), 123-134.

Reinitz, G. (1983). Relative effect of age, diet, and feeding rate on the body composition of young rainbow trout (*Salmo gairdneri*). *Aquaculture, 35,* 19-27.

Robinson, E. H. (1991). A practical guide to nutrition, feeds, and feeding of catfish. Mississippi Agriculture Forestry Experiment Station Bulletin 979, Mississippi State, MS.

Soivio, A., & Koskela, J. (1992). The effect of antiobiotics on the apparent digestibility of feed for rainbow trout. *Aquaculture, 100* (1-3), 235.

Storebakken, T., & Austreng, E. (1987). Ration levels for salmonids. I. Growth, survival, body composition and feed conversion in Atlantic salmon fry and fingerlings. *Aquaculture, 60,* 189-199.

Sumpter, J. P. (1992). Control of growth of rainbow trout *Oncorhynchus-mykiss*. *Aquaculture, 100* (1-3), 299-320.

Health of
Aquatic Animals

D isease is rarely a simple association between a pathogen and a host fish. Usually other circumstances must be present for active disease to develop in a population. These circumstances are generally grouped under the umbrella term stress. Management practices limiting stress help prevent disease outbreaks.

This chapter makes no attempt at covering all diseases of aquatic animals. Rather, this chapter presents the importance of good management practices and their interaction with some representative diseases.

Chapter 4, Aquatic Management Practices, provides some additional details for specific species.

Learning Objectives

After completing this chapter, the student should be able to:

In Aquaculture

- Outline fish health management
- List behavioral signs of sick fish
- List common physical signs of sick fish

■ List common stressors of fish

■ Outline general management measures for preventing disease outbreaks

■ List and compare treatment methods

■ Complete a list of general guidelines for treatment of fish diseases

■ Calculate treatments for fish ponds

In Science

■ Define terms associated with disease conditions

■ Discuss disease resistance

■ Define terms associated with severity of disease or condition

■ Discuss the role of stress in fish diseases

■ Describe the immunization of fish

■ List signs of stress and disease

■ Discuss common diseases caused by pathogenic viruses

■ Discuss common diseases caused by pathogenic bacteria

■ Describe a fungal infection

■ Name and describe a common pathogenic protozoan parasite

■ Name and describe a common pathogenic crustacean parasite

■ Describe a grub or fluke infection

■ Name and describe a common pathogenic worm parasite

■ List noninfectious diseases and give examples

■ Describe an infection of lice

Understanding of this chapter will be enhanced if the following terms are known. Many are defined in the text and others are defined in the glossary.

KEY TERMS		
Abnormalities	Antigenicity	Dip
Acidosis	Bacteria	Disease
Alkalosis	Bath	Flush
Antibiotic	Cellular	Fungus
Antibodies	Chronic	Hemoglobin
Antigen	Diagnosis	Hemorrhagic

KEY TERMS		
Host	Mortality	Stress
Hydrate	Mucus	Susceptible
Immune	Noninfectious	Therapeutic
Infectious	Parasites	Transmission
Infiltration	Pathogen	Vaccine
Injections	Prognosis	Virus
Inoculation	Secondary	Visceral
Lethargic		

HEALTH MANAGEMENT

Disease is any condition of an aquatic animal that impairs normal physiological functions. Fish disease outbreaks increase production cost because of the investment lost in dead fish, the cost of treatment, and decreased growth during convalescence. In nature we are less aware of fish disease problems because sick animals are quickly removed from the population by predators. In addition, fish are much less crowded in natural systems than in captivity. Parasites and bacteria may be of minimal significance under natural conditions but can contribute to substantial problems when animals are crowded and stressed under culture conditions.

Fish health management describes management practices which are designed to prevent fish disease. Once fish get sick, salvage is difficult.

Successful fish health management begins with prevention of disease rather than treatment. Good water quality management, nutrition, and sanitation prevent fish diseases. Without this foundation, outbreaks of opportunistic diseases are impossible to prevent. As Figure 6-1 illustrates, disease results from the interaction of the host fish with the environment and the pathogen.

The fish is constantly bathed in potential pathogens, including bacteria, fungi, and parasites. Poor water quality, poor nutrition, or immune system suppression generally associated with stressful conditions allow these potential pathogens to cause disease. Medications used to treat diseases buy time for fish and enable them to overcome opportunistic infections but are no substitute for proper animal husbandry.

Daily observation of fish behavior and feeding activity allows early detection of problems when they do occur so that a diagnosis can be made before the majority of the population becomes sick. Successful treatment starts early in the course of the disease while the fish are still in good shape.

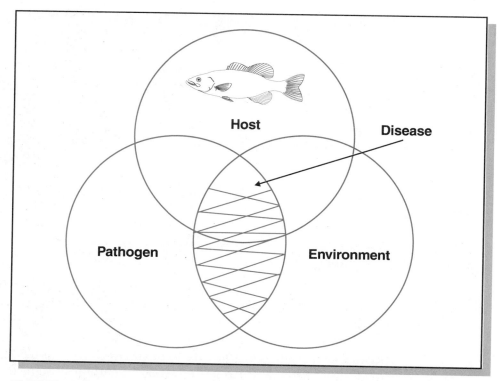

FIGURE 6-1 Disease rarely results from a simple contact between the host fish and the disease-causing organism (the pathogen).

FIGURE 6-2 Checking golden shiners for disease.

STRESS AND DISEASE

Stress is a condition in which an animal is unable to maintain a normal physiologic state because of various factors adversely affecting its well-being.

Stress is caused by placing a fish in a situation that is beyond its normal level of tolerance. Specific examples of things that can cause stress (stressors) are listed in Table 6-1.

Stress increases blood sugar. This is caused by a secretion of hormones from the adrenal gland. Stored sugars, such as glycogen in the liver, are metabolized. This creates an energy reserve that prepares the animal for an emergency action. Osmoregulation is disrupted because of changes in mineral metabolism. Under these circumstances, a freshwater fish tends to absorb excess water from the environment (over-hydrate). A saltwater fish will tend to lose too

TABLE 6-1 Stressors in Fish

Types of Stressor	Example	Causes
Chemical stressors	Poor water quality	Low dissolved oxygen, improper pH
	Pollution	Intentional or accidental like chemical treatments and insecticides
	Diet composition	Type of protein and amino acids
	Metabolic wastes	Nitrogenous—accumulation of ammonia or nitrite
Biological stressors	Population density	Crowding
	Other species of fish	Aggression, territoriality, lateral swimming space requirements
	Microorganisms	Pathogenic and nonpathogenic
	Macroorganisms	Internal and external parasites
Physical stressors	Temperature	Important influence on immune system of fish
	Light	
	Sounds	
	Dissolved Gases	
Procedural stressors	Handling	
	Shipping	
	Disease Treatments	

much water to the environment (dehydrate). This disruption requires that extra energy be used to maintain osmoregulation. Respiration increases, blood pressure increases, and reserve red blood cells are released into the circulation. The inflammatory response is suppressed by hormones released from the adrenal gland.

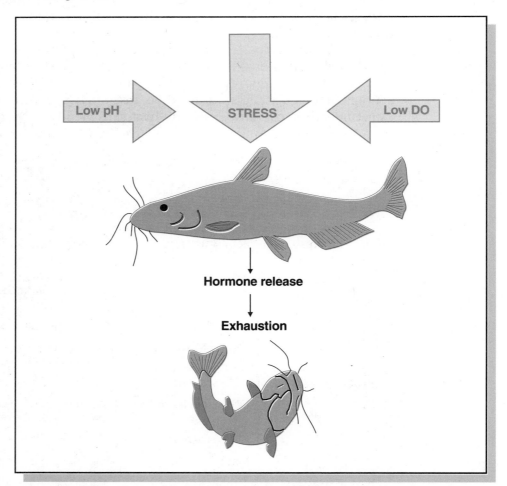

FIGURE 6-3 Stress triggers a chain of events that result in an "alarm reaction" (fight or flight response) by the fish, which then triggers a series of hormonal changes. As the fish tries to adjust to the insult it uses up energy reserves, but during this time it is able to resist or compensate for the insult. If the insult is not removed, its energy reserves become depleted and the fish becomes "exhausted." At this phase its ability to resist disease organisms, with which it is in constant contact, is severely compromised and the fish may become sick or die.

DISEASE RESISTANCE

All fish do not get sick and die each time a disease outbreak occurs. Many factors affect how an individual responds to a potential pathogen. The pathogen—bacteria, parasite, or virus—must be capable of causing disease. The host (fish) must be in a susceptible state, and certain environmental conditions must be present for a disease outbreak to occur, such as one or more of the stressors in Table 6-1.

Protective Barriers Against Infection

Fish possess four protective barriers to protect against infection:

1. Mucus
2. Scales and skin
3. Inflammation
4. Antibodies.

Mucus (slime coat) is a physical barrier that inhibits entry of disease organisms from the environment into the fish. It is also a chemical barrier because it contains enzymes (lysozymes) and antibodies (immunoglobulins), which can kill invading organisms. Mucus also lubricates the fish, which aids movement through the water, and it is also important for osmoregulation.

Scales and skin function as a physical barrier protecting the fish against injury. When these are damaged, a window is opened for bacteria and other organisms to start an infection.

Inflammation is a cellular, nonspecific response to an invading protein. An invading protein can be a bacteria, a virus, a parasite, a fungus, or a toxin. Inflammation is characterized by pain, swelling, redness, heat, and loss of function. It is a protective response and an attempt by the body to wall off and destroy the invader.

Antibodies, a specific cellular response, are molecules formed to fight invading proteins or organisms. The first time the fish is exposed to an invader—an antigen—antibodies form that protect the fish from future infection by the same organism. Exposure to sublethal concentrations of pathogens is extremely important for a fish to develop a competent immune system. An animal raised in a sterile environment will have little protection from disease. Young animals do not have an immune response that works as efficiently as the immune response in older animals. Figure 6-4 illustrates the formation of antibodies.

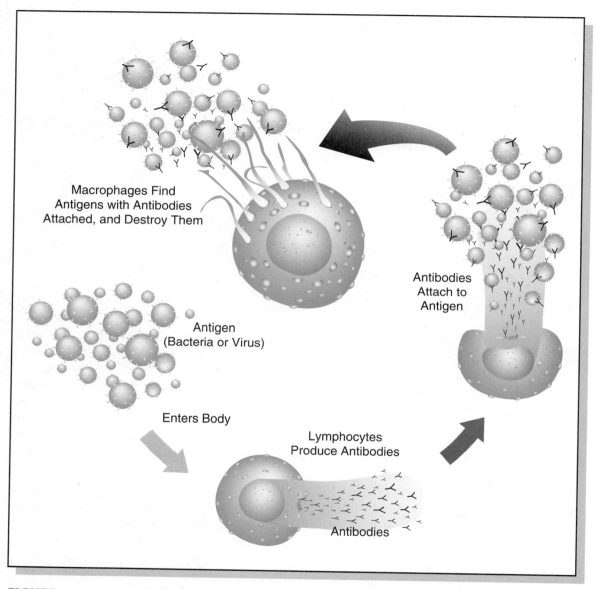

FIGURE 6-4 The formation of antibodies—the immune response.

Effect of Stress on Protective Barriers

Any stress causes chemical changes in mucus that decreases its effectiveness as a chemical barrier against invading organisms. Stress upsets the normal electrolyte—sodium, potassium, and chloride—balance, which results in excessive uptake of water by freshwater fish and dehydration in saltwater fish. The need for effective osmoregulatory (electrolyte balance) support from mucus components is increased.

Handling physically removes mucus from the fish. This decreases chemical protection, osmoregulatory function, and lubrication. The fish uses more energy to swim at a time when its energy reserves are already being used up metabolically. Handling also disrupts the physical barrier against invading organisms.

Chemical stress like disease treatment often damages mucus, resulting in loss of protective chemical barrier, loss of osmoregulatory function, loss of lubrication, and damage to the physical barrier created by mucus.

Scales and skin are most commonly damaged by handling stress. Any break in the skin or removed scale creates an opening for invasion by pathogenic organisms.

Trauma caused by fighting (reproductive stress or behavioral stress) could cause breaks in the skin or scale loss.

Parasitic infestations can result in damage to gills, skin, fins, and loss of scales, which could create breaks in the skin for bacteria to enter. Many times, fish that are heavily parasitized actually die from bacterial infections. The parasite problem, associated physical damage, and stress response allow the bacteria in the water to invade the fish, causing a lethal disease.

Temperature stress, particularly cold temperatures, can completely halt the activity of antibodies of the immune system, eliminating an important first defense against invading organisms. Excessively hot temperatures are also very detrimental to fish.

A sharp decrease in temperature severely impairs the fish's ability to quickly release antibodies against an invading organism. The time lapse required to mount an antibody response gives the invader time to reproduce and build up its numbers, giving it an advantage that may allow it to overwhelm the fish.

Prolonged stress severely limits the effectiveness of the immune system. This increases the opportunities for an invader to cause disease. Also, any stress causes hormonal changes that decrease the effectiveness of the inflammatory response.

Prevention of Stress and Disease

Good management prevents stress. This means maintaining good water quality, good nutrition, and sanitation.

Good water quality involves preventing accumulation of organic debris and nitrogenous wastes, maintaining appropriate pH and temperature for the species, and maintaining dissolved oxygen levels of at least 5 ppm. Poor water quality is a common and important stressor of cultured fish and precedes many disease outbreaks.

Good nutrition means feeding a high-quality diet that meets the nutritional requirements of the fish. Each species is unique, and the nutritional requirements of different species will vary. Chapter 5, Fundamentals of Nutrition in Aquaculture, provides information on nutrition and digestion.

Proper sanitation implies routine removal of debris from fish tanks and disinfection of containers, nets, and other equipment between groups of fish. Organic debris that accumulates on the bottom of tanks or vats is an excellent medium for reproduction of fungal, bacterial, and protozoal agents. Prompt removal of this material from the environment decreases the number of agents the fish is exposed to. Disinfection of containers and equipment between groups of fish minimizes transmission of disease from one population to another.

DISEASE TYPES

Two broad categories of disease affect fish, infectious and noninfectious diseases. Infectious diseases are caused by pathogenic organisms present in the environment or carried by other fish. They are contagious, and some type of treatment may be necessary to control the disease outbreak. In contrast, noninfectious diseases are caused by environmental problems, nutritional deficiencies, or genetic defects. Noninfectious diseases are not contagious and usually cannot be cured by medications.

Infectious Diseases

Infectious diseases are broadly categorized as parasitic, bacterial, viral, or fungal diseases.

Parasitic. Parasitic diseases of fish are most frequently caused by microscopic organisms called protozoa that live in the aquatic environment. A variety of protozoans that infest the gills and skin of fish cause irritation, weight loss, and eventually death. Most protozoan infections are relatively easy to control using standard fishery chemicals such as copper sulfate, formalin, or potassium permanganate.

Bacterial. Bacterial diseases are often internal infections and require treatment with medicated feeds containing antibiotics approved for use in fish by the Food and Drug Administration. Typically fish infected with a bacterial disease will have hemorrhagic spots or ulcers along the body wall and around the eyes and mouth. They may also have an enlarged, fluid-filled abdomen and protruding eyes. Bacterial diseases can also be external, resulting in erosion of skin and ulceration. Columnaris is an example of an external bacterial infection that may be caused by rough handling.

Viral. Viral diseases are impossible to distinguish from bacterial diseases without special laboratory tests. They are difficult to diagnose, and no specific medications are available to cure viral infections of fish. Immunization can protect fish from a viral disease, but vaccines do not exist for all viral diseases.

Fungal. Fungal diseases are the fourth type of infectious disease. Fungal spores are common in the aquatic environment but are not normally a problem in healthy fish. When fish are infected with an external parasite, bacterial infection, or injured by handling, the fungi can colonize diseased tissue on the exterior of the fish. These areas appear to have a cottony growth or may appear as brown matted areas when the fish are removed from the water. Potassium permanganate is effective against most fungal infections. Since fungi are usually a secondary problem, diagnosis and correction of the original problem are important.

Noninfectious Diseases

Noninfectious diseases can be broadly categorized as environmental, nutritional, or genetic. Environmental diseases are the most important in commercial aquaculture. Environmental diseases include low dissolved oxygen, high ammonia, high nitrite, or natural or man-made toxins in the aquatic environment. Proper techniques of managing water quality enable producers to prevent most environmental diseases.

PARASITIC DISEASES

Parasites obtain their food from the host. This reduces the performance of the host and leads to other diseases. Parasites produce large numbers of eggs and many spend different phases of their life cycle in different hosts.

Tapeworms

Causative organisms include the Asian tapeworm (*Bothriocephalus opsarichthydis*), the bass tapeworm (*Proteocephalus ambloplitis*), the catfish tapeworm (*Corallobothrium fimbriatum* and *Ligula intestinalis*), and others. The species of tapeworm varies with the species of host fish.

All fish are susceptible, especially black basses, Chinese carps, catfishes, sunfishes, and golden shiners.

Signs. Fish often show no outward indication, but fish may be listless, lose weight, or become sterile. In some cases, such as severe infections of Asian tapeworms in grass carp, the abdomen becomes distended and the intestine blocked. Microscopic verification determines presence of tapeworms.

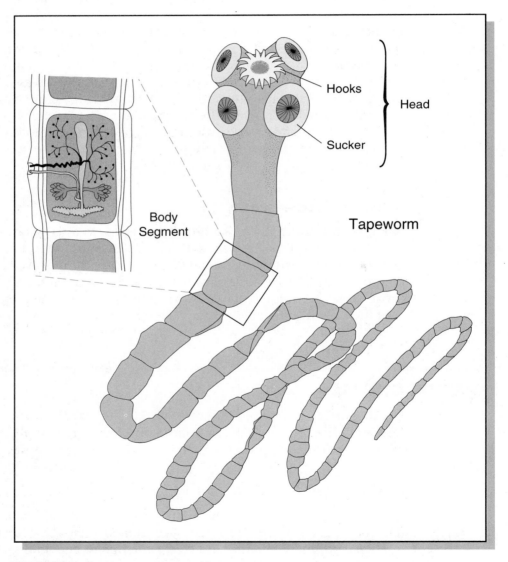

FIGURE 6-5 A tapeworm and its parts.

Contributing Factors. Use of brood fish infected with tapeworms, purchase of contaminated fry and fingerlings, or the use of surface water containing tapeworm-infested hosts contribute to an infection. Droppings of fish-eating birds in and near the pond can introduce tapeworms.

Prevention and Treatment. Aquaculturalists avoid maintaining or purchasing infected fry, fingerlings, and broodfish. They drain, dry, and disinfect ponds between fish crops and eliminate or reduce exposure to intermediate hosts. Currently, no **therapeutic** agents are available.

Life Cycle of the Asian Tapeworm
Bothriocephalus opsarichthydis

During their life cycle many parasitic worms require a number of animals as intermediate hosts. The Asian tapeworm, for example, requires a copepoda (a small, nearly microscopic free-swimming organism) as an intermediate host. The Asian tapeworm probably develops in any fish species that eats the infected copepoda. The tapeworm often causes problems in grass carp, golden shiner, fathead minnow, Colorado squawfish, and mosquitofish.

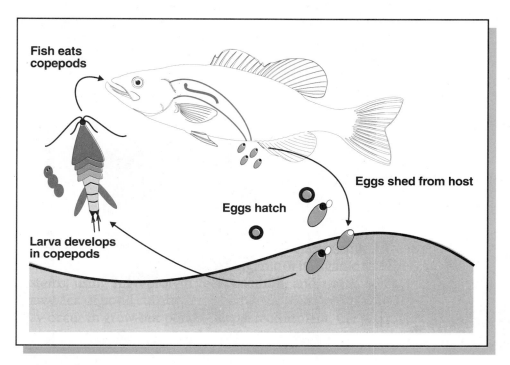

FIGURE 6-6 Life cycle of Asian tapeworms.

After introduction of a parasite into a pond or reservoir, only the removal of all fish and disinfection of the pond bottom to kill the copepoda or other intermediate hosts can ensure its eradication. The ponds should be filled with copepoda-free water—usually well water—and stocked with only parasite-free broodfish.

Trichodiniasis

Protozoa of the genus *Trichodina* causes trichodiniasis. All freshwater fish are susceptible.

Signs. Affected fish are listless and lethargic. Fish also eat less and may have frayed fins.

Contributing Factors. Reduced water quality, including low dissolved oxygen and high concentrations of organic material, contribute to the disease. Fluctuating temperatures during fall and spring and malnutrition can debilitate the fish and accelerate the buildup of the parasite. Microscopic examination verifies the presence of *Trichodina*.

Prevention and Treatment. Maintain good water quality, including dissolved oxygen concentrations of 4 ppm or above. Offer the fish an adequate amount of a nutritionally complete feed, and avoid overcrowding fish, especially fingerlings.

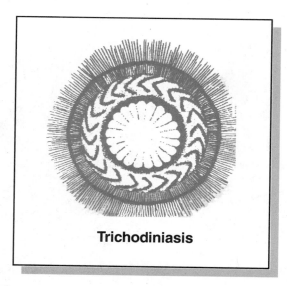

Trichodiniasis

FIGURE 6-7 Protozoan causing trichodiniasis.

Possible therapeutic agents include formalin, potassium permanganate, and copper sulfate.

Ichthyophthiriasis

Ichthyophthiriasis multifiliis, a ciliated protozoan, causes ichthyophtiriasis, commonly called Ich, and all freshwater fish are susceptible.

Signs. Parasites appear as small raised spots that resemble sprinkled table salt, over the entire body surface and fins. Affected fish may flash against the bottom or sides of a tank. Heavily infected fish often congregate at the intake or outlet of the pond or tank.

Contributing Factors. Poor water quality and malnutrition contribute to Ich. Susceptibility is increased when the water source is contaminated with wild fish and when the water temperature is 60° to 75°F (16° to 24°C).

Microscopic examination verifies the presence of the Ich protozoas.

Prevention and Treatment. Avoid contaminated water supply, nets, and other equipment. Provide water of good quality, and offer nutritionally adequate feeds.

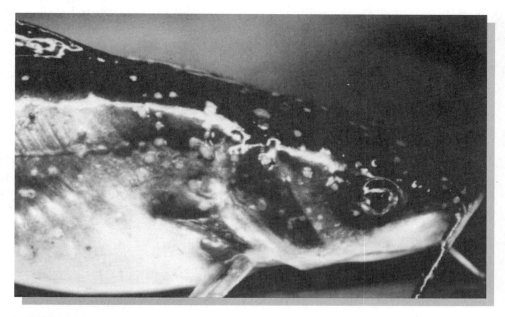

FIGURE 6-8 Tiny white spots on the fins and skin are noticeable after Ich reaches the mature stage. (Photo courtesy Chuck Weirich, Delta Research Center, Stoneville, MS)

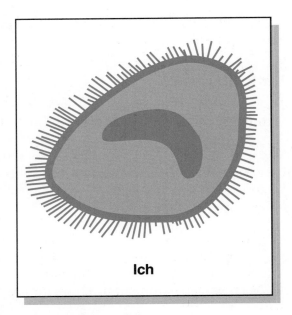

Ich

FIGURE 6-9 Ciliated Ich protozoan.

Formalin, table salt, copper sulfate, and potassium permanganate can be used as therapeutic agents.

Life Cycle of Ich Parasite

The life cycle of the Ich parasite involves several stages. Because treatments for one stage often do not affect the other stages, the disease frequently recurs. The adult stage of this protozoan emerges from the skin and mucus of the infected fish and drops to the bottom of the pond, where it forms a cyst. The organism then divides many times and produces many tiny juvenile forms. Under suitable conditions, these juveniles, called theronts, emerge and swim about, seeking a fish of any species to penetrate. When they find one, they burrow into it and develop to maturity. Only the free-swimming stages of this parasite are vulnerable to treatment.

Monogenetic Flukes

Monogenetic flukes include flukes of the genera *Gyrodactylus, Dactylogyrus,* and *Cleidodiscus.* Representatives of these genera are similar in size and general appearance.

All warmwater fish are susceptible.

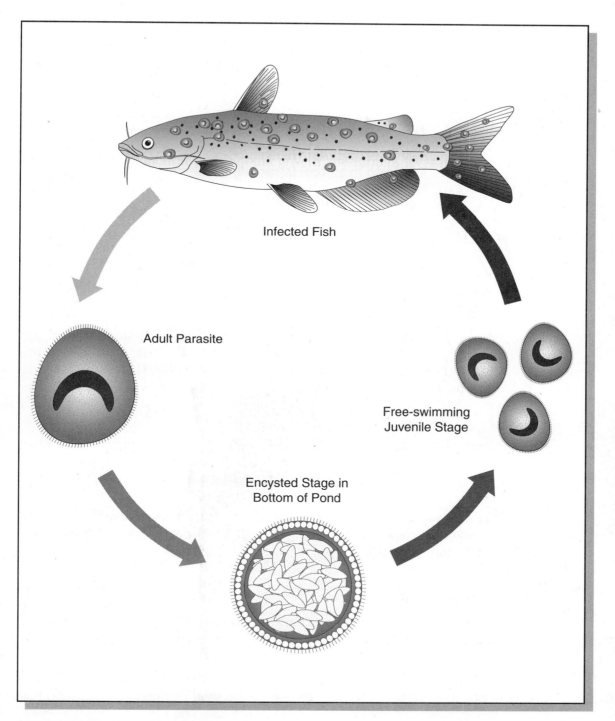

Infected Fish

Adult Parasite

Free-swimming
Juvenile Stage

Encysted Stage in
Bottom of Pond

FIGURE 6-10 Life cycle of Ich parasite.

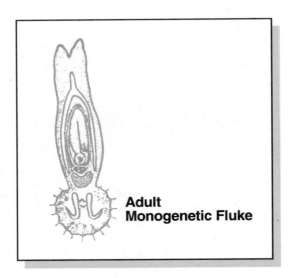

FIGURE 6-11 Adult monogenetic fluke.

Signs. Fish show discomfort, sometimes rubbing or flashing against the tank or pond walls and bottom, or becoming listless and staying near the edge of the pond. Gills may be flared on small fish. When flukes are abundant, their primary damage, combined with secondary bacterial infection, may cause death.

Contributing Factors. Poor water quality accompanied by inadequate nutrition, crowding, and fluctuating water temperatures all can contribute.
Microscopic examination verifies the presence of flukes.

Prevention and Treatment. Maintain good water quality, offer adequate feeds, and avoid overcrowding. Masoten (Dylox) formalin, potassium permanganate may be used as therapeutic agents.

Anchor Parasites

This disease is caused by a parasitic crustacean (copepoda) *Lernaea cyprinacea*. All freshwater fish, but especially baitfish, catfish, and carp are susceptible.

Signs. Signs include small reddish lesions on the external surface, often surrounded by fungus. The parasite resembles a shaft of a small barb (similar to a broom straw) inserted into the flesh of the fish. The anchor tends to prevent detachment of the parasite. Figure 6-12 illustrates an anchor parasite.

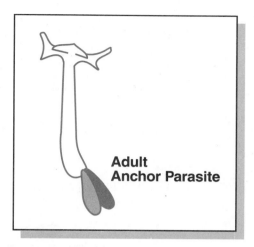

**Adult
Anchor Parasite**

FIGURE 6-12 Adult anchor parasite.

Contributing Factors. Disease-free fish become infected by stocking of fish contaminated with the anchor parasite into parasite-free populations. Movements of wildlife such as ducks or muskrats from pond to pond can spread the parasite.

The presence of anchor parasite can be verified by visual examination.

Prevention and Treatment. The best prevention is to avoid stocking parasite-infected stocks of fish. Masoten (Dylox) can be used as a therapeutic agent.

Costiasis Disease

Protozoa of the genus *Ichtyobodo* cause Costiasis disease. It is sometimes called Costia. All freshwater fish are susceptible.

Signs. Blue-gray film sometimes appears over the body surface. Fish do not feed and are often listless and lethargic. Gill filaments may appear ragged in gross examinations.

Contributing Factors. Overcrowded conditions aggravated by fluctuating water temperatures that are common in fall and spring contribute to susceptibility. Malnutrition can also contribute to the susceptibility to Costia.

Microscopic examination verifies the presence of the protozoan.

Prevention and Treatment. Good water quality and nutritionally complete feeds help prevent infections of Costia. Some therapeutic agents include table salt, formalin, or acetic acid, and copper sulfate followed by potassium permanganate.

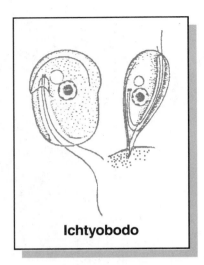

Ichtyobodo

FIGURE 6-13 Protozoan parasite Ichtyobodo causes costiasis.

Fish Lice

Fish lice are parasitic branchiurans of the genus *Angulus*. These are related to the anchor parasite. All freshwater fish are susceptible.

Signs. Infected fish will flash or rub against the tank or pond bottoms or sides. Also, they will be listless and show red spots. When infections are heavy fish start dying.

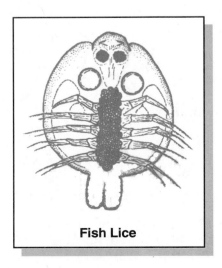

Fish Lice

FIGURE 6-14 Fish lice.

Contributing Factors. Stocking of lice-contaminated fish into parasite-free populations allow the lice to spread. Depending on size, examination by eye or microscope verifies the presence of fish lice.

Prevention and Treatment. Stocking of parasite-free fish is the best prevention. Masoten (Dylox) can be used as a therapeutic agent.

Fish Grubs (Larval Flukes)

Grubs of the genera *Crassiphiala* (black spot), *Clinostomum* (yellow grub), and *Posthodiplostomum* (white grub) cause the disease signs. Bluegills and other sunfishes, black basses, and most minnows are susceptible.

Signs. Signs include small pigmented nodules, either black, cream-colored, or white in the flesh and visceral cavity, and sometimes in the gills. Fish may show a loss of equilibrium and become deformed.

Contributing Factors. Infected fish and the presence of intermediate hosts—fish-eating birds—and snails. Microscopic examination verifies the presence of grubs.

Prevention and Treatment. Eradicating snails in ponds and removing bird roosts in the vicinity eliminates the intermediate hosts. No therapeutic agents are available.

Life Cycle of Black Spot Grub

The spots typical of black spot disease result from a reaction of the fish's body to the presence of larval parasitic worms in its flesh. The worms reach the adult stage only when an infected fish is eaten by a susceptible bird host. As shown in Figure 6-15, the life cycles have at least two free-swimming stages and involve three kinds of hosts: fish, snails, and birds. When an animal other than a bird eats an infected fish, no danger of infection exists.

Fungus

Fungi, usually of the genera *Saprolegnia* and *Achlya*, cause the disease signs. All freshwater fish can be affected.

Signs. Fish have a general cotton-like or fur-like appearance, usually associated with localized discolored areas or lesions. Fungus assumes the color of materials suspended in the water.

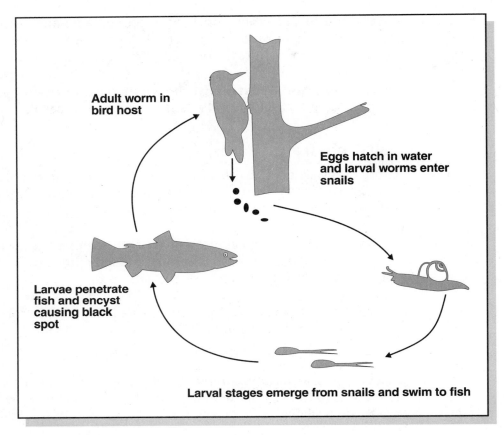

Adult worm in bird host

Eggs hatch in water and larval worms enter snails

Larvae penetrate fish and encyst causing black spot

Larval stages emerge from snails and swim to fish

FIGURE 6-15 Life cycle of the black spot grub.

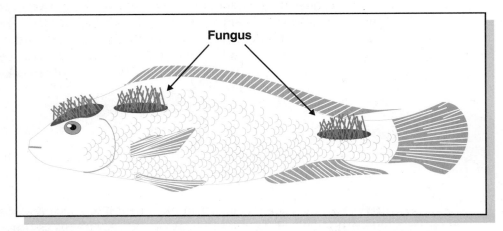

Fungus

FIGURE 6-16 Fish with fungus have a cotton- or fur-like appearance.

Contributing Factors. Fungal infections are generally secondary and indicate other adverse conditions. Fungal infections seldom become established on healthy fish unless they have been subjected to stress or injury. Stress conditions include prolonged periods of very low temperatures, malnutrition, and possibly low dissolved oxygen.

Microscopic examination verifies the presence of fungi.

Prevention and Treatment. Fish farmers maintain good water quality and feed nutritionally adequate feeds throughout the year. Feeding just before winter and in early spring is especially important.

Copper sulfate and potassium permanganate can be used as therapeutic agents.

BACTERIAL DISEASES

Bacteria are microscopic one-celled organisms having various shapes. They often cause disease though some bacteria are beneficial.

Bacteremia (Hemorrhagic Septicemia)

Bacteria like *Aeromonas hydrophila* or *Pseudomonas fluorescens*, and possibly other bacteria cause bacteremia. Bacteremia is literally bacteria in the blood. All fish can be affected.

Signs. Infected fish are listless and lethargic and eat less feed. They show shallow, irregular-margined reddish sores or ulcers on the sides, popeye, enlarged (swollen) fluid-filled belly, raised scales, red streaks in the fin rays and at the bases of the fins, and a reddened area around the anus.

Contributing Factors. Outbreaks may occur in spring when the water warms, particularly when the fish spawn, are handled, moved, or are overcrowded. Also, outbreaks occur when dissolved oxygen content of the water is low, and possibly when other conditions such as malnutrition and disease weaken the fish.

Laboratory culture confirms bacteremia.

Prevention and Treatment. Precautions include avoiding rough handling and overcrowding, especially during summer, maintaining good water quality, and providing a well-fortified feed containing greater than recommended levels of ascorbic acid (vitamin C).

Possible therapeutic agents include antibiotics like oxytetracycline (Terramycin) in the diet. The addition of oxytetracycline to fish transport water may retard the transfer of the bacterium but will not cure infected fish.

Columnaris Disease

The bacterium *Flexibacter columnaris*, sometimes called *Cytophaga columnaris* or *Chondrococcus columnaris*, is responsible for columnaris disease. All fish are susceptible.

Signs. Affected fish show discolored patches on the body with little or no hemorrhaging or sloughing of scales. Discolored patches and scale loss superficially resemble damage caused by fungus infections. Other signs include mouth and barbel erosion, fin erosion, tail loss, and decayed areas in gills.

Contributing Factors. Mechanical injury caused by rough handling, especially when water temperature exceeds 68°F (20°C), can contribute to an infection. Overcrowding in holding and transport facilities, poor water quality (low dissolved oxygen), fluctuating water temperatures, and malnutrition can also contribute to a columnaris infection.

Microscopic examination and laboratory culture verifies the presence of columnaris.

Prevention and Treatment. Preventive measures include avoiding rough handling and overcrowding, especially during summer, maintaining good water quality and providing a well-fortified feed containing greater than recommended levels of ascorbic acid (vitamin C).

Possible therapeutic agents include water treatments with potassium permanganate or Diquat. The medication of feed with oxytetracycline may be helpful if the infection is systemic (internal).

Enteric Redmouth

This infection in trout is caused by the bacterium *Yersinia ruckeri*. Sometimes the disease is called redmouth or Hagerman redmouth. Enteric redmouth (ERM) disease occurs in salmonids throughout Canada and much of the United States. The disease has been reported in rainbow trout, steelhead, cutthroat trout, coho, chinook, and Atlantic salmon. The bacteria was first isolated in 1950 from rainbow trout in the Hagerman Valley, Idaho.

Signs. ERM is characterized by inflammation and erosion of the jaws and palate. Trout typically become sluggish, dark in color, and show inflammation of the mouth, opercula, isthmus, and base of the fins. Reddening occurs in the body fat and the posterior part of the intestine. The stomach may fill with a colorless, watery liquid, and the intestine fills with a yellow fluid. ERM often produces sustained, low-level mortality, but can cause large losses.

Contributing Factors. Disease spread can be linked with fish movement. Large-scale outbreaks occur if chronically infected fish are stressed during hauling or exposed to low dissolved oxygen or other poor environmental conditions. Diagnosis of infections can be determined only by isolation and identification of the bacterium.

Prevention and Treatment. The best control is to avoid the pathogen. Fish and eggs should be obtained from sources known to be free of ERM. Vaccination is available and can be administered efficiently to fry for long-term protection.

Some drugs, such as sulfamerazine, may be required to stop mortality during an outbreak.

Enteric Septicemia of Catfish

The bacterium, *Edwardsiella ictaluri*, causes enteric septicemia. Because of one of the signs of the disease, it is sometimes called hole-in-the-head disease. It affects channel catfish.

Signs. Fish may have a "hole-in-the-head" lesion between the eyes. This may appear as a white or reddish raised area before the hole appears. Fish will also have pimple-like lesions over the general body surface, a typical dropsy appearance, bloody-looking internal organs, or yellowish or reddish fluid in the body cavity. Fish may cease feeding, become listless, hang tail-down in the water, and spasmodically swim rapidly in circles.

Contributing Factors. Low dissolved oxygen, high ammonia and nitrite concentrations, and water temperatures between 70° and 82°F (21° and 27.8°C) appear to be the conditions usually associated with the onset of the disease. Laboratory culture verifies the existence of the columnaris bacteria.

Prevention and Treatment. Aquaculturalists should maintain good quality water, keep dissolved oxygen level above 4 ppm, and provide good quality fed containing supplemental ascorbic acid (vitamin C). Also, the aquaculturalist should avoid using brood fish that have a history of the disease.

Possible therapeutic agents include antibiotics such as oxytetracycline in the feed.

BACTERIUM OR VIRUS— WHAT'S THE DIFFERENCE?

A bacterium is a microscopic single-celled organism, and very different from a virus. The plural of bacterium is bacteria. Bacteria occur everywhere life exists. They possess a tough, rigid outer cell wall through which they absorb their food. Some bacteria have a slimy outer capsule and some may have a whiplike flagella to propel them through liquids. If flagella are positioned all around the bacterium, it is called peritrichous, but if flagella are at each end it is called lophotrichous. Some bacteria may simply drift in air or water currents.

Bacteria generally reproduce by splitting into two. This is called binary fission and it may occur once every 15 to 30 minutes. Under favorable conditions one bacterium could form over 150 trillion bacteria in 24 hours! This usually does not happen. Bacteria are very numerous and very tough. A pinch of soil contains millions, and some bacteria can survive freezing, intense heat, drying, and some disinfectants. To survive adverse conditions bacteria form spores, which can remain active for years.

Bacteria can be classified by their shapes. Bacteria shaped like a sphere are called cocci. Those shaped like rods are called bacilli. Spirillum are in a spiral shape. Vibrio are comma shaped. Mycobacteria are very small rods. Flexibacter form long thin rods. The figure below illustrates some of the common shapes.

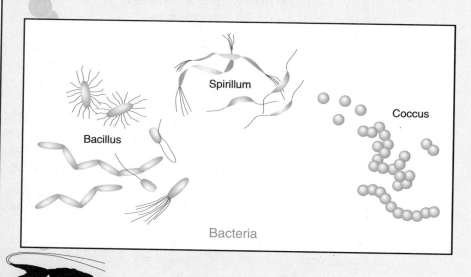

Spirillum

Coccus

Bacillus

Bacteria

Many bacteria perform useful functions for humankind. For example, helpful bacteria include those responsible for decay, sewage treatment, cheese and yogurt production, and those responsible for the nitrification process. Some bacteria cause disease. These are called pathogenic. Bacterial infections can be treated with antibiotics or similar drugs and some can be prevented by vaccination.

A virus is smaller and simpler than a bacterium. In fact a virus is so small and simple that it is on the borderline between a living organism and an inanimate particle. A virus is so small that it cannot be seen with an ordinary light microscope, but requires the use of an electron microscope.

Viruses live and reproduce inside of other living cells. They depend on living cells to reproduce, but some can live for quite some time outside cells of the body and some can even survive freezing and drying. Because viruses live inside of the cells of plant and animals, chemical treatment is often out of the question since this would kill the host cell. Some drugs relieve the symptoms that viruses produce, but the only effective way of controlling a viral infection is to remove the infected individual. The affected individual's own immune system must produce antibodies and counteract the infection. Vaccination provides a means of preventing viral infections.

Viruses take on many different shapes and forms. The figure below illustrates some of these.

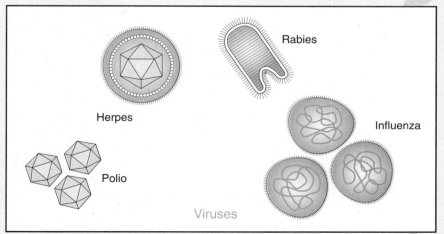

Viruses

Ulcer Disease of Goldfish

The warmwater strain of *Aeromonas salmonicida* causes ulcer disease of goldfish. Obviously, goldfish are the susceptible species, especially the broodfish. Sometimes it is called goldfish furunculosis.

Signs. Fish show ulcers or lesions with irregular margins on the sides of the fish. The lesions start as small white spots and progress to large hemorrhagic (bloody) sores. Scales are lost at the site of the ulcer.

Contributing Factors. Stress associated with handling, transportation, and spawning contribute to the development of the disease. The condition is aggravated by water temperatures of 55° to 75°F (12.8° to 23.9°C), and poor nutrition.

Laboratory culture verifies the presence of the bacteria.

Prevention. Fish farmers should collect, handle, and transport the brood fish in winter when the water temperature is less than 55°F (12.8°C), use young brood fish, and offer a nutritionally complete feed fortified with ascorbic acid (vitamin C). No therapeutic agents are available.

Chilodonelliasis

In cold water, the bacterium, *Chilodonella cyprini* and in warm water, the bacterium *C. hexasticha* cause chilodonelliasis. Most freshwater fish are susceptible.

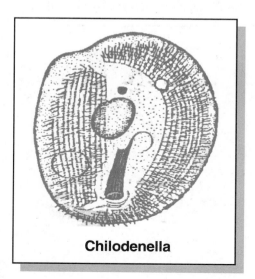

Chilodenella

FIGURE 6-17 The bacteria that cause chilodonelliasis.

Signs. Fish with chilodonelliasis have bright red gills that sometimes bleed when touched.

Contributing Factors. Water of reduced quality containing especially high levels of organic matter contribute to infections. The condition is aggravated by crowding, malnutrition, and by water temperatures of 40° to 70°F (4° to 21°C).

Verification requires microscopic examination.

Prevention and Treatment. Maintain good water quality, and offer the fish adequate amounts of good quality feed, especially in the late winter and early spring.

Formalin and potassium permanganate can be used as therapeutic agents.

VIRAL DISEASES

Viruses are submicroscopic pathogens that invade and destroy living cells, causing the release of large numbers of new particles—viruses—identical to the original. Viruses are essentially a protein coat surrounding a core of nucleic acid—genetic material.

Channel Catfish Virus Disease

Channel catfish virus causes channel catfish virus disease (CCVD). Channel catfish are susceptible during their first summer, usually when they are less than 5 in (12.7 cm) long.

Signs. Infected fish show a swollen abdomen with clear yellow fluid in the body cavity, popeye, erratic swimming—whirling or spiraling on the longitudinal axis—hemorrhage at the bases of the fins and through the skin on the ventral surface, and dark red spleen.

Contributing Factors. Low dissolved oxygen, high ammonia concentrations, and water temperatures above 68°F (20°C) and especially above 85°F (29°C) contribute to CCVD. Rough handling and stress from excessive chemical treatments may also be involved.

Tissue culture is necessary to isolate the virus.

Prevention and Treatment. Aquaculturalists maintain good water quality, and especially maintain dissolved oxygen concentration at 4 ppm or

FIGURE 6-18 Channel catfish virus disease affects fingerlings. The bloated abdomen is a characteristic sign of the disease. (Photo courtesy Chuck Weirich, Delta Research Center, Stoneville, MS)

higher. They do not overcrowd the fish, and handle them only when necessary during the first summer and avoid the use of chemical prophylactics. Disinfecting nets, tubs, and other equipment used when fish are handled or transported and purchasing fry from virus-free broodstock help prevent CCVD. No therapeutic agents are available for CCVD.

Infectious Hematopoietic Necrosis

A virus causes infectious hematopoietic necrosis (IHN). Coho salmon seem resistant to the virus, but rainbow trout are severely affected.

Signs. Affected fish show abdominal swelling, pale gills, dark coloration, lethargy, and hemorrhages at the base of the fins.

Contributing Factors. Survivors of the disease are carriers. The main avenues for infection are fish-to-fish and fish-to-eggs. Feeding byproducts made from infected fish also transmits the disease. Mortality is high in young fish but resistance to the virus appears to increase with age.

Prevention and Treatment. The only prevention is to avoid infected eggs, fish, and feed. No therapeutic agents are available. When an infection occurs, infected fish must be destroyed and the facility disinfected.

Infectious Pancreatic Necrosis

All salmonids and goldfish are susceptible to the virus that causes infectious pancreatic necrosis (IPN).

Signs. Affected fish may show an overall darkening of the body, popeyes, abdominal swelling, and hemorrhages at the base of the fins. Fish may be observed spiraling along the long axis of the body. Also, the death rate increases.

Contributing Factors. IPN virus spreads through feces, eggs, and seminal and ovarian fluids from parent to progeny. Fish surviving the disease become carriers. Susceptibility decreases with age; first-feeding fry are most susceptible.

Prevention and Treatment. Fish farmers do not use fish infected with the IPN virus. No treatment is known for IPN. Infected fish must be destroyed and the facility disinfected.

NONINFECTIOUS DISEASES

Infectious diseases are transmitted from fish to fish by an identifiable pathogenic agent. One fish can give the disease to another. Noninfectious diseases result from an animal's reaction to environmental changes. One affected fish cannot pass the condition on to others.

Oxygen Starvation

Reduced dissolved oxygen levels cause oxygen starvation.

Signs. Affected fish gather at the water inflow or outlet. Also, fish will be observed gasping at the water surface. Oxygen starvation may be noted as sudden mortality.

Prevention and Treatment. Producers monitor dissolved oxygen levels and try to predict any sudden drops in dissolved oxygen. Treatment is some form of aeration.

Alkalosis

Water that becomes too basic for the species causes alkalosis. In other words, the pH increases to a level higher than the species can tolerate.

Signs. When the pH is high for an extended period, fish die. Alkalosis can also cause corroding of the skin and gills or a milky turbidity of the skin.

Prevention and Treatment. Aquaculturalists must monitor pH level and maintain the pH in a range optimal for the species being cultured. When the problem continues to reoccur, the farmer must determine the cause and correct it. In some cases, the pH level can be reduced by adding alum or agricultural gypsum.

Acidosis

Acidosis is caused by a drop in the pH to a level too low for the species.

Signs. Affected fish shoot through the water with sudden rapid fin movements. Fish gasping for air sometimes jump out of the water. The skin may have a milky turbidity or be red and inflamed. Also, brown deposits will be noted on the gills. Death of fish can occur very quickly or it can take a slow course.

Prevention and Treatment. Monitoring the pH and maintaining the pH at an optimal range is the best prevention. Liming raises the pH and total hardness of the water. The farmer must determine the cause of the imbalance and prevent its reoccurrence.

Nutritional Deficiency

A nutritional deficiency can be caused by unsuitable food, too much food, too little food, or a vitamin deficiency. Nutritional diseases can be very difficult to diagnose. A classic example of a nutritional disease of catfish is "broken back disease," caused by vitamin C deficiency. The lack of dietary vitamin C contributes to improper bone development, resulting in deformation of the spinal column. Another important nutritional disease of catfish is "no blood disease," which may be related to a folic acid deficiency. Affected fish become anemic and may die. The condition seems to disappear when the deficient feed is discarded and a new feed provided.

Signs. The signs vary depending on the type of deficiency. Some common signs include body deformities; slow, weak movements; loss of appetite; slow growth; and a hollow-bellied profile. Chapter 5, Fundamentals of Nutrition in Aquaculture, Table 5-7, lists the effects of various deficiencies.

Prevention and Treatment. Fish farmers must feed a nutritionally balanced ration designed for the species being cultured. Also, a farmer must be

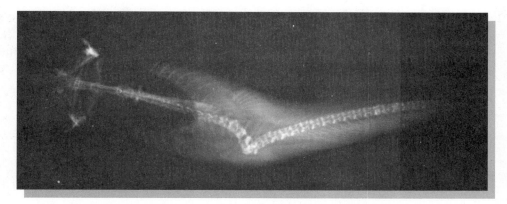

FIGURE 6-19 Nutritional deficiency of vitamin C causes "broken back" disease or spinal curvature in catfish. (Photo courtesy Department of Fisheries and Aquaculture, Auburn University, Auburn, AL)

careful not to overfeed or underfeed fish. Calculating the feed conversion ratio provides information for adjusting feed levels. Some conditions caused by nutritional deficiency cannot be reversed. For those conditions that can be reversed, the type, amount, or time of feeding should be changed.

Poisoning

Any toxic substance or toxic levels of a substance in water can poison fish.

Signs. Obviously, the signs vary with the poison. The most noted sign may be sudden mortality.

Prevention and Treatment. Before building ponds, aquaculturalists should test the soil and water for pesticide contamination or other toxic substances. Water should be monitored for ammonia, nitrate, iron, and other potential hazards. Treatment varies depending on the toxin present. Emergency measures require dilution of the toxin with fresh, clean water or a total water change. Some conditions cannot be reversed.

Brown Blood Disease

High nitrite levels in the water oxidize hemoglobin in the blood to methemoglobin.

Signs. Signs of brown blood disease include loss of appetite, topping, and literally brown blood. Fish may die suddenly.

Prevention and Treatment. Brown blood disease can be prevented by monitoring nitrite levels in water and anticipating high nitrite levels when water temperature and pH rise. Sodium chloride, or common table salt, when applied at a rate of 5 ppm per A-ft effectively reverses the effects of nitrite.

Anemia

Poor nutrition or a chronic disease can cause anemia in fish. Anemia is the inability of the blood to carry oxygen.

Signs. Anemic fish are lethargic, lose their appetite and color, and have pale gills. Eventually, they die.

Prevention and Treatment. The best prevention and treatment is feeding a nutritionally complete diet and controlling disease.

Gas Bubble Disease

Water supersaturated with oxygen or nitrogen causes gas bubble disease. This situation is found naturally in well and spring water when ice melts or when air is introduced into water lines or pumps.

Signs. Affected fish show bubbles under the skin and in the gill tissues. Also, fish rustle when they are taken out of the water.

Prevention and Treatment. Fish farmers monitor dissolved oxygen levels and maintain these levels at optimum for the species. Also, farmers control algae growth and avoid algae blooms. The situation can be corrected by mechanical aeration.

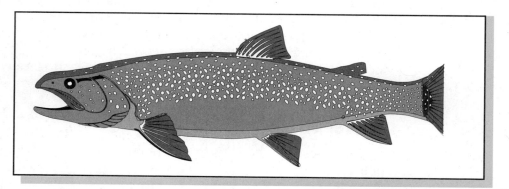

FIGURE 6-20 Genetic abnormalities such as no tail are rare and not very significant to the industry.

Genetic

Genetic **abnormalities** include conformational oddities such as lack of a tail or presence of an extra tail. Most of these are of minimal significance. Using unrelated fish as broodstock every few years minimizes inbreeding and the expression of genetic abnormalities.

DETERMINING THE PRESENCE OF DISEASE

The most obvious sign of sick fish is the presence of dead or dying animals. The careful observer can usually tell that fish are sick before they start dying because sick fish often stop feeding and may appear lethargic. Healthy fish should eat aggressively if fed at regularly scheduled times. Pond fish should not be visible except at feeding time. Fish that are observed hanging listlessly in shallow water, gasping at the surface, or rubbing against objects indicate something may be wrong. These behavioral abnormalities indicate that the fish are not feeling well or that something is irritating them.

In addition to behavioral changes, some physical signs should alert producers to potential disease problems in their fish. These include the presence of sores such as ulcers or hemorrhages, ragged fins, abnormal body conformation like a distended abdomen or dropsy, and exopthalmia or popeye. When these abnormalities are observed, the fish should be evaluated for parasitic or bacterial infections.

When fish are suspected of getting sick, the first thing to do is check the water quality. Low oxygen is a frequent cause of fish mortality in ponds, especially in the summer. High levels of ammonia are also commonly associated with disease outbreaks when fish are crowded in vats or tanks. In general, it is appropriate to check dissolved oxygen, ammonia, nitrite, and pH during a water quality screen associated with a fish disease outbreak.

Ideally, daily records should also be available for immediate reference when a fish disease outbreak occurs. These should include the dates fish were stocked, size of fish at stocking, source of fish, feeding rate, growth rate, daily mortality, and water quality. This information is needed by an aquaculture specialist to solve a fish disease problem. Good records, a description of behavioral and physical signs exhibited by sick fish, and results of water quality tests provide a complete case history for the diagnostician.

Fish submitted to a diagnostic laboratory should be collected live, placed in a freezer bag without water, and shipped on ice to the nearest facility. Small fish can be shipped alive by placing them in plastic bags that are partially filled—30 percent—with water. Oxygen gas can be injected into the bag prior to sealing it. Insulated containers for shipping live, bagged fish minimize tem-

AquaFood Farms Health Records

Week of _____

Observation	Date	Time	Loc#1	Loc#2	Loc#3
DO level					
pH					
Total alkalinity					
N					
Total ammonia N					
Unionized ammonia					
Chloride					
Total hardness					
Temperature					
Observed signs of disease					
External parasites					
Feeding behavior					
Number of mortalities					

FIGURE 6-21 Fish records are important to determining disease.

perature fluctuations during transit. In addition to fish samples, a water sample collected in a clean jar should also be submitted.

DISEASE TREATMENT

The following questions help determine whether or not treatment is warranted:

1. What is the prognosis? Is the disease treatable, and what is the possibility of a successful treatment?

2. Is it feasible to treat the fish where they are, considering the cost, handling, **prognosis**, and other factors?

3. Is it worthwhile to treat, or will the cost of treating exceed the value of the fish?

4. Can the fish withstand the treatment considering their condition?
5. Does the loss rate and the disease warrant treatment?

Before any treatment is started, the following four factors must be known:

1. The volume of water in the holding or rearing unit to be treated.
2. Fish species and ages, since different species and ages will react differently to the same drug or chemical.
3. The toxicity of the chemical to the particular species of fish to be treated and the effect of water chemistry on the toxicity of the chemical should also be known.
4. The disease being treated. Although this factor appears to be self-evident, it is often forgotten.

Methods of Treatment

Various methods of treatment and drug application control fish diseases.

Dip. A strong solution of a chemical is used for a relatively short time. This method can be dangerous because the solutions used are concentrated. The difference between an effective dose and a killing one is usually very slight. Fish

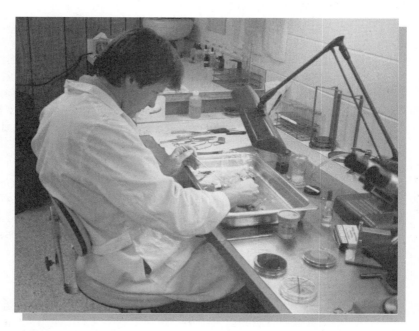

FIGURE 6-22 After fish are submitted to a diagnostic laboratory, tissue samples are taken to isolate the disease.

are usually placed in a net and dipped into a strong solution of the chemical for a short time, usually 15 to 45 seconds, depending on the type of chemical, the concentration, and the species of fish being treated.

Flush. This method is fairly simple and consists of adding a stock solution of a chemical to the upper end of the unit to be treated, then allowing it to flush through the unit. An adequate water flow must be available so the chemical can be flushed through the unit or system in a short period of time. This method cannot be used in ponds.

Prolonged. There are two types of prolonged treatments: a short term, or bath, and an indefinite prolonged treatment.

Bath. The required amount of chemical or drug is added directly to the rearing or holding unit and left for a specified time, usually one hour. The chemical or drug is then quickly flushed with fresh water. Several precautions must be observed with this treatment to prevent serious losses. Although a treatment time of one hour may be recommended, fish should be observed during the treatment period. At the first sign of distress, fresh water is added quickly. Use of this method requires extreme caution to ensure that the chemical is evenly distributed throughout the unit to prevent the occurrence of a hot spot of the chemical.

Indefinite. Usually this method is used for treating ponds or hauling tanks. A low concentration of a chemical is applied and allowed to dissipate naturally. This is generally one of the safest methods of treatment. One major drawback is the large amounts of chemicals required, which can be prohibitively expensive. As in the bath treatment, the chemical must be distributed evenly throughout the unit to prevent hot spots.

Feeding. In the treatment of some diseases, the drug or medication must be fed or in some way introduced into the stomach of the sick fish. This can be done by either incorporating the medication in the food or by weighing out the correct amount of drug, putting it in a gelatin capsule, and then using a balling gun to insert it into the fish's stomach. This type of treatment is based on body weight.

Injections. Large and valuable fish, particularly when only small numbers are involved, can at times be treated best by injecting the medication into the body cavity—intraperitoneal (IP)—or in the muscle tissue—intramuscular (IM). Most drugs work more rapidly when injected IP than IM. IP injections require caution to ensure that no internal organs are damaged. The easiest location for IP injections is the base of one of the pelvic fins.

For IM injections the best location is usually the area immediately next to the dorsal fin.

CALCULATING TREATMENTS

Calculation of treatment levels, units of measure, terminology, and treatment levels used in prescribing treatment rates are often confusing, not only to the fish farmer, but also to many biologists. Even though most people are familiar with pounds, ounces, gallons, acres, and feet, it can be confusing to convert these to kilograms, grams, liters, acre-feet, and meters.

Most aquaculture systems, including ponds, tanks, and raceways, eventually require some type of chemical treatment to combat a disease or aquatic plant problem or improve water quality. Some treatments, like fertilizers or lime, are based on the surface area of water, but most treatments include the total volume of water in the production or holding unit.

All commercial aquaculturists should know how to calculate treatment rates, determine the amount of chemical or material needed, and apply the treatment. Before any treatment is applied, the water, fish, chemical, condition of the problem, and method of treatment should be known and understood. Producers can experience high economic losses when treatment rates are not properly calculated.

Before any calculation is made, the units of measurements should be determined. The unit of measurement selected should be familiar and convenient for the specific situation. For example, the large volume of water in ponds is usually expressed as acre-feet, while the volume of a small tank may be expressed in gallons or cubic feet. Another decision is whether to use the English or metric system of measurement. A working knowledge of both systems is important because reports or publications may use either one. Metric units are easiest to work with when small volumes or weights are involved.

To make calculations easy, refer to Tables A-1 and A-2 in the Appendix for conversion factors and parts per million equivalents.

Basic Formula

Most treatments can be calculated using this basic formula:

$$\text{Amount of chemical needed} = V \times C.F. \times \text{ppm desired} \times \frac{100}{\%AI}$$

Where: V   =   The volume of water in the unit to be treated.

C.F.   =   A conversion factor that represents the weight of chemical that must be used to equal one part per million (1 ppm) in one unit of the volume (V) of water to be treated. The unit of measurement for the results is the same as the unit used for the C.F. (pounds, grams, or other unit). Table A-2 in the Appendix contains a list of these conversion factors for various units.

ppm   =   The desired concentration of the chemical in the volume (V) of water to be treated expressed in parts per million.

$\dfrac{100}{\%AI}$ =   100 divided by the percent active ingredient (AI) contained in the chemical to be used.

Most chemicals are 100 percent AI unless otherwise specified, so this value is usually 1. The percent AI is found on the label of most fisheries chemicals. Chapter 7, Water Requirements for Aquaculture, provides example problems using the basic formula.

IMMUNIZATION

Vaccines given to fish cause the fish to produce antibodies to a specific disease, giving them immunity to the disease. Technology in fish vaccination has advanced rapidly in the past few years for four reasons:

1. The list of antibacterial drugs that could legally be used is extremely small and the prospects for enlarging the list were dim.

2. The effectiveness of the few available antibacterial drugs was rapidly being diminished because of the development of antibiotic resistance among the bacterial fish pathogens.

3. The danger that this antibiotic resistance might be transmissible to microorganisms of public health concern, and the real possibility that drugs now approved for use in fish culture would have their approval revoked.

4. The viral infections in fish could not be treated with any of the antibiotics available.

The advantages of immunizations include—

■   Does not produce antibiotic resistance bacteria

■   Can be applied to control viral and bacterial diseases

■ Vaccination is convenient and economical

■ Requirements for licensing vaccines are less stringent than those for antibiotics.

Vaccination Methods

Unfortunately, the biggest single factor working against the widespread use of fish vaccination was the lack of a safe, economical, and convenient technique for vaccinating large numbers of fish. Attempts at oral vaccination have been unsuccessful, and alternative procedures devised include mass inoculation infiltration, and spray vaccination.

The mass inoculation method works well with fish in the 5 to 25 gm range, and individual operators are able to vaccinate 500 to 1,000 fish per hour. Cost of the technique seems reasonable, but the number of fish that can be treated is limited by the human resources available for short-term employment and by the size of the inoculating tables.

The infiltration method or hyperosmotic immersion allows vaccination of up to 9,000 fish (1,000 per lb) quickly and safely in approximately four minutes. The method uses a specifically prepared buffered hyperosmotic solution. Through osmosis, fluid is drawn from the fish body during its immersion in the buffered prevaccination solution. The fish are then placed into a commercially prepared vaccine that replenishes the body fluids and simultaneously diffuses the vaccine or bacterin into the fish.

Fish are spray vaccinated by removing them briefly from the water and spraying them with a vaccine from a sand-blasting spray gun. Antigenicity— ability to produce antibodies—of the preparations is markedly enhanced by the addition of bentonite, an absorbent.

In 1976, two bacterin were licensed for sale and distribution by the U.S. Department of Agriculture. These products are enteric redmouth and *Vibrio anguillarum* bacterin.

SUMMARY

Disease is any condition of an aquatic animal that impairs normal physiological function. Prevention of disease is better than treatment. Prevention of fish diseases is enhanced by good management of the water environment and the feed and by observing fish frequently for signs of disease.

Management practices must minimize stress on fish to decrease the occurrence of disease outbreaks from parasites, bacterial, viral, or fungal pathogens. When disease outbreaks occur, the underlying cause of mortality should be identified as well as underlying stress factors that may be compromising the natural survival mechanisms of the fish. Correction of stressors like poor water quality and crowding should precede or accompany disease treatments.

Stress compromises the fish's natural defenses so that it cannot effectively protect itself from invading pathogens. A disease treatment is an artificial way of slowing down the invading pathogen so that the fish has time to defend itself with an immune response. Any stress that adversely effects the ability of the fish to protect itself will result in an ongoing disease problem. As soon as the treatment wears off, the pathogen can build up its numbers and attack again. Rarely would a treatment result in total annihilation of an invading organism. Disease control depends upon the ability of the fish to overcome infection as well as the efficacy of the chemical or antibiotic used. Effects of the disease and treatment can be very costly to the producer.

For some diseases, vaccinations immunize the fish. The continued development of vaccines is important. Few treatments are approved for fish diseases. Vaccines are effective for bacterial and viral infections and are easier to approve.

STUDY/REVIEW

Success in any career requires knowledge. Test your knowledge of this chapter by answering these questions or solving these problems.

True or False

1. Disease is any condition of an animal that impairs normal function.

2. All bacteria cause disease.

3. Fish can get lice.

4. Fish can be vaccinated.

5. Only a pathogen is necessary to cause disease.

Short Answers

6. In aquaculture, why is fish health management more important than the treatment of disease?

7. List two general practices that help prevent fish diseases.

8. Give an example of a chemical stressor.

9. Give an example of a biological stressor.

10. Give an example of a physical stressor.

11. Which of the following is not one of the four protective barriers protecting fish against an infection?
    a. antibodies
    b. dorsal fin
    c. mucus
    d. scales and skin

12. Handling, crowding, and poor water quality are examples of _____ that decrease the fish's ability to resist disease.

13. Name the two broad categories of diseases that affect fish.

14. An example of an infectious disease is—
    a. acidosis
    b. alkalosis
    c. Ich
    d. brown blood

15. What three things interact to produce a disease in fish?

16. Name four protective barriers that normally protect fish against infection.

17. Give three examples of stress affecting the normal protective barriers.

18. List the four broad categories of infectious diseases.

19. Name a disease caused by a ciliated protozoan.

20. What disease gives fish a cotton-like or fur-like appearance?

21. What's another name for hole-in-the-head disease?

22. When fish receive a treatment, what ways can the treatment be administered?

23. Name three methods of vaccinating fish.

Essay

24. Define health management.

25. Name four types of stressors and give an example of each.

26. What controls osmoregulation?

27. List three good management practices that prevent stress.

28. Distinguish between bacterial, viral, and fungal diseases.

29. List three broad categories for noninfectious diseases and give an example of each.

30. Describe the life cycle of a fluke.

31. When describing a disease in animals, why should the disease be described in terms of signs instead of symptoms?

32. Explain why treatment of viral diseases is difficult.

33. What causes acidosis?

34. Describe four behavioral patterns that indicate fish are healthy.

35. Name five physical abnormalities that indicate the presence of disease.

36. Why are records important to fish health management?

KNOWLEDGE APPLIED

1. Based on the information in this chapter, develop a checklist of things to observe when checking fish for signs of disease. Make the checklist complete enough that it could be given to a new employee.

2. Visit or contact a disease diagnostic laboratory. Find out which species the lab will perform diagnostic procedures on. Determine some of the diseases the lab dealt with in the past year. Ask what the requirements are for submitting an animal for diagnosis and how much the diagnosis costs. Also, ask what diseases they have to report.

3. Using the formula for calculating the amount of chemical needed, develop a series of problems for calculating the treatment of fish in ponds, raceways, or tanks. For example, how much potassium permanagate is needed to treat a pond 700 ft long × 700 ft wide × 4 ft deep with a concentration of 2 ppm? Potassium permanagate is 100 percent active ingredient. Or, for example, how much formalin is needed to treat a circular tank that is 4 ft in diameter and has a water depth of 3 ft with 250 ppm? Formalin liquid is 100 percent active. (More sample problems are available in the resources listed at the end of the chapter.)

4. Visit a veterinary supply store. Ask the veterinarian or sales representative to discuss the types of vaccines sold. Also, find out what types of restrictions are placed on the sale of vaccines, antibiotics, and other medicines. As an alternative, ask a veterinarian or sales representative to visit class and discuss vaccines, antibiotics, and other medicines.

5. From a biological supply house or the biology department at a college or high school, obtain prepared microscope slides of bacteria. Observe the different shapes—rods, spheres, or spirals. Also, if available, obtain prepared slides of microscopic parasites, such as some of the protozoans that infect fish.

6. Research and report the role of bacteria in human health and commerce. Some bacteria cause disease while other bacteria have a role in human welfare.

LEARNING/TEACHING AIDS

Books

Adams, S. (1990). *Biological indicators of stress in fish.* Bethesda, MD: American Fisheries Society.

Andrews, C., Exell, A., & Carrington, N. (1988). *The manual of fish health.* Morris Plains, NJ: Tetra Press.

Dupree, H. K., & Huner, J. V. (1984). *Third report to the fish farmers: The status of warmwater fish farming and progress in fish farming research.* Washington, DC: U.S. Fish and Wildlife Service.

Ellis, A. (1988). *Fish vaccination.* San Diego, CA: Academic Press.

Piper, R. G., McElwain, I. B., Orme, L. E., McCraren, J. P., Fowler, L. G., & Leonard, J. R. (1982). *Fish hatchery management.* Washington, DC: U.S. Department of the Interior, Fish and Wildlife Service.

Sindermann, J. C., & Lightner, D. V. (1988). *Disease diagnosis and control in North American marine aquaculture.* New York: Elsevier Publishing.

Stolen, J. S. (1993). *Techniques in fish immunology, FITC1* (2nd Ed.). Fair Haven, NJ: SOS Publications.

Stoskopf, M. K. (1993). *Fish medicine.* Philadelphia: W.B. Saunders.

Walker, S. S. (1990). *Aquaculture.* Stillwater, OK: Mid-America Vocational Curriculum Consortium, Inc.

Wolf, K. (1988). *Fish viruses and fish viral diseases.* Ithaca, NY: Comstock Publishing Associates.

Bulletins

Technical Information Service provides nine bulletins on fish diseases. They also maintain a collection of research publications on fish diseases. These can be requested from Technical Information Service, Leetown Science Center, 1700 Leetown Road, Kearneysville, WV 25430, phone 304-725-8461 or FAX 304-728-6203.

The Regional Aquaculture Centers also provide a number of bulletins about specific disease. Addresses for these centers are listed in the Appendix, Table A-11.

Submitting Samples for Fish Disease Diagnosis explains how the fish farmer as well as the diagnostician is responsible for obtaining an accurate disease diagnosis. This booklet is published by the U.S. Fish and Wildlife Service, Fish Farming Experimental Station, Stuttgart, AR.

Treatment Tips for Fish Producers defines many of the variables to be considered when the producer is treating fish diseases. It also offers information that should help make treatments safe and beneficial. This booklet is published by the U.S. Department of the Interior, Fish and Wildlife Service.

Water Requirements for Aquaculture

H igh-quality water and plenty of it is the primary consideration for any aquaculture facility. This is true for finfish, shellfish, and crustaceans. Water provides oxygen, food, serves as an excretory site, helps regulate body temperature, and may harbor disease-causing organisms. Gaining an understanding of water helps aquaculturalists become more productive.

Learning Objectives

After completing this chapter, the student should be able to:

In Aquaculture

- Explain why water is important in aquaculture
- Explain the quality features of water for aquaculture
- Define terms related to water quality management with their definitions
- Calculate water needs and filling time
- Calculate treatments for volumes of water
- List causes of dissolved oxygen loss
- List signs of dissolved oxygen efficiency

■ Select facts about the prevention of oxygen depletion
■ List methods of correcting dissolved oxygen deficiency
■ Know what causes turbidity
■ List the purposes of liming
■ Discuss aquatic plant control methods with their descriptions
■ List ways to dispose of wastewater

In Science

■ Describe the properties of water
■ List cations and anions found in water
■ Describe why and how aquatic solutions change
■ Explain how changes in water affect aquatic life
■ Match compounds and elements with their chemical formulas and symbols
■ Discuss the importance of oxygen in water quality management
■ Discuss the role of temperature in oxygen management

FIGURE 7-1 Water, lots of it, for aquaculture.

- ▪ List chemicals, compounds, and elements that are detrimental to water quality
- ▪ Understand the importance of nitrogen compounds in water quality management
- ▪ Complete statements about pH and water quality
- ▪ Select from a list methods of managing the pH cycle
- ▪ Select general guidelines for water chemistry management

Understanding of this chapter will be enhanced if the following terms are known. Many are defined in the text and others are defined in the glossary.

KEY TERMS		
Acid	Grams per liter (g/L)	Parts per thousand (ppt)
Aeration	Heat capacity	
Alkaline	Heavy metals	pH
Anaerobic	Hydroponics	Salinity
Anions	Ions	Salts
Beer's Law	Liming	Saturation
Buffers	Logarithmic	Settling pond
Cations	Microgram	Solubility
Colorimetric	Microohms	Thermal stress
Conductance	Milliequivalents per liter (meq/L)	Titrant
Dissolved oxygen		Titrimetric
Effluent	Milligram	Turbidity
Equivalents per million (epm)	Organic	Wastewater
	Parts per billion (ppb)	Water hardness
Fertilization	Parts per million (ppm)	

WATER QUALITIES, MEASUREMENTS, AND ALTERATIONS

Hydrogen and oxygen associate to form water, H_2O. In nature, water is not a pure substance. It consists, to some degree, of a number of dissolved and suspended substances. Unless the water is particularly muddy, these substances consist predominately of dissolved ions. Ions are elemental forms or groups that carry an electrical charge in solution. Positive ions are called cations, and examples are calcium, potassium, magnesium, and sodium. Negative ions or anions are substances like carbonate, bicarbonate, sulfate, and chloride.

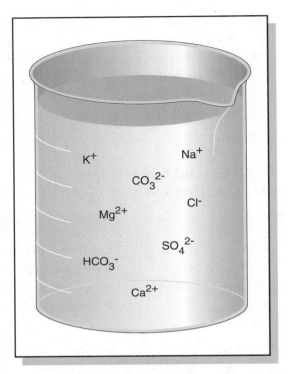

FIGURE 7-2 Cations or positive ions and anions or negative ions in water.

Substances that dissolve readily in water to form simple ions are called salts. Because many of the dissolved ions in water would combine upon evaporation to form salt compounds, the content of simple ions in solution is frequently referred to as salt content.

A sample of natural surface water also contains organic matter. Organic matter consists of living material, its excretions, and decomposing material. This is usually only a small percentage of a residue. Organic matter can affect water chemistry by requiring oxygen for living processes as the complex organic matter is built up from and torn down into simple compounds.

Gases are also present in water, either dissolving at underground pressures in the case of groundwater or under normal pressure at the surface from the gas mixture of the air. Some gases enter the water as they are formed by living agents in water or by chemical release from muds in the bottom.

Units of Measure

The most common unit of measure of trace elements, gases, ions, and pesticides is parts per million (ppm). A part per million is the same as a milligram per liter (mg/L). Smaller concentrations are sometimes reported as parts per

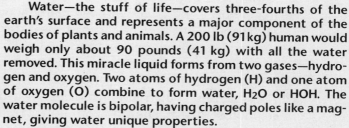

HOH—WATER!

Water—the stuff of life—covers three-fourths of the earth's surface and represents a major component of the bodies of plants and animals. A 200 lb (91 kg) human would weigh only about 90 pounds (41 kg) with all the water removed. This miracle liquid forms from two gases—hydrogen and oxygen. Two atoms of hydrogen (H) and one atom of oxygen (O) combine to form water, H_2O or HOH. The water molecule is bipolar, having charged poles like a magnet, giving water unique properties.

Depending on the temperature, water exists in three forms. It is a liquid, between 32° and 212°F (0° and 100°C). It is a gas or vapor at temperatures above 212°F (100°C), and water becomes a solid (ice) at temperatures below 32°F (0°C).

Of all the naturally occurring substances, water has the highest specific heat. This makes it a good coolant in biological systems and makes it resist rapid temperature changes. Specific heat is the amount of heat required to raise the temperature of a substance 1°C.

Water is the universal solvent, dissolving almost everything. It is powerful enough to dissolve rocks yet gentle enough to hold an enzyme in a fragile cell. As a solvent, it acts as a medium for biochemical reactions and carries waste products and nutrients.

This miracle liquid supports life.

billion (ppb), which is the same as a microgram per liter (mcg/L). Larger concentrations are sometimes reported as parts per thousand (ppt) or grams per liter (g/L).

Sometimes, and particularly with ionized components of salts, a measure called equivalents per million (epm) is used. Equivalents per million is the same as the expression milliequivalents per liter (meq/L). Standard multiplication factors convert epm to mg/L.

Rarely, the measure grains per gallon is used. One grain per gallon equals 17.1 parts per million.

pH

The pH—one of the most common water tests—is a measure of hydrogen ions in the water. The pH scale spans a number range of 0 to 14 with the number 7 being neutral. The pH scale is logarithmic, so every one-unit change in pH represents a ten-fold change in acidity. Measurements above 7 are basic and below 7 are acidic. The farther a measurement is from 7, the more basic or acidic is the water. Acid and alkaline (basic) death points for fish are approximately pH 4 and 11. Growth and reproduction can be affected between pH 4 and 6 and pH 9 and 10 for some fishes. Also, pH affects the toxicity of other substances, such as ammonia and nitrite.

The pH of some ponds may change during the course of a day and is often between 9 and 10 for short periods of afternoons. Fish can usually tolerate such rises that result when carbon dioxide, an acidic substance, is used up by plants in photosynthesis. The most common pH problem for pond fish is when water is constantly acidic. The nature of the bottom and watershed soils is usually responsible. Water with a stable and low pH is only correctable with liming.

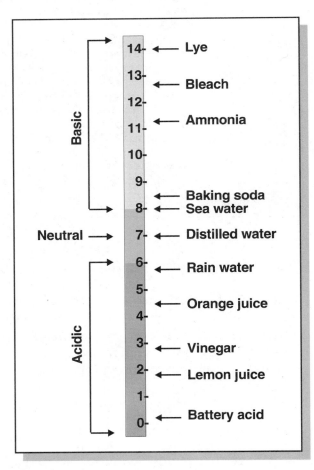

FIGURE 7-3 The approximate pH of some common substances.

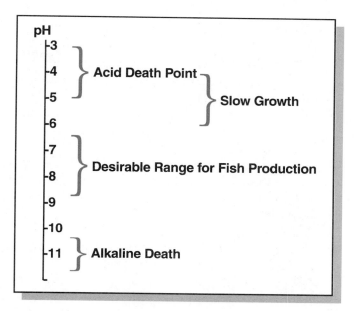

FIGURE 7-4 The effect of pH on fish survival.

Temperature

Water temperature helps determine which species may or may not be present in the system. Temperature affects feeding, reproduction, immunity, and the metabolism of aquatic animals. Drastic temperature changes can be fatal to aquatic animals. Not only do different species have different requirements, but optimum temperatures can change or have a narrower range for each stage of life.

All species tolerate slow seasonal changes better than rapid changes. Thermal stress or shock can occur when temperatures change more than 2° to 3°F (1° to 2°C) in 24 hours.

The heat capacity of water is very high, making it resistant to changes in temperature. This moderates the daily and seasonal climatic changes in temperature. But cooling is often impractical and heating is possible but costly.

Water temperature is an important variable in many chemical tests and electronic measures for water quality. Many determinations require adjustment for the temperature or noting the temperature at the time of sampling.

Water temperature affects many biological chemical processes. Spawning is triggered by temperature. Temperature differences between the surface and bottom waters help produce vertical currents moving nutrients and oxygen throughout the water column. Temperature influences the solubility of oxygen and the percentage of unionized ammonia in water.

Salinity

Salinity is the measure of the total concentration of all dissolved ions in water. Sodium chloride (NaCl) is the principal ionic compound in sea water, but most inland ponds contain substantial concentrations of other ionic compounds (salts) such as compounds of sulfate and carbonate. Salinity also can be measured according to the density the salts produce in the water, the refraction they cause to light or by electrical **conductance**. The result in all cases is reported in parts per thousand (ppt) salinity.

Standard sea water is 35 ppt or more. Well waters sometimes accumulate high amounts of dissolved ions due to ionization of compounds of underground minerals or as a result of leaching from the high salt content of arid land surfaces.

Dissolved ionic substances can be measured by electrical conductance. On laboratory reports this may be shown as specific conductivity. Conductivity is reported as **microohms** per cm or $EC \times 10^6$. From such an electrical measure, tables can be used to derive tons per A-ft., parts per million, grains per gallon, and other measures. Conductivities in natural surface water measure from 50 to 1,500 microohms per cm.

Freshwater fish such as catfish, carp, tilapia, and trout tolerate some salinity and still grow and survive.

FIGURE 7-5 Test kits provide useful, quick information about water quality. (Aquaculture Test Kit, FF-3. Photo courtesy Hach Company, Loveland, CO)

Total Dissolved Solids

Another measure reporting the presence of dissolved ionic constituents is total dissolved solids. This measurement is made by weighing the residue of an evaporated sample after passing it through filter paper. If the sample is not filtered the reported value is total solids (TS) instead of total dissolved solids (TDS). Because ionic compounds dominate the content of dissolved substance in most water samples, TDS reflects closely the numerical quantity derived in the measure of salinity.

TDS is reported in mg/L of whole sample. In both surface and ground waters of inland areas where TDS exceeds 2,000 mg/L the principal anions are sulfate and chloride. The measure of these ions in mg/L when added together usually approximate one-half the measure of TDS. Although not the same as salinity, TDS sometimes serves as the only measure on which to make decisions regarding dissolved salts. As mentioned for salinity, 2,000 mg/L (2 ppt) is known to adversely affect sensitive species or younger stages of some species. Catfish handle 6,000 to 11,000 mg/L salinity quite well, depending on acclimation. Besides sea water, water that flows through limestone and gypsum dissolves calcium, carbonate, and sulfate with resulting high levels of TDS.

Major Dissolved Ionic Components

Helpful tests determine various ionic constituents of water. Calcium (Ca) and magnesium (Mg) compounds are preferred as major ionic components. Solubility of liming compounds is affected by the presence of sodium. Aquatic species are more affected by total ionic presence than individual ion concentrations.

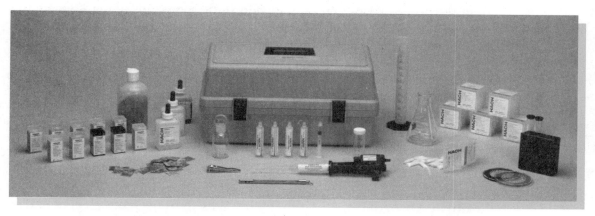

FIGURE 7-6 Application of Aquaculture Portable Lab Test Kit, DREL/2000. (Photo courtesy Hach Company, Loveland, CO)

Carbonate (CO_3^{2-}) and bicarbonate (HCO_3^-) are present in both surface and groundwater supplies at levels consistent with their solubilities. Measurements normally range below 300 mg/L and are considered harmless to fish life. Carbonate and bicarbonate act as pH **buffers**, resisting changes in pH.

Sulfate (SO_4^{2-}) and chloride (Cl^-) are expected to range higher than carbonate/bicarbonate in well and surface waters where the water comes in contact with appropriate rocks such as those composed of aluminum sulfate. Normal surface waters contain less than 50 mg/L of sulfate and chloride, but some well waters far from coastal areas reach 1,000 to 2,000 mg/L sulfate and several times that amount in chlorides.

Nitrate (NO_3^-) is generally nontoxic to fishes and can be expected to occur at less than 2 mg/L in natural surface water. Fish can tolerate several hundred mg/L. In some recycled waters or where feeding causes enrichment, nitrate could climb to several mg/L.

Phosphate (PO_4^{3-}), fluoride, and silicate are minor constituent anions. Phosphate, like nitrate, is usually present in slight amounts (less than 0.1 mg/L) in natural surface and well water. Aside from promoting unwanted growth of algae in ponds it is considered harmless.

Fluoride concentrations in surface water would be considered normal at less than 0.5 mg/L, high at 1 to 2 mg/L, and rare at over 10 mg/L. Fish react differently to fluoride according to overall water conditions and species. In some cases, 3 mg/L causes major losses, but normal populations have been recorded in lakes where concentrations reach over 13 mg/L.

Silica is rather unreactive and harmless and is normally present in pond waters at less than 10 mg/L.

Potassium (K^+) usually represents a very low percentage of surface water cations. Calcium (Ca^{2+}) and magnesium (Mg^{2+}) are typically greater and vary from site to site in proportion to sodium. Sodium (Na^+) is much more soluble than the other cations and can range into the thousands of mg/L where the others are limited to hundreds of mg/L. Water with 5 or less mg/L calcium is considered very low and 10 mg/L or less low in calcium. Problems to fish life from maximum cation amounts are typically associated with intolerance by fish to extremely high total dissolved solids (salts).

Some waters fed by wells have calcium and magnesium ions dissolved to perhaps twice that which is considered possible for natural surface water and yet will maintain living populations of fishes and other aquatic animals. Excess sodium has a detrimental influence on liming efforts and can cause liming to have little positive effect.

Sodium Absorption Ratio. This test is used to check the alkali hazard of irrigation water. It compares sodium concentration with calcium and magnesium to assess the potential for sodium buildup in cropland.

FIGURE 7-7 Obtaining a water sample that reflects the condition of all the water is difficult.

Sodium Percentage. This is another irrigation water test, also known as sodium hazard, that compares the amount of sodium to all cations present. The effect of sodium on aquatic microplant production is not well understood. On crops, the sodium percentage should be less than 60.

Total Alkalinity and Total Hardness. This is a measure of the basic substances of water. Because in natural water these substances are usually carbonates and bicarbonates the measurement is expressed as mg/L of equivalent calcium carbonate. In some cases, such as many groundwater and western ponds, sodium carbonate is the predominant basic substance. These basic substances resist change in pH (buffering) and where an abundance of calcium and magnesium bicarbonate is dissolved the pH will stabilize between 8 and 9. If an abundance of sodium carbonate is present the pH may exceed 9 or 10. Some laboratory forms report carbonate (CO_3^{2-}) and bicarbonate (HCO_3^-) in addition to total alkalinity. These are typically derived from alkalinity measurements by multiplication by standard conversion factors.

Total hardness is the measure of the total concentration of primarily calcium and magnesium expressed in milligrams per liter (mg/L) of equivalent calcium carbonate ($CaCO_3$). Calcium and magnesium are usually present in association with carbonate as calcium carbonate or magnesium carbonate. Total hardness relates to total alkalinity and indicates the water's potential for stabilizing pH. Waters can be high in alkalinity and low in hardness if sodium and potassium are dominant. Table 7-1 lists water hardness classifications as ppm of $CaCO_3$.

TABLE 7-1 Water Hardness Levels

Water Hardness	As ppm $CaCO_3$
Soft	0 to 20
Moderately soft	20 to 60
Moderately hard	61 to 120
Hard	121 to 180
Very hard	>180

Fish do best when measurement of hardness or alkalinity measures between 20 and 300 mg/L. Below 20 mg/L can result in poor production. Cases where total hardness is considerably below the measure for total alkalinity is also not desirable. Liming increases total hardness and total alkalinity.

Lime Requirement. This test determines the amount of liming needed to neutralize acid because of leached-out cropland cations. Sediment (mud) is removed and analyzed to determine the amount of ground limestone needed to bring the pH of soils to an acceptable level and hold it there for two to five years. For aquaculture use, the test determines the amount of lime necessary to raise the pH of the bottom mud to 5.8 and raise the water hardness to acceptable levels. The lime requirement of the bottom mud must be satisfied before lasting effects can be expected in the water column. The lime requirement is reported in lbs per A and may amount to one to several tons of lime per acre. Liming efforts may be futile where sodium presence is great or muds are extremely acid.

Suspended Solids and Turbidity

Suspended solids (unfiltered residue) measure less than 2,000 mg/L in muddy pond waters. Many times this amount is needed to directly affect fingerlings and adult fishes. Muddiness can affect natural food production at 250 mg/L suspended solids by shutting out sunlight. Also, turbidity interferes with reproduction of some fish at less than 500 mg/L. Sometimes turbidity is used as a measure of suspended solids and is given in turbidity units. Turbidity units are derived by light transparency. Turbidity in fertile surface waters is largely due to organic material, particularly algae. An expected measurement for water is 2 units, while algae-rich waters measure 200 units.

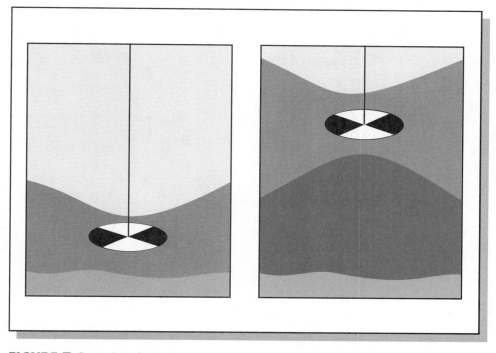

FIGURE 7-8 A Secchi disk can be used to determine turbidity.

Trace Elements

Trace elements include ionic constituents of water that dissolve to a small extent. Some elements are routinely included in reports of irrigation water and soil tests because of the impact on plants. Boron is most common in this category. More generalized water reports include certain metals that could have toxic effects, and some laboratories offer metal analyses as special order items.

Aquatic organisms of all types are sensitive to metal poisoning when the concentration of these reaches a certain level in the water column. Certain fish groups tend to be more sensitive than others to particular metals. Copper, for example, is more toxic to rainbow trout than to channel catfish. Exact levels of tolerable metals in solution that are considered safe for aquatic life are the subject of much discussion and disagreement. The toxic nature of metals is influenced by the hardness of the water. A metal may poison fish in very soft water at a rate that would have to be increased ten-fold to produce the same effect in hard water. Because the concentration of a metal required to produce toxicity may differ according to the overall water chemistry, a clear statement on the toxicity of the various metals is difficult to find.

Trace elements normally present in unpolluted surface waters at concentrations of less than one mg/L include:

Aluminum (Al)
Arsenic (As)
Barium (Ba)
Beryllium (Be)
Cadmium (Cd)
Chromium (Cr)
Cobalt (Co)
Copper (Cu)
Iron (Fe)
Lead (Pb)
Manganese (Mn)
Mercury (Hg)
Molybdenum (Mo)
Nickel (Ni)
Selenium (Se)
Silver (Ag)
Zinc (Zn)

Availability of trace elements at toxic levels is affected by water hardness, dissolved organics, and suspended clays. Toxic action is also influenced by the form, for example, free ion or bound in organic compound. While trace elements can be toxic, some are essential for health of aquatic life.

Except for aluminum, arsenic, barium, and iron, these elements should be considered potentially harmful when present in concentrations above 0.1 mg/L.

Dissolved Gases

Dissolved gases determine the basic suitability of water for fish survival. These gases include oxygen (O_2), carbon dioxide (CO_2), nitrogen (N_2), ammonia (NH_4^+ and NH_3), hydrogen sulfide (H_2S), chlorine (Cl_2), and methane (CH_4). Dissolved gases are usually not found on water analysis report forms because the manner by which samples are collected and shipped can cause gas measurements to be inaccurate. Ammonia is the most stable of the group and if a sample is processed within a day after collection it should measure fairly accurately. Other measures are best taken at the water site using appropriate meters or chemical test procedures.

Dissolved Oxygen. Aquatic life requires dissolved oxygen (DO). It varies greatly in natural surface water and is characteristically absent in groundwaters. Most aquatic animals need more than a 1 ppm concentration for survival. Depending on culture circumstances, aquatic animals need 4 to 5 ppm to avoid stress. Concentrations considered typical for surface water are influenced by temperature but usually exceed 7 to 8 mg/L (ppm). In ponds, dissolved oxygen fluctuates greatly due to photosynthetic oxygen production by algae during the day and the continuous consumption of oxygen due to respiration. Dissolved oxygen typically reaches a maximum during the late afternoon and a minimum around sunrise. Cloudy weather, rain, plankton die-offs, and heavy stocking and feeding rates result in low levels of dissolved oxygen, which can stress or kill fish. As Table 7-2 indicates, the gains and losses of the dissolved oxygen in ponds are very close.

Oxygen is only slightly soluble in water. Water may be frequently supersaturated with oxygen in ponds with algae blooms. For example, at sea level at a temperature of 77°F (25°C), pure water contains about 8 ppm of oxygen when 100 percent saturated, but during the afternoon hours, levels of 10 to 14 ppm in ponds with healthy algae blooms are not uncommon.

As water warms, is raised to higher altitudes, or becomes more saline, its oxygen holding capacity declines. Water saturated with oxygen at 59°F (15°C) contains about 9.8 ppm, while water at 86°F (30°C) is saturated at about 7.5 ppm.

Aquaculturalists measure dissolved oxygen with oxygen meters or chemical test kits, which give results in mg/L. Guidelines for oxygen management usually report that oxygen levels should be maintained above 4 mg/L (ppm) to avoid stress. Most warmwater fish experience significant oxygen stress at levels of 2 mg/L, and that levels of less than 1 mg/L (ppm) may result in fish kills. While these guidelines are accurate, fish actually respond to the percent satu-ration of oxygen rather than the oxygen content in water. A reading of 1 mg/L at 30°C (13.3 percent saturated) is a higher concentration than 1 mg/L at 15°C (10.2 percent saturated) and represents more available oxygen.

TABLE 7-2 Dissolved Oxygen Ledger for Ponds

Source	Gain (ppm)	Loss (ppm)
Photosynthesis by phytoplankton	6–20	—
Diffusion	1–5	—
Plankton respiration		5–15
Fish respiration		2–6
Diffusion		1–5
Respiration by other organisms		1–3
Total	7–25	9–29

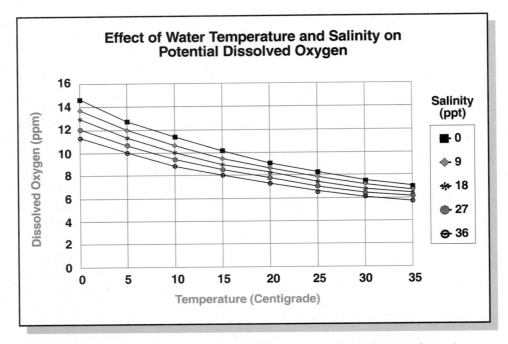

FIGURE 7-9 Increasing the water temperature or the salinity reduces its oxygen holding capacity.

If dissolved oxygen reaches low levels, fish will show signs including—

■ Not eating and acting sluggish

■ Gasping for air at the surface

■ Grouped near water inflow pipe

■ Slow growth

■ Outbreaks of disease and parasites.

Proper water management prevents the problems from depletion of dissolved oxygen. Management techniques include—

■ Monitoring dissolved oxygen at critical times

■ Avoiding overfeeding

■ Proper stocking level

■ Avoiding overfertilization

■ Controlled plant growth

■ Some form of aeration

■ Keeping water circulating.

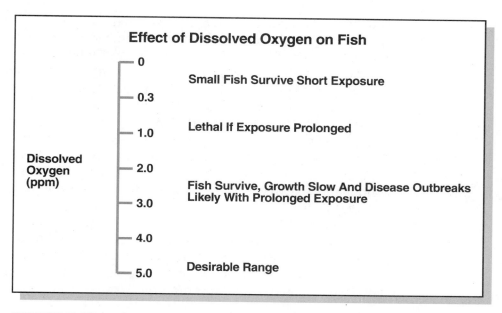

Effect of Dissolved Oxygen on Fish

Dissolved Oxygen (ppm)

- 0 — Small Fish Survive Short Exposure
- 0.3
- 1.0 — Lethal If Exposure Prolonged
- 2.0
- 3.0 — Fish Survive, Growth Slow And Disease Outbreaks Likely With Prolonged Exposure
- 4.0
- 5.0 — Desirable Range

FIGURE 7-10 Fish need dissolved oxygen at 5 ppm. Less stresses fish and can be lethal.

Carbon Dioxide. Carbon dioxide (CO_2), a minor component of the atmosphere, is highly soluble in water. Most carbon dioxide in pond water occurs as a result of respiration. Levels usually fluctuate inversely to dissolved oxygen, being low during the day and increasing at night, or whenever respiration

FIGURE 7-11 Pond aeration increases the dissolved oxygen.

occurs at a greater rate than photosynthesis. Carbon dioxide is present in surface water at less than 5 mg/L (5 ppm) concentrations but may exceed 60 mg/L (60 ppm) in many well waters and 10 mg/L where fish are maintained in large numbers. Some aquatic animals, including fish, can endure stress and survive at up to 60 mg/L. If oxygen is lowered into its stress-causing range, carbon dioxide limitation is reduced to 20 mg/L.

Carbon dioxide interferes with the ability of the aquatic animal to extract oxygen from water, contributing to stress of fish during periods of low oxygen. Aerating water to improve its oxygen content drives off excess carbon dioxide.

Adding quick lime ($Ca(OH)_2$) to water rapidly removes carbon dioxide without affecting oxygen content. This improves the ability of fish to use the available oxygen.

Carbon dioxide acts as an acid in water, lowering pH as it increases in concentration. Carbonate buffers in water neutralize carbon dioxide and stabilize pH fluctuations within the range tolerated by fish. Waters low in alkalinity and hardness may experience extremes of pH due to its poor buffering against changes in carbon dioxide concentrations.

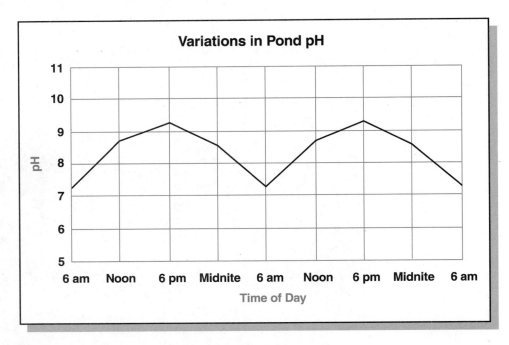

FIGURE 7-12 Carbon dioxide (CO_2) content of the water affects the pH of water in a pond during the course of a day. Respiration raises the CO_2 and lowers the pH, while photosynthesis removes CO_2 and raises the pH.

Hydrogen Sulfide. Hydrogen sulfide (H_2S), rotten-egg gas, is present in some well waters but is so easily oxidizable that exposure to oxygen readily converts it to harmless form. Its toxicity depends on temperature, pH, and dissolved oxygen. Any measurable amount after providing reasonable aeration could be considered to have potential to harm fish life.

Hydrogen sulfide occurs in ponds as a result of the anaerobic decomposition of organic matter by bacteria in mud. Hydrogen sulfide is toxic to fish and interferes with normal respiration. Toxicity is increased at higher temperatures and a pH less than 8 when the largest percentage of hydrogen sulfide is in the toxic unionized form. Vigorous aeration or splashing is usually sufficient to remove hydrogen sulfide from well water.

In ponds, hydrogen sulfide can be released from anaerobic mud when the bottom is disturbed by seining and harvest activities. Liming ponds raises mud pH and reduces the potential for the formation of hydrogen sulfide. Potassium permangenate at 2 to 6 mg/L (2 to 6 ppm) removes hydrogen sulfide from water and reverses the effects of its toxicity to fish.

Ammonia. Ammonia is present in slight amounts in some well and pond waters. As fishes become more intensively cultured or confined, ammonia can reach harmful levels. Any amount is considered undesirable, but stress and some death loss occurs at more than 2 mg/L (2 ppm), and at more than 7 mg/L (7 ppm) fish loss increases sharply.

Ammonia is a waste product of protein metabolism by aquatic animals (Figure 7-13). In water, ammonia occurs either in the ionized (NH_4^+) or unionized (NH_3) form, depending upon pH. Unionized ammonia is considerably more toxic to fish and occurs in greater proportion at high pH and warmer temperatures. For example, at 82.4°F (28°C) and pH 8, 6.55 percent of the total ammonia is present in unionized form. At pH 9, 41.23 percent of the ammonia is unionized. Unionized ammonia is stressful to warmwater fish at concentrations greater than 0.1 mg/L and lethal at concentrations approaching 0.5 mg/L. Concentrations of 0.0125 ppm cause reduced growth and gill damage in trout.

Test kits for determining ammonia in water measure total ammonia. To determine if a large percentage of the ammonia is in unionized form, pH is also measured. A pH above 8 in the presence of ammonia concentrations above 0.5 mg/L is cause for concern.

Algae use ammonia as a nitrogen source for making proteins. Concentrations usually remain low in ponds with phytoplankton blooms. The greatest concentration of ammonia often occurs after plankton die-offs, at which time pH is low due to high levels of carbon dioxide, and the majority of ammonia is present in the relatively nontoxic ionized form.

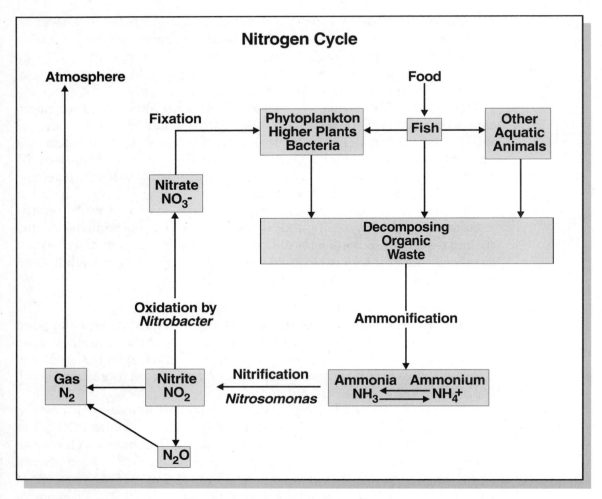

FIGURE 7-13 Nitrogen cycle. Feed and living tissue all contain nitrogen. Eventually, the nitrogen is released and converted to several forms.

Management is the key to preventing problems from ammonia. The management techniques are similar to those for preventing oxygen depletion—

- Avoid overcrowding
- Avoid overfeeding
- Add freshwater
- Control plant growth
- Monitor the pH
- Remove fecal material.

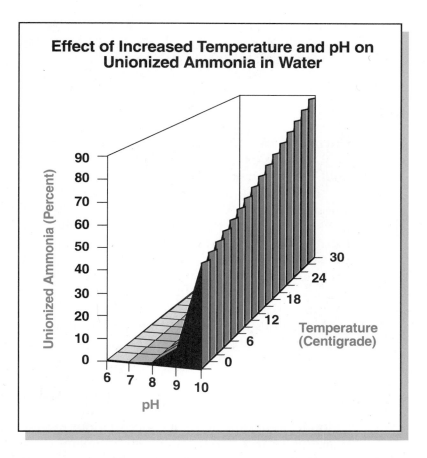

Effect of Increased Temperature and pH on Unionized Ammonia in Water

FIGURE 7-14 As temperature and pH increase, so does unionized ammonia.

Nitrite. Nitrite (NO_2-) is one of the basic products of organic matter decomposition. It acts as an intermediate stage in the conversion of ammonia to nitrate. These reactions occur in soils, mud, and water. Nitrite changes quickly to nitrate if oxygen is present. In culture ponds that are rich from feeding, a temporary accumulation of this chemical in harmful amounts sometimes occurs. Nitrite measurements of more than 1 mg/L (1 ppm) should be suspect in causing fish deaths.

The binding of nitrite with the hemoglobin molecule gives blood a chocolate brown color. Fish farmers call this condition brown blood disease. In medical terms it is known as methemoglobinemia. The toxicity of nitrite to fish is lessened by the presence of chlorides in water. The addition of salts, either calcium chloride or sodium chloride at rates of 20 mg/L for each 1 mg/L of nitrite, is a standard treatment for preventing nitrite poisoning in freshwater ponds. Most warmwater fish can tolerate at least 0.4 mg/L of nitrite in freshwater without treatment if oxygen levels remain above 4 mg/L.

Nitrite should be monitored frequently if a problem is suspected since its concentration may increase rapidly in pond water, especially during spring and fall, or when algae blooms suddenly die.

Chlorine and Chloride. Chlorine is usually present at approximately 1 mg/L in municipal water supplies as a result of chlorination. Fish succumb quite easily at these levels.

Some aquaculturalists may mistake the difference between chlorine and chloride. Chlorine (Cl_2), in gaseous form or as hypochlorites, is widely used for disinfection of water supplies and in aquaculture for sterilization of equipment, tanks, and standing water in drained ponds. Chlorine acts as a powerful oxidizing agent and it is toxic to fish at concentrations of less than 0.05 mg/L. Residual chlorine in municipal water supplies is normally between 0.5 and 2.0 mg/L. Water used for fish culture should not contain any residual chlorine to be considered safe. Chlorine can be removed from water by extended periods of aeration before use, or more rapidly by the addition of sodium thiosulfate at a rate of 7.0 mg/L for each 1 mg/L of chlorine. Sodium thiosulfate is not considered toxic to fish at concentrations required to remove chlorine from water.

Chloride is a by-product of chlorine dissociation in water but is also widely associated with numerous other compounds that are highly water soluble. Chloride occurs in a range of concentrations in water and is often used as an indicator to characterize aquatic environments as saline or fresh. Fish, too, are classified as freshwater or marine as determined by their physiological adaptation to salinity. Chloride is important to fish in osmoregulation and other physiological process and is regarded as non-toxic within the tolerance range of each species. Tests to determine the chlorine or chloride content of water require completely different chemical procedures. The distinction between the two compounds and their significance in water is obvious, but occasionally confused.

Nitrogen and Methane. Nitrogen (N_2) and methane (CH_4) are considered to play a critical role only at abnormally high levels. For example, nitrogen may be driven to high levels in waters that are plunged deeply at dam outfalls or through pump cavitation. As total gas concentrations exceed 115 percent of normal, fish are affected by bubble formation in the blood, called gas bubble disease.

Total gases is a measure sometimes used in aquatic analyses but is not often seen in laboratory reports.

Organic Material and its Breakdown Products

In water, organic material consists of living organisms and various dissolved organic chemicals of their excretion and dead matter and with its associated decomposition products. Organic chemicals or tissues consist mostly of the elements carbon, oxygen, hydrogen, nitrogen, and sulfur. Organic material decomposes constantly into its ultimate breakdown products of carbon dioxide, sulfide, ammonia, nitrate, hydrogen ion, and water.

Carbon is always a component of organic chemicals. In the analysis of organics, the characteristics of decomposition potential and carbon presence are used for measurement. Measurement of organics can give the pond owner an idea of overall enrichment and because decomposition requires oxygen, a general idea of what will be demanded of a pond's oxygen content.

Some organic materials such as pesticides will act as toxins to fish at very low levels. For low level detection of these specific organic constituents sophisticated instrumentation techniques such as chromatography are used.

Total Organic Carbon. This test is commonly reported on laboratory forms and is given in mg/L. Natural surface water would be expected to contain 10 mg/L, and water that receives regular feeding could build up to over 30 mg/L. Water with decomposing plant life could also be expected to have high organic carbon. The measure of total organic carbon includes dissolved organic carbon and suspended material, commonly called particulate organic carbon.

COD. Chemical oxygen demand (COD) is a speedy and reliable estimate of organic load that is reported in mg/L. A normal measure would read less than 10. A measure of 60 would be considered rich.

BOD. Biochemical oxygen demand (BOD) is a standard test for organic material. It is reported in mg/L per hour or per total test time. On a per-hour basis, 0.5 would be considered rich. BOD is usually measured over a five-day test period.

Chlorophyll. The measure of this photosynthetic pigment gives an estimation of plant life suspended in the water column. Unfertile ponds range up to 20 micrograms per liter (mcg/L), and fertile ponds with rich phytoplankton blooms range from 20 to 150 mcg/L.

Oil and Grease. Most oils and grease float at the surface. Aquatic animals suffer at rates above 0.1 mg/L where the oily products are not those of the natural release of plants or other living agents of the pond. Roughly, sudden die-off of fishes would be expected to occur when such concentrations exceed 100 mg/L. An oil slick is usually apparent in cases of pollution, but a sudden die off of an algae population will cause a slick from oil material of the plants.

Pesticides. Pesticides are not a natural part of the environment. Their presence is determined by past activities of humans. A knowledge of the historical application of pesticides in the watershed gives a clue to which chemicals should be tested for. Test laboratories offer scans of pesticides known to retain their integrity months and years after application. New pesticides disappear from detection soon after application.

Insecticides are more toxic to crustaceans than fish because of their closer kinship to insects. Pesticide groups include—

- Organochlorides like DDT and toxaphene at 0.01 to 0.5 mg/L
- Organophosphates like methyl parathion and malathion at 1 to 10 mg/L
- Carbamates like carbaryl and methomyl at 0.1 to 10 mg/L
- Pyrethroids like permethrin at 0.1 to 10 mg/L
- Fungicides like benomyl and captan at 0.05 to 5 mg/L
- Herbicides at 0.05 to 500 mg/L (aquatic and terrestrial toxicity varies greatly).

OTHER FACTORS

Several factors relate to the water requirements for aquaculture. These include the soil type, the effect of rainfall, the rate of evaporation, the amount of light, the odor of the water, and the color of the water.

Soil

The soil must hold water, so clay-type soils are desirable. Soil core samples at various places around the site determine if adequate clay is present to prevent excess seepage. The local Soil Conservation Service Office can provide assistance. Table 7-3 shows seepage in various soil types. Ponds can be built in soils that have high percolation rates if they are lined with a layer of packed clay or plastic.

TABLE 7-3 Soil Types and Seepage Losses

Soil Type	Seepage Losses (inches/day)
Sand	1–10
Sandy loam	0.52–3
Loam	0.32–0.8
Clayey loam	0.1–0.6
Loamy clay	0.01–0.2
Clay	0.05–0.4

Once water is added to a pond the soil in the pond bottom becomes an important part of the system. Nutrient and oxygen fluxes between the pond bottom (mud) play an important role in determining water quality. The mud receives much of the organic matter entering the pond. Organisms produced in and on pond bottoms become food for crustaceans and fish. Some fish lay eggs in the pond bottom.

Rainfall

Rainfall varies from place to place, year to year, and season to season. Rainfall influences the amount of water in a pond. Also, precipitation falling on the land may become runoff. As runoff it dissolves various substances that may affect the water used for aquaculture. Rain gauges are inexpensive but provide one more piece of helpful information for making management decisions. Rainfall affects runoff pollution from land, the temperature, pH, and the total dissolved solids of surface water.

Evaporation

Evaporation tends to follow the amount of sunlight (solar radiation) and the air temperature. As the solar radiation and air temperature increase, evaporation increases. Humidity and wind speed also affect evaporation. Low humidity and high wind speed increase evaporation. Ponds need compensation for evaporative losses.

Light

The amount of light striking a pond influences the amount of photosynthesis. As the light increases, photosynthesis increases. Clouds, roughness of the water surface, and turbidity decrease the amount of light available for photosynthesis.

Odor

Odor of water affects its recreational value and the taste of fish and other aquatic foods. Odor comes from sources like municipal and industrial waste discharges, decomposing vegetable matter, and microbial activity. The human nose is the best instrument for detecting odors in water. Table 7-4 can be used to classify and record odor from a water sample.

Color

Water color results from dissolved substances and suspended material. It can provide useful information about the water's source and content. Transparent

TABLE 7-4 Odor Classifications

Description	Nature of Odor	Examples of Odor
Aromatic (spicy)	Camphor, cloves, lavender, lemon	Same
Balsamic (flowery)	Geranium, violet, vanilla	Same
Chemical	Industrial wastes or treatments	
	Chlorinous	Chlorine
	Hydrocarbon	Oil refinery wastes
	Medicinal	Phenol and iodine
	Sulfur	Hydrogen sulfide (rotten egg)
Disagreeable (pronounced, unpleasant)	Fishy	Uroglenopsis, Dinobryon (dead algae)
	Pigpen	Anabaena algae (visit a pig farm to sample this distinctive odor)
	Septic	Stale sewage
Earthy	Damp earth	
	Peaty	Peat
Grassy	Crushed grass	Same
Musty	Decomposing straw	
	Moldy	Damp cellar
Vegetable	Root vegetables	Same

water with low amounts of dissolved or suspended substances indicates low productivity and appears blue. Dissolved organic materials add a yellow or brown color. Some algae and dinoflagellates produce reddish or deep yellow colors in water. Water rich in phytoplankton appears green. Soil runoff produces various red, brown, yellow, and gray colors.

Methods like the Forel-Ule Color Scale and the Borger Color System attempt to standardize the descriptions of the apparent color of water. Color determinations should be made against a white background. The white quadrants of a Secchi disk can be used.

OBTAINING WATER

The prime consideration for any aquaculture facility is a plentiful supply of high-quality water. This is true whether the organisms to be grown are fish or shellfish. Also, it is an economic necessity that the water be relatively inexpensive. A secondary consideration is whether the soil in the area will hold water. Suitable alternatives cost proportionately more.

For crawfish and shrimp the depth of water required is usually 1.5 to 3 ft (0.5 to 0.9 m) while for finfish the depths recommended are from 3.5 to 5 ft (1.1 to 1.5 m). Each A-ft of water contains about 326,000 gal (1,234,044 L). A 1-A pond 5 ft deep requires over 1.6 million gallons for a single filling. Some flushing of the pond is usually desirable and practiced, which means an additional 0.4 million gal per A per year. Planning any type of aquaculture facility requires estimating the water needs (Figure 7-15).

Calculating Water Needs

If water supply is inadequate, then a facility may not meet its production goals. A suitable water supply in both volume and quality is essential for the intensive, commercial production of fish.

The following examples will illustrate the steps and calculations required to estimate the water requirements for various production units.

EXAMPLE 1. A producer wants to construct four ponds in the same area and service all with one water well. The ponds vary in size and the owner wants to be able to fill any pond within seven days. The sizes of the ponds are 6 A, 4 A, 5.5 A, and 3.5 A. The average water depth in each pond is 5 ft. What flow rate in gallons per minute (gpm) is required from the service well to fill any pond in at least seven days?

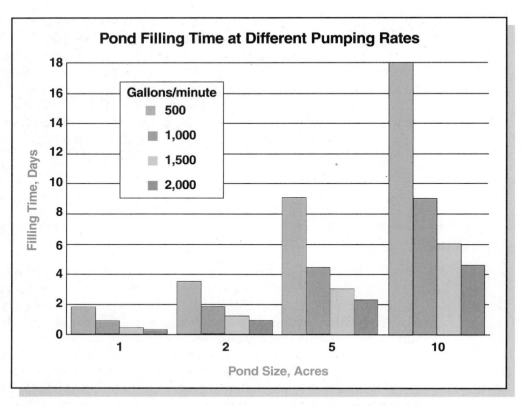

FIGURE 7-15 The effect of pumping rate (gallons per minute) becomes very evident in larger ponds.

STEP 1. Determine the volume of water in the largest pond. If the largest pond can be filled in seven days, then any smaller ponds will fill in seven days or less.

Volume in A-ft = 6 A × 5 ft average depth = 30 A-ft

STEP 2. Convert 30 A-ft to gallons:

30 A-ft × 325,850 gal/A-ft = 9,775,500 gal

STEP 3. Determine the minimum flow rate needed to fill the pond in seven days. In this case, we are not including any adjustment for seepage, evaporation, or rainfall.

$$\frac{9{,}775{,}500 \text{ gal}}{7 \text{ days} \times 24 \text{ hours/day} \times 60 \text{ minutes/hour}} = 970 \text{ gpm}$$

To be on the conservative side, a 1,000-gpm well should be adequate.

EXAMPLE 2. With the 1,000-gpm well from Example 1, what would be the filling time in days for the smallest pond of 3.5 A and 5 ft average water depth?

STEP 1. Determine the volume of water in the pond:
Volume in A-ft = 3.5 A × 5 ft average depth = 17.5 A-ft

STEP 2. Convert 17.5 A-ft to gallons:
17.5 A-ft × 325,850 gal/A-ft = 5,702,375 gal

STEP 3. With a flow rate of 1,000 gpm, the filling time in minutes is:

$$\frac{5,702,375 \text{ gal}}{1,000 \text{ gal/min}} = 5,702 \text{ minutes}$$

STEP 4. Convert 5,702 minutes to days:
5,702 minutes ÷ 60 minutes/hr = 95 hours
95 hours ÷ 24 hours/day = 3.96 or 4 days to fill pond

EXAMPLE 3. A fish hatchery facility is being planned that will include six fish holding tanks each 4 ft wide and 30 ft long. The average water depth in each is 3 ft. The water supply for these tanks has to provide at least two complete water exchanges per hour in all tanks at the same time. The facility will also have 20 troughs each 2 ft wide and 10 ft long with an average depth of 1 ft. A water supply of 5 gpm is required for each trough and all may need water at the same time. What is the minimum water requirement in gallons per minute (gpm) for this facility?

STEP 1. Determine the water requirement for the six holding tanks.

Determine the volume of water that the six holding tanks contain.
Volume per tank = 4 ft (width) × 30 ft (length) × 3 ft (depth)
= 360 ft³/tank

Convert 360 ft³ to gallons:
360 ft³ × 7.48 gal/ft³ = 2,693 gal/tank
2,693 gal/tank × 6 tanks = 16,158 gal capacity for tanks

STEP 2. The desired filling time for all tanks is 1 hour or 60 minutes. The gpm required to fill tanks is calculated as follows:

$$\frac{16,158 \text{ gal capacity of all tanks}}{60 \text{ minutes}} = \frac{270 \text{ gpm required to fill all tanks}}{\text{in 60 minutes}}$$

STEP 3. Determine how much water is required to supply two complete water exchanges per hour to all tanks. Two exchanges per hour means that the total volume of water in all tanks needs to be completely replaced every 30 minutes.

$$\frac{16{,}158 \text{ gal in tanks}}{30 \text{ minutes}} = \frac{540 \text{ gpm for two water exchanges per hour}}{\text{in all tanks}}$$

The water requirement for the six holding tanks is 540 gpm. It is not 540 gpm plus 270 gpm because the maximum requirement of 540 gpm for continuous flow use can be reduced to 270 gpm to fill all tanks within 60 minutes. If fewer tanks are filled at one time or the flow rate is increased, then the filling time will decrease.

STEP 4.   Now, determine the water requirements for the 20 troughs. The water requirement for these troughs needs to be added to the water requirement for the holding tanks because all facilities may be in use at the same time. First, determine the total volume of water in the troughs:

Volume per trough = 2 ft (width) $\times$ 10 ft (length) $\times$ 1 ft (depth)
$= 20 \text{ ft}^3/\text{trough}$

Convert 20 ft^3 to gallons:
$20 \text{ ft}^3 \times 7.48 \text{ gal/ft}^3 \times 150 \text{ gal/trough}$

Total water volume in troughs:
$150 \text{ gal/trough} \times 20 \text{ troughs} = 3{,}000 \text{ gal}$

The desired filling time for all troughs is a maximum of 60 minutes. To determine the water requirement, do the following calculations:

$$\frac{3{,}000 \text{ gal capacity all troughs}}{60 \text{ minutes}} = \frac{50 \text{ gpm required to fill all troughs}}{\text{in 1 hour}}$$

STEP 5.   Determine the flow rate required for all troughs once they are full of water. The required continuous flow rate for each tank is 5 gpm.

$5 \text{ gpm/trough} \times 20 \text{ troughs} = 100 \text{ gpm}$

For the troughs, the water requirement is 100 gpm because only 50 gpm is needed to fill the tanks within the desired time of 1 hour.

STEP 6.   To determine the total water requirement for the facility, the water requirements for tanks and troughs are added. If more water is needed for other purposes that would occur when all other facilities are operating, then these water requirements are also calculated and added to the total requirement. The total water requirement in gpm for facility is 540 gpm for six holding tanks plus 100 gpm for 20 troughs equals 640 gpm.

Sources

Water for aquaculture comes as surface water or ground water. Either source may require some pumping.

Surface Water. If water from a stream or reservoir is used to fill the pond, the water should be screened or filtered. A series of screened boxes or a gravel filter bed are useful at the intake source. Unfiltered water at the pond inlet should run through a fine mesh screen box or saran filter sock. The primary purpose of the screens and filters is to prevent undesirable wild fish, fish eggs, and predaceous invertebrates from entering the pond.

Ground Water. Water pumped from wells is often low in oxygen and may contain high levels of undesirable gases such as hydrogen sulfide and carbon dioxide. It is beneficial to aerate this water prior to its entering a pond or tank, especially if the in-flowing water rapidly replaces the pond or tank water. A splash board or other device that splashes, sprays, or agitates the water as it enters the pond is normally used.

Pumping. The final decision maker in any aquaculture production facility is operating cost. Pumping costs contribute significantly to operating costs. The costs to pump an A-ft of water includes, using a diesel-powered pump, diesel fuel, depreciation, interest, and maintenance costs of the well and pumping equipment. Cost of wells is extremely variable and impossible to relate to pumping lift or the amount of fuel required to pump a given quantity of water. Well costs must be estimated on an individual basis.

FIGURE 7-16 Pumping water contributes significantly to the cost of an operation. (Photo courtesy Chuck Weirich, Delta Research and Extension Center, Stoneville, MS)

MANAGING WATER

Managing water requires maintaining the quantity, testing the quality, and maintaining the quality. Other factors include aquatic life, evaporation, and seepage.

Quality Measurements

A practical understanding of water quality is essential to the aquaculturalist to allow an evaluation of environmental conditions and implementation of effective management strategies. Often, aquaculturalists do not have extensive training in water chemistry and as a result may misinterpret or misapply information about water quality and its management. While the chemistry of water is a complex subject, most aspects of general importance to fish farmers can be simplified to allow for easier understanding and practical approaches to management.

Pond water is tested for many reasons. The selection of water tests would be different for each. Reasons for testing include—

■ A catastrophe with sudden die-off of fish and no clue to a cause.

■ A catastrophe with sudden die-off of fish and a good idea of cause.

■ Little or no die-off but with knowledge or suspicion of some accident or malicious action in the vicinity of the pond.

■ Evaluation of suitability of pond water or water source.

If a probable cause is known, the test requirements can be easily narrowed down. If a pesticide is suspected it is very helpful for the pond owner to do an initial investigation to determine what type of pesticide was used. Otherwise the testing may be no more than searching for a "needle in a haystack" because of the great variety of manufactured pesticides.

When the cause of the catastrophe is a mystery, or where water is merely being evaluated, a general spectrum of tests that evaluate parameters known to commonly affect aquatic life can be conducted. Table 7-5 lists common tests, sample selection, and methods.

Test Methods

Whether the water is tested in a laboratory or in the field by an aquaculturalist, general test methods include titrimetric, colorimetric, and electronic meters. For many water quality tests, companies sell test kits complete with all of the chemicals and standards.

TABLE 7-5 Summary of Water Management Methods

Parameter	Procedure	Method
Air Temperature—see Temperature, Air		
Alkalinity	Collect samples in plastic or glass bottles. Fill completely and cap tightly. Avoid excessive agitation and prolonged exposure to air. Keep samples cool in refrigeration unit or ice chest, and analyze within 24 hours. Warm to room temperature before analyzing.	Test kit; titrimetric
Ammonia	Collect samples in clean glass or plastic bottles. Samples not analyzed immediately may be preserved by reducing the pH to 2 or less with sulfuric acid. Refrigerate samples and analyze within 24 hours. Warm to room temperature and neutralize before analysis.	Test kit; electronic colorimeter
Biochemical Oxygen Demand (BOD)—see Oxygen, Biochemical Demand (BOD)		
Carbon Dioxide	Collect sample in clean glass or plastic bottles. Fill completely and cap tightly. Avoid excessive agitation or prolonged exposure to air. Analyze as soon as possible but sample can be stored at least 24 hours by cooling to a temperature lower than the source.	Test kits; titrimetric
Chemical Oxygen Demand (COD)—see Oxygen, Chemical Demand		
Chloride—see Salinity		
Depth, Pond	Install staff gauge in each pond and read at same time each day, before restoring to specified depth.	Read and record
Dissolved Oxygen—see Oxygen, Dissolved		
Evaporation and Inflow	Surface inflow/outflow and evaporation should be determined.	Read and record
Hardness	Collect samples in plastic or glass bottles that have been washed with detergent, rinsed with tap water and rinsed with 1:1 nitric acid solution and deionized water. Store only if prompt analysis is not possible.	Test kits; titrimetric and colorimetric, electronic
Nitrate	Collect samples in clean plastic or glass bottles. Store at 39°F (4°C) if sample analyzed in 24 to 48 hours. Warm to room temperature before running test.	Test kits; colorimetric, visual and electronic; Electronic meter

(continued)

TABLE 7-5 Summary of Water Management Methods *(continued)*

Parameter	Procedure	Method
Nitrogen, Total Kjeldahl	Collect samples in clean glass or plastic bottles. Samples should be refrigerated and analyzed within 24 hours. Samples preserved by adjusting the pH to 2 or less with sulfuric acid can be stored in a refrigerator for up to 28 days.	Digestion required; colorimetric, electronic
Oxygen, Biochemical Demand (BOD)	Determined by measuring the dissolved oxygen in a freshly collected sample and comparing it to the dissolved oxygen level in a sample collected at the same time but incubated at 20°C for five days. The difference between the two oxygen levels is the BOD.	Test kits
Oxygen, Chemical Demand (COD)	Collect samples in glass bottles. Use plastic bottles only if they are known to be free of organic contamination. Test as soon as possible. To store sample, bring pH to 2 or less with sulfuric acid.	Test kits; titrimetric or colorimetric
Oxygen, Dissolved	Measure directly in water source or collect samples. Sampling and sample handling important for meaningful results. Dissolved oxygen changes due to many variables.	Test kit; Winkler titration Electronic meter
pH, Water	Take readings at 10 inches (25 cm) below the water surface. Collect samples in clean plastic or glass bottles. Fill bottles completely and cap tightly. If a probe is used, calibrate using a precision thermometer. Calibrate meter with standard buffers at pH 7 and pH 10.	Test kit; colorimetric; visual method

Calibrated electronic meter |
| Phosphorus, Total | Collect one sample by pooling three samples. For most reliable results analyze immediately. Samples could be refrigerated and analyzed within 24 hours. | Test kit; colorimetric, electronic |

Pond Soil Characteristics—see Soil, Characteristics

Pond Temperature—see Temperature, Water

Rainfall	Install three rain gauges at site. Read and empty at 24-hour intervals, or more frequently to prevent gauge overflow. Report average of three readings.	Read rain gauge

(continued)

TABLE 7-5 Summary of Water Management Methods *(concluded)*

Parameter	Procedure	Method
Salinity	Collect sample pooled from three levels if collected from a pond. Use clean glass or plastic bottles. Samples can be stored for up to 28 days in sealed containers.	Test kit; titration Test kit; electronic colorimeter Meter; conductivity Electronic hydrometer refractometer
Secchi Disk Visibility—see Visibility, Secchi Disk		
Seepage	Determine seepage from a 24-hour water balance, preferably when there is no rainfall, inflow, or outflow.	Read and record
Soil, Characteristics	Determine soil type and particle size.	Soil triangle and International system of particle sizes
Solar Radiation	Install Solar Monitor and Quantum Sensor and read the cumulative radiation each day and at end of each time interval.	
Temperature, Air	Install three maximum-minimum thermometers in the shade near ponds; read at 24-hour intervals and report average maximum and average.	Thermometers
Temperature, Water	Take readings at 10 inches (25 cm) below the water surface. Ideally, in a pond take readings also at mid-water and 10 inches (25 cm) above the bottom. If probe is used, calibrate using a precision thermometer.	Certified thermometer Electronic meter
Total Dissolved Solids (TDS)	Collect sample, measure volume, dry, and weigh dried sample. A convenient alternative is to test the conductivity that can be used to estimate the TDS.	Gravimetric Electronic meter
Total Kjeldahl Nitrogen—see Nitrogen, Total Kjeldahl		
Total Phosphorus—see Phosphorus, Total		
Wind Speed	Install totalizing anemometer, read at 24-hour intervals.	Anemometer
Visibility, Secchi Disk	At two locations in each pond, calculate Secchi Disk Visibility.	Secchi disk

Titrimetric. Titrimetric analyses use a solution of known strength—the titrant—which is added to a known or specific volume of sample in the presence of an indicator. The indicator produces a color change, indicating the titration is complete. To calculate the results, the amount of titrant used is measured. A microburette or precision pipet adds the titrant.

Colorimetric. Beer's law states that the higher the concentration of a substance the darker the color produced in a test reaction. This law provides the basis for determining the concentration of many substances in water samples. Known chemical reactions produce typical colors. The concentration that these colors represent is determined visually by comparing the color obtained from sample to a set of standards. Since the human eye interpretation can be quite subjective, electronic colorimeters provide a more accurate indication of the color intensity. Colorimeters consist of a light source passing through a sample that is measured on a photo detector providing an analog or digital readout. Electronic colorimetric readings also compare the sample value to readings from a set of standard (known) readings.

Electronic Meters. Modern electronics provides the aquaculturalist with a variety of electronic meters designed to measure specific water quality factors including pH, total dissolved solids, conductivity, dissolved oxygen, temperature, and turbidity. Like the chemical methods, standard solutions are important. They are used to calibrate the electronic meter.

Using test kits or electronic meters, aquaculturalists regularly check the oxygen, pH, carbon dioxide, and ammonia of water in production.

Flesh Analysis. Fish flesh is often taken in large scale fish kills to analyze for possible toxicants. Pesticides and heavy metals are the most common items investigated. Some metals will normally be found in much greater concentration in the flesh than the water in which that fish swims. Some pesticides characteristically also accumulate in quantities much greater than are present in the water.

Sediments. Sediments or muds from pond bottoms are usually checked for heavy metals and pesticides. Measurements of heavy metals in sediments are typically much higher than those measured in the water column. Usual measures of lake sediments in mg/L include—

Cadmium (Cd), 0.1 to 1.5
Cobalt (Co), 4 to 40
Chromium (Cr), 50 to 250
Copper (Cu), 20 to 90
Lead (Pb), 10 to 100
Lithium (Li), 15 to 200

Manganese (Mn), 100 to 1,800
Mercury (Hg), 0.1 to 1.5
Nickel (Ni), 30 to 250
Zinc (Zn), 50 to 250
Iron (Fe), 11,000 to 70,000.

Sediment pesticides are often checked when establishing a particular site as suitable for fish culture. The very persistent pesticides are the object of such checks and organochlorines are usually tested as "scans." Sediments are also checked for pesticides when they are suspect in a catastrophe situation. The testing laboratory should be advised as to which type of pesticide has been used. Otherwise many useless tests may be conducted that result in a very high expense to the pond owner.

Other Management Factors

Other water management factors include pond fertilization, insect control, vegetation control, evaporation control, and seepage.

Pond Fertilization. Ponds stocked with fish eggs, fish fry, or fingerlings benefit from fertilization. The nutrients provided stimulate the productivity of the natural food chain, resulting in abundant zooplankton, the preferred natural food of most young fish. A mix of organic and inorganic fertilizers are used most often. Quantities of organic fertilizers used are dependent on quality and nutrient content. Cottonseed meal or chicken manure at 300 to 500 lbs per A (340 to 568 kg per ha) will help promote development of the zooplankton bloom. Chapter 4, Aquatic Management Practices, provides examples of using animal manure to fertilize ponds.

Phytoplankton growth is stimulated by adding 8 to 10 lbs per A (9 to 11 kg per ha) of phosphorus in a liquid or granular fertilizer mix. The most commonly used inorganic fertilizers are ammonium phosphate or liquid polyphosphates. When using a liquid fertilizer, application rates are usually about 1 to 1.5 gal per A (9.5 to 14 L per ha).

Natural fertility of the water determines the need for fertilization in any case. Older ponds with nutrient rich sediments may respond without fertilization. Fertilization should take place when or shortly after the pond is filled. At the time the fish are fed a prepared diet, the need for fertilization is eliminated.

Aquatic Insect Control. The larvae and adults of many insects prey on small fish, particularly their eggs and fry. Newly filled ponds attract insects as sites for egg laying. As a result, ponds newly prepared for stocking young fish also grow a crop of predacious insects. These insects reduce survival of fish to larger sizes. Insects and their larvae can be eliminated by applying chemicals to the water that are specifically approved for this use. State Departments of Agriculture list legal chemicals.

When selecting a chemical for use in fish ponds, it must be labeled for use in ponds, and labeled specifically for the type of fish that is being grown, such as foodfish, baitfish, or goldfish. Chemicals that are toxic to insects may also be toxic to some fish. Treatments are generally more effective at pH below 9 and water temperatures between 65° and 85°F (18.3° and 29.4°C). The appropriate chemical is mixed with water and sprayed or otherwise well dispersed around the pond edge. The product label contains a great deal of information about the product and should be read thoroughly and carefully before each use.

A major drawback of using chemicals is that many are also toxic to desirable zooplankton, an important food for young fish. To benefit from insect control and zooplankton production, timing of application is critical. Ponds should be treated for insect control five to seven days before stocking fish fry. If eggs are stocked, application should occur five to seven days before eggs are expected to hatch. This treatment will control insects but allow zooplankton to develop because their recovery time is shorter. Following initial control, applications should not be repeated until approximately three weeks following the initial treatment or until fish have depleted the zooplankton supply and are feeding on prepared diets. In ponds used for producing small fish or fingerlings, regular control of insects at two- to three-week intervals will enhance survival if ponds receive supplementary feeding to compensate for depletion of zooplankton populations. Fish longer than 1 in (2.5 cm) are much less susceptible to predation by insects.

In food fish ponds where these chemicals cannot be used, a mixture of 4 gal (15 L) of diesel fuel and 1 qt (1 L) of motor oil per surface A can be applied to kill air-breathing predaceous insects. This treatment clogs the breathing tubes of the insects when they rise to the surface for air. To minimize damage to zooplankton populations, the diesel fuel should be aerated for 24 hours prior to application. It is approved for use on food fish.

Aquatic Vegetation Control. Proper preparation, filling, and fertilizing of ponds helps prevent aquatic weed problems. A pre-filling application of herbicide can prevent weed and algae growth. It also reduces phytoplankton bloom development. A phytoplankton bloom established in a pond out-competes aquatic weeds. If weeds become established, they prevent development of a satisfactory bloom. For small weed infestations, manual removal is best. Once weeds are removed, development of the phytoplankton bloom will prevent their reestablishment. If extreme weed problems develop, aquatic herbicides may be required and the advice of a weed specialist should be considered as to the type and amount of herbicide to use.

In aquaculture, the grass carp is often used for biological weed control at densities dependent upon the extent of weed infestation. For rapid control, grass carp of about 8 to 12 in (20 to 30 cm) in length should be stocked at about 100 per A (250 per ha). Larger carp can be stocked at lower densities. As the grass carp control the weeds, they should be removed since they compete with other fish for feed added to the pond. At harvest time, due to their large size, they can physically damage small fish if confined in the same net. A large-mesh seine to remove the grass carp solves this problem. Grass carp can be used at densities as low as 1 to 2 per A (3 to 5 per ha) for maintenance after weed control has been achieved. In states where grass carp are legal, triploid (sterile) grass carp may be required.

Evaporation and Rainfall. In some areas, evaporation from a pond surface exceeds the amount of rainfall that enters the pond directly. And as most aquaculture ponds are not constructed to receive water from an outside drainage area, additional water is required for maintenance of the desired depth of water in the aquaculture ponds.

Seepage. Ponds should have sufficiently high clay content to hold water well. Proper pond construction in addition to a careful selection of the pond site is imperative.

CALCULATING TREATMENTS

Most aquaculture systems, including ponds, tanks, and raceways, eventually require some type of chemical treatment to combat a disease or aquatic plant problem or improve water quality conditions. Some treatments, like fertilizers or lime, are based on the surface area of water, but most treatments include the total volume of water in the production or holding unit.

Aquaculturalists should know how to calculate treatment rates, determine the amount of chemical or material needed, and apply the treatment. Before any treatment is applied, the water, fish, chemical, condition of the problem, and method of treatment should be known and understood. Producers can experience high economic losses when treatment rates are not properly calculated.

Before any calculation is made, the units of measurements should be determined. The unit of measurement selected should be familiar and convenient for the specific situation. For example, the large volume of water in ponds is usually expressed as A-ft, while the volume of a small tank may be expressed in gallons or ft^3. Another decision is whether to use the English or metric system of measurement. A working knowledge of both systems is important because reports or publications may use either one. Metric units are easiest to work with when small volumes or weights are involved.

To make calculations easy, refer to Tables A-1 and A-2 in the Appendix for conversion factors and parts per million equivalents.

Basic Formula

Most treatments can be calculated using this basic formula:

$$\text{Amount of chemical needed} = V \times C.F. \times \text{ppm desired} \times \frac{100}{\%AI}$$

Where: V   = The volume of water in the unit to be treated.

C.F.   = A conversion factor that represents the weight of chemical that must be used to equal one part per million (1 ppm) in one unit of the volume (V) of water to be treated. The unit of measurement for the results is the same as the unit used for the C.F. (pounds, grams, or other unit). Table A-2 in the Appendix contains a list of these conversion factors for various units.

ppm = The desired concentration of the chemical in the volume of water expressed in parts per million.

$\frac{100}{\%AI}$ = 100 divided by the percent active ingredient (AI) contained in the chemical to be used.

Most chemicals are 100 percent AI unless otherwise specified, so this value is usually 1. The percent AI is found on the label of most fishery chemicals. Most treatments can be calculated using this basic formula:

To understand how the basic formula is used to calculate treatments, review the following examples that represent a variety of practical situations.

EXAMPLE 1.   How much copper sulfate is needed to treat a 10-A pond with an average water depth of 4 ft with a 1.3 ppm treatment?

Select the unit of measurement and then determine the volume of water in the pond. Volume (V) of water in the pond can be expressed as A-ft.

10 A × 4 ft = 40 A-ft

The conversion factor (C.F.) for A-ft is 2.7 lbs. This weight is required to give 1 ppm in 1 A-ft of water.

The parts per million (ppm) or concentration of copper sulfate desired is 1.3.

Copper sulfate is 100 percent active (AI), so when 100 is divided by 100, the result is 1.

The amount of copper sulfate needed is determined by using the correct numbers in the basic formula as follows:

$$\text{Weight of chemical needed} = V \times C.F. \times \text{ppm desired} \times \frac{100}{\%AI}$$

lbs of copper sulfate = 40 A-ft × 2.7 lbs × 1.3 ppm × 1 = 140.4 lbs

EXAMPLE 2. How much formalin is needed to treat a holding tank 20 ft long by 4 ft wide and has a water depth of 2.5 with 250 ppm?

Determine volume (V) of water in tank by multiplying length × width × depth in ft.

$$\text{Volume in ft}^3 = 20 \text{ ft} \times 4 \text{ ft} \times 2.5 \text{ ft} \times 200 \text{ ft}^3$$

The conversion factor (C.F.) for ft^3 is 0.0283 gm, or the weight of chemical needed to give 1 ppm in 1 ft^3 of water.

The parts per million (ppm) desired is 250.

Formalin is regarded as 100 percent active (AI) for treatment purposes, so 100 divided by 100 equals 1.

$$\text{Formalin needed} = 200 \text{ ft}^3 \times 0.0283 \text{ gm} \times 250 \text{ ppm} \times 1 = 1,415 \text{ gm}$$

Formalin is a liquid, so the weight in grams must be converted to a volume unit. This is done by dividing the weight, 1,415 gm, by 1.08, the specific gravity of formalin.

$$\frac{1,415 \text{ gm}}{1.08 \text{ gm/cm}^3} = 1,310 \text{ cm}^3$$

To convert cubic centimeters to another unit of volume for measuring in fluid ounces, refer to Appendix Table A-3. Multiply the conversion factor of 0.0338 × 1,310 cm^3 to obtain volume in fluid ounces, 44.3 fl. oz.

EXAMPLE 3. How much potassium permanganate is needed to treat a circular tank that is 6 ft in diameter and has a water depth of 4 ft with 10 ppm?

Volume of circular tank is: $V = 3.14 \times r^2 \times d$

$$\text{Volume} = 3.14 \times (3 \text{ ft} \times 3 \text{ ft}) \times 4 \text{ ft} = 113 \text{ ft}^3$$

Conversion factor (C.F.) for ft^3 is 0.0283 gm.

The desired ppm is 10.

Potassium permanganate is considered to be 100% AI.

The values are substituted into the basic formula to obtain grams of potassium permanganate needed:

$$113 \text{ ft}^3 \times 0.0283 \text{ gm} \times 10 \text{ ppm} \times 1 = 32 \text{ gm}$$

DISPOSING OF WATER

Each year agricultural practices come under closer scrutiny for evidence that they may deteriorate the environment. Some claims against agriculture are valid, some are not. Agriculturalists and aquaculturalists respond to valid claims by changing practices.

Clean, plentiful water is the life blood of successful aquaculture. Wastewater is the major pollutant from aquaculture.

Water used in aquaculture carries with it any uneaten feed, fish wastes or excrement, chemicals, dead fish, sediments, algae, and escaped fish. This wastewater is known as effluent, and it is produced by hatcheries, rearing tanks, processing plants, ponds, raceways, flushing, harvesting drawdown, and haul tanks. Aquaculturalists recognize that wastewater harms the environment and must be creatively dealt with. Improper disposition of wastewater can cause damage to the aquatic species being cultured. Finally, wastewater disposal is regulated by laws, regulations, and agencies.

Pollution

Wastewater pollution can cause a number of problems for the environment. Dead fish look unsightly and smell. Nutrients in the wastewater cause the aging or eutrophication of a body of water. The wastewater increases sedimentation and may contain weed seeds. Wastewater smells because of decaying materials and the gases released. Finally, wastewater may contain a species of fish that could genetically damage wild fish or become a trash fish in the wild.

Treating and Disposing

Along with plans for adequate clean water, disposal of wastewater receives considerable attention. Basically, six methods are used to treat wastewater—

1. Settling ponds or vats
2. Irrigation water
3. Percolation ponds

4. Filtering systems
5. Hydroponics
6. Chemical additives.

Settling Ponds or Vats. Wastewater moves into the settling area where the solids settle to the bottom and the top water is released into the environment or a stream. Solid material needs to be removed occasionally.

Irrigation Water. As long as the wastewater contains no chemicals harmful to a field crop, it can be used for irrigation. Some of the substances in the wastewater serve as fertilizer for the crop.

Percolation Ponds. These ponds are similar to settling ponds, but the bottoms of the ponds are porous. Water absorbs into the ground. This method has the potential of polluting groundwater supplies.

Filtering Systems. Elaborate filtering systems prepare wastewater for release into the environment. These filtering systems are often expensive and require a high level of maintenance.

FIGURE 7-17 Settling ponds allow solid material to settle to the bottom. Eventually, they have to be cleaned.

Hydroponics. Wastewater from aquaculture provides the nutrified water for raising plants hydroponically. Hydroponics is the growing of land (field) crops with their roots dangling in nutrified water. Often, the wastewater from aquaculture contains nutrients below the level required by the plants.

Chemical Additives. Chemicals or biochemicals added to wastewater can remove pollutants. This method requires care since the improper addition of chemicals to the water could also create pollution.

Aquaculturalists deal with wastewater by using a modification or a combination of the six basic methods. What they choose to do could be influenced by the regulations that apply to wastewater in their area.

Regulations and Agencies

Federal, state, and local governments all establish regulations for wastewater disposal. The federal Environmental Protection Agency (EPA) requires any aquaculture facility that produces more than 100,000 lbs (45,455 kg) or discharges wastewater for more than 30 days to obtain a National Pollution Discharge Elimination System (NPDES) permit. The EPA sets minimum standards to be met and delegates specific regulations to the states.

FIGURE 7-18 Wastewater from aquaculture production facilities can be disposed of by irrigation.

Another federal agency, the U.S. Army Corps of Engineers, has jurisdiction in some areas over streams and rivers and the water in them.

The specific responsibilities of inspection and regulation often fall to the state and local agencies. Aquaculturalists should contact these agencies for complete details. Agencies to contact include the Soil Conservation Service and the Cooperative Extension Service. Ignorance of the regulations is no defense, and failing to follow regulations can result in fines.

SUMMARY

Successful aquaculture depends on a reliable source of high-quality water. Water provides oxygen, food, serves as an excretory site, regulates body temperature, and may harbor disease-causing organisms. Water quality must be monitored and maintained.

Water contains cations and anions. Their amounts and types affect the water quality. One major change influenced by ions is the pH. Water too acidic or too alkaline stresses aquatic life.

Water contains dissolved gases. The most important of these is oxygen. Low oxygen levels also stress fish and can result in a rapid fish kill. Increased water temperature decreases the level of oxygen in water. When oxygen levels threaten fish, supplemental aeration is used. Carbon dioxide in water comes from the respiration of plants and animals. High levels of carbon dioxide interfere with the fish's ability to obtain oxygen and also decreases the pH of the water.

Nitrogen in water comes primarily from the metabolic wastes of the animals. Nitrogen compounds in the water include ammonia, nitrates, and nitrites. Levels of ammonia are especially detrimental to the fish culture even at very low levels in the unionized form. Unionized ammonia increases as the pH increases.

Water for aquaculture supports a very complex system. Successful aquaculture requires a thorough understanding of water chemistry, constant observation, and testing. Common tests to monitor water quality use colorimetic, titrimetric, and electronic methods.

Management of water for aquaculture includes consideration of soil types used in ponds, rainfall, evaporation, light, odor, and color. Management of water also includes math skills to determine volumes and treatment levels.

After the water is used in aquaculture, the final problem is disposing of the wastewater. Aquaculturalists choose the best method of disposal for their needs.

STUDY/REVIEW

Success in any career requires knowledge. Test your knowledge of this chapter by answering these questions or solving these problems.

True or False

1. An increase in pH means the water is more acidic.

2. A Secchi disk is used to measure dissolved oxygen content.

3. Phytoplankton increases the oxygen content of the water in a pond.

4. Cations are negatively charged ions.

5. Plants produce most of the nitrogen in a pond.

6. Electronic meters can determine the dissolved oxygen content of water.

Short Answer

7. Which of the following is not a cation?
   a. chloride
   b. calcium
   c. sodium
   d. potassium

8. Fifteen parts per million (ppm) is the same as:
   a. 15 grams per pound
   b. 10 ounces per quart
   c. 15 milligrams per liter
   d. 15 grams per liter

9. A pH of 7 would be:
   a. basic
   b. neutral
   c. acidic
   d. 7 ppm

10. Respiration by fish adds carbon dioxide to the water. This _____ the pH of the water.
    a. increases
    b. decreases
    c. does not effect
    d. neutralizes

11. Name three processes affected by water temperature.

12. Name three cations and three anions found in water.

13. What is the most common measurement used to describe the amount of gases, ions, pesticides, and elements in water?

14. How does pH affect unionized ammonia?

15. What effect does increased water temperature have on oxygen content?

16. List three ways to measure salinity.

17. What is the relationship between water hardness—carbonate and bicarbonate—and pH changes?

18. TDS closely reflects what other measure?

19. What simple instrument is used to measure turbidity?

20. What is the purpose of liming and what is used to lime a pond?

21. Name four potentially harmful trace elements that may be found in water.

22. List five gases that could be found in a water sample.

23. What is the optimal level of dissolved oxygen for fish?

24. Name four events that reduce the dissolved oxygen in water.

25. What processes in water produce carbon dioxide and what process uses it?

26. What is the optimal pH range for aquatic animals?

27. How can hydrogen sulfide be removed from water?

28. Name four forms in which nitrogen may exist in water.

29. What chemical element is always a part of organic compounds?

30. An earthen pond is 10 A with an average depth of 5 ft. How many gallons and how many A-ft of water will be required to fill this pond? Also, if the well used to fill this pond delivers 1,500 gal per minute, how many days and hours will it take to fill the pond?

31. A concrete raceway 5 ft wide and 45 ft long with an average depth of 3 ft holds how many gallons?

32. List four heavy metals that might be found in the bottom soil of ponds.

33. A tank needs a potassium permanganate treatment of 10 ppm. How much potassium permanganate is needed to treat a rectangular tank 10 ft long and 2 ft wide with an average water depth of 1.5 ft? Assume the potassium permanganate to be 100 percent AI.

34. List three reasons why wastewater from aquaculture facilities is considered a source of pollution.

35. Name four methods used to treat wastewater from aquaculture facilities.

Essay

36. Why does photosynthesis in a catfish pond cause the pH to increase?

37. Does increased carbon dioxide in water raise or lower the pH? Why?

38. List three management practices that prevent or help overcome low dissolved oxygen levels in water.

39. What form of nitrogen is the most harmful to fish? Why?

40. What poor management practices cause high levels of ammonia in water?

41. Explain the difference between chlorine and chloride.

42. Why is soil type an important consideration in pond water management?

43. How would pesticides get into water used for aquaculture?

44. What is the best instrument for detecting odor? Name six odor descriptions commonly used.

45. Other than routine testing, give three other reasons that may cause an aquaculturalist to test water quality.

46. Explain the difference between a titrimetric test and a colorimetric test.

47. Why are ponds fertilized?

KNOWLEDGE APPLIED

1. Take a field trip to a local municipal wastewater facility to observe its procedures. Find out what water quality measurements/tests they use and compare these to aquaculture.

2. Plans for simple hydroponic projects can be easily obtained. Construct a simple hydroponic demonstration project. Determine what nutrients the plants require in the water. Compare this to the nutrients expected in aquaculture wastewater.

3. Construct a Secchi disk. Use a 12-in by 12-in piece of sheet metal. Mark an 8-in (20 cm) circle on the sheet metal. Make a small hole in the center. Paint the top of the disk with flat white paint. Divide the circle into equal quadrants. Paint two opposite quadrants flat black. Insert an eye bolt

through the hole in the center and place a lead weight on the bolt before attaching the nut. Attach a calibrated line to the eye bolt.

Use the Secchi disk to take readings on a pond. Lower the disk into the water until it just disappears. Record the measurement on the calibrated line. Lower the disk until it disappears. Then, raise it until it reappears. Record this measurement and use the average of the two.

4. From one of the suppliers listed below in Learning/Teaching Aids, obtain test kits for dissolved oxygen, ammonia, and pH. Practice using these test kits on various water sources. If possible, obtain an electronic pH meter and a dissolved oxygen meter and compare their readings to those obtained from the kits.

5. Take measurements of various shapes of containers that hold water. Use the measurements to determine the volume of water they will hold. Convert English volume measurements to metric volumes. For raceways or ponds, practice stepping off the measurements and using this to estimate the volume.

6. Obtain soil samples of clay, loam, and sand. Examine the properties of each in dry, damp, and wet states. On a field trip, practice identifying soil types.

7. In a fairly large body of water such as a pond, determine the temperature at various depths. For example, record the temperature at 1-ft intervals. Explain any differences.

8. Since ponds and other water-holding facilities are found in many shapes, estimating the surface area and volume provides an excellent chance for applied geometry. For example, ponds can be a combination of a rectangle and a triangle. Some tanks are round.

9. Track and record the weather for several months. Record the wind speed, hours of daylight, amount of cloud cover, amount of precipitation, and the maximum and minimum temperatures.

LEARNING/TEACHING AIDS

Books

American Public Health Association, American Water Works Association, & Water Pollution Control Federation. (1989). *Standard methods for the examination of water and wastewater* (17th ed.). Washington, DC: APHA.

Boyd, C. E. (1990). *Water quality in ponds for aquaculture.* Birmingham, AL: Birmingham Publishing Company.

Council for Agricultural Education. (1992). Module III: Using water. *Aquaculture.* Alexandria, VA: Council for Agricultural Education.

Dupree, H. K., & Huner, J. V. (1984). *Third report to the fish farmers: The status of warmwater fish farming and progress in fish farming research.* Washington, DC: U.S. Fish and Wildlife Service.

Hach Company. (1992). *Hach water analysis handbook.* Loveland, CO: Hach.

LaMotte Company Staff. (1992). *The monitor's handbook.* Chestertown, MD: LaMotte Company.

Tennessee Valley Authority. (1991). *Catalog of water quality educational materials.* Chattanooga, TN: Tennessee Valley Authority.

U.S. Department of Agriculture. *Water: The yearbook of agriculture 1955* (House Document No. 32). 84th Congress, 1st Session.

Walker, S. S. (1990). *Aquaculture.* Stillwater, OK: Mid-America Vocational Curriculum Consortium, Inc.

Videos

USDA Regional Aquaculture Centers. (N.D.). *Introduction to Water Quality Testing.*

USDA Regional Aquaculture Centers. (N.D.). *Procedures for Water Quality Management.*

USDA Regional Aquaculture Centers. (N.D.). *Water Quality Dynamics.*

(*Note:* Addresses for Regional Aquaculture Centers are listed in Appendix Table A-11.)

Equipment and Supplies

Argent Chemical Laboratories, 8702 152nd Avenue NE, Redmond, WA 98052
Carolina Biological Supply Company, 2700 York Road, Burlington, NC 27215
Hach Company, P.O. Box 389, Loveland, CO 80539
LaMotte Company, P.O. Box 329, Chestertown, MD 21620
NASCO, 901 Janesville Avenue, Fort Atkinson, WI 53538-0901
Royce Instrument Corporation, 13555 Gentilly Road, New Orleans, LA 70129
YSI, Inc., P.O. Box 279, Yellow Springs, OH 45387

Aquatic Structures and Equipment

P onds and raceways are the most common structures for raising fish. With water supplies reaching maximum and new species being seriously cultured, many other structures for culture are being used. These include tanks, silos, cages, and recirculating or closed systems.

Aquaculture has developed some unique equipment and some unique uses for equipment. This chapter cannot cover all aquaculture structures and equipment, but some additional equipment is discussed in related chapters. For example, Chapter 7, Water Requirements for Aquaculture, discusses testing equipment, and Chapter 4, Aquatic Management Practices, discusses equipment used for harvesting.

Learning Objectives

After completing this chapter, the student should be able to:

In Aquaculture

- Distinguish between four types of ponds
- Identify factors in pond site selection
- Explain important pond construction requirements

- ■ Define tank and raceway culture
- ■ List advantages and disadvantages of tank and raceway culture
- ■ Define cage culture
- ■ List advantages and disadvantages of cage culture
- ■ Describe cage design requirements
- ■ Identify site-specific factors that determine costs
- ■ List factors to consider when planning pond size
- ■ Identify layout and design considerations
- ■ List advantages of small versus large ponds
- ■ Identify four types of aerators
- ■ Describe the seines used on ponds
- ■ Describe a live-car or sock
- ■ Identify the need for boats, tractors, trucks, and pumps

In Science

- ■ Identify species for pond, cage, raceway, tank, or silo culture
- ■ List steps in determining a site's water quality
- ■ Determine whether soil is suitable for pond construction
- ■ Describe the biological concerns in a recirculating or closed system
- ■ Compare some of the biological concerns with cages and closed systems

Understanding of this chapter will be enhanced if the following terms are known. Many are defined in the text and others are defined in the glossary.

KEY TERMS		
Biofilters	Levee	Relift pump
Biological filtration	Live-cars	Seine
Borings	Mud line	Silos
Cage	Net pens	Slopes
Feeding ring	Paddlewheel	Sock
Flow index	Percolation	Stratification
Freeboard	Raceways	Topography
Impoundment	Recirculating	Watershed

PONDS

With the large production of catfish in the United States, ponds are the most common type of structure for raising fish.

Four types of ponds are natural, impoundment, excavated, and levee. Each type can be used for fish production. Natural ponds are located in low areas that fill with ground water or act as catch basins for runoff and springs. An impoundment pond is built by constructing a dam across a ravine or stream bed retaining the water behind the dam. The simplest and most common type of constructed pond is the excavated pond, which is made by digging a pit. Levee-type ponds are constructed by pushing up dikes on flat land and filling them with water from wells.

Harvesting techniques in ponds depend on the size range of the animals, the intensity of production, and the species cultured. Technology exists for harvesting most cultured aquatic animals. Some harvest systems, such as those used in intensive catfish pond culture, are well developed and modern. For other species, harvesting equipment and techniques need to be improved. Animals reared for recreational fishing have special handling and transporting requirements that are not readily adaptable to mechanized harvesting methods and remain labor-intensive and expensive.

Table 8-1 lists characteristics of each pond type.

Design Standards and Construction

Well-designed ponds save money and time in construction and operation. The requirements of the fish and methods of harvest are factors that will determine

FIGURE 8-1 Levee-type ponds constructed side-by-side are popular all over the Mississippi Delta.

TABLE 8-1 Fish Pond Types

Pond Type	Recommended Soils	Recommended Fish Species	Advantages	Limitations
Natural	—	Sunfish Crappie Bullhead Bass	No construction cost; presence of natural foods.	Not drainable; no control of pond design or location; difficult to harvest; difficult to manage or renovate; possible wild fish or weed problems.
Impoundment	Clay	All species	Drainable, allowing control of water level; easy to manage, renovate, and harvest; pond bottom can be dried for removal of weeds, wastes, disease, and wild fish.	High construction cost; often relies on runoff; possible siltation or pollution problems.
Excavated	Clay Sand Peat Gravel	Sunfish Crappie Bullhead Bass	Moderate construction cost; applicable to many types of soils and topography.	Usually not drainable; difficult to harvest; size limitation; renovation difficult; bottom uneven making seining ineffective; maintenance difficult.
Levee	Clay	All species	Bottom even; drainable; harvesting easy with correct equipment.	Expensive construction; flat land required; high level of management required.

the size and depth of the pond. Another consideration in design is whether the pond will be used for spawning, fish-out, and single- or multi-year production.

When more than one pond is planned, the ponds should be located so that each pond can be drained and filled separately from the others. This will avoid the spread of disease and allow more timely harvests of various ponds independently.

Size

Fish can be grown in ponds of any size. Fish ponds range in size from 0.1 A to 20 A (0.04 to 8.1 ha). The site conditions, limitations of construction equipment, and harvest methods will determine pond size. Ponds of less than 1 A are easier to manage and harvest. The pond should not be so large as to require stocking more fish than the aquaculturist is prepared to purchase. Fingerlings are expensive and can make up a large part of the investment in fish farming.

The site and design of the pond may be the most important factors controlling profitability. Ponds that leak, have irregular bottoms, or routinely suffer from a shortage of water will not produce a consistent crop of fish.

Ideally, levee ponds built on flat land and filled with groundwater or surface water are more suitable for commercial catfish production. Some terrain is rolling and not conducive to this kind of construction. In hilly terrain, pond builders must take advantage of the natural formations by constructing dams across valleys between hillsides so that runoff from rainfall on the watershed will be stored behind the dam.

Water Supplies

Water to fill and maintain watershed ponds usually comes entirely from runoff, though groundwater (wells) and surface water (springs, streams, and reservoirs) can also be used as supplemental sources. The ratio of watershed to water surface acreage should be large enough so that ponds fill and sometimes overflow during rainy months but drop no more than 2 ft (0.6 m) during drier

FIGURE 8-2 Small earthen excavated ponds for trout. (Photo courtesy Terry Patterson, College of Southern Idaho, Twin Falls, ID)

months. The best ratio of watershed to water surface varies according to the type of land on which the pond is built. For a watershed of heavy clay soil on open land, the best ratio is 5 A (2 ha) of land for each surface A (0.4 ha) of pond. For a sandy watershed in a wooded area, the best ratio is 30 A (12 ha) or more of land for each surface A (0.4 ha) of pond.

When a watershed is too small and unable to supply enough water to the pond, or an outside source of water is needed for filling during dry periods, water from wells, streams, or rivers can be pumped into the pond. Water containing wild fish should be filtered to avoid introducing these fish into the pond.

When ponds are built in series in a valley, less watershed is needed to maintain an A of water. Before harvest, water can be pumped or drained from one pond to another for storage. This procedure not only allows a producer to refill using the stored water immediately after harvest, but it also eliminates the possibility of draining nutrients into nearby natural waters.

Soil Characteristics

Good-quality soil that is at least 20 percent clay is necessary for building the core of dams. This includes clay, silty clay, and sandy clay soils. Soil should be sampled by frequent **borings** along a proposed dam site to determine if the clay foundation is large enough to build the dam.

Borings for soil samples should also be taken from the proposed dam site and the shoreline to be sure there is enough clay to build the dam. Usually a good source of clay can be found in the hillside near the dam site. If such a source is available, using it to build the dam can add to the size of the pond. If removing the clay will uncover rock formations, sand, or gravel in the pond bottom, it is best to leave the clay in place.

Pond construction in limestone areas can be especially risky because of the possibility of underlying cracks and sinks that may cause the pond to leak. In areas where the soil of the proposed pond bottom could leak, soils should be bored to check for quality. Approximately four borings per A are sufficient, unless there are variations in soil type in the pond bottom. Figure 8-3 shows the parts of a dam, including the core and drainage system.

Topography

Topography affects the size and shape of a watershed pond. Generally, steep **slopes** in V-shaped valleys require dams of larger volume per water surface A than sites with gently sloping hills and wide, flat valleys. So, ponds built in steep terrain usually cost more per pond A than those built in gently rolling terrain.

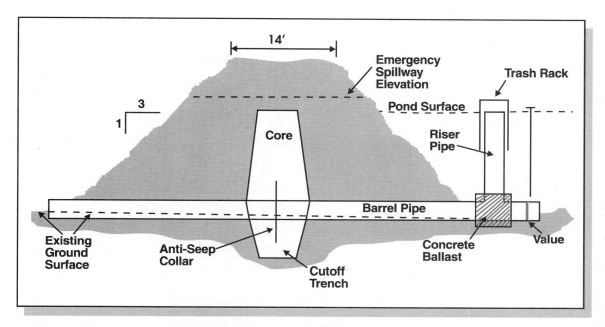

FIGURE 8-3 Cross-section of a typical dam at the drainpipe.

Ideally, watershed ponds should be less than 10 ft (3 m) deep at the drain. This depth allows the producer to harvest the pond without draining it. Deep ponds must be drained of much of their water before they can be seined for a complete harvest.

Some sites with gentle slopes and large flood plains allow for the construction of two-sided and three-sided watershed ponds (Figure 8-4). These ponds are usually constructed parallel to hills bordering a creek. Runoff is used as a water source, but the dam does not cross a hollow or draw. The great advantage of this kind of pond is that it does not have to be drained for harvest.

Other Considerations

The site for a watershed pond should be selected so that pipes and valves can be installed to drain the pond completely. The proposed shoreline should be excavated to provide a depth of at least 3 ft (0.9 m) around the edge of the pond. Pond bottoms should be smooth and slope gently to the drain pipe. A poorly constructed pond with an uneven bottom will cause incomplete harvests.

Floods from nearby rivers must not flow over the dam, and floods within the watershed must not weaken the dam. Ponds constructed in flood plains should be located so they will not cause damage to adjacent property if flooding does occur. Information on floods and their 100-year potential is available from the USDA Soil Conservation Service field office in each county.

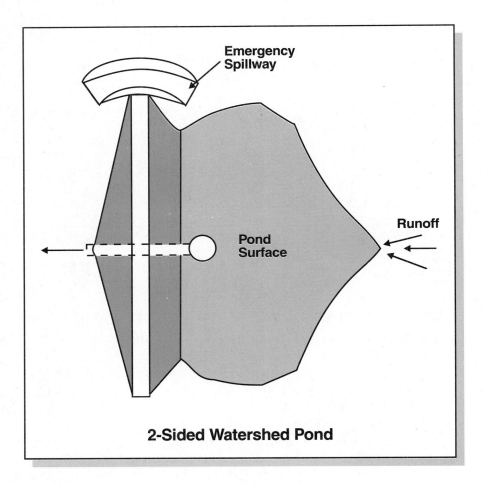

FIGURE 8-4 Diagram of conventional hill ponds.

After deciding on a dam site, the permanent water line and the potential flood-stage water line of the proposed pond should be marked off to make sure that water will not encroach on other property. Also, if the pond site contains 1 A (0.4 ha) or more of wetland, the U.S. Army Corps of Engineers must grant a permit before the pond can be constructed.

Cost of Construction

Cost estimates include clearing, earthfill, excavation, pipe and drain, concrete, seeding, and road gravel. Each pond site is unique. In general, a large, shallow, one-sided watershed pond is relatively inexpensive to construct. A three-sided pond may cost about twice as much as a one-sided pond.

Chapter 8: Aquatic Structures and Equipment ■ 397

CLAY

Gravel, sand—very coarse to very fine—silt, and clay make up the soil. Clay is the smallest of these soil particles. An International System of soil classification compares clay with the other soil particles.

Name of Particle	Diameter limits (mm)
Gravel	above 2.00
Coarse sand	2.00 to 0.20
Fine sand	0.20 to 0.02
Silt	0.02 to 0.002
Clay	below 0.002

Clay particles are plate-shaped and are often composed of complex compounds such as kaolinite, illite, and montmorillonite. Clay is sticky and capable of being molded—highly plastic—when wet and very hard and cloddy when dry. Because of the small particle size of clay and because all particles have the same negative charge, clay possesses some unique characteristics.

Clays in water forms a colloid. Since all clay particles have the same kind of electrical charge, they repel one another. Because the clay particles are so closely packed they cannot move freely but maintain the same relative position. If a force is exerted upon the clay, then the particles slip by each other. This permits clay to take on a shape as in pottery or brick making. When the force is removed, the particles retain their new position, since the same electrical forces as before act upon them. Clays become permanently hard when baked or fired.

In ponds, clay can cause the water to remain muddy. The negatively charged clay particles repel each other and will not settle out. The addition of some positively charged particles is necessary to cause coagulation and precipitation of the clay particles. Compounds used to do this include limestone, alum, and gypsum.

Clay is important in pond construction. It keeps water in ponds. Without clay in the construction of a pond, water losses by seepage are excessive and expensive. A pond made of fine sandy loam or sandy clay loam can lose over 14 in of water a day. A pond made of clay or clay loam but not properly packed can lose over one in of water a day. Ponds constructed of properly compacted heavy clay lose only about 0.002 in of water in one day.

When clay is not available for pond construction, bentonite is incorporated into the pond. Bentonite is a clay product that swells when wet and fills the minute holes in the soil. Also, organic matter such as manure, or a plastic liner can be used to help prevent seepage.

Site Selection of Levee-Type Production Ponds

Considerable thought and planning should go into selecting sites for commercial fish production ponds. Construction costs, ease and cost of operation, and productivity can be greatly affected by the site selected.

Water Availability. Water for filling levee-type ponds must come from a well, spring, reservoir, or stream since there is no watershed for runoff water to enter the pond. One of the first considerations in selecting a site for commercial fish ponds is to make sure that an adequate supply of suitable quality water is available for the size of farm planned.

Usually one well with a capacity of 2,000 to 3,000 gallons per minute (gpm) (7,571 to 11,356 L/min) is adequate for four 20-A (8.1 ha) ponds, or a minimum of 25 gpm per A (238 L/ha) of pond surface. Enough water must be available to fill the pond completely within ten days, otherwise problems with vegetation and water quality management will occur.

A local well drilling company, a groundwater geologist, or the closest office of the U.S. Geological Survey should be able to tell if a well of the desired capacity can be developed at the site. Information on the quality of groundwater at the site should be available from the U.S. Geological Survey or a local groundwater geologist.

FIGURE 8-5 Ponds require wells that can supply about 25 gal per min per A of pond surface.

Water Sources. Water for levee ponds for commercial fish production can come from a well, spring, stream, or reservoir. Of the four choices, a well is usually the best for several reasons. A spring is almost as good. Streams and reservoirs should be used only as last resorts because they usually contain various species of wild fish that can get into the pond and cause severe management problems. Streams and reservoirs can also become contaminated with pesticides or industrial chemicals that could kill fish. During droughts water levels become so low in the streams and reservoirs that they can no longer be used as sources of needed pond water. Wild fish in streams and reservoirs serve as constant sources of reinfection with various infectious diseases. The quality of water in streams and reservoirs also can change enough during droughts or floods as to be unusable.

Soil Characteristics. The soil must hold water, so clay-type soils are desirable. Soil cores taken at various places around the site ensure adequate clay is present to prevent excess seepage. The local Soil Conservation Service Office can provide assistance. Ponds can be built in soils that have high percolation rates if they are lined with a layer of packed clay or plastic, but costs are almost prohibitive.

Topography. The topography or lay-of-the-land determines the amount of dirt that has to be moved during pond construction. Pond construction on flat land requires less dirt moving than building on rolling or hilly land. Levee-type ponds built on flat land usually require about 1,100 to 1,200 yd^3 of soil to be moved per A (2,103 to 2,293 m^3 per ha).

Wetlands. If the site is classified as a wetland, a permit is required from the U.S. Army Corps of Engineers before clearing or building on wetlands. In some states, permits are needed from one or more state agencies before any clearing or building can take place on wetlands.

Draining. For management purposes and for harvesting, ponds must be drained. The site should be selected with this in mind. Further, ponds need to drain by gravity flow. Draining of ponds should not cause flooding on a neighbor's land or block the drainage from a neighbor's land.

Flooding. The site under consideration should not be subject to periodic flooding. This can be checked at the local U.S. Geological Survey Office or Soil Conservation Service Office.

Utility Right-of-Ways. Before building ponds over pipelines or under power lines, producers first should check locations of right-of-ways with the utility company to avoid possible legal problems later.

FIGURE 8-6 For management purposes ponds must be drained and dried. Here a small research pond is being drained.

Pesticides. If row crops were ever grown on or adjacent to the site being considered, it should be checked for pesticide residues. Three areas within a site that must be checked for pesticide residue levels are—

- Low areas where run-off collects
- Any area where spray equipment, either aerial or ground, was filled with pesticides
- Any area in the site where pesticides were disposed of or were stored.

Analysis of soils for pesticide residues can be done by commercial analytical laboratories or, in some states, by the state chemical laboratory.

CONSTRUCTION OF LEVEE-TYPE PONDS

Construction of ponds for the commercial production of aquatic animals is one of the most expensive and important aspects in developing a fish farm. Unless careful consideration is given to the design and the cost of pond construction, the producer may find the layout is not suitable for the species of fish desired and the cost of building makes it impossible to receive a profit. As discussed, site selection is the first step in construction.

Size

Most commercial catfish ponds in the Mississippi Delta are built on a 20-A (8.1 ha) land unit, which gives a surface area of about 17.5 A (7.1 ha) of water depending on the slope of the inside (wet) levee, the top width of the levee, and the height of the levee above the normal water level, often referred to as freeboard. The 20-A (8.1 ha) size is a compromise between ease of management and cost of construction. A larger pond is much more difficult to manage, and ponds smaller than this are more expensive to construct because more valves and inflow and drain pipes must be used.

In addition to costing more to construct, smaller ponds decrease the amount of water area available for fish production. If a 20-A (8.1 ha) land unit is made into two 10-A (10 ha) pond units with 16-ft (4.9 m) tops, 3:1 slopes and 1.5 ft (0.5 m) of freeboard, about 0.34 surface A (0.14 ha) of water will be lost due to the increased amount of levee needed. Assuming an average annual production of 4,200 lbs (1,905 kg) of catfish per A, this loss of water due to the increased amount of levees equals a loss of about 1,428 lbs (648 kg) of catfish production annually per 20-A (8.1 ha) pond unit.

Decisions on actual pond size depend on what the producer wants to raise, the topography of the land, and the amount of land available for pond construction. Before deciding on pond size, the aquaculturalist should look at the cost of building several different size ponds and go with the size that is most economical and consistent with production goals and degree of management planned. A typical layout of levee-type catfish ponds is shown in Figure 8-7.

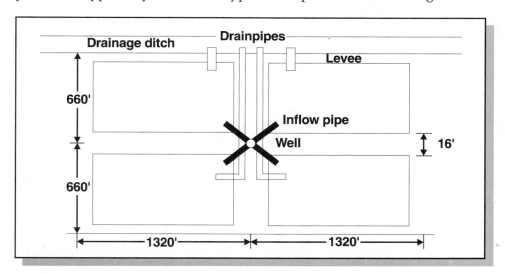

FIGURE 8-7   Layout of a typical levee-type catfish pond. (Source: Thomas L. Wellborn, Construction of Levee-type Ponds for Fish Production, Southern Regional Aquaculture Center, Publication No. 101)

Shape

Pond shape is largely determined by the topography and by property lines. Most commercial levee-type fish ponds are rectangular because of the greater ease and economics in harvesting and feeding, although square ponds are cheaper to build. A square 20-A pond requires 3,596 ft of levee, but a rectangular 20-A pond that is 660 ft by 1,320 ft requires 3,822 linear ft of levee, a difference of 226 ft. A rectangular 10-A pond (467 ft × 933 ft) requires 2,729 linear ft of levee, whereas a square 10-A pond (660 ft × 660 ft) requires 2,569 ft of levee, a difference of 160 ft.

Ponds with a curving, irregular shape are difficult to harvest unless drained. They are also extremely difficult to manage with respect to water quality. Irregular-shaped ponds should not be constructed before carefully weighing the advantages and disadvantages of cost and management.

Levee Width

Levees should be at least 16 ft (4.9 m) wide at the top. The main levees where the wells are located should have a 20-ft (6.1 m) top width to allow an easier flow of traffic for feeding, harvesting, and water quality management. Levees with top widths less than 16 ft (4.9 m) require more maintenance than do those with top widths of 16 ft (4.9 m) or greater. Narrow levees also are a greater hazard to employees and equipment in wet weather.

If the soil type is such that the levees become impassable in wet weather except to four-wheel-drive vehicles, at least two sides of each pond should be graveled to permit all-weather access for feeding, harvesting, disease treatment, and water quality management.

Freeboard and Depth

Freeboard is the height of the levee above the normal water level. The amount of freeboard should not exceed 2 ft (0.6 m) nor be less than 1 ft (0.3 m). Levees with a freeboard in excess of 2 ft (0.6 m) are expensive to build. Excess freeboard makes it difficult to get equipment into and out of the pond. Levees with a freeboard of less than 1 ft are subject to erosion, thus increasing maintenance costs.

Depth of the pond is important because of management implications. The depth of the pond at the toe of the slope at the shallow end should never be less than 2.5 ft (0.8 m) nor greater than 3.5 ft (1.1 m). At the deep end of the pond, the maximum depth at the toe of the slope should not exceed 6 ft (1.8 m), with a 5-ft (1.5 m) depth preferred. The pond bottom must be flat, free of all roots, stumps, and debris that might interfere with seining, and slope from the shallow end to the deep end at a rate of 0.1 to 0.2 percent along the long axis of the pond.

Slope of Levees

Slope is expressed as a ratio of the horizontal distance in feet for each 1 ft of height. For example, a 3:1 slope extends out 3 ft (0.9 m) horizontally for each ft in height and a 6:1 slope extends 6 ft (1.8 m) horizontally for each ft of height.

The actual slope of the levees will depend on the type of soil at the pond site, but for most soil types a 3:1 slope is satisfactory if properly compacted. Increasing the slope to 4:1 or 5:1 substantially increases the amount of soil required for the levees. This increases the construction. For example, a levee 6 ft (1.8 m) high, with a 16-ft (4.9 m) top width and 3:1 slope contains 7.6 yds^3 of soil per linear foot. A levee with the same dimensions except for a 4:1 slope requires 8.9 yds^3 of soil, and a levee with 5:1 slope needs 10 yds^3 of soil per linear ft. The cross-section of a typical levee for a catfish pond is shown in Figure 8-8.

Orientation

The direction in which the long axis of the pond is oriented depends mostly on the topography of the site and the property lines. Some say ponds should be oriented with the long axis parallel or at right angles to the prevailing winds. Levees of ponds with the long axis parallel to the prevailing winds are subject to erosion because of increased wave action, but the ponds are better aerated because of this increased wave action. Ponds oriented at right angles are subject to less levee erosion and are not as well aerated.

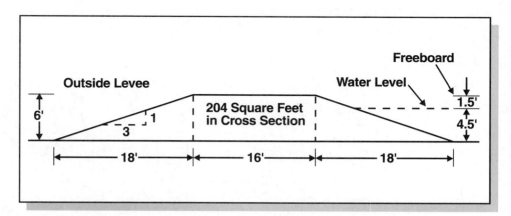

FIGURE 8-8 Cross-section of a typical levee for a commercial catfish pond. A levee with these dimensions contains 7.6 ft^3 fill material in each linear ft. (Source: Thomas L. Wellborn, Construction of Levee-type Ponds for Fish Production, Southern Regional Aquaculture Center, Publication No. 101)

Site Preparation and Construction

All existing vegetation, roots, stumps and topsoil must be removed from the site before starting levee construction to allow a good bond between the foundation soil and the fill material. After completion of the levee, several inches of topsoil are put back on the top and outside slopes so a vegetative cover can be established quickly.

Because of their speed, self-loading pans are the most efficient equipment to build pond levees. They also give the best compaction of the fill material when complete wheel track coverage is made over each layer of fill placed in the levee. For proper compaction the soil must have at least 12 to 15 percent moisture. If the soil is dry during construction, each layer of fill dirt placed on the levee must be wetted before compaction. Laser-controlled pans increase the accuracy and speed of the cuts and fills.

Bulldozers can be used to build pond levees but they do not give good compaction. If bulldozers are used, using a sheepsfoot roller gives complete coverage over each layer of fill.

The area in the levee where the drain is located is left open during construction for drainage of any storm water during construction. As soon as the pond bottom and drain site are excavated to grade, a drainpipe of appropriate diameter is installed. A small, sloped ditch, about one-third of the diameter of the drainpipe, is dug to give uniform support for the pipe. Fill material is placed around the side and over the top of the drain and is hand compacted. This fill material must be moist during hand compaction to insure a watertight seal around the drain pipe. The hand-compacted fill over the drain gives protection from heavy equipment during the completion of the levee construction.

FIGURE 8-9 Laser-controlled pan used to build catfish ponds.

Drains

The ponds must be sited and constructed so they can be drained by gravity flow. Obviously, the lowest part of the pond must be higher than the ditch into which it is to be drained. The pond bottom must be flat and slope from the shallow to the deep end with a slope of 0.1 to 0.2 percent. A flat, sloping bottom is necessary for harvesting and draining.

A harvesting basin inside or outside the pond is not built in levee ponds. Harvest basins do have a place in the production of certain types of fish, though they are expensive to construct. In the early years of commercial catfish production, harvest basins were routinely built but were quickly discarded because of problems and expense associated with them.

Levee-type ponds use several types of drain structures. The best type of drain structure to use is the one that most closely meets the producer's needs based on the pond design selected and estimated cost. The so-called inside swivel drain or modified Canfield outlet is the most common drain used in levee-type ponds. It is located at the lowest part of the pond. The depth of water in the pond is determined by the length of the upright drainpipe, and water level is adjusted by pivoting the upright drain up or down on its swivel joint. It must be held securely in position to prevent unplanned drainage. This can be done with a chain from the end of the upright drain to a post on the bank.

Swivel joints require grease. Maintenance of the swivel joints can be a problem since work has to be done under water or when the pond is drained.

FIGURE 8-10 The modified Canfield outlet or inside swivel drain controls the water level by pivoting the drainpipe up or down on its swivel joint.

FIGURE 8-11 Swivel drain in a newly constructed pond. Drainpipe is completely lowered and the walkway provides access to a hand winch to raise and lower the drain pipe.

Many farmers who raise catfish in levee-type ponds now use outside drains. The drainpipe is laid through the levee at the lowest point in the pond with the inside end extending at least 5 to 10 ft (1.5 to 3 m) out from the toe of the slope to prevent clogging by dirt sloughing from the levee. This inside end is screened to prevent loss of fish.

The outside end of the drainpipe should extend 5 ft (1.5 m) past the toe of the slope to prevent excess erosion of the levee during draining. The outside end of the pipe is fitted with a T and a standpipe of the height necessary to maintain the desired normal water level in the pond. The end of the T is fitted with a valve for water level manipulation and complete draining if needed.

Another method is to have the outside standpipe 24 in (61 cm) high, rather than the height of the normal water level in the pond, and fitted with a valve. The end of the T is capped. During rain, the valve is partially opened to remove excess water. This system permits rapid draining of 3 to 4 ft (0.9 to 1.2 m) of water with slight danger of wild fish entering the pond through the drainpipe. The pond can be completely drained by removing the cap at the end of the T.

The discharge end of the drainpipe needs to be at least 24 in (61 cm) above the surface of the water in the drainage ditch. By doing this, wild fish can be prevented from entering the pond through the drain.

Size of Drain

The size of the drainpipe needed is dictated by the size of the pond and how rapidly the farmer wants to drain the pond. Regardless of the pond size, a drain should allow the pond to be completely drained in no more than seven days, and preferably in five days.

Design Assistance

Local Soil Conservation Service technicians provide assistance in planning the design of fish ponds. In some areas of the United States, they are experienced in pond design and can help prevent severe design flaws. They can also help calculate the cubic yards of dirt that must be moved during construction using a computer program such as *Auto Pond II*.

RACEWAYS AND TANKS

Most raceways in the United States are made of concrete, but a few are constructed of other materials such as earth, stone, metal, concrete, or plastic, including fiberglass. Sites suitable for production in raceways are limited primarily by the water source. Most hatcheries in which raceways are used are in the Snake River Valley of Idaho, where high-quality flowing water is available.

FIGURE 8-12 Raceways at the world's largest commercial trout production facility, Clear Springs in Buhl, Idaho.

The number of raceways in the effluents of electric power plants could increase as aquaculturists seek new sources of high-quality, heated water and the economics of production justify the cost of development and operation.

Raceways receiving water from power plant effluents have been used to culture marine and estuarine organisms, channel catfish, striped bass, rainbow trout, American eel, and other species.

The density or weight of fish that can be stocked per raceway depends on fish size, raceway size, and the available flow of quality water. These relations can be expressed as follows—

$$F = \frac{W}{L \times I}$$

where F = the **flow index**, W = weight of fish in ponds, L = length of fish in inches, and I = water inflow in gallons per minute.

The flow index is derived practically by the maximum weight of fish that can be raised in a raceway. Once the flow index has been established, the formula can be rewritten and used as follows—

$$W = F \times L \times I$$

If all measures are metric, then flow index and maximum weight can be figured in metric. This form of the equation is used to calculate the weight of fish that can be safely maintained as fish length increases or water flow changes. Most hatcheries of the U.S. Fish and Wildlife Service are operated with a flow index equivalent to 10 to 25 when F is calculated in metric units (1.0 to 2.5 when F is calculated in English units). Hatchery efficiency declines at flow indexes below 10, and water quality deteriorates at indexes above 25. Most raceways for salmonids have been designed for fish loads of 2 to 3 lb/ft^3 (32 to 48 kg/m^3).

Many who are interested in commercial fish farming strike upon the idea of raising catfish or other warmwater species in raceways. This idea seems appealing because yields per unit of growing area may be theoretically increased and the need for large ponds eliminated or reduced. Still some factors make the raceway culture of warmwater fish rather risky. Some unique situations exist where raceway systems may be practical and profitable.

Raceways are generally constructed in a ratio of 5:1 (or greater) length to width, and with a depth of 3 to 5 ft (0.9 to 1.5 m). Water should flow evenly through the system to eliminate areas of poor water circulation where waste materials or sediment may accumulate. Raceways may be constructed above ground or in the ground from cement or fiberglass, and even wood has been used. Fish cultured in raceways require a large quantity of good-quality water, preferably supplied by gravity flow from artesian wells or higher elevations. If pumping is required, operating costs may be high and risks increased due to possible failure of pumps or power supply.

Raceways should be considered only if an abundance of good-quality water is available. On the average, 1 to 3 gal/min (3.8 to 11 L/min) of flow should be available for each ft³ (28 L) of raceway volume at densities of 3 lbs of fish per ft³ (1.4 kg/29 L). If supplemental aeration is used, the water requirement may be somewhat reduced. Water flow should be sufficient to keep solid waste material from accumulating in the raceway and to dilute liquid waste, primarily ammonia, excreted by fish.

For many intensively cultured species, water quality, not physical space, limits the number of animals that can be reared in a raceway or similar system. Water must be continuously added to flush away wastes and to maintain a quality environment.

In some raceways, effluents from downstream sections are pumped back upstream and reused. The effluent may first be improved by aeration, filtration, sedimentation, ozonation, or a combination of these and other processes before recycling. In single-pass raceways, where water is used only once before being discharged, incoming water quality depends on the source.

Effluent quality from intensive systems is a function of the number and size of fish or other organisms, the feeding rate and water flow rate.

Silos

Silos are deep cylindrical tanks similar in operation to horizontal raceways. Water is exchanged at a rapid rate to maintain adequate oxygen levels and to remove waste products. Water quality in silos is usually no better than that of

FIGURE 8-13   Small rectangular tank used for trout fry.

FIGURE 8-14 Fiberglass circular tanks used in a research facility.

the incoming water, but dissolved oxygen may be increased by aeration and agitation, or in some units by the addition of gaseous or liquid oxygen. Silos may be constructed with sediment basins to remove solid waste before water is discharged.

Since rectangular raceways typically have length:width:depth ratios of 30:3:1, they occupy considerably more surface area than do circular tanks of similar capacity with a diameter:depth ratio of about 1:2. These deep circular tanks or silos require a relatively large flow of water. Silos were initially developed in Pennsylvania for the culture of trout in areas where there was an adequate flow of water of high quality to provide a gravitational head sufficient to flush solid wastes from the tank.

Silos have been incorporated into water recirculation culture systems. Such systems require biological, mechanical or chemical filters, or a combination of these, to remove waste metabolites and allow filtered effluent water to be recycled through the culture tanks. These systems have been used to culture striped bass, channel catfish, and salmonids.

Circular Tanks

Circular tanks constructed of plastic, concrete, and steel are widely used throughout the United States to culture aquatic species. Small 25- to 500-gal (100 to 2,000 L) tanks are used to spawn fish, to maintain fry and fingerlings, and to culture brine shrimp and other forage organisms. Circular tanks are used

FIGURE 8-15 Concrete silos used for trout production.

as flow-through units and also in water recirculation systems. Tank size corre-lates with growth, food conversion efficiency, and survival of rainbow trout.

CAGES AND PENS

Cage culture of fish uses existing water resources but encloses the fish in a cage or basket that allows water to pass freely between the fish and the pond.

Modern cage culture began in the 1950s with the invention of synthetic materials for cage construction. In the United States, universities did not begin conducting research on cage rearing of fish until the 1960s.

Today, cage culture receives more attention by both researchers and com-mercial producers. Factors such as increasing consumption of fish, declining wild fish stocks, and a poor farm economy produced a strong interest in fish production in cages. Many of America's small or limited-resource farmers look for alternatives to traditional agricultural crops. Aquaculture appears to be a rapidly expanding industry and one that may offer opportunities even on a small scale. Cage culture also offers the farmer a chance to use existing water resources which, in most cases, have only limited use for other purposes.

FIGURE 8-16 Cage culture offers a farmer the chance to use existing water resources and to start on a small scale.

Cage culture of fish is not foolproof or simple. Cage production can be more intensive in many ways than pond culture and should probably be considered as a commercial alternative only where open pond culture is not practical.

Like any production scheme, cage culture of fish has advantages and disadvantages.

Advantages

Cage culture has some distinct advantages, including—

- Many types of water resources can be used, including lakes, reservoirs, ponds, strip pits, streams, and rivers, which could otherwise not be harvested. Specific state laws may restrict the use of public waters for fish production.

- A relatively low initial investment is all that is required in an existing body of water.

- Catching is simplified.

- Observation and sampling of fish is simplified.

- Allows the use of the pond for sport fishing or the culture of other species.

These advantages are appealing. A potential fish farmer can produce fish in an existing pond without destroying its sportfishing and try fish culture with reasonable risks.

Disadvantages

Cage culture also has disadvantages, including—

- ■ Feed must be nutritionally complete and kept fresh.
- ■ Low dissolved oxygen syndrome (LODOS) is an ever-present problem and may require mechanical aeration.
- ■ The incidence of disease can be high and diseases may spread rapidly.
- ■ Vandalism or poaching is a potential problem.

Good management of cage-cultured species prevents the disadvantages causing the failure of this production option.

Aquatic organisms may be confined at very high densities in cages or pens placed in ponds, lakes, rivers, and estuaries. Mollusk cages may be suspended from piers, long lines, or rafts. Water quality within cages and pens depends primarily upon both the quality of water in the surrounding area and the rate of exchange by circulation through these aquaculture units. Cages and other devices placed in estuarine areas with tidal flows are flushed by tidal action. In lakes and ponds the swimming and feeding activity of fish cultured in the cages, and of wild fish around the cages, can generate sufficient water movement to permit water exchange in cages. Wind-generated waves also help to circulate and exchange water around cages and pens placed in ponds and lakes with little or no natural current.

Cages placed in rivers or industrial effluents take advantage of the flowing water to remove wastes. Cages, rafts, and other off-bottom systems may be moved from one site to another to take advantage of better environmental conditions. For example, caged fish cultured in the intake canal of a power plant during the spring and summer may be moved to the discharge canal to take advantage of the thermal effluent during the fall and winter.

Many factors may seasonally alter water quality, for example, nitrogen gas supersaturation in thermal effluents, agricultural run-off, industrial discharges, and reservoir drawdown for irrigation or flood control. These must be known and understood by the aquaculturist before placing cages or pens in a body of water.

Harvesting is a problem in cage and pen culture. Cages or pens are usually located away from land and access is provided by boat or barge. Pond and raceway harvesting techniques are unworkable in this setting. Most cage- and pen-reared fish must be dipped by hand and placed in containers, a labor-intensive operation. **Net pens** or cages do not permit partial or selective harvest.

Cage Construction and Placement

Cages for fish culture have been constructed from a variety of materials and in practically every shape and size imaginable. Basic cage construction requires that cage materials be strong, durable, and non-toxic. The cage must retain the fish yet allow maximum circulation of water through the cage. Adequate water circulation is critical to the health of the fish, to bring oxygen into the cage and remove metabolic wastes from the cage. Location of the cage in the pond may be critical to proper circulation through the cage. Mechanical circulation and aeration through the cage may be necessary if stocking densities are high, cages are large, or water quality deteriorates during production.

Cage Materials. Cage components consist of a frame, mesh or netting, feeding ring, lid, and flotation. Cage shape may be round, square, or rectangular. Shape does not appear to affect production. Cage size depends on the size of the pond, the availability of aeration and the method of harvest. Most fish farming supply companies sell manufactured cages, cage kits, or materials for constructing cages. Common cage sizes used in ponds are:

- Cylindrical—4 × 4 ft
- Square—4 × 4 × 4 ft and 8 × 8 × 4 ft
- Rectangular—3 × 4 × 3 ft and 8 × 4 × 4 ft

The frame of the cage can be constructed from wood—preferably redwood or cypress—iron, steel, aluminum, fiberglass, or PVC. Frames of wood, iron, and

FIGURE 8-17 Cages positioned in a pond. (Photo courtesy Chuck Weirich, Delta Research Center, Stoneville, MS)

steel—unless galvanized—should be coated with a water-resistant substance like epoxy or an asphalt-based or swimming pool paint. Bolts or other fasteners used to construct the cage should be of rust-resistant materials.

Mesh or netting materials that can be used include galvanized wire, plastic-coated welded wire, solid plastic mesh, and nylon netting (knotted or knotless). Solid plastic (polyethylene) mesh is most commonly used for small cages because of its durability. Mesh size should be no smaller than $\frac{1}{2}$ in (1.3 cm) to assure good water circulation through the cage. A larger mesh size can be used if large fingerlings are stocked. The feeding ring or collar can be made of $\frac{1}{8}$- or $\frac{3}{16}$-in (0.3 to 0.5 cm) mesh and should be 12 to 15 in (30.5 to 38.1 cm) in width. The feeding ring keeps the floating feed from washing through the cage sides.

All cages should have lids to prevent fish from escaping and to keep predators—including people—out of the cages. Lids may be made from the same material as the rest of the cage or can be from other materials, such as plywood, masonite, aluminum, or steel. Plywood, masonite, and steel will need to be painted with exterior or epoxy paint. The manager of the cage needs to be able to observe feeding behavior and have easy access to the cage to remove uneaten feed and any dead fish.

Flotation of the cage can be provided by inner tubes, styrofoam, waterproofed foam rubber, capped PVC pipe, or plastic bottles. Cages can also be suspended from docks. Plastic bottles should be made of sturdy plastic, like antifreeze or bleach bottles and should have their caps waterproofed with silicon sealer. Floats should be placed around the cage so that it floats evenly with the lid about 6 in (15.2 cm) out of the water.

Cage Construction. Figure 8-18 shows the construction of a simple cage. The simplest cage design to construct is a 4 × 4 ft (1.2 × 1.2 m) cylindrical cage fashioned from $\frac{1}{2}$-in (1.3 cm) plastic mesh. The mesh comes in a roll 4 ft wide by 50 ft long (1.2 by 15.2 m). A total of 21 ft (6.4 m) of plastic mesh is used per cage. Thirteen feet (4.0 m) of mesh is used for the cylinder with two 4-ft (1.2 m) panels for the bottom and lid. The plastic mesh is easily cut with tin snips. The cylinder is formed around two metal, PVC, or fiberglass hoops at the top and bottom of the cage. A third hoop is used to form the lid.

The cage can be laced together with 18-gauge bell wire—plastic coated solid copper wire—stainless steel wire, hog rings, or black plastic cable ties. White cable ties deteriorate in sunlight.

The basic design and construction principles apply to all cages—

■ Attach the netting or mesh securely to the frame.

■ Lace carefully leaving no gaps or holes.

■ Lids and feeding rings are essential.

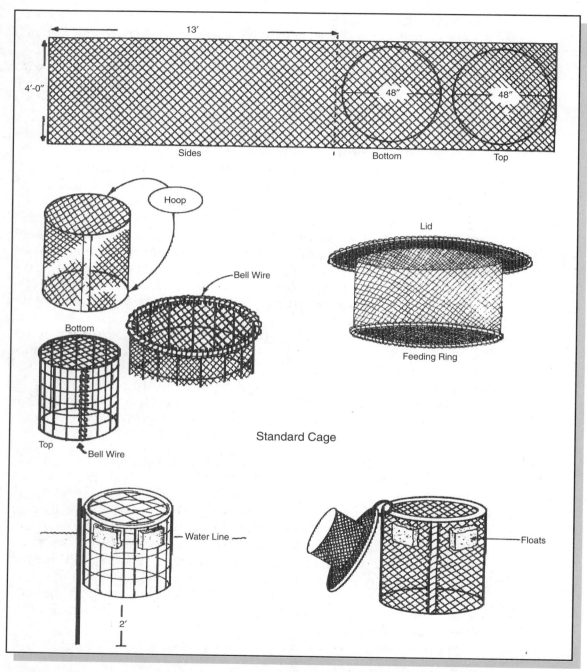

FIGURE 8-18 Construction of a 4 ft × 4 ft cylindrical cage. (Source: Michael P. Masser, Cage Culture: Cage Construction and Placement, Southern Regional Aquaculture Center, Publication No. 162)

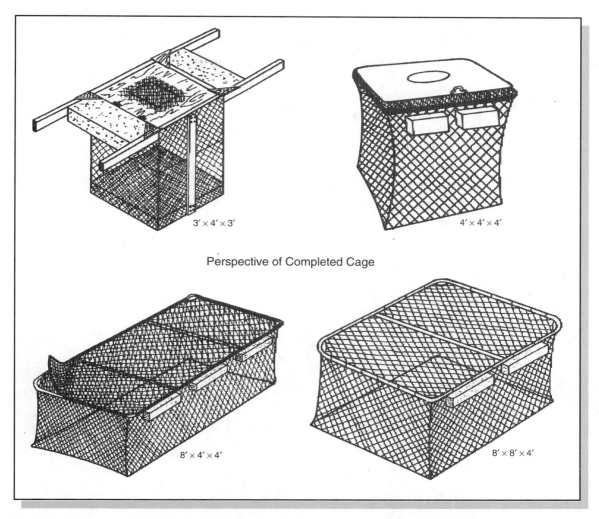

Perspective of Completed Cage

FIGURE 8-19 Some other common cage designs. (Source: Michael P. Masser, Cage Culture: Cage Construction and Placement, Southern Regional Aquaculture Center, Publication No. 162)

Cage Placement. Location of the cage in the pond may be critical to its success. Two factors to consider in cage placement are: (1) access to the cage and (2) maintenance of water quality. Daily feeding and management of the cage require easy access under almost any weather condition. Access may be by pier or by boat. Critical factors for locating cages in an area to maximize water quality are—

■ Windswept. It is important that the cage be in an area where it will receive maximum natural circulation of water through the cage. Usually, this is in an area that is swept by the prevailing winds.

◼ At least 6 ft (1.8 m) of water depth. A minimum of 2 ft (0.6 m) of water is needed under the cage to keep cage wastes away from the fish.

◼ Away from coves and weed beds. Coves, weed beds, and overhanging trees can reduce wind circulation and potentially cause problems.

◼ Away from frequent disturbances from people and animals like dogs and ducks. Disturbances from people frequently walking on the dock, fishing, or swimming near the cage, or from animals that frequent that area of the pond will excite the fish and can cause stress, injury, reduced feeding, and secondary disease.

◼ At least 10 ft (3 m) from other cages. Cages should not be too close together. Close proximity to other cages may increase the likelihood of low dissolved oxygen syndrome (LODOS).

Access to electricity or to a location where a tractor-driven paddlewheel, irrigation pump, or other aeration device can deliver aerated water to the cages should be considered before locating cages. Aeration devices should move oxygenated water through the cages.

FIGURE 8-20 A 40,000 gal recirculating tilapia system using two Model PBF-30 propeller washed bead filters for biofiltration and solids capture. Each unit provides 12,000 ft² of surface area for bacterial attachment. (Photo courtesy Douglas Drennan and Dr. Ron Malone, Department of Civil and Environmental Engineering, Louisiana State University)

Large Cages. Larger cages have been built and used, particularly in large lakes, reservoirs, rivers, bays, and estuaries. Many times large cages are called net pens because they are constructed from nylon netting. Some research and demonstrations include large cages for ponds that have built-in aeration devices. Apparently large cages can be designed that maximize the number of fish sustainable by the pond and actually support increased densities.

CLOSED SYSTEMS OR RECIRCULATING SYSTEMS

Recirculating water systems are designed to minimize or reduce dependence on water exchange and flushing in fish culture units. These systems have practical applications in commercial aquaculture hatcheries, holding tanks, and aquaria systems, as well as small scale aquaculture projects. Water is typically recirculated when a specific need exists to minimize water replacement, to maintain water quality conditions that differ from the supply water, or to compensate for an insufficient water supply. There are many designs for recirculating systems. Most will work effectively if they accomplish—

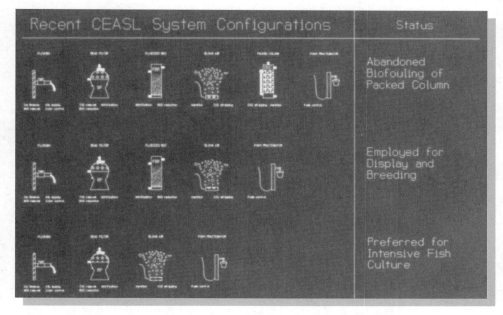

FIGURE 8-21 Recent recirculating treatment configurations proposed by CEASL (Civil Engineering Aquatic Systems Laboratory). (Photo courtesy Douglas Drennan and Dr. Ron Malone, Department of Civil and Environmental Engineering, Louisiana State University)

- Aeration
- Removal of particulate matter
- **Biological filtration** to remove waste ammonia and nitrite
- Buffering of pH.

These processes can be achieved by a simple composite unit such as an aquarium filter, or in larger systems, by several interconnected components.

Aeration

Water must be aerated to maintain adequate dissolved oxygen concentrations for fish and for proper functioning of the biological filter. Aeration is usually applied in the fish culture tank and again before or within the biological filter, that portion of the recirculating system where organic waste products are broken down through bacterial decomposition. Trickling filters and revolving plate **biofilters** are designed to be self-aerating. Air

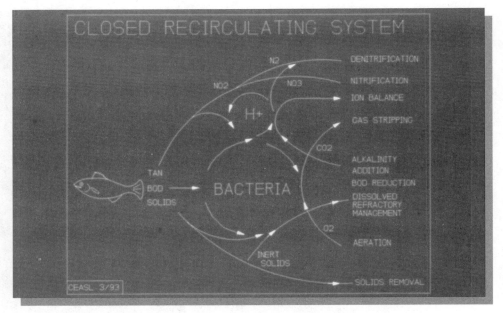

FIGURE 8-22 Recirculating system finfish waste wake. Diagram depicts typical waste products generated by finfish and typical treatment processes used to remove the waste from a recirculating system. (Photo courtesy Douglas Drennan and Dr. Ron Malone, Department of Civil and Environmental Engineering, Louisiana State University)

lift pumps are often used to move water through the tanks, accomplishing both aeration and pumping. Superintensive systems may use pure oxygen injection, though this is seldom economical. The level of aeration should be sufficient to sustain dissolved oxygen levels above 60 percent saturation throughout the system.

Removal of Particulate Matter

Solids resulting from fish waste and uneaten feed contribute a portion of the oxygen demand and toxic ammonia in the system and should be concentrated for removal. This is done in a settling basin with reduced water turbulence, or by mechanical filtration through porous material such as sponge, screen, sand, or gravel. Solids that accumulate will gradually be mineralized—broken down—by bacterial action and their volume reduced. Although this process adds additional oxygen demand to the system, it reduces the need for frequent cleaning if the solids do not become resuspended or interfere with normal water flow. Mechanical filters require regular cleaning since they are prone to clogging when dirty. To prevent excess amounts of solids from accumulating in the biofilter, small particles are usually removed before or as the first component of biofiltration.

FIGURE 8-23 Biofilter of hollow plastic balls to increase surface area.

Biological Filtration

Fish and other aquatic organisms release their nitrogenous wastes primarily as ammonia (NH_3) excreted across the gill membranes. Urine, solid wastes, and excess feed also have undigested nitrogen fractions and other sources of ammonia. Ammonia is toxic to fish and can exert sublethal stress at concentrations of less than 0.05 mg/L ammonia nitrogen (NH_3^-), resulting in poor growth and lower resistance to disease.

To control ammonia levels in recirculating water systems, extensive surface area is provided for bacteria that biologically oxidize ammonia into relatively harmless nitrate (NO_3^-). Bacterial nitrification is a two-stage process resulting first in the transformation of ammonia to nitrite (NO_2^-), then a further oxidation of nitrite to nitrate. Nitrite is also toxic to fish at low concentrations, so both reactions must occur for successful biofiltration.

The bacteria responsible for these reactions occur widely in soil and water environments and can be easily inoculated into biofilters from natural sources or with material from established filters. To ensure bacterial populations are sufficient to remove ammonia and nitrite at rates required during operation, a biofilter is typically conditioned for several weeks by adding ammonia and monitoring its breakdown before stocking fish.

Media for biofilters can be virtually any substrate that provides maximum surface area for bacterial growth, for example, oyster shell, gravel, nylon netting, plastic rings, corrugated fiberglass panels, and sponge foam pads. In designing biofilters, the principal concerns should be—

■ Maximum surface area for bacterial growth

■ High dissolved oxygen levels

■ Uniform water flow through the filter

■ Sufficient void space to prevent clogging

■ Proper sizing to ensure adequate ammonia removal capability.

The required size of a biofilter is difficult to predict since filter surface area, fish density, and water flow are important considerations. A 3:1 ratio of tank volume to biofilter volume is usually more than sufficient.

Management

Recirculating water systems should be designed for simplicity of operation and economic feasibility. Conditioning of the biofilter requires enough time before introducing fish. Ammonia and nitrite concentrations must be checked frequently. Dissolved oxygen should be sustained above 60 percent of saturation and periodically verified. Alkalinity, hardness, and pH need to be measured and adjusted, if necessary, at regular intervals. Filters should be inspected and cleaned. Medications used to treat fish diseases may be toxic to bacteria in the biofilter. An ability to isolate fish tanks for disease treatment should be provided. Frequent monitoring of the performance of the recirculating system allows the manager to improve and refine its operation over time.

Buffering of pH

Fish metabolism and bacterial nitrification result in the formation of acids that lessen the buffering capacity of water and lower the pH. Most fish can tolerate a pH range of 5 to 10. A range of 6.5 to 8.5 is preferred for most aquaculture species. To replace lost alkalinity and sustain the buffering capacity of water, carbonate (CO_3^{2-}) in the form of limestone, bicarbonate of soda, or other common sources is added. Often, biofilter media (oyster shell) or some other component of the system (concrete tanks) serve as a source of carbonates. Depending on the species cultured, frequent monitoring of water hardness, alkalinity, and pH may be required.

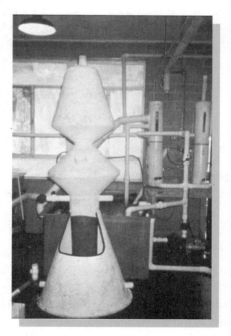

FIGURE 8-24 A combination fluidized bed/bubble washed bead filter (Model CBF-4). The opaque section at the bottom is used to determine the percent fluidization of the sand bed. A fluidized sand bed effectively doubles the specific surface area available for biofiltration. (Photo courtesy Douglas Drennan and Dr. Ron Malone, Department of Civil and Environmental Engineering, Louisiana State University)

FIGURE 8-25 Propeller washed bead filter (Model PBF-6) on a 1,000 gal recirculating tilapia system. The filter is used for solids capture and biofiltration, and the tank is capable of holding 500 lbs of fish. (Photo courtesy Douglas Drennan and Dr. Ron Malone, Department of Civil and Environmental Engineering, Louisiana State University)

OTHER MAJOR EQUIPMENT

Aquaculture uses a wide range of devices, such as aeration equipment, nets and seines, boats, pumps, tractors, and trucks.

Aeration Equipment

Aerators work by increasing the area of contact between air and water. Aerators also circulate water so fish can find areas with higher oxygen concentrations. Circulation reduces water layering—stratification—and increases oxygen transfer efficiency by moving oxygenated water away from the aerator.

Fish farmers have used emergency aerators powered by tractor power take-offs (PTOs) for many years. With intense production, the need for PTO aerators can be quite expensive because each aerator requires a tractor. More electric aerators are being used. Large tractor-powered aerators are still used as back-ups during severe oxygen depletions, equipment failure, or power outages.

Each producer decides which aeration device should be purchased or built. This decision is important and should be made with the specific application and associated costs of energy and equipment in mind. Most aerators are in one of the following categories: surface spray or vertical pump, pump sprayer, paddlewheel, diffused air, and propeller aspirator pump.

Surface Spray or Vertical Pump. Surface spray aerators have a submersible motor that rotates an impeller to pump surface water into the air as a spray. They float, are lightweight, portable, and electrically powered. Units of 1 to 5 hp with pumping rates of 500 to 2,000 gpm (1,893 to 7,571 L/min) are available.

They are designed to be operated continuously during nighttime, cloudy weather, or when low dissolved oxygen concentrations are expected. Surface spray aerators have prevented fish kills when used at 1.5 to 2 hp/A. They are usually of little use in large ponds because of relatively low oxygen transfer rates and their inability to create an adequately large area of oxygenated water.

Pump Sprayer. Pump sprayer aerators are found on many fish farms. Most are powered by tractor power takeoffs or electricity. Some units are engine driven and require mounting on a trailer frame for transport. Pump sprayer aerators are equipped with either an impeller suction pump, an impeller lift pump, or a turbine pump. Some have a capped sprayer pipe or "bonnet" with outlet slits attached to the pump discharge. Others discharge directly through a manifold which has discharge slits on top and outlets at each end. Water is sprayed vertically through the discharge slits from each end of the manifold.

FIGURE 8-26 Surface spray or vertical pump aerator.

This type is commonly referred to as a T-pump or bankwasher and directs oxygenated water along a pond bank where distressed fish often go. Pump sprayers typically have no gear reduction, which reduces mechanical failure and maintenance. These units do not erode the pond bottom, and minimum operating depth is reached when the intake is covered with water.

Paddlewheel Aerators. Paddlewheel aerators have been used on catfish farms for many years. Farm-made paddlewheels are usually made from $3/4$-ton truck differentials and vary with drum size and configuration, shape, number, and length of paddles. Units are powered by PTOs or driven by self-contained diesel engines. The self-contained units are usually on floats and attached to the pond bank or held in place by steel bars secured in the bank or pond bottom.

Increasing either the speed of the drum rotation (rpm) or paddle depth generally increases aeration capacity. Paddle depth affects oxygen transfer rates more than does the speed of rotation. This increase in capacity is not cost-free because horsepower requirements increase and oxygen transfer efficiency may decrease. The maximum rotational speed of a tractor-powered paddlewheel aerator for extended operation is limited by the tractor and the gear reduction of the paddlewheel.

FIGURE 8-27 Portable paddlewheel aerators can be powered by a tractor PTO, mounted engine, or an electric motor.

The shape of the paddles is also important. For example, U, V, or cup shapes are more efficient than flat paddles. Paddlewheels create vibrations that can be reduced when paddles are arranged in a spiral pattern.

The oxygen transfer rate and power requirement increase with paddle immersion depth and the diameter of the paddlewheel drum. The size of the spray pattern likewise increases. The power required to operate a paddlewheel aerator at any given speed and paddle depth is constant. Fuel consumption and operating costs depend on the power source.

Most producers do not have enough paddlewheel aerators for all ponds and move these units from pond to pond. When fish are stressed with low dissolved oxygen, they often go to shallow areas of the pond near the banks. A paddlewheel, though mobile, can be difficult to situate in the pond properly so that it is effective without damaging itself or the tractor.

Electric Paddlewheel. Electric paddlewheel units are 4 to 12 ft (1.2 to 3.7 m) long with paddles of triangular cross section and a total drum diameter of about 28 to 36 in (71 to 91 cm). Paddlewheel speed is usually 80 to 90 rpm with a paddle depth of about 4 in (10 cm), enough to load the motor. The correct paddle depth can be determined in the field as the depth needed to draw the rated amperes of the motor. To extend the service life of the motor, the motor should draw only 90 percent of its full-load ampere rating unless the

FIGURE 8-28 Side wheel style paddlewheel aerator. This aerator requires the PTO of a 45 hp tractor.

FIGURE 8-29 Electric-powered brush aerator. This aerator is anchored to the shore, and two drums on each side float the aerator.

manufacturer recommends differently. Motor sizes range from 0.5 to 19 hp and larger.

Methods used to reduce the motor speed to the desired aerator shaft speed include V-belts and pulleys, chain drive and gears, and gearboxes. Shafts of most electric motors run at 1,750 rpm and most units are mounted on floats.

Diffused Air Systems. Diffuser aerators operated by low pressure air blowers or compressors forcing air through weighted aeration lines or diffuser stones release air bubbles at the pond bottom or several feet below the water surface. Efficiency of oxygen transfer is related to the size of air bubbles released and water depth. The smaller the bubble and the deeper it is released, the more efficient this type aerator becomes. When tested at normal catfish pond depths, these aerators were found to be inefficient compared to other devices.

Limited studies in commercial catfish ponds showed no improvement in fish production when a diffused aeration system was used. One of the biggest problems with diffused-air systems is clogging of the air lines and diffusers so that periodic cleaning is required. Also, the air lines interfere with harvesting.

Diffused air systems are used for supplemental oxygen in trout facilities.

Propeller-aspirator Pump. These aerators consist of a rotating, hollow shaft attached to a motor shaft. The submerged end of the shaft is fitted with an impeller that accelerates the water to a velocity high enough to cause a drop in pressure over the diffusing surface, which pulls air down the hollow shaft. Air passes through a diffuser and enters the water as fine bubbles that are mixed

FIGURE 8-30 Small diffused air system being used to maintain catfish in a small laboratory tank.

into the pond water by the turbulence created by the propeller. These are electrically powered and models range from 0.125 to 25 hp.

Seines

For every 2 ft (0.6 m) of pond width to be seined, 3 ft (0.9 m) of seine length is required. The same ratio applies to pond depth. Floats can be made of styrofoam or plastic attached on 18-in (46 cm) centers.

Most catfish seines have a **mud line** on the bottom of the net. A mud line is made of many strands of rope or a roll of menhaden netting bound together. As the seine is drawn across the pond bottom, the mud line stays on top of the mud, eliminating the digging effect of lead-weighted lines.

Seines should be made of polyethylene or nylon. Catfish spines will not catch in polyethylene material. Nylon netting requires a net treatment to prevent spines from entangling.

The mesh size to be used varies according to the minimum size of the fish to be captured. Buying the proper mesh seine for an operation allows capture of only fish that are large enough for the market. The size of the fish caught varies somewhat with the mesh width and the condition and activity of the fish. Fish do not grade as well when water temperatures are cold.

FIGURE 8-31 After seines are made, they must be treated so they do not rot in water. This "ferris wheel" is used to wrap the treated seine around while it dries.

Live-cars or Socks

Holding live fish is sometimes necessary if the market cannot take all the catch in one day, or because of a delay between capture and hauling fish to market. Often, catfish producers may want to sell fish directly to consumers. In these cases, catfish can be held in **live-cars** or **socks**. Live-cars are net enclosures that can be placed in a pond to temporarily hold the fish (Figures 8-32 and 8-33). They are made of the same material as seines.

Holding fish in live-cars requires caution. Diseases, oxygen stress, weight loss, and poaching are common problems. Aeration may be necessary near the live-car at night, particularly in warm weather. Producers limit the time the fish are held to only a few days to reduce weight loss and prevent disease. Disease can occur in holding devices during any season but is much more prevalent when water temperatures are highest. Poachers can easily steal fish from unguarded holding facilities.

Typical live-cars are 20 to 40 ft (6.1 to 12.2 m) long, 10 ft (3 m) wide, and 4.5 ft (1.4 m) deep, and are constructed of a mesh size that retains harvestable fish and allows the escape of the smaller fish. The escape of market-sized fish is prevented by attaching a line of floats to the upper edge of the live-car, or sometimes by sewing an apron of netting, 1 to 2 ft (0.3 to 0.6 m) wide, with supporting floats, inside the top of the live-car. Stakes anchor the unit in the pond.

FIGURE 8-32 Live-car or sock being made.

FIGURE 8-33 About 40 ft of netting will be "sewn" onto the rope with a float every few feet.

FIGURE 8-34 A small live-car or sock staked for visitors to observe.

Live-cars can be used advantageously by the catfish producer. They can be coupled with a harvesting seine to serve as a temporary holding container and grader. Hauling trucks can then be scheduled accurately because the quantity of fish is known, and the truck can be loaded without delay. A filled live-car can

FIGURE 8-35 Live-car or sock positioned in a pond. Note stakes in the foreground.

be moved from the seine landing area to an area more accessible to the hauling truck. Hauling efficiency is also improved because the fish essentially empty their digestive tracts overnight, and can be precooled by positioning the live-car near the well outlet. Live-cars can be used to hold small quantities of fish for processing or as repositories for selected fish during sorting.

Loading is simplified because fish cannot escape by swimming under the seine. Boom-mounted loading baskets can be lowered into the live-car and filled by crowding fish into them, eliminating time-consuming dipnetting. Workers stand outside the live-car to load fish and are much less likely to be injured by spines of catfish fins.

The fish capacity of a live-car depends on unit size, pond depth and temperature, length of time fish are to be held, and whether a well discharge or a circulating pump is available.

Other Equipment

For harvesting fish, producers may also need—

- A seine reel for hauling in and storing the seine
- Seine stakes
- Tractors

FIGURE 8-36 Boom-mounted brailer lowered into a live-car and filled by crowding the catfish, eliminates time-consuming dipnetting and hand-carrying up the levee to the transport truck.

■ Sturdy dip nets

■ Baskets

■ Boots or chest waders

■ Scales

■ A boom

■ A boat and motor.

Transporting

Transporting live fish requires maximum care to avoid fish losses. In transport, fish are crowded into a relatively small amount of water. Agitators, blowers, compressed oxygen, compressed air, or liquid oxygen can be used individually or in combination to keep the fish alive. Transport containers are usually made of wood, fiberglass, or aluminum. Many types are commercially available.

Generally, the dissolved oxygen content in the water is the factor that determines whether the fish live or die. Fish should not be fed for at least 24 hours before transport so that excessive fish wastes do not accumulate during transport. Fish wastes and regurgitated feed consume large quantities of oxygen and can produce ammonia and carbon dioxide problems.

FIGURE 8-37 Large live-haul truck used to haul catfish to the processing plant. Note the oxygen tanks on the front to supply oxygen to each of the units that holds fish.

Transporting fish in cool weather or in cool well water increases fish survival. Cool water holds more oxygen than warmwater, and fish consume less oxygen at lower temperatures. Also, large fish consume less oxygen by weight than small fish. An oxygen probe in the hauling tank and the meter in the cab of the truck to monitor oxygen concentrations during transport is a good practice.

Fish health and survival depend on the producer's ability to limit stress. Stress from netting, loading, hauling, and stocking weakens the fish and makes them more susceptible to disease and water-quality problems.

Successful hauling begins with the use of good pond management practices and careful harvesting. Fish stressed during production or harvest cannot be transported successfully. Fish are not fed for two to three days before harvesting and hauling because they can be handled much better if the intestinal tract is empty of feed. After harvesting, a common procedure is to haul the fish in cool water, and then temper the hauling water to the receiving water before the fish are unloaded. Preferred water temperature for hauling is 60° to 65°F (16° to 18°C) in summer and 45° to 50°F (7° to 10°C) in winter.

The loading level in the transport unit varies with the size of fish to be transported, length of the haul, and temperature of the hauling water. For example, catfish of 1 to 2 lbs (0.4 to 0.9 kg) can be hauled at the rate of 4 lbs of fish per gallon of transport water if the trip requires less than 12 hours.

Tanks for hauling fish from the harvest area to the holding facilities must be fitted with water-aerating devices suitable for the size and species of fish. Very small fish are sometimes injured by compressed air and agitators, but tolerate much better the small bubbles of (bottled) oxygen coming from carbide aerator stones. Other fish are damaged little by agitators, compressed air, or sprayed water. About 1 lb of fish per gallon of water can be moved in cold weather, but this amount should be reduced to about 0.67 lb in warm weather.

Oxygen Testing Equipment

Intensive culture of fish requires periodic checks of dissolved oxygen levels. During certain times of the year these dissolved oxygen checks must be made several times in each 24-hour period. For small operations, a chemical test kit will suffice. Chemical test kits for 100 oxygen determinations can be purchased. Large operations use electronic oxygen meters to save time and labor in making all of the dissolved oxygen checks required. At least one backup oxygen meter is needed because they can easily be damaged. Additionally, a chemical oxygen test kit is needed to check the accuracy of oxygen meters.

FIGURE 8-38 The convenience of electronic oxygen testing meters taken one more step. The meter probe is on a long pole and the technician quickly takes readings from many ponds.

Tractors

Depending on the operation, at least one tractor—90 to 100 hp—is needed for such things as pulling a feeder, providing power for a paddlewheel aeration device, a **relift pump**, and pulling a seine reel. For commercial catfish production, at least two tractors are needed for four 17.5- to 20-A (7.1 to 8.1 ha) ponds.

FIGURE 8-39 Tractors are used to pull the seine reel and the seine through the pond.

FIGURE 8-40 PTO relift pump for aeration and quick exchange of water in ponds.

PTO Relift Pump

Catfish producers need at least one relift pump for every four 17.5- to 20-A (7.1 to 8.1 ha) ponds. When the discharge end is capped and slotted, this pump is the second-most efficient emergency aeration device. Enough pipe should be available to pump water from one pond to an adjacent pond. Pumping good water from a pond into an adjacent one with low oxygen levels can be a good way to keep fish alive until the problem can be corrected. Also, at times it may be necessary to remove water rapidly from a pond to correct certain water quality problems.

Boats

Boats are used for dispensing certain chemicals directly into the water for the control of diseases, aquatic weeds, and water quality problems. The boat should be powered by an outboard motor of about 10 hp. A trailer may be used for transporting the boat and motor from one pond to the next.

FIGURE 8-41 Boats are used in commercial catfish production during harvest to push the seine and to dispense chemical treatments.

FIGURE 8-42 Typical boat used to harvest crawfish. The wheel serves to hold the boat straight in the wind. This is a so-called "go devil" rig. (Photo courtesy J.V. Huner, Director, Crawfish Research Center, University of Southwestern Louisiana, Lafayette, LA)

Crawfish producers use a boat for setting traps while other aquaculturalists use boats to capture wild broodstock to produce seed.

Trucks

One or more ½- to ¾-ton pickup trucks are necessary for routine work around the farm. The number needed depends on the size of the farm, the type of farm, the number of ponds, and the number of employees.

FIGURE 8-43 Trucks are used for many purposes. This truck is fitted with a small tank for transporting fish.

SUMMARY

Although pond culture still predominates, the use of raceways, tanks, silos, cages, and recirculating systems has increased. As aquaculture production increases and technology improves, these additional culture systems can be expected to become more widely used. Regardless of the culture system, planning is essential for successful aquaculture. Next, the aquaculturalist must completely understand the type of production facility being used.

Aquaculture has created new uses for traditional equipment and new equipment for new jobs. Each aquaculturalist needs to select carefully the best equipment for their facility and be knowledgeable in the use of the equipment.

STUDY/REVIEW

Success in any career requires knowledge. Test your knowledge of this chapter by answering these questions or solving these problems.

True or False

1. In the United States, raceways are the most common type of structure for raising fish.

2. For a potential pond site, the Soil Conservation Service can provide information about soil types.

3. Most commercial catfish ponds on the Mississippi Delta are built on 5-A units.

4. Paddlewheels are used to aerate raceways.

5. A live-car or sock sits on the back of a semi truck and hauls fish to market.

Short Answer

6. Name three types of ponds that could be used for aquaculture.

7. Levee-type ponds are constructed by:
   a. digging a hole
   b. building a dam across a ravine
   c. pushing up dikes on flat land
   d. pouring concrete walls above the ground

8. What size of pond can fish be grown in?
   a. 1 A
   b. 5 A
   c. 0.1 A
   d. any size

9. Two general sources for water to fill ponds or raceways include _____ and _____ water.

10. _____ is an essential soil type for pond construction and the core of dams.

11. Why it is important that the bottom of a pond be level?

12. When drilling a well, who should be able to tell if the well will produce the desired capacity of the facility?
   a. U.S. Geological Survey
   b. U.S. Fish and Wildlife Service
   c. a local well-drilling company
   d. a and c

13. Why is a well usually the best choice for a levee pond for commercial production?

14. Levee-type ponds should be constructed:
   a. on hilly land
   b. on flat land
   c. between mountains
   d. along the coastline

15. If crops were grown on or adjacent to a potential site, what should be checked before proceeding?
   a. pesticide residues
   b. crop residues
   c. organic matter
   d. depth of cultivation

16. List three sources of water for ponds.

17. Name three areas within a pond site that need to be checked for pesticide residues.

18. What is the best shape for a levee-type pond for commercial catfish production?

19. Give two reasons for making main levees 20 ft wide at the top.

20. How does the slope of a levee affect the amount of soil to be moved?

21. Name or describe two types of drains used in levee-type ponds.

22. For a raceway, what is the ratio of the length to the depth and the width?

23. What species can be used for cage culture?

24. Give three basic design principles for all cages.

25. List four problems all recirculating systems must address.

26. Name four types of aerators.

27. A pond is 300 ft wide. What length must the seine be?

28. List two guidelines for live transportation of fish.

29. Name three uses for a tractor in aquaculture.

Essay

30. Describe why clay is essential for some pond construction.

31. Define topography and indicate how it affects pond construction.

32. Define freeboard.

33. Compare raceway construction to levee pond construction.

34. What is a silo?

35. Why would cage culture be beneficial for the novice aquaculturist?

36. What are three disadvantages of cage culture?

37. Why do cages need lids?

38. Why must water 6 ft deep be used for a cylindrical cage that is 4 ft × 4 ft?

39. Why should cages be placed 10 ft apart?

40. What is the role of bacteria in a recirculating system?

41. What is a live-car and how do catfish producers use them?

42. How are boats used in aquaculture?

KNOWLEDGE APPLIED

1. Visit a construction company that specializes in earth moving. Observe equipment used and ask someone to explain how it is used. Also, ask what the cost per yard is for moving soil.

2. With the help of information from the Soil Conservation Service and a good soils manual, determine the type of soil available in your area. Also, using a sample of clay and a sample of sand, mix each with water and observe the different behavior. If possible, invite a Soil Conservation official to visit the class and discuss and demonstrate soil types.

3. If water resources permit, use information resources at the end of this chapter and construct a small cage. If a cage cannot be built, use the information resources to design a cage and develop a list of materials. Then calculate the cost to purchase the materials.

4. Obtain a test kit for measuring dissolved oxygen. Design some simple aeration methods for water samples. Using the test kits determine the effectiveness of these aeration methods.

5. With paper or a computerized drafting program, design a levee-type pond. Determine the slope of the levee, size of the pond, size of the levee, and calculate how many yards of soil will need to be moved to make the levees.

6. Closed systems or recirculating systems use some of the same technology as water treatment or sewage treatment plants. Visit a water or sewage treatment plant and compare the technology being used with that necessary for a recirculating system in aquaculture. As an alternative, invite an official from the local sewage treatment plant to visit class and discuss water quality.

LEARNING/TEACHING AIDS

Books

Beem, M. (1987). *Building cages for fish farming.* Brookings, SD: South Dakota State University.

Beveridge, M. (1987). *Cage aquaculture.* Farnham, Surrey, England: Fishing News Books.

Chakroff, M. (1982). *Freshwater fish pond culture and management.* Washington, DC: Peace Corps, U.S. Government Printing Office.

Dupree, H. K., & Huner, J. V. (1984). *Third report to the fish farmers: The status of warmwater fish farming and progress in fish farming research.* Washington, DC: U.S. Fish and Wildlife Service.

Klontz, G. W. (1991). *Fish for the future: Concepts and methods of intensive aquaculture.* Moscow, ID: Idaho Forest, Wildlife and Range Experiment Station College of Forestry, Wildlife and Range Sciences, University of Idaho.

Lee, J. S., & Newman, M. E. (1992). *Aquaculture: An introduction.* Danville, IL: Interstate Publishers, Inc.

Tchobanoglous, G., & Burton, F. (1991). *Wastewater engineering: Treatment, disposal and reuse* (3rd ed.). New York: McGraw-Hill.

Walker, S. S. (1990). *Aquaculture.* Stillwater, OK: Mid-America Vocational Curriculum Consortium, Inc.

Wheaton, F. (1977). *Aquacultural engineering.* Malabar, FL: Krieger Publishing Company.

Booklets

Backyard aquaculture in Hawaii. (1989). Kaneohe, HI: Windward Community College.

Chamberlain, G. W. (1985). *Texas shrimp farming manual: An update on current technology.* Corpus Christi, TX: Texas Agricultural Extension Service.

Helfrich, L. A., & Libey, G. (N.D.). *Fish farming in recirculating aquaculture systems (RAS).* Blacksburg, VA: Virginia Tech.

Jensen, G. (1989). *Handbook for common calculations in finfish aquaculture.* Baton Rouge, LA: Louisiana State University Agricultural Center.

Lock, J. T., Griffin, W., & Steinbach, D. W. (1988). *Inland aquaculture handbook.* College Station, TX: Texas Aquaculture Association.

Losordo, T.M. (N.D.). *An analysis of biological, economical, and engineering factors affecting the cost of fish production in recirculating aquaculture systems.* Raleigh, NC: North Carolina State University.

Losordo, T.M. (N.D.). *Engineering considerations in closed recirculating systems.* Raleigh, NC: North Carolina State University.

South Carolina Sea Grant Consortium. (1989). *Production guidelines for crawfish farming in South Carolina.* Clemson, SC.

Texas A&M University Sea Grant College Program. (1991). *Practical manual for semi-intensive commercial production of marine shrimp.* College Station, TX.

Wellborn, T. L. (1987). *Catfish farmer's handbook.* Mississippi State University, Cooperative Extension Service.

Williams, K., Schwartz, D. P., Gebhart, G. E., & Maughn, O. E. (1987). *Small scale caged fish culture in Oklahoma farm ponds.* Langston, OK: Cooperative State Research Service, Langston University, and Stillwater, OK: Oklahoma Cooperative Fish and Wildlife Research Unit.

Fact Sheets

Beem, M., & Beghart, G. (N.D.). *Cage culture of rainbow trout.* Langston, OK: Langston University.

de la Bretonne, L. W. (1990). *Crawfish culture: Site selection, pond construction and water quality.* Southern Regional Aquaculture Center.

Jensen, J. W. (1990, August). *Watershed fish production ponds.* Southern Regional Aquaculture Center.

Masser, M. P. (1988). *Cage culture: Cage construction and placement.* Southern Regional Aquaculture Center.

Masser, M. P. (1988). *Cage culture: Harvesting and economics.* Southern Regional Aquaculture Center.

Masser, M. P. (1988). *Cage culture: Site selection and water quality.* Southern Regional Aquaculture Center.

Masser, M. P. (1989). *Cage culture: Species suitable for cage culture.* Southern Regional Aquaculture Center.

Rakocy, J. E. (1989). *Tank culture of tilapia.* Southern Regional Aquaculture Center.

(*Note:* Fact sheets are available from any of the five Regional Aquaculture Centers listed in Appendix Table A-11.)

Software

Aquaculture recirculating systems. (1991). Washington, DC: Aquaculture Information Center, National Agricultural Library, USDA.

Videos

USDA Regional Aquaculture Centers. (N.D.). *Cage Culture: Raising Fish in Cages.*

USDA Regional Aquaculture Centers. (N.D.). *Introduction to Water Quality Testing.*

USDA Regional Aquaculture Centers. (N.D.). *Procedures for Water Quality Management.*

USDA Regional Aquaculture Centers. (N.D.). *Water Quality Dynamics.*

Aquaculture Business

U nder the right conditions and with careful preparation, aquaculture can be profitable, both financially and emotionally. For someone poorly prepared and uninformed, aquaculture can be a disaster. Beginners should consider starting with small, simple systems. For example, much practical and relatively inexpensive experience can be gained by initially growing fish in a few floating cages or in an existing pond or with a small shellfish plot. As experience in production and marketing is gained, the business may expand into larger and more complex operations. Another option is to work with someone who successfully operates an aquabusiness.

Learning Objectives

After completing this chapter, the student should be able to:

- Define terms related to aquacultural business management
- List reasons for keeping records
- Distinguish between basic kinds of records
- List guidelines for building and maintaining a good credit standing
- List factors that a lender looks for in a borrower
- List factors that a borrower looks for in a lender

■ Identify indicators of good loan repayment ability

■ List sources of credit for aquacultural enterprises

■ List the essential components of all budgets

■ Select budgeting principles from a list

■ Define related management terms

■ Describe functions in the management process

■ Identify management considerations in planning an aquabusiness

■ Explain important skills of managers

■ Describe the importance of records and reports

■ Explain important human relations skills

Understanding of this chapter will be enhanced if the following terms are known. Many are defined in the text and others are defined in the glossary.

K E Y T E R M S		
Accounting	Enterprise budget	Liquidity
Accrual accounting	Enterprises	Net worth
Assets	Equity	Partnerships
Balance sheet	Estate	Principal
Break-even analysis	Evaluation	Profitability
Capital	Fixed costs	Risks
Cash-basis accounting	Goals	Shareholders
Cash flow	Income	Sole proprietorship
Collateral	Input costs	Solvency
Corporations	Interest	Strategic planning
Economic	Liabilities	Variable costs

COUNTING THE COST

No one should enter an aquaculture business without counting the costs. These costs can be personal or social considerations that may affect the success of the business venture as well as the actual costs of getting into and staying in business.

Personal/Social Costs

Some of the personal and social costs to consider when entering an aquaculture business come from the following checklist:

1. Are you willing to work long, hard, and irregular hours, for example, 16 hours a day, 7 days a week?

2. Do you get along well and communicate effectively with people? (Small producers not only grow fish or shellfish, they must also promote and market themselves and their product.)

3. Are you comfortable with mathematical problem-solving and mechanical troubleshooting?

4. Will you seek help when needed?

5. Do you personally have the technical expertise with fish or shellfish to manage the operation?

6. Can you afford to hire an experienced technician?

7. Do you know others in the business who will provide help or information?

8. Does your state have an aquaculture association you can join?

9. Are you willing to learn of current practices and new developments?

10. Are you familiar with legal issues of marketing your product?

11. Do you have the resources to construct and operate an approved facility if fish will be processed, for example, dressed and filleted?

12. Is the proposed culture site an unrestricted area, for example, not a right-of-way or wetland?

13. Is the prospective culture site located near the market and processing facilities?

14. Is the proposed site suitable for aquaculture?

15. Do you live close enough to the culture site to visit and monitor as needed and to ensure security?

16. Is an adequate supply of high quality water available and suitable for aquaculture production?

17. Can you control water to, from, and within your system?

18. Can you effectively manage waste produced by your operation?

19. Will your neighbors and other user groups, for example, recreational and commercial fisheries, accept the aquaculture operation?

20. Have you discussed your planned operation with the appropriate local, state, or federal agencies?

21. Have you identified the permits required to construct and operate an aquaculture operation?

22. Have you determined what species you want to culture, and do you know its biology?

23. Have you explored the different production technologies available and identified one that satisfies your interests and resources?

24. Do you have the resources—financial, technical, and spacial—needed?

25. Are disease diagnostic services and dependable technical assistance readily available?

26. Do you have access to a dependable workforce for physical labor?

Actual Dollar Costs

Anyone planning to enter an aquaculture business should make an economic evaluation of the investment. Making a realistic evaluation on paper improves the chances of success once money is committed and reduces the possibility of unpleasant surprises. Then, as the business matures, records are used to continually evaluate the economics of doing business.

Even before a single fish, crayfish, or clam is harvested, many input costs must be considered. These input costs include such items as—

Fry or fingerling (seed)
Feed mixtures
Water chemicals
Structures
Equipment
Water testing
Harvesting
Collection
Aeration
Heater
Supplies
Transportation
Processing and storage
Electricity or fuel power
Antibiotics and sanitation
Labor
Capital
Land
Insurance
Recordkeeping

Income taxes
Interest payments
Depreciation
Licenses.

Table 9-1 lists the inputs for catfish production.

Enterprise Budgeting

Estimating the costs and returns for a particular activity is called developing an enterprise budget. This procedure reflects the economic value of producing a specific output using a given set of inputs by following specific production practices. Profitability can be estimated by subtracting all the costs from the expected revenues.

TABLE 9-1 Selected Nondepreciable Inputs Used in Producing Catfish for Food, Delta of Mississippi

Item	Unit
Catfish floating feed (32% protein)	Ton
Fingerlings	Each
Diesel fuel	Gallon
Electricity	Kwh
Gasoline	Gallon
Earth moving	Cubic yard
Chemicals:	
Medicated feed (Romet-30)	Ton
Potassium permanganate	Pound
Copper sulfate	50-lb bag
Triple superphosphate	50-lb bag
Sodium chloride	50-lb bag
Test kits:	
Alkalinity	Each
Ammonia	Each
Carbon dioxide	Each
Chloride	Each
Nitrate/nitrite	Each
pH measurement	Each
Total hardness	Each

Two types of costs should be considered in developing enterprise budgets: variable and fixed. Variable costs are the expenses that vary based on production output, such as feed, fingerlings, etc. Fixed costs are the expenses that do not change, regardless of whether production occurs, such as depreciation, interest on investment, insurance, taxes, etc. Figure 9-1 shows that variable costs make up the largest portion of the total costs of fish production. In an examination of variable costs alone, feed comprises almost two-thirds of the costs, with fingerlings coming in a distant second, as shown in Figure 9-2.

Tables 9-2, 9-3, and 9-4 provide example enterprise budgets for catfish, trout, and crawfish.

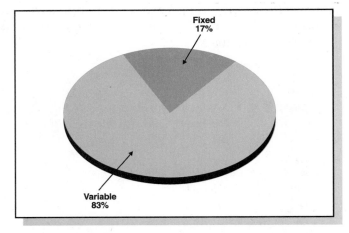

FIGURE 9-1 Variable costs make up the largest portion of the total cost of fish production.

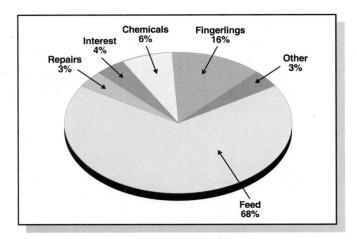

FIGURE 9-2 Feed comprises almost two-thirds of the variable costs, with fingerlings coming in a distant second.

TABLE 9-2 Catfish Budget for an Open Pond[1]

Item	Weight Each	Unit	Quantity	Price or Cost/Unit	Value or Cost
1. Gross Receipts					
Catfish	1.00	Lbs	32,463.00	0.65	21,100.95
2. Variable Costs					
Fingerlings (4-inch)		Each	35,000.00	0.07	2,450.00
Floating Feed (32%)		Ton	34.68	250.00	8,670.70
Chemicals		Appl./A	1.00	100.00	1,000.00
Harvest Labor		Hr.	32.00	4.00	128.00
Tractor (Fuel, Oil, Lube)		Hr.	82.00	2.25	184.50
Electricity		Kwh	10,800.00	0.07	756.00
Miscellaneous		A	10.00	5.00	50.00
Machine and Equipment (Repair)		Dol.			358.28
Interest on Operating Capital		Dol.	5,155.48	0.12	618.66
Total Variable Costs					14,216.14
3. Income Above Variable Costs					6,884.81
4. Fixed Costs					
Interest on Building and Equipment		Dol.	9,424.50	0.12	1,130.94
Depreciation on Building and Equipment		Dol.			1,583.10
Other Fixed Charges on Building and Equipment		Dol.			133.30
Total Fixed Costs					2,947.34
5. Total of All Specified Expenses					17,163.48
6. Net Returns Above All Specified Expenses					3,937.47
Net Returns Per Acre:	Above Specified Variable Expenses				688.48
	Above Specified Total Expenses				393.75
Break-even Price (Per Cwt. Sold):	To Cover Specified Variable Expenses				43.79
	To Cover Specified Total Expenses				52.87

[1] Assumptions for 10 A include: Stocking in spring, custom harvest in fall; estimated annual costs/returns; using recommended management practices; 3,500 fish stocked per A; 20 lb/1,000 beginning weight; 2 lb of feed/lb of gain; 200 days in growing season; 1 lb ending weight; 7.25% death loss/unharvested fish. Net returns are to land, existing pond, operator's labor, and management. These estimates should be used as guides for planning purposes only.

TABLE 9-3 Costs and Returns for Commercial Trout Production[1]

	Total Expenses	Total Income
Returns		
17,600 lb fish @ $1.50		$26,400
Cash Costs		
Repairs	$ 1,216	
Repair labor 60 hours @ $5.00	300	
Eggs 66,000 @ $43/1,000	2,840	
Feed 22,000 lb @ $.25/lb	5,500	
Electricity 450 kwh, $186.00/month	2,232	
Labor (feeding, processing, marketing, etc.) 744 hours @ $5.00/hr	3,720	
Insurance and license	1,000	
Miscellaneous, 5% of cash costs	782	
Interest on cash costs @ 12%	745	
Total cash costs	18,335	
Return over cash costs		8,065
Fixed Costs		
Interest on average investment	2,139	
Depreciation on buildings and equipment	1,690	
Return to management		$4,236

[1] Average cash expenses outstanding December through October at 12%. (Source: Southern Regional Aquaculture Center (1990), *Budgets for Trout Production.*)

Each producer faces different situations when trying to analyze the economic feasibility of fish production. So, budget estimates in enterprise budgets should be used only as a starting point for planning.

MANAGING THE BUSINESS

Aquaculture is a business like any other agricultural business. It requires expert management. Management time is critical to the success of the business. Management involves—

TABLE 9-4 Estimated Annual Operating Costs Associated with a 40-A Crawfish Pond in Southwestern Louisiana

	Cost
Variable Costs—Items	
Forage	$ 1,641
Fuel/Well	1,834
Repairs and Maintenance	1,059
Labor ($5/hr)	2,918
Herbicides	157
Sacks	146
Bait ($0.16/lb)	5,835
Total Variable Costs	$13,590
Fixed Costs	
Depreciation	6,978
Interest (12%)	3,605
Total Fixed Cost	$10,583
Total Annual Cost	$24,173

- Setting goals and objectives
- Recognizing and identifying problems
- Responding and acting when problems occur
- Seeking, compiling, and using relevant information
- Considering and analyzing alternative courses of action
- Making specific decisions
- Carrying out decisions or taking action
- Accepting responsibility for these decisions
- Evaluating the results of these decisions
- Developing training programs for family members and employees
- Directing and evaluating family members and employees
- Making buy and sell decisions
- Controlling financial operations
- Organizing the use of resources

■ Establishing the timing of operations

■ Monitoring operations and check up on everything.

Although the list overlaps it could easily be expanded. The list does indicate a lot of things that managers do. That is why good full-time managers are crucial to most sizable businesses and why time must be set aside for management in any business, even if the principal laborer is also the manager.

Management Functions

At the core of management concepts is a set of goals and objectives for the business, developed, and understood with clarity by the owner, by management, and by labor. Expectations about levels of annual earnings and production, maintenance of farm buildings and grounds, tradeoffs between capital appreciation and current earnings, long-term growth, and achievements must be established. While these goals and objectives are not always formalized in writing, they need to be reasoned and discussed.

Figure 9-3 suggests five basic management functions or activities used to achieve the goals and objectives of a business.

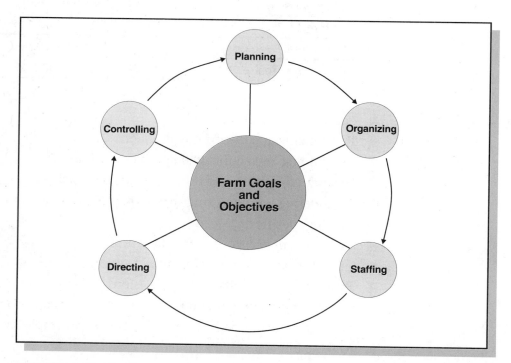

FIGURE 9-3 Five basic management functions or activities used to achieve the goals and objectives of a business.

Planning. While all five of the basic functions are important, planning is crucial because a good plan involves all the other functions. Planning involves—

- Setting daily priorities and schedules. What should be included in today's "To Do" List? Who should complete each priority activity?

- Recognizing problem areas and looking for alternative solutions. Why did production drop last month? Did we have the protein level right in the food? Did feed quality change? Should a consultant look at the ponds?

- Making a financial plan and cash-flow statement for the year, knowing when and how much credit must be obtained, and where the cash will come from to meet the regular obligations.

- Looking at alternative marketing plans.

- Establishing the overall enterprises for the business.

- Developing the business. How fast should the business grow? Is new staff needed? What professional development is needed for each manager?

Planning cannot be done just once a year. It is an ongoing process. Plans get revised when goals are not being reached. Most planning deserves undivided attention without interruptions. A well-spent hour with a banker, with the computer, or in discussion with a trusted neighbor or partner may save a lot of money, time, and energy.

Organizing. Organizing is establishing an internal structure of the roles and activities required to meet the farm's goals. The manager must define the positions to be filled, the duties, responsibilities, and authority attached to each, and coordinate efforts among people.

Organizing includes—

- Deciding who reports to whom. This is often referred to as the chain of command.

- Determining the functions in each position (job design), including the degree of authority.

- Establishing the work routines and standard operating procedures for each production enterprise.

Staffing. Staffing is as crucial a management function to a small or part-time business as it is to a much larger one. Often, the need to figure out how to get all the jobs done on time is even more critical. No business should try to operate without the possibility of hiring assistance when needed. Assistance can range from hiring a teenager after school to help with a few operations to contracting with an accountant to prepare tax records.

Recruiting and hiring workers. Whether a business needs one full-time worker, two or three part-time helpers on a regular basis, or hourly help for seasonal work, maintaining a competent force is essential. Labor management starts with obtaining qualified workers who understand what is expected of them. Written job descriptions ensure that the manager has thought through what is expected. Terms of compensation and benefits must be established—another reason for having something in writing.

Training and evaluating workers. Managing an aquabusiness means that someone takes responsibility for assigning tasks and making sure that workers understand how to do their jobs and what is expected of them. Incentives for high levels of achievement help. Telling people when they did something well may be even more important that telling them when they did something wrong. Both are necessary. This is even more important when all the workers are family members.

Directing. Directing is closely related to staffing. The smaller the business, the more the two are interlocked. Delegation of authority is often one of the most difficult things for the manager of a small business to accomplish. All the workers need to know their responsibilities and have a sense of when they can make decisions and when the boss must be involved. The lines of authority become more crucial with more employees.

Motivation is part of directing. Knowing what is going on and listening to employee concerns helps build communication and confidence. Creating a team spirit where every worker feels some responsibility for the success or failure of the operation is desirable. Openness and understanding by a manager are respected in close working relationships.

Controlling. Controlling is another key function. Control is part of business management that determines what new methods are needed to turn out positive results when an investment decision is proven to be less profitable than planned. Control requires keeping track of expenses and income. It forces a manager to monitor what is happening every day.

Some of the important activities that are a part of controlling include:

Monitoring the records and accounts of the operations. These records can always be kept by someone besides the manager, but the work must be done on a regular basis. The manager analyzes these records to know what is going on.

As with any business, good managers pay close attention to those factors that affect profits most. Table 9-5 summarizes the sensitivity of major price and production efficiency factors and their effect on profit potential. For example, for every 0.1 lb that the feed conversion rate can be lowered, the cost of production would decrease by $1.69 for every 100 lbs of catfish produced.

Fish production requires a great deal of money. The overall net worth and cash flow of the potential producer should be large enough to withstand both the start-up period and any unforeseen setbacks.

Comparing rates of production and levels of performance or productivity against established goals or generally accepted standards. Control ensures that these comparisons are made systematically and discussed with the people directly involved. Problems in production arising from natural causes need to be recognized and allowed for in good management.

Monitoring production processes and making changes as necessary. Adjusting when to treat, when to feed, and when to start and stop harvest are all results of control. Keeping track of the work routines and making sure that plans are accomplished (or revised) makes the difference.

Organizing and operating an aquaculture business requires a manager to make and carry out many decisions. Some decisions take time and study. Others cannot wait until tomorrow. Part of the satisfaction of operating a business is seeing a major change in enterprises work better than expected, solving a nutrition problem that was finally recognized, or knowing that a pond was harvested when conditions where best. Good management allows the other farm resources to be used effectively. Understanding the functions of management is one step in becoming a better manager.

TABLE 9-5 Sensitivity of Price and Production Factors and Their Impact on Profits for a 10-A Pond

Item	Unit Change	Dollars Per Cwt. Sold
Pond Construction	$100/A	0.42
Death Loss	1%	0.41
Stocking Rate	500/A	2.57
Feed Conversion	0.1 lb	1.69
Feed Price	$25/ton	2.24
Interest Rate	1 point	0.44
Fingerlings	1 cent each	1.13

PLANNING—THE SECRET OF BUSINESS SUCCESS

What separates a successful aquabusiness from an unsuccessful one? Numerous factors—quality of the land, water managerial skill, and sufficient equity capital—are all important. Yet, some businesses that seem to have these basics are less successful than other businesses that are not so well endowed.

An important attribute of good business management is to be able to step away from the immediate concerns to see the future. Strategic planning is analyzing the business and the environment in which it operates in order to create a broad plan for the future.

For smaller aquabusinesses, the most effective planning may take place at the kitchen table. To establish an appropriate atmosphere for strategic planning requires setting aside time away from day-to-day problems and interruptions so that the key participants—owners, managers, family members—can reach a common understanding about what they want to do in the next three to five years and how they want to do it.

Management needs to take a broad overview of the economy and the industry to determine the major opportunities and threats. Tactical planning is concerned with day-to-day and week-to-week decisions, such as what and how much pesticide to use, when to harvest fish, or whether to overhaul the old tractor or buy a new one. The results of strategic planning could lead to new enterprises, major capital investments, or perhaps even an exit from the business. This broader focus over a longer time distinguishes strategic planning from tactical planning.

FIGURE 9-4 Sometimes the decision must be to hold on to the old tractor instead of buying a new one.

Why Do Strategic Planning?

Strategic planning permits more profits, in the long run, by—

■ Establishing a clear direction for management and employees to follow

■ Defining in measurable terms what is most important for the firm

■ Anticipating problems and taking steps to eliminate them

■ Allocating resources (labor, machinery and equipment, building and capital) more efficiently

■ Establishing a basis for evaluating the performance of management and key employees

■ Providing a management framework that can be used to facilitate quick response to changed conditions, unplanned events, and deviations from plans.

Steps in Strategic Planning

Strategic planning involves the first seven steps shown in Figure 9-5. An eighth step—implementation—is strategic management.

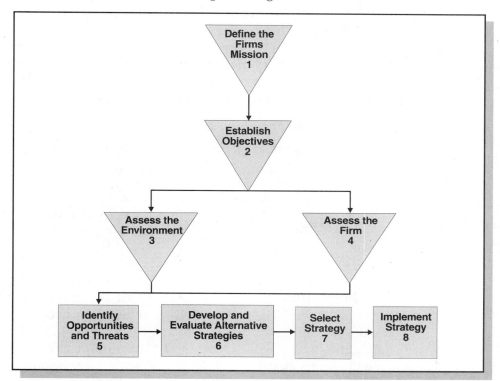

FIGURE 9-5 Strategic planning involves the first seven steps.

Step 1: Define the Mission. The mission statement defines the purposes of the organization and answers the question, "What business or businesses are we in?" Defining the aquabusiness's mission forces the owner-operator or manager carefully to identify the products, enterprises, and/or services toward which the organization's production is oriented. This statement answers the question, "What is our current situation?" For example—

- What markets are likely to produce the best opportunities?
- What type of agricultural commodities or services can we produce to take advantage of these opportunities?
- What, if any, other activities are we involved in, and what are the priorities of these activities?

Answering these questions will suggest goals that will help to clarify objectives in the next step.

A mission statement is not necessarily a long document. In fact, it should contain fewer than 100 words, and two or three sentences may be sufficient. Establishing strategic goals is the key element of the mission statement. The mission statement answers the question, "Why are we in business?"

Step 2: Establish Objectives. Goals, which are the more general, long-term desires of owner-operators or managers, clarify the aquabusiness's purpose. Objectives should translate the mission into concrete terms. Objectives should be measurable and straightforward statements such as the following—

- Increase sales by 100 percent in the next five years
- Reduce labor costs by 25 percent in the next three years
- Increase production per acre of pond by 30 percent in the next five years
- Add four more raceways in the next two years.

These objectives should be chosen in such a way that they contribute to reaching the goals identified in Step 1. Each objective is measurable and time-limited. This allows management to evaluate progress in implementing the plan.

Step 3: Assess the External Environment. Every agricultural firm faces uncertainties, threats, and opportunities that are beyond its control. Market forces may cause prices to plunge, either in the long-run or short-run. Overproduction, declining consumer demand, a strong dollar, high interest rates, changes in government policies, and regulation of labor and pesticides are external threats that can cut profits or make business more difficult. New market opportunities are created by demographic changes, changing consumer lifestyles, regional population growth, and technological breakthroughs.

The operator/manager must understand the economic, social, and technological forces that will affect the firm. Then reasonable expectations may be formulated about what will happen to product prices, interest rates, the rate of inflation, labor markets, and input prices over the next three to five years.

Step 4: Assess the Organization's Strengths and Weaknesses. The quality and quantity of resources within the control of the operator/manager is the first part of this assessment. What are the abilities and limitations of the operator/manager? What skills and abilities do the employees have? How modern and efficient is the physical plant? How large is the resource base? How much water is available? What is the cash position of the farm? These resources need to be compared against those of competitors.

Many farms have an unrealistic view of their own resources and operation because they do not compare themselves to others in the same business. The process of providing candid answers to these questions forces the operator/manager to recognize that every business is constrained in some way to its physical resources as well as its human skills and abilities.

Step 5: Identify Opportunities and Threats. This step combines the data gathered in Steps 3 and 4 to determine the threats and opportunities the business might encounter in the planning period. Difficulties in the external environment can present opportunities in another segment of agriculture. For example, concern about cholesterol in meat products created new markets for poultry and fish. Concern about carcinogens in the environment, including some pesticides, brought about new opportunities. Some firms have avoided problems by creatively turning an external threat into an opportunity.

Step 6: Develop and Evaluate Alternative Strategies. Steps 6 and 7 are at the heart of the strategic planning process. This is the point at which the business develops the alternative plans that describe the methods for achieving objectives and obtaining greater long-run profits.

In what ways can production agriculture firms gain a competitive advantage? The answer to this question depends to a great extent on whether or not the farmer is a price-taker. Farmers are traditionally price-takers. They take the price set by others who control the market.

Some types of strategies operator/managers use to gain a competitive advantage include—

1. Become more efficient. Increase profits by—

   ■ Reducing input use, holding product and price (quality) constant

   ■ Using more, or higher quality, inputs increasing revenue more than costs.

2. Seek out alternative enterprises.

3. Exploit quality differences. Obtain price premiums for quality that more than offset the additional costs involved in producing higher quality commodities.

4. Integrate horizontally. Farm more units, or add enterprises, enlarge enterprises to gain more complete use of existing resources. Acquire additional resources. Spread fixed costs over more units of output.

5. Integrate vertically. Obtain more profit by moving higher or lower into the marketing and distribution channels—add storage or packing facilities, trucks to haul products, and direct marketing; acquire resources to produce inputs that formerly were purchased.

6. Reduce risks through diversification and hedging.

7. Identify new markets or narrow (niche) markets.

Organizations that have some degree of control in the market because of fewer competitors or because of the possibility for differentiation of products and services have the potential for additional strategies to gain a competitive advantage.

Possibly, no identified alternatives will permit a particular farm family to attain its objectives. Nonfarming alternatives may need consideration. For example, to sell some farm assets, keeping part or all of the land and residence, and to seek off-farm employment. Nonfarming alternatives should not be neglected in selecting alternatives.

Once alternatives are developed, Step 6 is only half completed. These alternative strategies must be evaluated. In practice, management may develop a long list of possible alternatives. These can usually be whittled down with reasoning and logic. Once the obvious losers are eliminated, "pencil-pushing" is in order. No single or preferred method is used for evaluating alternatives, but some combination of the following may be used—

1. Budgeting alternatives, both profitability and cash flow

2. Break-even analysis

3. Projections of income, cash flow, and balance sheet statements

4. Computerized decision aids.

Table 9-6 shows a monthly cash flow budget for a trout farm. Income for the operation occurs only in November, yet expenses occur every month.

Step 7: Select a Strategy. From the analysis in Step 6, the firm selects a strategy—an alternative or a combination of alternatives—that will enable the operator/manager to achieve the desired objectives. Evaluating alternatives may show that the original objectives are not feasible. The operator/manager

TABLE 9-6 Cash Flow and Monthly Labor Requirements for Trout[1]

Item	Total	Jan	Feb	Mar	April	May	June	July	Aug	Sept	Oct	Nov	Dec
Cash Receipts 17,600 lbs @ $1.50	$26,400											$26,400	
Cash Expenses:													
Repairs	$1,216	$200				$200	$200	$200	$100	$100			$216
Eggs, eyes	2,840											$2,840	
Feed	5,500					400	600	700	900	1,100	$1,200	600	
Electricity	2,232	186	186	186	186	186	186	186	186	186	186	186	186
Labor, all	3,720	60	200	200	250	200	200	200	200	200	150	1,800	60
Insurance and license	1,000							1,000					
Miscellaneous	782	67	65	65	65	65	65	65	65	65	65	65	65
Interest on cash expenses	625						625						
Monthly Net	$8,485	(513)	(451)	(451)	(501)	(1,051)	(1,876)	(2,351)	(1,451)	(1,651)	(1,601)	$20,909	(527)
Labor Hours	744	12	40	40	40	40	40	40	40	40	30	360	12

[1] *Trout Farming in Washington*, Cooperative Extension, Washington State University.

NO PLAN, NO MONEY

A potential lender or investor wants to see a written business plan. The plan can be used as a guide in developing the business. Business plans can have different formats depending on the type and source of funding being sought or the general purpose. If funding is already in place, a plan can easily be used as a guide for strategy. Successful business plans contain certain elements.

Title page. The title page should minimally include four pieces of information: the name of the proposed project, the name of the business, the principals involved, and the address and phone number of the primary contact.

Table of contents. The table of contents should include the major topics of the body of the business plan and critical tables or figures that the investor should take particular notice. Every detail of the plan should not be included in the table of contents. Too much detail takes too much effort by the reader to try to find the critical aspects of the business. The main function of the table of contents is to guide the reader to the critical areas of the business plan.

Statement of purpose. This section is a brief mission statement of the aquaculture venture: what is to be accomplished, why this project was chosen, and how it is to be done. The statement should include outside funding requirements and describe the repayment plan and the source of repayment.

Executive summary. This section presents the key elements of the business plan to prospective lenders/investors. The length should be under five pages to increase its likelihood of being read. The summary should begin with a brief restatement of the purpose of the project. Next, the venture's specific products, markets, and business objectives should be described along with why the idea has a competitive advantage. The specifics of the management team should then be presented with the various strengths and weaknesses—and solutions for the weaknesses. Lastly, the financial aspects of the venture should be summarized showing projected returns, outside financing requirements, and timing of cash flows.

The business. The next section of the business plan provides details of the venture. The following points should be addressed:

- History. You should describe your present organization, including when it was founded, progress made to date, present form—partnership, corporation, etc.—past financing, and prior aquaculture successes and experiences.

- Description. The proposed venture should be laid out concisely. The location, size, products, facility design, and installation procedures need to be described. This section answers in detail what the venture intends to do, such as what type of production technique will be used, water conditions of the chosen site, etc. Certain justifications for the facility and products should be given when it is a critical assumption of the business. For example, the production technique—intensive, extensive, or semi-intensive—should carry a cost/benefit justification.

- Market. The business plan should include a description of the market in which the aquaculture product will be sold.

- Marketing. A section on how the venture will specifically cater its product to the market is necessary. This section should include a product description and how you will be taking your product to the market you described in the above section.

- Competition. An analysis of significant competitors should be presented. This section includes a listing of direct competitors such as other aquaculturists within the market segment and indirect competition such as fisheries products, imitation fish products, and fish imports. Possible reactions from competitors and the effects your fish or shellfish will have on the market should be addressed. Strategic moves by the venture in relation to competitors should be discussed and explained.

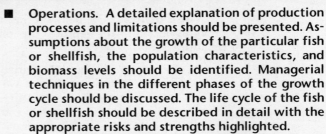

- **Operations.** A detailed explanation of production processes and limitations should be presented. Assumptions about the growth of the particular fish or shellfish, the population characteristics, and biomass levels should be identified. Managerial techniques in the different phases of the growth cycle should be discussed. The life cycle of the fish or shellfish should be described in detail with the appropriate risks and strengths highlighted.

  A suggested technique for describing operations is to divide the project into its separate functions. Each function can be explained in detail along with its importance. The functions can then be brought together to form the whole system of operations.

- **Management.** This section discusses how the venture will run. The organizational structure is described to show lines of authority and responsibility. An organizational chart should be presented depicting clear functional duties and communication lines. Each critical functional area—pond, hatchery, maintenance—requires managerial control and feedback descriptions. The type and style of management can be discussed. For example, the fish farm may be run by a sole proprietor who subcontracts most of the labor or a highly integrated team that performs all functions internally.

- **Research and Development (optional).** If the aquaculture venture is involved with new techniques or new fish or shellfish that have not been raised before, this section should be included. The discussion should include whether expertise is to be hired or developed from within. Details such as cost, expected accomplishments, and timing should be described.

- **Personnel.** The complete personnel plan is presented in this section to show the expected hiring needs and personnel policies over the planning horizon of the venture. The expected number of full- and part-time employees, the amount of over-

time, and any seasonal trends should be included. Also, fringe benefits and control policies need to be listed as well as whether it is critical to keep employees. A demonstration of the knowledge of any laws that may effect employees should be included.

■ Loan Application and Effects. This section presents any source of external financing and describes why it is necessary and what benefits are expected. Borrowing for equipment, land, feed, and other items should be listed with an explanation of the terms, critical aspects (such as options), the collateral required, and why borrowing makes sense. Any assumptions concerning financing should be spelled out.

■ Development Schedule. The timing of the venture development can be presented by chart or some other form showing critical dates. This section should include the decision milestones and guide the reader through each stage of the venture. An example is a calendar chart showing various stages of the venture and any completion dates. Any information such as dependence on governmental agencies, weather, or equipment manufactures should be included.

■ Summary. This section should present in abbreviated form all prior sections and information about the business. It is important to highlight the most important facts and assumptions and exclude the details.

Financial plan. This section includes sources and applications for capital, equipment list break-even analysis, pro forma balance sheet, pro forma income statement, pro forma cash-flow budget, historical financial statements, equity capitalization, debt capitalization, and supporting documents.

Still, those who fail to plan, plan to fail—especially without money.

may have to move back to Step 2 and select new objectives or reformulate combinations of alternatives. Selection of a final strategy may involve trade-offs among goals. An alternative is seldom likely to be superior to all other alternatives for attainment of each of the goals of the operator/manager and his or her family. The process of strategic planning should be recognized more as an art than a science.

Step 8: Implementation. Implementation is a crucial link in the strategic management chain. Management must periodically look back on the plan and determine how well the business is reaching its objectives. Assessing implementation will point to mid-course corrections. Assessment enables planners to understand the planning process. Perhaps objectives were set too optimistically or perhaps critical threats or opportunities were not recognized. Recognizing and correcting the plan's weaknesses will improve strategic planning the next time the process is undertaken.

Strategic planning should not be viewed as a formidable task resulting in detailed plans. It should be written, but a few pages will suffice. The process should include all the key players participating in the strategic management discussion. All individuals involved in managing the farm or ranch must understand where the business is going, how it plans to get there, and what problems or opportunities lie ahead.

SETTING GOALS FOR BUSINESS MANAGEMENT DECISIONS

Almost everyone is enthusiastic about goals. Most people like to discuss goals and some boast of having goals. Goals are definitively known and are important.

People who teach management also stress the importance of goals. Listen to almost any management guru to hear ideas like these—

■ Identify your goals. Manage to reach them.

■ Management is goal-directed.

■ Take charge of your life and work. Set goals and attain them.

■ Without goals, you cannot be a manager because you will not know what you want to achieve through your management decisions.

Almost everyone agrees on the importance of goals. The paradox of goals is this: Many people will publicly affirm that they have identified their goals and that goals are important. Most, though, cannot or will not record and communicate their desired outcomes in a goal statement that will guide their management decisions.

The communication, negotiation, and compromise required for goal identification yield additional important benefits. When goals are selected in a way that ensures that each person does work that he or she enjoys, motivation increases and management performance improves. Perceptions of reality are modified as participants gain a greater understanding of each other's roles, interests, and activities. Identifying goals has both immediate and long-term payoffs—the quality of daily management outcomes and focus of long-term decisions are improved.

Those who regularly set and write down goals report benefits like—

■ Communication among family members improved.

■ Management decisions and work activities effectively focused on priority concerns.

■ Cash-flow management in the production unit and household improved as impulse buying of production inputs and household items declined.

■ Borrowing, risk, and interest expense reduced.

■ Conflict reduced, and working relationships improved.

■ Expenses were kept under control, and profits increased.

■ Anxiety and concern over the present and future reduced.

■ A better balance between production activities and family life was achieved.

Goals and commitment, a combination that cannot be beaten, ensures that the aquaculture business will grow, change, and remain profitable.

BUSINESS AND RISKY DECISIONS

Aquaculture is a high-stress industry. The management of the business is fraught with risk and uncertainty. Aquacultural managers must consider the **risks** associated with the ever-changing political, social, economic, and ecological environment in which they operate.

Types of Risks

Identifying the different events or sources of risk that affect the outcome of a decision is a crucial step in the decision-making process. The relative importance of the sources of agricultural risk differs among enterprises and changes over time. Risks in aquaculture include market, production, financial, obsolescence, casualty, legal, and human.

Market Risk. The variability and unpredictability of the prices that farmers receive for their products and what they pay for production costs are market risks. Fluctuating supply and demand conditions result in price variations.

Production Risk. This source of risk is a result of the variability in production caused by such unpredictable factors as weather, disease, pests, genetic variations, and timing of practices. Examples include variations in yields, machinery breakdowns, and feed conversion efficiencies.

Financial Risk. Financing assets that the business controls creates risk. The increased use of borrowed capital leaves the operator vulnerable to not having enough cash to meet obligations or of not having adequate credit. Other examples of this source of risk include the possibility of losing the lease on the land and the ultimate disaster—bankruptcy.

Obsolescence Risk. The rapid development of new technology can make current production methods obsolete shortly after important investments have been made. The possibility of adopting new technologies too soon or too late is a risk farmers face.

Casualty Loss Risk. This a traditional source of risk referring to the loss of assets as a result of such events as fire, wind, hail, flood, and theft.

Legal Risk. Governmental laws and regulations are a growing source of uncertainty for farmers. Changing social attitudes have resulted in laws and

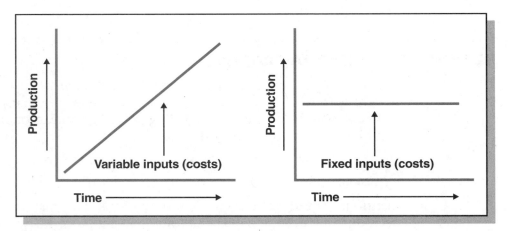

FIGURE 9-6 Managers understand the financial risks of increasing productivity when variable inputs increase and fixed inputs remain flat.

regulations governing environmental protection, water quality, food safety, and other farm-related matters. In addition, there is a risk of lawsuits resulting from accidents and other events.

Human Risk. The character, health, and behavior of individuals are unpredictable and contribute to the risk in farm management. The possibility of losing a key employee during a critical production period is one example of this type of risk. Dishonesty and undependability of business associates are other examples. Also, family needs and goals change, sometimes unpredictably.

Psychological studies suggest that business managers tend to overlook risk considerations as they make decisions. They do not deal with risk. When it comes to making decisions in today's risky agricultural climate, the wise manager must explicitly consider various sources of risk.

Managers respond to risk in different ways. Just as we classify people as being optimistic or pessimistic, conservative or liberal, we can also classify people according to their attitudes about taking risks—risk avoiders or risk takers.

The risk takers are the plungers, the more adventurous types who willingly make risky decisions. They are willing to accept greater risk in return for the small chance of a higher income.

Risk avoiders are the more conservative types who have a preference for less risky decisions. Risk avoiders are willing to sacrifice the small chance of higher income for less risk.

The framework for making risky decisions means managers must choose among alternative actions, the outcomes of which depend on events beyond their control. The outcome of each combination of choices and events is known as a pay-off.

For some risky decisions, managers use intuitive thought processes. Many decisions are too complex and important to be handled by intuition alone. A more formal approach such as a matrix or statistical comparison provides the discipline to ensure that all available information has been used.

Risk analysis does not simplify decision making or eliminate the agony of making difficult choices. Risk analysis does not eliminate risk, but it can help the manager select the right risks to take in the often uncertain world of aquaculture business.

BUSINESS STRUCTURES

A business entity is the legal structure under which a farm or any business is organized and operated. Aquaculture owners can establish their businesses as sole proprietorships, partnerships, or corporations. Whether individuals inherit or receive a farm through gifts, they must decide on the type of business structure for the farm. The selected structure often changes as the farm grows or new individuals enter the business. Individuals who originally owned and operated their business as a sole proprietorship, for example, may choose to shift to a corporation, a partnership, or a multiple business organization.

The sole proprietorship is the most common form of business organization since most small businesses are owned and operated by a single individual. Sole proprietorships have a common law origin and can be easily established and operated because the business structure is an extension of an individual's rights and responsibilities in property ownership and commercial transactions. Partnerships also have a common law origin, and thus have many of the characteristics of a sole proprietorship.

By contrast, incorporation has a statutory origin, which means state laws prescribe a corporation's structure, procedures, and conditions of organization and operations. Incorporating a farm business requires a series of legal steps. Corporate activities are closely regulated.

Sole Proprietorship

Personal and business objectives help decide the best organization structure for a small business. The sole proprietorship is usually best suited for a beginning business because it is the simplest and least regulated of all business types. No legal papers must be filed to establish and maintain the business. Since the proprietor owns and operates the business as an individual, records and planning are limited to those needed to reach management objectives, to file personal income tax returns, and to comply with laws and regulations common to all business ventures. The major drawback is that sole proprietorships, unlike some types of corporations, do not offer protection from personal liabilities.

Although the sole proprietorship is the simplest business structure, financial management considerations and family objectives may make this structure inappropriate as an enterprise becomes larger and more complex. The advantages of partnerships and corporations over sole proprietorship are too complex to warrant broad generalizations. The decision to shift to a new business structure must be made on a case-by-case basis. Each farm owner must decide if and when to move to a more complex and formal organizational structure.

Multiple Ownership

Increasing capital requirements and the economies of scale available to large operations led to the evolution of multiple ownership of agricultural enterprises. Increased capital requirements make it difficult for young people to start their own enterprises, and many young people enter into the ownership and management of their parents' businesses.

Because a sole proprietorship is, by definition, organized and operated by one individual, the transfer of a business to the next generation requires a business organization that accommodates multiple owners—either a partnership or a corporation. In some cases, separate proprietorships—with joint ownership of equipment and labor exchanges—may be established between parties. Separate proprietorships may be more feasible for some enterprises because of their large capital investment in facilities and equipment.

Shared Management

A large business with multiple owners, whether a partnership or a corporation, offers a chance to divide management responsibility among the partners or stockholder employees. Joint management decision making provides excellent on-the-job management training for less experienced managers. Partnership and corporate structures are equally flexible in the development of a management team that meets the needs of each business.

Income Sharing

Multiple ownership and management in a partnership or corporate structure offer many avenues for distributing income among the respective parties. A partnership pays no income taxes because the individual partners assume their own tax liabilities. Income can be shared through drawing accounts and distribution of residual income. If partners lease assets to the partnership, lease payments can compensate owners for their resources.

The corporate structure can distribute income among stockholder-employees in the form of salaries, dividends, or interest on debentures. Payments to stockholders must be reasonable and based upon services rendered, but there is much flexibility in sharing income among stockholders and employees.

Capital Transfer and Estate Planning

Capital transfer among common property owners in a partnership or corporation is a significant consideration when family members decide to continue the enterprise as an operating unit beyond the retirement of the present owners. With proper planning, the partnership and corporate structure can be used to reserve resources for retirement, transfer property to family members, and minimize expenses and transfer taxes.

Regardless of the business structure—sole proprietorship, partnership, or corporation—it is possible to develop a sound **estate** plan. The capital transfer through the estate can be handled with jointly held property ownership, wills, and trust arrangements. Although the partnership or corporate structures do not in themselves solve estate transfer problems, they can make capital transfer somewhat easier.

Attracting Capital

The traditional sources of capital for small farms are the equity provided by family members, reinvestment of retained earnings, lease agreements, and loans. Capital sources are the same regardless of the organization's structure. The sole proprietorship may be the most limited in terms of capital acquisition because only one family is involved in the operations. Multiple ownership through a partnership or corporation allows the combining of funds from more than one family, which results in a larger business.

Table 9-7 indicates the capital needed to start a small trout operation. Depending on location, this can be even more.

TABLE 9-7 Establishment Costs for a Trout Farm Producing Less Than 100,000 Pounds[1]

Category	Units	Price	Quantity	Value
Site preparation[2]				$187.50
Concrete floor[3]	Yd^3	$54.00	9.07	489.78
Concrete walls[3]	Yd^3	$54.00	7.39	399.06
Reinforcing steel	Pair			232.50
Drainpipe	Pair			232.50
Screening	Pair			37.50
Tank forms	Pair			137.50
Snapties and wedges	Pair			250.00
Labor	Hourly	$5.00	197.50	987.50
Water intake assembly	Pair			812.50
Miscellaneous	Pair			625.00
Total Establishment Cost				$4,391.34

[1] Estimated costs for establishment of one pair of up to ten pairs of tanks constructed in series. More than one series may be constructed. Expected annual production of 6,000 lbs per pair. (Source: Southern Regional Aquaculture Center (1990), *Budgets for Trout Production*.)
[2] Each tank based on dimensions of 35 ft long by 6 ft wide.
[3] Floor and walls at 6 in thick.

Federal Income Taxes

A sole proprietor's business pays no federal income tax. Instead, the taxable income of the business is included in the proprietor's personal income, and taxes are paid at the individual tax rates. Federal income taxes for a partnership are treated in a similar manner. The partnership files an information return showing the income and expenses, the names of the partners, and how the partnership earnings will be divided among the partners. The profits, losses, capital gains and losses, and tax credits are allocated to partners according to the terms of the partnership agreement. The partners pay taxes as individuals on their respective shares of partnership income.

Federal income tax savings may occur if a business incorporates and becomes subject to federal income taxation under Subchapter C of the Internal Revenue Code. Because a corporation is considered a separate taxpayer, the corporation can divide income among the corporation, owner-operator employees, and shareholders. The corporation pays individuals associated with the corporation for their contributions. Owner-employees receive a salary for their labor, and management and shareholders receive dividends for their capital investment. Residual income after all expenses are paid is taxed to the corporation at corporate income tax rates. Whether federal income taxes will be lower after incorporation depends upon the corporation's earnings level, the tax rates for individuals versus that for corporations, and the allocation of earnings.

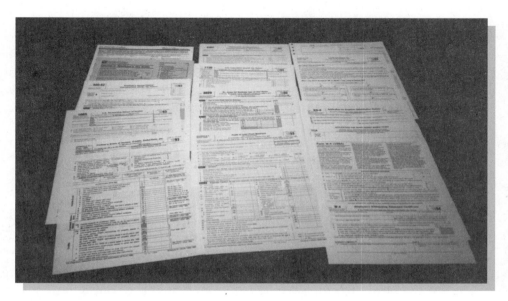

FIGURE 9-7 Federal income tax is a fact of life of all businesses.

When the corporation is owned primarily by a family, the tax objective is to minimize the family's total annual income tax burden. This means that the total taxes paid by the corporation, in addition to the personal income taxes paid on the stockholder-employee's salary, and any other personal income should be less than the total personal income taxes paid by the owners before incorporation.

Another tax advantage of incorporation is the increased business deductions available because the owners who work for the corporation become employees of the corporation. In addition to the employee's salary, the corporation can take a deduction for fringe benefits such as group life insurance plans, medical and hospital plans, pension and profit sharing plans, and others. It permits the corporation to use pre-tax dollars to pay for benefits received by a stockholder that the same individual not in a corporation would acquire by using after-tax dollars. This results in more after-tax total income available to the stockholder-employees.

A disadvantage of Subchapter C corporations is that double taxation is possible. It occurs when corporations pay dividends to their shareholders. Dividends are distributed from the corporation's after-tax income, and shareholders must include dividends in their taxable income. Thus, shareholders are in effect paying taxes a second time on the same profits.

If a corporation elects to be taxed under the special tax option or Subchapter S method, the corporation is not a taxpayer for income tax purposes. That is, the corporation itself is not taxed on income. The income of the corporation "flows through" to the shareholders and each shareholder pays a tax on the individual's prorated share of the corporation's earnings when filing an individual income tax return. All income is taxed the year it is earned whether it is retained or distributed. Subchapter S rules are similar to partnership rules in that an information return is filed annually on behalf of the corporation.

Thus, corporate earnings in a Subchapter S corporation are taxed only once—to the shareholder. This avoids the double taxation possibility present with Subchapter C corporations.

Just because federal income taxes may be reduced by incorporation, not all taxes and costs will necessarily be reduced. Rather, many increased costs and taxes exist with corporations. All of these must be examined in arriving at the total savings possible by incorporation.

Payroll Taxes

After incorporation, the sole proprietor or partner changes status from employer to employee. The business has at least one additional employee, if not more, which results in increased payroll taxes.

Social Security taxes are increased since the combined employee and employer rates under the corporate structure are higher than for self-employed individuals—partners or sole proprietors.

Stockholders-employees of corporations are also subject to Worker's Compensation charges on their salaries and are entitled to benefits under the Act. This is not true of sole proprietors or partners in a partnership. A stockholder-employee's salary may also be subject to the unemployment compensation tax.

Owners and operators of corporations and sole proprietors have to file personal income taxes through quarterly estimates or withholding rather than as a lump sum.

Structure Must Fit Objectives

The initial business organizational type for a small-scale family business is usually a sole proprietorship. When circumstances surrounding the operation suggest a partnership or corporation, an in-depth analysis needs to be made. An analysis of the organizational characteristics and the objectives of the family is perhaps the most important, but still the most neglected, phase of the process.

Usually, the decision does not need to be rushed. It is relatively easy and inexpensive to incorporate or form a partnership, but it may not be so easy and inexpensive to dissolve a corporation or partnership. Those thinking about changing business organizations should take enough time to weigh the advantages and disadvantages of each structure for their particular situation.

RECORDS IMPROVE PROFITABILITY

Managers need a complete and accurate farm records system in order to make informed management decisions that help maintain or improve business profitability. Records systems serve four functions—

1. As a service tool to assist in reporting to the Internal Revenue Service and other taxing entities, creditors, other asset owners, and to others who have a vested interest in the financial position of the business

2. As an indicator of progress

3. As a diagnostic tool for identifying strengths and weaknesses

4. As a planning tool.

With the proper records managers can determine the component cost of annual ownership and the annual operating cost per pound of product as shown in Tables 9-8 and 9-9. This information helps the manager make management decisions by comparing these costs to averages for the industry or historical information for the aquabusiness.

Records can also help the manager plan and implement farm business arrangements and do estate and other transfer planning. Managers can use records to determine efficiencies and inefficiencies, measure progress of the business, and plan for the future.

Aquaculture business managers do not need to be accomplished accountants or experts on taxes and law. They do need to know how to keep the required records for their businesses; they must realize that all business decisions

TABLE 9-8 Estimated Annual Ownership Cost Components Expressed in Cents Per Pound of Catfish Harvested, Delta of Mississippi[1]

Item	Cents Per Pound
Depreciation:	
Ponds	1.77
Water supply (well, pumps, motors, and outlet pipes)	0.23
Feeding (feeder with electronic scales and storage)	0.12
Disease, parasite and weed control equipment (boat, motor, and trailer)	0.04
Miscellaneous equipment	2.65
Total Depreciation	4.81
Interest on Investment:	
Land	2.23
Pond construction	0.97
Water supply (wells, pumps, motors, and outlet pipes)	0.24
Feeding (feeder with electronic scales and storage)	0.10
Disease, parasite and weed control equipment (boat, motor, and trailer)	0.01
Miscellaneous equipment	1.15
Total Interest on Investment	4.70
Taxes and Insurance	0.33
Total Annual Ownership Cost	9.85

[1] Source: *Economic Analysis of Farm-Raised Catfish Production in Mississippi.*

TABLE 9-9 Estimated Annual Operating Cost Components Expressed in Cents Per Pound of Catfish Harvested, Delta of Mississippi[1]

Item	Cents Per Pound
Repairs and Maintenance:	
Vegetative cover	0.23
Water supply (wells, pumps, motors, and outlet pipes)	0.13
Feeding (feeder with electronic scales and storage)	0.02
Disease, parasite and weed control equipment (boat, motor, and trailer)	0.02
Miscellaneous equipment	1.31
Total Repairs and Maintenance	1.72
Fuel:	
Mowing	0.05
Feeding	0.31
Outboard motor	0.02
Electric floating paddlewheels	0.86
PTO-driven aerators and low-lift pump	0.45
Pumping	1.28
Transportation	0.50
Total Fuel	3.46
Chemicals	2.38
Telephone expenses	0.19
Test kits	0.04
Fingerlings	6.45
Feed (32% protein)	24.50
Labor:	
Operations management	2.74
Hired labor	4.85
Total labor	7.59
Harvesting and hauling	4.00
Liability insurance	0.25
Interest on operating capital	2.35
Interest on fish inventory	0.25
Total Annual Operating Cost	53.19

[1] Source: *Economic Analysis of Farm-Raised Catfish Production in Mississippi.*

have income tax consequences; and they must be able to evaluate the accounting and legal professionals who serve their businesses.

Choosing a Records System

Records systems range from simple, hand accounting systems using pencil and paper to sophisticated double-entry computer accounting systems. Some require a mix of hand and computer operations.

A system should not only meet the accounting and planning needs of the operation, but it should also satisfy income tax, legal, and other outside reporting requirements. Programs should be selected with good detailed instructions for use.

Accounting Methods

Two types of accounting methods are used in farming—cash basis and accrual.

Cash-basis Accounting. This method is used primarily for income tax reporting purposes in service industries. Generally in cash-basis accounting, income is recorded as income when it is received and expenses are recorded as expenses when they are paid. Cash-basis accounting is simple and can provide some income tax advantages for businesses that are heavily dependent on inventory changes.

This method also has drawbacks. Cash-basis accounting can grossly distort the financial position, profitability measures, and operational results of the farm business. It is necessary to convert cash-basis accounting to accrual accounting for analysis and decision-making purposes.

Accrual Accounting. This method is required for tax purposes for most trading and manufacturing businesses. In accrual accounting, expenses are considered expenses when they are accrued (or committed) and income is counted as income when it is earned. This includes changes in inventories. This method does not depend on how the cash moves in the business. Expenses incurred are matched with related income to determine net income. This approach provides a better continuous picture of profitability. An assessment of cash flow is still needed to determine the financial feasibility of the business.

Basic Recordkeeping

Recordkeeping need not be a complex managerial activity if some simple rules are followed. A well-designed farm record system makes the job easier as well as more efficient.

Tips for Better Recordkeeping. Six suggestions for better recordkeeping in an aquaculture business include—

1. Always record the gross or total amount. Never, never net it out.
2. Always go through all the steps for each transaction.
3. Run everything through a checking account.
4. Separate business income and expenses from personal income and expenses.
5. Do periodic accuracy checks.
6. Staple your calculator tape to each page as you total your book so you can refer back to it. Do not do your work twice—once is enough!

Items one and two fit together as do items three and four.

Tax Records

The Internal Revenue Service requires a set of farm records to show all taxable income and expenses that are deductible. This can be done in many different formats. The manager or recordkeeper must maintain accounts to show the three different types of farm income: sale of "resale" (purchased) items, other ordinary income, and sale of capital items. Records must also be kept of the two types of expenses—ordinary expenses and capital expenses—along with some

FIGURE 9-8 Whether records are maintained by hand or with a computer, they are important for success.

expenses that could be classified in either category. Included in the expense category is the annual depreciation record.

The record system chosen should support items on a tax return. The records must provide evidence of the types of income and expenses. This requires sales slips, invoices, receipts, deposit records, and canceled checks. Income and expenses should be clearly identified. Records of loans, debt repayment, and interest expenses must be kept as long as they have any income tax or legal ramifications.

Other required records might include capital item records, Social Security records, Occupational Safety and Health Administration (OSHA) records, Federal Unemployment Tax records, worker's compensation, retirement plans, health insurance, operating agreements, carryovers and carrybacks, net operating losses, and income tax credits.

Balance Sheet

This statement summarizes three of the five accounts in a complete accounting system. The general accounting equation for the balance sheet is: assets equal debt plus equity. Phrased another way, assets minus debt equals equity.

The balance sheet is divided vertically into the left part, called assets (what the business owns), and the right part, called liabilities (what the business owes). The total of the two parts must be equal. Two kinds of liabilities are included: (1) debt or outside capital and (2) equity (net worth) or inside capital. The debt represents claims lenders have on the assets while equity represents claim owners have on the assets.

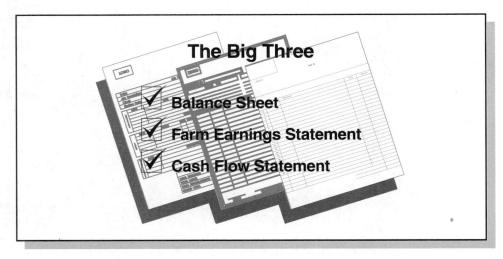

FIGURE 9-9 The big three.

Horizontally, the balance sheet can be broken into three categories.

1. **Current Assets.** The first category, current assets, contains those assets that are in cash or are usually turned into cash during the course of the year. For tax purposes, they are assets that would be considered ordinary income if sold or ordinary expenses if purchased.

2. **Intermediate Assets.** The second category includes intermediate assets. They are not true current assets but neither are they true long-term assets. They are assets used in the production of income and are generally viewed as non-real estate property, such as machinery and productive animals.

3. **Long-Term Assets.** The third asset category is composed of long-term assets. These generally include real estate property used for producing income.

An asset's length of life is sometimes used to distinguish between the three asset types. For example, assets that last less than one year are current, those that last from one to ten years are intermediate, and those that last more than ten years are long-term. Some accountants use only two categories, current and long-term.

A space for notations should be provided at the end of the balance sheet. Here, contingent liabilities and assets can be noted along with other salient information, such as the asset valuation methods used, that helps explain the data in the balance sheet.

FIGURE 9-10 Ponds represent a long-term asset. (Photo courtesy Chuck Weirich, Delta Research and Extension Center, Stoneville, MS)

Asset value. Determining the appropriate asset values is the biggest challenge when developing a balance sheet. The values selected depend on their use. It is best to have a double asset column balance sheet. Then, two sets of values can be shown for analysis purposes. One value should be the market value, which is what a willing buyer would pay a willing seller (given adequate time and sufficient knowledge). Credit worthiness and loan soundness are measured using this column. Another value sometimes included is the adjusted tax basis, but this data is readily available from properly kept tax records.

The balance sheet shows where the money is invested and how the business is financed. It provides a snapshot of the financial position of the business at a particular point in time. It shows the financial and credit soundness of the business. The balance sheet provides comparative data that can be used for evaluating the business and for developing the farm earnings statement.

Farm Earnings Statement

The second of the "big three" statements focuses on current activity. It shows the income earned by the business before taxes and contains the other two of the five accounts in a complete accounting system. The general accounting equation is: sales minus cost of goods sold minus operating expenses, plus or minus inventory and capital adjustments equals income before taxes. Or, simply, revenue minus expenses equals income before taxes.

The earnings statement is divided into three sections—

1. Cash operating statement
2. Adjustments for inventory
3. Adjustments for capital items

The first section shows all cash income and cash expenses and produces a figure called net cash farm income. The second section shows the inventory adjustment, which results in a figure called adjusted net farm operating income. The inventory adjustment is the difference between the ending current assets and beginning current assets, adjusted for changes in accounts payable. The third section shows the capital account adjustment, which results in net farm earnings—or the return to unpaid labor, unpaid management, and equity capital. The capital adjustment is the difference between the intermediate and long-term assets at the end of the year and the intermediate and long-term assets at the beginning of the year.

The earnings statement ties together the information from the balance sheet with cash-basis income tax accounting data. The bottom line is an excellent measure of the profitability of the farm business.

FORM 8A

BALANCE SHEET

NAME:

Market   Depreciated Cost   DATE:

CURRENT FARM ASSETS		Line No.	Value
Cash, checking balance		14	
Prepaid expenses and supplies		15	
Growing crops		16	
Accounts receivable		17	
Hedging accounts		18	
		19	

Crops held for sale or feed	Line No.	Crop Code	Quantity	
	20			
	21			
	22			
	23			
	24			
	25			
	26			
	27			
	28			
	29			

Crops under govt. loan	Line No.	Crop Code	Quantity	
	35			
	36			
	37			
	38			
	39			
	40			

Livestock held for sale	Line No.	Lvstk Code	Quantity	
	45			
	46			
	47			
	48			
	49			
	50			
	51			

Total Current Farm Assets   60

INTERMEDIATE FARM ASSETS

Breeding livestock	Number

Farm machinery and equipment

Total Intermediate Farm Assets   65

LONG TERM FARM ASSETS

Farm real estate	Acres

FLB stock and co-op equity

Total Long Term Farm Assets   70

TOTAL FARM ASSETS

NONFARM ASSETS

Vehicles	
Household goods	
Cash value of life insurance	
Stocks and bonds	

Total Nonfarm Assets   75

TOTAL ASSETS

CURRENT FARM LIABILITIES	Line No.	Amount
Farm accounts payable and accrued expenses		

Judgments and liens

Estimated/Accrued Taxes:
    Property
    Income Tax and Social Security

Accrued Interest: Current
                 Intermediate
                 Long term

Subtotal accounts payable and accrued expenses   79

Current farm notes payable	Due Date	Interest Rate	Annual Installment	Amount Delinquent	Principal Balance

Total Current Farm Liabilities   80

INTERMEDIATE FARM LIABILITIES

Description	Due Date	Interest Rate	Annual Installment	Amount Delinquent	Principal Balance

Total Intermediate Farm Liabilities   85

LONG TERM FARM LIABILITIES

Description	Due Date	Interest Rate	Annual Installment	Amount Delinquent	Principal Balance

Total Long Term Farm Liabilities   90

TOTAL FARM LIABILITIES

NONFARM LIABILITIES

Nonfarm accounts payable and accrued expenses

Nonfarm notes payable	Due Date	Interest Rate	Annual Installment	Amount Delinquent	Principal Balance

Total Nonfarm Liabilities   85

TOTAL LIABILITIES

NET WORTH

FIGURE 9-11 The balance sheet.

Cash-Flow Statement

The most action-oriented of the "big three," the cash-flow statement shows how cash moves into and out of the business. The general accounting equation is: inflows equal outflows. A complete cash-flow can also serve as a cash accuracy check.

Many different formats for developing a cash-flow statement are available. One way is to divide the cash-flow into four sections:

1. Income, which is the marketing plan
2. Operating expenses, which is the production plan
3. Capital purchases, which is the investment plan
4. Principal, interest, and additional borrowing, which is the debt service plan.

This type of organization gives a better perspective on total cash-flow and aids in planning and control.

Three columns are necessary for each accounting period. Then, one set of these columns can be for each month or at least for each quarter. The first column would be called projected, the second column would be called actual, and the third column would be called variance. In this fashion, the cash-flow statement can be used as a financial management control tool. In cash-flow planning for income, operating expenses, and investment, the business manager is asking, "How much am I going to sell or buy? At what unit price am I going to buy or sell? At what time am I going to buy or sell?"

Debt-service information can be obtained from credit records and the balance sheet. A two- to three-year cash-flow history is useful. Then, the manager can find out how this year is going to differ from previous years. This helps make budgeting easier and more accurate.

The cash-flow statement is useful as an evaluation, control, and planning tool. But used by itself, it can relay false information because it only considers cash. For best results, the cash-flow statement should be used with the balance sheet and earnings statement. Used together, the "big three" provide a complete set of financial statements.

Other Key Accounts

Several other accounts feed into or supplement the five accounts in the "big three" financial statements. These include income accounts, expense accounts, capital item accounts, depreciation records, enterprise accounts, labor records, marketing records, feed records, experimental records, individual machine records, and family records.

Cash-Flow Worksheet

FOR YEAR → 1995	JAN	FEB	MAR	APR	MAY	JUNE	JULY	TOTAL/YR
BALANCE ON HAND								
DESCRIPTION								
INCOME:								
CROP SALES								
FISH SALES								
CAPITAL ITEMS								
CUSTOM WORK								
INTEREST INCOME								
OTHER INCOME								
TOTAL INCOME								
EXPENSES:								
HIRED LABOR								
TAXES								
INSURANCE								
LEASE RENT								
LOAN PAYMENT								
CHEMICALS								
PESTICIDES								
FINGERLINGS								
FUEL, OIL, ETC.								
REPAIRS								
CUSTOM WORK								
FEED PURCHASES								
OTHER EXPENSES								
SUPPLIES								
OTHER EXPENSES								
TOTAL EXPENDITURES								
NET INCOME								
LIVING EXPENSES								
ENDING BALANCE								
AMOUNT TO BORROW								

FIGURE 9-12 A sample cash flow for six months used for projecting or tracking cash, income, and expenditures each month.

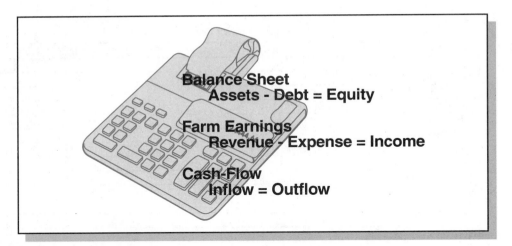

FIGURE 9-13 Formulas for the big three.

Accuracy Checks. Single-entry cash-basis accounting can result in significant errors. It is best to balance the checkbook against the record book on a monthly basis. Then at the end of the year, the manager can make three accuracy checks—

1. Cash-flow
2. Profit/net worth
3. Liabilities.

When these three accuracy checks balance, the business manager can proceed to file income tax and use the records to analyze and manage the business.

USING AN ACCOUNTING SYSTEM FOR ANALYSIS

Before decisions can be made or analyzed, the information necessary for the decisions must be available. The primary goal of any farm or ranch accounting system should be to provide business management analysis and control. The accounting system should be geared toward the farm or ranch manager. If the accounting system is not used, it is worthless. Many uses of the accounting system relate to individuals other than the manager, so the system must be able to provide financial information for them too. The accounting system supports major management functions by providing the information necessary for making decisions. The accounting system should supply three types of information—

■ Scorekeeping, or evaluating performance (generally, a retrospective look available in the financial statements)

■ Attention directing, to flag ongoing operating problems, inefficiencies, and opportunities (identified through analysis of the financial statements)

■ Problem solving or analyzing the relative merits of alternative courses of action.

The accounting system provides information the farm manager needs for external reporting for tax and credit purposes. Also accounting provides financial control of routine operations, business management analysis, and reporting to multiple owners.

Tax Requirements

The Internal Revenue Service (IRS) and most state income tax authorities require that enough business records be kept to justify all income and expense claims reported on an income tax return. The lack of standardized requirements for a minimum acceptable set of records has led agribusinesses to store their cash register receipts, invoices, bank statements, and canceled checks in a box or file drawer and do little more. Legally, such records are sufficient. This system can become an extremely expensive one during an IRS examination.

Other Taxes and Investments

Complete and accurate records minimize problems with estate, gift, and property taxes. The ability to participate in investments outside the normal business activities can be enhanced by having the information readily available to determine whether a particular investment opportunity is financially feasible.

Credit Application

Lenders stress repayment capacity of loans as well as **collateral** security. Most borrowers now need to show that the investment for which the loan is intended will be able to generate enough income to pay back the interest and principal owed within the specified time period.

Financial Control of Routine Operations

Astute business managers concern themselves with cash-flow management. How much to borrow, either in long-term credit or in operating credit, is only half the story. When and how much to pay back is just as important as tight control of cash reserves. Paying operating money back after a sale may not be the wisest operation if it puts the business in a cash-flow bind later. Interest charges must be analyzed in addition to liquidity needs of the business and of the family. Preparing a realistic cash-flow budget is one of the vital steps in the annual recordkeeping process. A cash-flow budget is a projection of anticipated cash receipts and cash expenditures, by category, for a future time period, typically 12 months. Borrowing and repayment plans are included. Cash budgeting involves all the steps required in the whole farm planning process: marketing, including price projections for inputs as well as outputs, yield projections, and enterprise combinations.

Despite the difficulty of preparation, the cash-flow budget helps document managerial abilities and loan repayment capacities. The cash-flow budgeting process can be extended one more step to provide an extremely effective financial control device. If monitored monthly or quarterly, the cash-flow budget can indicate potential problems before they arise. This ability to foresee problems allows the manager to adjust before the fact rather than react afterward.

Business Management Analysis for Strategic Planning

If a farmer is disciplined enough to develop and maintain a records system to meet income tax reporting and credit application needs, then virtually all the needed information will be available to meet what is probably the most important goal of a farm or ranch records system—business management analysis. Good business managers know exactly what their variable and total costs of production are. They know whether they are meeting the goals of their marketing plans or their cash-flow budgets. They have analyzed their strengths and weaknesses, both in physical terms and financial terms. They know where their business has been, where it is now, and where it is going.

Corporations and Partnerships

Multiple-owner forms of business organization require more detailed records because of more intricate tax reporting requirements, state corporation laws, and additional documentation needs of lenders. Perhaps the most important need for more detailed records in partnerships and corporations comes from the likelihood of problems and potential conflicts among the individuals involved.

Lease and Family Distributions

Individuals involved in informal family partnerships, joint ventures, and share leases need to rely on a detailed records system to ensure fairness in distribution of profits and contributions.

Uses and Interpretations of the Statements

Just doing the scorekeeping—producing the financial statements—is not enough. It takes interpretation and analysis of the financial information to meet the attention-directing and problem-solving needs.

Interpretation begins by evaluating net worth, a key measure of financial wealth. On a market value basis, net worth shows what would be left if all assets were converted to cash and all liabilities paid. Next, the net income should be sufficient to meet withdrawals for family consumption. More cannot be taken out of the business than is earned. On a cost basis, the change in net worth from one year-end balance sheet to the next equals net income minus withdrawals. The next step is to use data from the financial statements for a systematic financial analysis of the operation.

Financial Analysis. The first step in financial analysis is to identify appropriate criteria that will facilitate a comprehensive analysis, and then measures for each criterion must be established. For each of the following five criteria, one or more measures are suggested.

Liquidity. A concept describing a firm's ability to meet short-run obligations when due without disrupting the normal operation of the business. The ratio of current assets to current liabilities is a common measure.

Solvency. A longer-run concept relating to capital structure and a firm's ability to pay all obligations if assets were liquidated. The focus is on total debt in relation to equity. It is a financial risk measure because the risk of not being able to repay borrowed capital and interest increases as the proportion of debt to net worth increases. Another equally useful measure is debt as a percentage of total assets.

Profitability. Relates to revenue less expenses, called net farm income. But a dollar measure of net farm income is not sufficient because the size of business is not considered. Furthermore, net farm income is typically a return to unpaid labor, management, and capital in contrast to other businesses, where it is a return only to capital. Return on assets and return on equity are two common measures of profitability. Net farm income is typically adjusted to get a return to capital expressed as a ratio to total assets.

Financial efficiency. A measure of the efficiency of a business in generating profit out of gross production. The secret of a business is to maximize the dollar value of profit out of each $1,000 value of farm production—a measure of gross production. Net farm income divided by value of farm production is

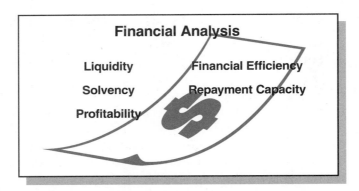

FIGURE 9-14 Criteria for financial analysis.

one useful measure. Similarly, operating expenses, interest, and depreciation can individually be evaluated as a proportion of the value of farm production.

Repayment capacity. An assessment of the firm's ability to repay debt. Ability to repay capital debt and interest is a major concern. One measure is all interest plus principal payments on capital debts, expressed as a percentage of the value of farm production. A non-ratio method—capital debt repayment capacity—is calculated as net income plus depreciation less withdrawals.

How Are We Doing?

Financial measures and ratios can be interpreted three ways: (1) comparative analysis, (2) trend analysis, and (3) actual versus budgeted. Comparative analysis is a comparison of one operation's results with those of operations of comparable size and type. For example, if an operation's debt to asset ratio is 40 percent, how does this compare with the debt level of other successful operators? Trend analysis compares results in one year with results achieved in past years. A trend analysis shows strengths and weaknesses and helps focus attention on areas where further strengthening is needed. Comparison of actual performance with the cash-flow budget requires developing an operational plan for the year ahead and then comparing monthly or quarterly performance with projections. Management should focus on variances, or the differences between budgeted and actual performance.

COMPUTERS AND MANAGEMENT DECISIONS

Agribusiness has never been more dynamic and competitive than it is today. Each decision a manager makes, or fails to make, can significantly impact the business. In some cases, a decision can affect a single production cycle of one enterprise. In others, a decision can change the direction of an entire operation. In this fast-paced, high-risk climate, computers can play an important part in helping managers make crucial decisions about their farms.

Computer programs help managers make a wide range of management decisions. Some programs are designed for strategic management, which is concerned with positioning the farm for success by matching the business's long-range direction with resources, management capabilities, and the economic environment of the industry. Other programs address tactical management, which focuses on the day-to-day, season-to-season activities needed to carry out the long-range strategic plans.

The success of strategic or tactical decision making depends to a large degree on managers' access to relevant information and their ability to use that information effectively in making decisions. Today's computer programs can help gather important data, provide a framework for analyzing options and perform calculations thoroughly and accurately in a fraction of the time it would take to do the same thing with pencil and paper.

Farmers' most important strategic management decisions deal with deciding the long-range direction of their farm businesses. Each must decide what enterprise, or combination of enterprises, offers the best long-term potential, how big the business should be, the type of financing needed, the amount of debt that can be handled, and how to ensure adequate profit

FIGURE 9-15 Once a luxury, now computers are essential to successful business management.

from the business now and in the future. The most effective way to approach these and other strategic questions is to identify a wide range of options and then narrow the field to the most feasible plans. The decisions made must be consistent with both business and family goals, available resources, the management ability available, and the risk-bearing capacity of the business and the people involved.

Tactical Decisions

While strategic management addresses long-range plans and objectives, tactical management focuses on the activities that move the farm toward those goals. Computer programs are available to help aquabusiness managers monitor or analyze production practices, develop financing plans, establish labor schedules, and create marketing plans.

Computer software can help managers develop marketing plans, keep pond records, track breeding programs, and many other tactical decisions. As with strategic planning, computers can speed up the decision-making process, reduce mathematical errors, and help the farm manager think critically about alternative courses of action.

Once a manager develops annual and long-range plans and answers, the how and when questions, the next step is to compare what is actually happening on the farm with expectations for production, marketing, and finances. Computers can help carry out the important and time-consuming task of monitoring farm operations on a daily basis. They can also help managers make adjustments when performance fails to meet expectations.

Having access to such a system depends on what records the manager is willing and able to keep on a day-to-day or week-to-week basis. Whether they are kept in a hand or a computerized system, they must be such that they can be summarized and analyzed at any point in time. The computer clearly has the advantage here in terms of quickly recalling structured data, calculating the desired measures, and detailing comparisons of plans to the actual outcomes in a timely fashion.

Computers for Decision Making

Of course, the manager must have the right kind of information for each decision, whether strategic or tactical. This includes information about the world at large as well as about the farm itself. For strategic decision making, the manager must keep abreast of general economic conditions, world supply and demand, credit policy, and so forth. In the area of tactical planning, the manager must keep up to date on current and future market prices, weather information, and other factors. While much of this information can be gleaned from the general news media, managers can often get information tailored specifically for their concerns through commercial computerized agricultural information networks.

When it comes to information about farm operations, many managers have been content to keep only the data necessary for income tax preparation. But in today's dynamic and competitive world, that approach is no longer sufficient. At a minimum, the manager must maintain production data in a useful format as well as information on assets, liabilities, and all credit transactions.

As with any technology, the value of the new computerized management tools depends on how conscientiously and wisely they are used. For the system to reach its potential, the manager must be willing to record vital data on a regular basis and use the resulting information and analyses in making crucial business decisions.

OBTAINING CREDIT

Sound use of agricultural credit is a two-way street affecting both borrower and lender. The individual seeking credit must be prepared to demonstrate to the lending institution that the proposed financing is feasible.

In any borrower-lender relationship, the borrower supplies up-to-date financial and production records to provide an understanding of the business. Financial records include a balance sheet and an income statement, as well as historical and projected cash flows. If possible, three to five years of financial and production data are desirable. Many lenders today are asking for income tax returns for three years.

On the other hand, it is the lender's responsibility to analyze these documents in a logical and systematic manner. This results in a timely decision on the borrower's credit worthiness. While good financial management is the primary responsibility of the borrower, both lender and borrower must use sound credit practices.

Selecting a Lender

Selecting a lender or lenders is a critical aspect of financial management. An owner/operator should shop for credit and investigate several sources before making a final decision. The borrower must be prepared to make judgments as well as be judged. Five guidelines to use in rating the quality of the credit service are—

1. Select a knowledgeable lender who understands aquaculture and agriculture today. Agriculture, undergoing rapid technological change, is

beset with unique problems and opportunities. The lender must be able to demonstrate up-to-date knowledge of problems, trends, and modern aquaculture practices specific to the particular enterprise and geographic region. Lenders need a track record that shows an understanding of its customers and a genuine interest in and concern for the customer's welfare and financial progress.

2. Select a lender who has experience in agricultural credit and a commitment to agriculture. In some years, a depressed agricultural economy caused some lending institutions to exit agriculture. As a borrower, examine the lender's farm loan experience. Check its reputation by asking other aquaculturalists. Evaluate the lending institution's commitment to agriculture and service by looking at its track record during periods of adversity.

3. Choose a lender who is willing to discuss lending policies and terms and provide prompt action to credit requests. While investigating a source of credit and related services, compare the terms of credit with other available sources. Total credit charges are more important than interest rates alone. Credit expenses, such as up-front charges for farm loan application and closing fees, can add substantially to credit cost. Examine fixed- and variable-rate interest options and determine the associated costs, benefits, and risks.

   Arrange repayment terms flexible enough to prevent undue hardship in case of special needs or emergencies. Payment schedules should mesh with the anticipated cashflow generated by the business. Investigate the privilege of prepaying without penalty.

   Timely action on loan requests should be a high priority in selecting a lender. A delayed credit decision can hinder crop and production schedules and ultimately affect cash flow and profits. The lender must have adequate information to make a sound credit decision.

4. Choose a lender who has the capacity to meet anticipated credit needs. Agricultural businesses frequently need large sums of capital. This could create a roadblock in the credit process. Some institutions may have a statutory limit placed on the amount of credit they can extend to any one individual or business.

5. Select a lender who has a reputation for honesty and integrity. In our market-based economy, customer service is essential in doing business. Seek a lender familiar with and skilled in financial and production analysis. Periodic visits by the lender to an operation show sincerity and concern and enhance the lender's understanding of the business. Lenders should explain all services offered in practical and understandable terms.

A lender with a reputation for honesty will judge potential borrowers on the same basis. A strong borrower/lender relationship is one of mutual confidence. Maintaining confidentiality of information and objectively evaluating a situation—backing credit decisions with facts—are strong attributes to consider in selecting a lender.

Preparing for the Lender

Five tips help prepare for the credit request and negotiate a financial package with the lender.

1. As a borrower, you must provide current, accurate financial statements and supporting records. A current balance sheet with supporting schedules and inventories is essential. A record of earnings (usually an income statement) and a projected cash flow for your business are also needed. A good set of records showing production plans, short- and long-range goals and procedures for implementation and evaluation enhance the chances of obtaining credit. Financial information that a particular bank wants may vary.

    Most lenders are accustomed to balance sheets and income statements prepared or compiled by a certified public accountant when dealing with commercial and industrial borrowers. An increasing number of agribusinesses are using professional services for their records. A common and significant difference between records prepared by an accountant and records prepared by an individual is valuation of assets on the balance sheet. Accountant-prepared balance sheets commonly value assets at cost less depreciation, while individuals commonly use cost or an estimate of market value. The issue of asset value is important because asset value variations also show up on the right-hand side of the balance sheet as changes in net worth and they also affect leverage ratios such as percentage equity or debt/worth. If accountant-prepared statements are not used, most lenders will desire a conservative estimate of asset values.

    A number of excellent farm record systems are available, many of which run on a microcomputer. Some will coordinate and reconcile income statements to balance sheets, ensuring consistent valuation and better tracking of financial progress.

2. Arrange credit in advance. Lenders do not like surprises. Do not inform the lender of a major decision after the fact. This can destroy trust and credibility and make future credit more difficult or impossible to obtain.

3. Allow your lender time to review your plans and make suggestions. Major purchase decisions are sometimes made on the basis of emotion

rather than profitability. A lender can provide objectivity and counsel in reviewing your credit request. Explaining goals and plans builds confidence and trust, which strengthens the working relationship.

4. Keep your lender informed. Even the best of businesses face adversity that reduces the ability to repay. Inform your lender as soon as possible of changes in plans or unforeseen problems that will interfere with making loan payments. Communication is the key element in the initial request and throughout the credit process.

5. Maintain a high level of integrity. If you expect a lender to be honest and aboveboard at all times, then the same will be expected of you. Inaccurate information and failure to honor commitments will jeopardize the borrower-lender relationship.

Managing Credit Use

Once credit is obtained, properly managing the credit becomes a major challenge in the business. Three basic financial statements—the balance sheet, income statement, and cash-flow statement—are tools used to monitor the financial strength of the business. When compiled and supported by accurate financial information, these tools can provide the support needed for many of the strategies and financial decisions faced.

Any business—whether it is an agribusiness firm, farm, corporation, or small business—must meet certain criteria to be successful, particularly if credit is used. A successful business must exhibit strength in repayment ability and capacity, liquidity and solvency, and profitability and financial efficiency. Coincidentally, the lender's cornerstones of sound credit, the five Cs, encompass the same qualities—

1. Character (honesty, integrity, and management ability)
2. Capacity (repayment ability and profitability)
3. Capital (liquidity and solvency)
4. Collateral (minimizing risk to the lender)
5. Conditions (for granting and repaying the loan)

Both producer and lender can determine the financial status of the business with these criteria.

Any analysis of the use of credit is only as strong as the quality of financial and other information provided. Circumstances such as size and mix of enterprises, costs, values, commodity prices, collateral values, type of business entity, and time of year can all affect interpretation. Do not base final interpretation on any one factor but rather on a balanced, comprehensive approach. Comprehensiveness is the number one factor in developing any valid analytical process.

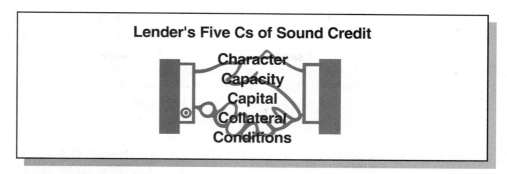

FIGURE 9-16 The five Cs.

The Lender's Viewpoint

Many lenders use a systematic approach to analyzing credit. They use some or all of the following guidelines and yardsticks—

- Annual earnings summary
- Earnings-coverage ratio
- Debt-payment ratio
- Business operating efficiency
- Current ratio
- Percentage equity
- Collateral position.

Credit Management. Once the debt and repayment structure is in place, constant monitoring and management of credit is essential. Debt structure and repayment terms, tracking of security, and marketing progress of repayment are frequent problems if numerous creditors are involved.

Sound credit analysis may include periodic review of open accounts with merchants, dealers, and suppliers. A check on personal credit card balances can be useful to analyze personal accounts. A strong sign of cash flow and credit management is when accounts payable, after initial billing, average less than 5 percent of revenue. If unpaid bills average more than 10 percent of revenue, it is a sign of pending credit problems. Any sharp increase in accounts payable or a general trend upward will be carefully scrutinized.

Production Management and Profitability. Production management of an operation can directly link to revenues and profits. Since production rates and efficiency can vary for a given area and enterprise, a general analysis is used to evaluate management relating to production, for example, top 20 percent, 20 to 50 percent, or below average. Lenders may want to evaluate the management skills of a loan applicant in areas such as production efficiency, cost control, and profitability. They may require an analysis for more than one year to determine trends.

Comparisons of production factors and financial ratios are revealing and benefit both borrower and lender. Comparison data may be available from the land-grant university in the area, the state extension service, farm organizations, or other groups offering farm records and analysis.

Farm businesses that are considered to be very profitable would have a return greater than comparable investments such as savings and other non-farm investments. Historically, income returns on farm investments have been relatively low, averaging in the 3 to 4 percent range. Return on equity has been even lower. A commercial operation can survive in the short run with small or negative returns, but both lender and farmer must carefully scrutinize the business if this scenario continues over the long run.

Individual and Farm Resources. Evaluating the financial situation and management of an agricultural business frequently involves more than analysis of the basic financial statements.

1. Personal Characteristics and Habits. A lender will look at the health and age of the individual requesting credit as well as that of the entire family. The stability of family relationships and evidence of estate planning or transfer of farm assets and short- and long-term goal setting are prime considerations. Education and practical experience should be observed, as well as how management techniques are applied to the farm operation.

2. On-site Visit. A good credit analysis will include on-site investigation of the overall resources—land, buildings, improvements, fish, and machinery—and personal living habits. These formal inspections can be used to monitor inventory and possibly do appraisals of real estate, fish, and machinery, as well as a check of security and collateral. The quantity and quality of land, including explanation of leases, land contracts, and other pending situations, will be examined. On-farm visits should include general observation of the amount and condition of fish, feed inventories, machinery, and buildings. Available storage and state of repair of machinery often can be a clue to future needs that could be classified under business development.

Lenders will critically evaluate the effects that economic and market trends have on the business. Forecasts, outlook, and other projections of costs and expenses related to various farm enterprises can help determine the overall health of the farm business and the customer's needs, desires, and strategies for success.

HUMAN RESOURCES

Effective human resource management begins with planning. Using a plan requires that personnel be recruited and then managed effectively. Managing personnel involves the major functions of work scheduling, training, motivation, evaluation, and discipline.

Personnel Planning

Effective personnel planning starts with a self-assessment by personnel managers. Their personal characteristics, attitudes, strengths and weaknesses, and supervisory skills directly affect the working relationships among employees and others in the farm business.

Personnel needs depend on the work (tasks) to be done, the types of products grown, and the machinery and technology of each operation. An analysis of personnel needs should result in a statement of the kind and amount of work to be done, which, in turn, provides a basis for determining the number and types of workers needed.

Matching current personnel—family and nonfamily—with tentative job descriptions is a critical step in developing job descriptions for new employees. Identifying mismatches between job descriptions and current responsibilities may help point up training needs, adjustments in job descriptions, shifts in responsibilities and, most important, tasks that cannot be adequately handled with existing personnel.

Hiring Employees

For a team of family and hired workers to function efficiently and effectively, one or more supervisors must carry out the following five personnel management functions: work scheduling, training, motivation, evaluation, and discipline.

Work Scheduling. Work planning and scheduling increase labor efficiency. Waiting for instructions, searching for a supervisor, duplicating the work of another employee, waiting for equipment to be available, doing maintenance work during critical periods of the production season, and wasting harvesting time because equipment was not ready for the season are examples of inefficiencies caused by poor work scheduling.

Work scheduling should be based on a list of tasks to be accomplished, the machinery and equipment needed for the tasks, the people available to do the tasks, and the time in which the work must be done. A task list identifies what needs to be done within the next period or periods of time. The work schedule accompanying the task list identifies the workers and equipment for the tasks. Providing instructions to workers about the tasks they are to do and when and where they are to do them is the final element of the work schedule. The instructions do not have to be given every day if employees are well trained and well supervised.

Training. Farm managers who hire workers with little aquaculture work experience must provide extensive training to new employees. The complexity of many farm tasks, the risk of injury to untrained workers and the labor inefficiencies that result from undirected, on-the-job stumbling make training essential.

Hiring experienced workers is sometimes considered an alternative to carefully planned and implemented training programs. In fact, all employees require training. Experienced employees may require considerable training to change poor work habits, inefficient practices, and lax attitudes toward safety that can endanger themselves and fellow workers. Some employers even prefer to hire inexperienced workers for some tasks because

FIGURE 9-17 Being a manager means working with groups of people—managing the human resources.

training can focus on the skills that are needed and not on retraining or changing old habits.

Motivation. Employees—family members included—do not change their behavior simply because someone tells them to do so. In fact, threats, bribery, and other types of manipulation may make little difference in an employee's work habits or attitude. The challenge for the farm manager is to balance workers' needs for job satisfaction with the farm's overall business goals. To do this, the farm manager must identify employees' most important unsatisfied needs and then determine the feasibility of satisfying those needs through work itself or conditions at the workplace.

A person working primarily to satisfy a need for social interaction may care little about labor productivity or sales. Can the person satisfy social needs at break times, before and after work, or through casual conversation during work? Or must the worker be disciplined for wasting time on the job?

Evaluation. A formal evaluation program lets employees know where they stand on a regular basis and includes guidelines for wage increases. The evaluation should tell employees how they are doing, identify areas where improvement occurred, and offer constructive suggestions for work improvement. Specific plans for training and job improvement should be discussed. Workers should also have the opportunity to make suggestions, raise questions, and air frustrations and complaints.

In addition to ongoing daily or other regular communications with workers, at least one formal evaluation meeting should be conducted with each employee every year. This meeting provides opportunities to review performance and progress during the past year and to establish performance goals for the coming year.

Compensation should be discussed during the evaluation meeting. Any changes in compensation should be consistent with the strengths and weaknesses discussed in the evaluation meeting. Merit increases should go only to those who have earned them, and employees should understand why they are or are not getting a raise.

Discipline. Workers function best when the rules are clear and they know the consequences of breaking them. Discipline problems can be minimized through careful employee recruitment and training, clear communication of work rules, and proper attention to human needs. When discipline is necessary, the supervisor should not sidestep the responsibility. Failure to provide discipline sends wrong and confusing messages to workers.

WHAT IS YOUR MANAGEMENT PHILOSOPHY?

Management is the art of successfully pursuing desired results with the resources available to the organization. Management is—

- People-oriented
- An art, not a science
- An ability to establish and meet prescribed goals
- Working with available resources, stressing efficiency.

Some describe management as a division of areas of responsibility, for example, finance, marketing, production, and personnel. Others view it as coordinating a series of resource inputs, for example, money, markets, material, machinery, methods, and manpower—the Six M concept. Other concepts of management divide it into industrial engineering, organizational, behavioral, and functional approaches.

Industrial engineering approach. This approach scientifically analyzes work processes and strives for increased productivity. Frederick Taylor, the "Father of Scientific Management," was a major proponent of this management concept. Job description, time-and-motion studies, and production standards form the basis of the industrial engineering approach to management. This school of management asserts that if these relationships and tasks are carefully designed, productivity is the natural result. The use of power and organizational authority will result in maximum effectiveness.

Organizational approach. This approach studies the ways in which power and authority may be distributed in order to increase productivity. The organizational approach focuses on such areas as specialization, division of labor, ways in which power and authority are distributed throughout the organization, staff relationships, span of control, for example how many employees can be controlled by one supervisor, and span of attention, for example how many different operations can be controlled by one manager. This school of management asserts that if the tasks are carefully designed, productivity is the natural result.

Behavioral approach. This approach stresses managing human resources in order to better the working environment both for the employer and employee viewpoints which, in turn, increases overall productivity. The behavioral approach urges the manager to enlarge and enrich jobs to give individual workers more responsibility and authority and to provide a working environment in which employees can satisfy their own needs to be recognized, accepted, and fulfilled. Douglas McGregor, Abraham Maslow, and Frederick Herzberg were leaders in developing this approach to management.

Maslow's need hierarchy, one of the most widely used models for human needs, was developed by Abraham Maslow. It is based on the idea that different kinds of needs have different levels of importance to individuals, depending on the individual's current level of satisfaction. Needs basic to human survival take priority over other needs, but only until survival has been assured. After that point, other needs form the basis for the individual's behavior—

- ■ Survival—The most basic human concern is for physical survival, such things as food, water, warmth, shelter.

- ■ Safety—Once immediate survival has been assured, humans are concerned about the security of their future physical survival. Today, this may take the form of income guarantees, insurance, and retirement planning.

- ■ Belongingness—After safety is assured, people become concerned with their social acceptance and belonging.

- ■ Ego status—With a comfortable degree of social acceptance, most individuals become concerned with status in their group. Group respect and the need to feel important depend heavily on the responses of other group members.

- ■ Self-actualization—The highest level of need, the feeling of self-worth, may be achieved through creative activities such as art, music, helping others in community activities, or building a business.

Finally, another popular concept views management as a series of functions. This school of thought describes management as PODCC—planning, organizing, directing, coordinating, and controlling. Two other functions should be added—communicating and motivating—since these functions underlie the success or failure of the first five functions. Under this approach, the best of all schools of management philosophy can be combined.

BUSINESS MANAGERS OF TOMORROW

The changing agricultural environment creates wide-reaching implications for managers of the future. A turbulent business environment arising from increased integration of the agricultural sector into the national and world economies, technological change, projected changes in government regulations, and expected changes in weather patterns will create new challenges for managers.

Other changes occurred in the agricultural sector that will affect the types of skills farm managers need. These include the movement toward fewer and larger commercial farms, a proliferation of part-time farmers near urban areas, increased vertical integration, and increased involvement of lending institutions in management and ownership. These changes affect the way operations are managed, the knowledge tomorrow's managers will need, and the forms their training will take.

Tomorrow's managers will be fewer in number, better educated, and more diverse than those of yesterday. They will use a broader set of managerial skills to meet the challenges of the turbulent business environment of the future. They will have access to new information and management skills that go beyond their formal education. This continuing educational process will take many forms, ranging from technology-based information transfer to intensive management development programs.

Tomorrow's Managers

The challenges of managing the aquaculture of tomorrow will be met by a wide and diverse group of managers. Farming the water is a business, and the successful farm operation will be managed as a business. The farms may be owned and operated by families, partnerships, or corporations, but the management will rely on business skills for success.

While experience is one method of developing these business skills, continuing education and training combined with farm experience is a faster, less risky means of developing sound business skills and practices.

Skills Managers Will Need

Tomorrow's managers will need an expanded set of managerial skills to succeed. Managerial skills in three areas—communication, business and economics, and technology—will be developed through formal university education, business experience, and continuing education.

Managing Innovation and Change. Changes in technology, information, and marketing systems occur at an increasing rate. As this rate of change continues, managers will be forced to adopt new practices and employ strategic thinking to survive. Changing consumer demand also will require increased innovation to fill existing or developing market niches.

Managing Risk. The growing exposure to global competition in production and financial markets, changing government policies related to trade, supply, and the environment, and changes in climate require new managerial skills in dealing with risk. Tomorrow's farm and ranch managers will be forced to use the futures and options markets, contractual arrangements, and other risk-shifting tools to manage these risks.

Designing Effective Organizations. As the structure of farms and ranches evolves over the next several years, managers will confront many organizational challenges, including the need to develop organizations that use labor effectively and that can take advantage of new relationships with buyers, suppliers, and competitors.

Designing Information Systems. Increasing information and rapidly evolving information technologies create a distinctive set of challenges for farm and ranch managers. Practical information acquisition systems and computer decision support systems will continue to be developed to aid managers in this area.

Managing Human Resources. As the average size of commercial farms increases and the number of part-time farmers grows, human resource problems will become more important to farm and ranch managers. These problems may be increased by absentee land owners and managers. The challenges of dealing with more seasonal employees, reliance on specialized personnel and expanded interactions with suppliers, buyers, and processors are likely to occupy more and more of the farm or ranch manager's time. Part-time farmers will face the need to balance farming demands with off-farm employment.

Strategic Planning

In addition to developing these five specific skills, managers will have to be strategic thinkers, capable of dealing with a turbulent environment by using the techniques and tools of strategic planning. These techniques will aid in managing technological innovations and in dealing with changing governmental policies, markets, weather, and business. Although based on theory and technical knowledge, these skills will best be developed through case-study learning experiences.

Global Marketplace

The managers of tomorrow will need to understand agricultural production and marketing in the global marketplace. This implicitly includes an understanding of consumers and their evolving needs. As the firms that farm and ranch managers deal with as buyers, suppliers, and competitors become increasingly global, the managers of the production operation will need to understand their needs to better develop working relationships. Exciting opportunities may exist for cooperation between individual operations or among groups, particularly in satisfying consumer needs by filling niche markets. Managers will likely learn about changing consumer needs through nontraditional study, including internships, study abroad programs, and various conferences and institutes.

Computers and Management Tools

Managers of the future must also be computer literate. In addition to serving as a valuable information-management tool, the computer will become increasingly important to the manager as a decision aid and as a means of communicating with other producers, trade associations, private firms, and government agencies. Short courses and home study will likely provide important training, as will expanded use of computer simulations in traditional classroom settings.

Agriculturists have always made decisions in a risky and uncertain environment. Increasing factors will affect that uncertain environment. Computer programs help managers analyze the risks associated with their decisions. Almost all college graduates today are exposed to, understand and accept computer technology and know it can be used to sort vast quantities of information and aid in making decisions. As past generations of managers are followed by a computer-literate generation, demand will increase for continuing education on ways to use the latest information and technology for short- and long-term decision making.

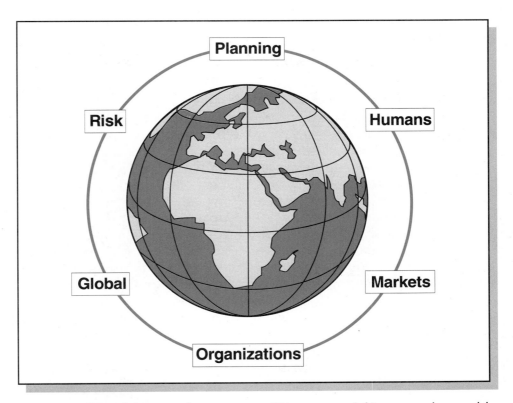

FIGURE 9-18 Managers of tomorrow will be successful in a complex world.

Multiple Training Sources

To train tomorrow's managers, existing educational programs are likely to be modified—with a great deal of support from nontraditional sources. The following will be key sources of training for tomorrow's managers—

- Formal college education
- Management development programs
- Extension education
- Home study through new communication methods
- Professional and trade associations.

The manager of the future will face an increasing need to understand the behavioral aspects of markets, employees, and competitors. Increased understanding of the liberal arts, as well as business concepts and practices, will be important. Effective networks among peers will also help. These needs will be best met through a variety of training sources. The successful business managers will take advantage of all sources.

SUMMARY

No one should enter an aquaculture business without counting the personal, social, and financial costs. Management skills are necessary to operate the business successfully. These skills are gaining in importance as the economy changes and the workplace changes. Management involves the best use of human resources of the business and the best use of the financial resources of the business. Records are essential to good financial management.

Record systems can be simple or complex, hand-kept or computerized. Besides meeting tax requirements, they provide an effective part of planning and evaluating the business. When a business needs to borrow money, records support the need and ability to repay the debt.

Obtaining and using credit in an aquabusiness is a two-way street. Borrowers look for fair, understanding, and knowledgeable lenders. Lenders lend money to honest, knowledgeable borrowers who can demonstrate a plan and an ability to repay the loan. Both the borrower and lender manage the credit.

Business managers of the future face challenges. More will be required for success—more planning, more technology, more knowledge, and more human resources. Managers of the future will need to know how to learn and seek training and education from a variety of sources.

STUDY/REVIEW

Success in any career requires knowledge. Test your knowledge of this chapter by answering these questions or solving these problems.

True or False

1. People and businesses always write down their goals.

2. Most aquaculture businesses keep records at least for income tax purposes.

3. A balance sheet is a type of accounting method.

4. A planning document is one of the "big three" statements.

5. Cost of feed represents a fixed cost.

Short Answer

6. List six input costs for starting an aquaculture facility.

7. Name six variable costs associated with catfish production.

8. Give three examples of fixed costs associated with any aquaculture operation.

9. List five basic management functions used to achieve the goals and objectives of an aquabusiness.

10. List the eight steps of strategic planning.

11. List three benefits of setting business goals.

12. Identify eight general business risks.

13. What is the most common form of business organization?

14. Name two forms of multiple business ownership.

15. List the four functions for records.

16. Name two accounting methods.

17. List the three categories of a balance sheet.

18. What is the general accounting equation for a cash-flow statement?

19. Name three accuracy checks for a single-entry accounting system.

20. List five steps to financial analysis of an aquabusiness.

21. List the five Cs that are the cornerstone of sound credit.

22. Name three guidelines or yardsticks lenders use to systematically analyze credit worthiness.

23. Identify five functions of human resource management.

24. List six managerial skills needed by the manager of the future.

Essay

25. Indicate four personal and social costs associated with starting an aquacultural business.

26. Explain why enterprise budgeting is useful in establishing an aquaculture business.

27. Explain the difference between fixed cost and variable cost.

28. Define management.

29. Define the following: organizing, staffing, directing, and controlling.

30. Explain why motivation is an important part of management.

31. Explain the difference between tactical planning and strategic planning.

32. Define assets.

33. Define a farm earning statement.

34. A good accounting system supplies what three types of information?

35. Give three reasons why the computer helps make better management decisions.

36. Describe three guidelines for selecting a lender.

37. Give three tips for negotiating a loan.

KNOWLEDGE APPLIED

1. Invite a successful business person from the community—an entrepreneur—to class to describe start-up costs and planning for a new business. Discuss the variable costs and the fixed costs of the business.

2. Invite an accountant who specializes in farm accounts for aquaculture or agriculture to visit class. Ask about the types of records provided to the farmer or aquabusiness. Find out how the records are generated. Discuss the value of the "big three" financial statements.

3. Determine the total cost to obtain the following production equipment for a catfish operation. Also, describe more completely the specifications of each piece of equipment. Use the table on the following page.

Production Equipment

Item	Cost
Tractors (50 hp), units × cost/unit	
Feeder, units × cost/unit	
Feed bin (10 tons) with pad, units × cost/unit	
Trucks (1 ton), units × cost/unit	
Aluminum boat (14 ft)	
Boat motor (25 hp)	
Boat trailer	
Transport tank (125 gal) and equipment	
Fixed electric aerators, units × cost/unit	
Portable aerators, units × cost/unit	
Storage and service building (30 ft × 50 ft)	
Side-mount mower (6–7 ft)	
Oxygen meter and accessories, units × cost/unit	
Miscellaneous farm shop equipment (repair tools)	
Waders, units × cost/unit	
Scales	
Low lift pump (PTO driven), units × cost/unit	
Dip nets (¼ in and 1 in), units × cost/unit	
Water quality test kit	
Battery and charger	

4. Visit with a banker who specializes in agricultural or aquacultural loans. Find out what is required to obtain a loan and how the loans are managed by the lending institution. Discuss the roles of production knowledge and trust in the lending process.

5. Prepare an enterprise budget. Collect realistic values for the budget. Use mortality figures and feed conversion rates normal for the species chosen. Show fixed costs and operating costs computed on an annual basis. Estimate the break-even and the payback. Use the table provided below.

Budget Analysis for the Annual Commercial Production of Fish

Item	Unit	Price or Cost/Unit	Quantity	Value or Cost Per Pound	Your Cost
1. EXPECTED RECEIPTS					
Fingerlings					
Food Fish					
Other					
Total Value:					
2. OPERATING EXPENSES					
Fingerlings					
Feed					
Chemicals					
Fuel					
Pumping					
Aeration					
Feeding					
Transportation					
Repairs and Maintenance					
Ponds					
Water Supply					
Feeding Equipment					
Harvesting Equipment					
Miscellaneous Equipment					
Labor					
Manager					
Hired Full-Time					
Hired Part-Time					

Item	Unit	Price or Cost/Unit	Quantity	Value or Cost Per Pound	Your Cost
Labor *(cont'd.)*					
Telephone					
Advertising					
Subtotal:					
Depreciation					
Pond Construction					
Water Supply					
Feeding Equipment					
Harvesting Equipment					
Miscellaneous Equipment					
Subtotal:					
3. INTEREST ON INVESTMENT					
Loan No. Purpose					
1 ( )					
2 ( )					
Subtotal:					
4. TAXES					
Property					
FICA					
State					
Federal					
Subtotal:					
5. INSURANCE					
Equipment					
Liability					
Life					
Subtotal:					
Total Costs:					

From your values, determine the following:

a. Income above Operating Expenses: _____
   Total Value Item 1 – Subtotal Item 2

b. Net Return to Land, Management, and Risk: _____
   Total Value Item 1 – Subtotal Items 2 + 3 + 4 + 5

c. Break-even Cash Price: _____
   Per Pound Value Item 1 – Per Pound Subtotal Item 2

d. Total Break-even Price: _____
   Per Pound Subtotals for Items 2 + 3 + 4 + 5

6. Visit with several local business owners and discuss the problems encountered in human relations and human resources in the workplace. Discuss the roles of leadership training and managing people.

7. Collect case studies of business problems, financial or management. Suggest solutions for these problems. Match your skills against those of local business managers and owners.

8. Visit with a representative of a local small business association. Discuss the types of ownerships for a business—sole proprietorship, corporation, or partnership. Develop a list of advantages and disadvantages for each.

LEARNING/TEACHING AIDS

Books

Chaston, I. (1988). *Managerial effectiveness in fisheries and aquaculture.* Farnham, England: Fishing News Books.

Chaston, I. (1991). *Business management in fisheries and aquaculture.* Cambridge, MA: Blackwell Scientific Publications.

Dixon, R. L. (1982). *The McGraw-Hill thirty-six hour accounting course* (2nd ed.). New York: McGraw-Hill Book Company.

Garrard, A. B., Fuller, J. J., & Keenum, M. E. (1988). *Economic analysis of small-scale processing for Mississippi farm-raised catfish.* Mississippi State: Mississippi Agricultural and Forestry Experiment Station.

Huner, J., & Dupree, H. K. (1984). *Methods and economics of channel catfish production and techniques for the culture of flathead catfish and other catfishes.* Washington, DC: U.S. Fish and Wildlife Service.

Jolly, C. M., & Clonts, H. A. (1993). *Economics of aquaculture.* Binghamton, NY: Haworth Press.

Keenum, M., & Waldrop, J. E. (1988). *Economic analysis of farm-raised catfish production in Mississippi.* Mississippi State: Mississippi Agricultural and Forestry Experiment Station.

Walker, S. S. (1990). *Aquaculture.* Stillwater, OK: Mid-America Vocational Curriculum Consortium, Inc.

Software

Regional Aquaculture Centers listed in Appendix Table A-11 can provide leads on software developed for financial evaluations and management decisions.

Federal, State, and International Agencies and Regulations

$\mathbf{A}$s aquaculture has assumed a greater economic importance, more federal, state, and local agencies have assumed roles to protect the consumer or the culturalist. Also, some agencies have developed roles to support aquaculture-related activities. Whatever the reason for involvement from federal, state, local, or international agencies, the aquaculturalist does not act in a vacuum. For example, a report on the regulatory process in California indicated that a single aquaculture venture could require up to 42 federal, state, and local agency permits. This chapter describes those agencies and their involvement in aquaculture. The aquaculturist will need to seek specifics when starting or changing an operation. Also, the aquaculturist needs access to information. This chapter describes agencies providing information.

Since aquaculture activities involve so many government departments and agencies in the United States, the government established the Joint Subcommittee on Aquaculture (JAS) in the late 1970s. The JAS was created to increase the effectiveness and productivity of federal aquaculture programs by improving the communication and coordination among agencies. The JAS is composed of the secretaries of the Departments of Agriculture, Commerce, Interior, Energy, and Health and Human Services; the administrators of the Environmental Protection Agency, and the Agency for International Development; the Chief of Engineers of the U.S. Army Corps of Engineers; the Chair of the Tennessee Valley Authority; and other agencies as appropriate.

Learning Objectives

After completing this chapter, the student should be able to:

In Aquaculture

■ List six agencies of the U.S. Department of Agriculture involved in aquaculture

■ Name six agencies concerned with the production of food

■ Identify acronyms used frequently by agencies

■ Name the service that provides diagnostic assistance to aquaculture producers

■ Name two sources of operating capital or start-up capital for aquaculture businesses

■ Describe the role of the Soil Conservation Service in planning an aquaculture facility

■ Indicate the government act that gives the Army Corps of Engineers authority

■ Describe the role of the National Sea Grant College Program

■ Describe the role of the FDA in aquaculture

■ Describe the role of the U.S. Fish and Wildlife Service in aquaculture

■ Explain the role of the EPA in aquaculture

■ List three programs of the Tennessee Valley Authority related to aquaculture

■ Give the state location of the five Regional Aquaculture Centers

■ Cite two international laws that preceded the Laws of the Sea

■ List four individual responsibilities when starting or changing an aquaculture operation

In Science

■ Identify agencies that support scientific research in aquaculture

■ Provide examples of research conducted by government agencies

■ Give the location of the aquaculture research programs

■ Name four programs and agencies that provide research information and data to aquaculture

■ Name five environmental issues addressed by the EPA

Understanding of this chapter will be enhanced if the following terms are known. Many are defined in the text and others are defined in the glossary. Key words are in color the first time they appear in the text. In this chapter, agency names are also in color the first time they appear in the text.

KEY TERMS		
Affiliate	Estaurine	Maritime
Archiving	Exclusive Economic	Matching funds
Biocontrol	Zone	On-line
Biologics	Export	Pollutant
Database	Grants	Treaty
Discharge	Import	Wetlands

U.S. DEPARTMENT OF AGRICULTURE

Established in 1862, the U.S. Department of Agriculture is headed by the Secretary of Agriculture. This department administers loans, grants-in-aid, and assists in production. Thirteen agencies or services under the U.S. Department of Agriculture are involved some way in aquaculture. These include—

FIGURE 10-1 USDA in Stoneville, MS.

1. Agricultural Marketing Service
2. Agricultural Research Service
3. Animal and Plant Health Inspection Service
4. Cooperative State Research Service
5. Economic Research Service
6. Extension Service
7. Farmers Home Administration
8. Foreign Agricultural Service
9. National Agricultural Library
10. National Agricultural Statistics Service
11. Office of Aquaculture
12. Office of International Cooperation and Development
13. Soil Conservation Service

Table 10-1 shows other federal government agencies and their regulatory responsibilities that can affect aquaculture.

TABLE 10-1 Federal Government Agencies and Their Regulatory Responsibilities to the Food Industry

Agency	Responsibility
Department of Agriculture (USDA)	Enforcement of standards of identity, plant sanitation inspection, veterinary inspection of meat, grading of meat, dairy, fruit, and vegetable products.
Environmental Protection Agency (EPA)	Control and enforcement of food processes resulting in pollution of waters, land, or air; control of pesticide application; control of water supplies.
Department of Commerce (National Marine Fisheries Service)	Regulation of catch, imports, and processing of fish products.
Department of Commerce and Department of Agriculture	Regulatory powers of international food trade.
Department of Health and Human Services (USPHS and FDA)	Enforcement of food legislation regarding food wholesomeness and aspects concerning human health and food process sanitation; food served by transportation companies; advisory capacity in milk, food, and shell-fish sanitation.
Federal Trade Commission (FTC)	Enforcement of all food legislation related to packaging, labeling, and advertising of food products.

Agricultural Marketing Service

The Agricultural Marketing Service of the U.S. Department of Agriculture provides assistance to aquaculture through the Federal-State Marketing Improvement Program (FSMIP). This program seeks to improve the marketing of aquaculture commodities and to reduce marketing costs for the benefit of producers and consumers. Projects include work on innovative marketing techniques, testing study findings in the marketplace and developing state expertise to provide service to marketers of agricultural products. FSMIP is a matching funds program exclusively available on a competitive basis through State Departments of Agriculture or other state agencies for projects aimed at improving marketing services in the states and/or regional projects.

The Livestock and Seed Division of the Agricultural Marketing Service, U.S. Department of Agriculture, is responsible for purchasing meat and fish products in order to stabilize market conditions and furnish nutritious food to meet the needs of the department's domestic feeding programs. Under certain excess supply conditions, aquaculture products may qualify for this program.

The Transportation and Marketing Division of the Agricultural Marketing Service, U.S. Department of Agriculture, conducts research on the market opportunities for aquaculture products. This research helps determine the economic feasibility of developing an aquaculture program based on the potential opportunities for marketing the product. This work applies to those already engaged in commercial production and to those planning new production and marketing organizations.

FIGURE 10-2 Logo for the USDA.

Agricultural Research Service

Aquaculture research through the Agricultural Research Service (ARS) includes marine shrimp in the Pacific region, cold, freshwater species in the Northeast region, and warm, freshwater species in the Midsouth region of the United States. Research is conducted on genetics, breeding, nutrition, disease diagnostics and control, water quality, and use and production systems to increase production capacity and technology transfer. Improved product quality and marketing are supported with research on processing, off-flavors, food texture and taste, packaging, food safety, and value-added products. Research programs are conducted at—

■ Stoneville, Mississippi
■ Southern Regional Research Center, New Orleans, Louisiana
■ Animal Parasite Research Laboratory, Auburn, Alabama
■ University of Hawaii, Honolulu, Hawaii
■ Lane, Oklahoma

Cooperative research programs are conducted at—

■ Oceanic Institute, Waimanalo, Hawaii
■ Spring and Groundwater Resources Institute, Shepherdstown, West Virginia
■ Mississippi State University, Mississippi State, Mississippi
■ Northern Crops Institute, Fargo, North Dakota

FIGURE 10-3 Mississippi State University, Delta Branch Experiment Station at Stoneville.

Animal and Plant Health Inspection Service

The **Animal and Plant Health Inspection Service** (APHIS) provides a broad range of cooperative animal and plant health protection services to livestock and crop producers through a field force located in all 50 states and in foreign countries. APHIS provides several services to both plant and animal agriculture.

Current services include **import** requirements for aquatic plants to prevent the importation and dissemination of plant pests and diseases and noxious aquatic weeds into and within the United States. APHIS also participates in a joint state-federal program to control noxious aquatic weeds in the United States and helps develop control methods, including use of **biocontrol** organisms. For example, grass carp can be used at densities as low as one to two per acre for maintenance after achieving weed control in a pond. Some states require permits to use grass carp.

Aquaculture producers experiencing problems with migratory birds and other animals receive on-site assistance from APHIS' **Animal Damage Control Division**. Most birds causing serious problems are protected under the federal Migratory Bird **Treaty** Act (MBTA). The MBTA is often confused with the endangered species laws. Under the MBTA, migratory birds may not be killed or trapped without permits.

APHIS licenses veterinary **biologics**—vaccines and diagnostic kits, for example—for prevention, diagnosis, and treatment of diseases of animals, including aquatic animals. Several fish vaccines are presently licensed by APHIS.

APHIS' **National Veterinary Services Laboratories** (NVSL) in Ames, Iowa, provides a limited amount of diagnostic assistance to aquaculture producers, mostly in problem cases.

Cooperative State Research Service

The **Cooperative State Research Service** (CSRS) works with the state agricultural experiment stations, forestry schools, 1890 land-grant colleges, Tuskegee Institute, and colleges of veterinary medicine. CSRS allocates money to states for maintaining high-quality research programs in agriculture. CSRS awards grants in aquaculture on a competitive basis through the Aquaculture Special Grant Program and the National Research Initiative. It also awards grants in aquaculture on research problems that Congress believes are important to the nation. In cooperation with the extension service, CSRS administers five Regional Aquaculture Centers for the performance of aquacultural research and extension and demonstration projects. CSRS houses the Office of Aquaculture, responsible for coordination of all aquaculture programs in the U.S. Department of Agriculture.

USDA STARTED IN THE PATENT OFFICE

Henry Leavitt Ellsworth (1791–1858) could be called the founder of the Department of Agriculture. Mr. Ellsworth was a lawyer, farmer, leader of an agricultural society, and head of an insurance company in Hartford, Connecticut. In 1839, he became head of the newly established Patent Office. Often, Mr. Ellsworth received seeds and plants from naval and consular officers overseas. He distributed these to farmers without any government authority or help.

Mr. Ellsworth pleaded constantly for government aid to agriculture. In 1839, he got $1,000 from the government to collect and distribute new and valuable seeds and plants, to carry on agricultural investigations, and to collect agricultural statistics. Mr. Ellsworth had a great insight to the importance of agricultural improvement—

> If the application of the sciences be yet further made to husbandry, what vast improvements may be anticipated! Mowing and reaping will, it is believed, soon be chiefly performed on smooth land by horse power. Some have regretted that modern improvements make so important changes of employment—but the march of the arts and sciences is onward, and the greatest happiness of the greatest number is the motto of the patriot.

In 1849, Daniel Lee, M.D., a former editor of the *Genesee Farmer* and professor of agriculture, was hired as a "practical and scientific agriculturist" to supervise agricultural matters in the Patent Office. Dr. Lee also prepared separate annual reports on agriculture. The first of these was dated 1849 and it eventually became the *Yearbook of Agriculture*, which was published every year until President Clinton ended the practice in 1993. Dr. Lee preached soil conservation, and he issued reports on tillage, runoff, drainage, insects, fertilizers, the improvement of farm animals, rural science, and the statistics of weather. Dr. Lee considered agricultural progress to come from scientific inquiry, practical wisdom, intellectual honesty, and forthright advocacy.

Eventually the staff expanded in the Division of Agriculture of the Patent Office. Townend Glover (1813–1888), a British entomologist, was hired to collect statistics and other information on seeds, fruits, and insects in the United States. He thoroughly studied the insects of the southern United States and gathered an extensive collection of insects, birds, models of fruit, and herbarium plants. In 1865, Mr. Glover represented U.S. agriculture at an exposition in Paris. A botanist and a chemist also joined the staff.

Isaac Newton (1800–1867) grew up in Pennsylvania, where he managed a model dairy farm. He sold butter to many special customers, including the White House. After the formation of the United States Agricultural Society, he was delegate to many of its meetings. Here he repeatedly introduced a resolution calling on Congress to establish a Department of Agriculture. In 1861, he was appointed Superintendent of the Agricultural Division of the Patent Office. When the Department of Agriculture was organized May 15, 1862, his friend President Lincoln made him the first Commissioner of Agriculture. Mr. Newton established an agricultural library and a museum. He also selected the grounds for the Department of Agriculture. One year after Mr. Newton took office, the new Department had a horticulturist, a chemist, an entomologist, a statistician, an editor, and 24 other staff members. Congress appropriated $100,000 for an office building in 1867.

Even before the Department was established, its advocates urged that it be made an executive department, headed by a secretary who would be a member of the Cabinet. Advocates argued that agriculture was the single most important economic activity in the nation. Finally, in 1889, the Congress elevated the Department of Agriculture to Cabinet status. Then the department employed 488 people with an annual appropriation of $1.1 million. Today the Department of Agriculture employs about 123,000 people with an appropriation of about $56.4 billion.

Economic Research Service

The mission of the Economic Research Service (ERS) is to provide economic and other social science information and analysis for improving the performance of agriculture and rural America. ERS produces such information as a service to the general public and to help Congress and the administration develop, administer, and evaluate agricultural and rural policies and programs.

The *Agriculture Situation & Outlook* is part of a series of reports that analyzes the production and demand for agricultural commodities, food and fiber products, and production resources. The *Agriculture Situation & Outlook* provides information on the supply, demand, pricing, and trade for aquacultural and related wild-harvested fisheries products. Information is also provided on regulatory and public policy issues.

Extension Service

The Extension Service is the primary educational arm of the U.S. Department of Agriculture. Through the Cooperative Extension System's 74 land-grant universities located in all 50 states, 6 territories and the District of Columbia, programs are implemented in partnership with other federal agencies and

FIGURE 10-4 Reading the *Agriculture Situation & Outlook* and other publications provides insight into the business.

state and county governments. The system functions as a nationwide educational network and includes professional staff in nearly all of the nation's counties. It transfers research, technology, and management information through educational programs and technical assistance to help people improve their lives. The Cooperative Extension System relies on information generated by research and helps interpret research results to speed the application and dissemination of this information to the public.

Many states have developed extension educational and service programs in aquaculture. They provide programs such as workshops for new fish farmers, short courses in management and fish diseases, aquaculture demonstrations, farm visits, field days, in-service training programs, 4-H youth programs, and written and videotape educational materials on aquaculture development. Many educational programs include timely newsletters on a variety of aquaculture topics.

Farmers Home Administration

The **Farmers Home Administration** (FmHA) makes and guarantees farm ownership and operating loans and provides technical management assistance to family farmers and ranchers. Farm ownership loans may be used to buy, improve or enlarge farms, including buildings, ponds, wells, and water systems. Farm operating loans may be used to pay for items needed for a successful operation, such as farm and home equipment, feed, fuel, chemicals, and hired labor.

FIGURE 10-5 The FmHA provides loans for agriculture and aquaculture and the SCS office can provide planning assistance to aquaculture.

Foreign Agricultural Service

The Foreign Agricultural Service (FAS) represents the United States' agricultural interests overseas, including aquacultural interests. With more that 80 overseas offices covering over 100 countries the FAS assists exporters of U.S.-origin products in developing foreign markets. The FAS can also advise on foreign importer and government contacts by providing information on export credit guarantee programs available through the USDA Commodity Credit Corporation (CCC).

The FAS Trade Assistance and Planning Office (TAPO) focuses on counseling new exporters. The TAPO shows potential exporters how to establish contacts with foreign buyers. Also, the TAPO provides advice on the best marketing and distribution approach for a product. The TAPO provides exporters with a kit that contains FAS publications and other background information on the FAS.

National Agricultural Library

The National Agricultural Library (NAL) established an Aquaculture Information Center, mandated by the National Aquaculture Improvement Act of 1985. The Aquaculture Information Center serves as a repository for national aquaculture information and provides document delivery service for many of its materials through the Lending Branch. Staff of the Center publish bibliographies of interest to potential and practicing aquaculturists. Also, the staff conduct on-line computerized searches of aquaculture-related databases and make referrals to specialist or other contact sources. The Center networks with states, Regional Aquaculture Centers, libraries, and the private sector to enhance information exchange.

National Agricultural Statistics Service

The National Agricultural Statistics Service prepares and publishes reports from monthly surveys of catfish processed, end-of-the-month inventories, prices paid to catfish producers, and prices received by processors. The 16 major catfish-producing states collect and publish producer inventory and sales data. The four largest catfish-producing states—Alabama, Arkansas, Louisiana, and Mississippi—collect data quarterly. The other 12 states collect data twice annually. Trout producer sales and losses data are collected and published for the 15 major producing states. These survey results are published each year in September.

FIGURE 10-6 Logo of National Agricultural Statistics Service.

Office of Aquaculture

The Office of Aquaculture coordinates for department-wide aquaculture activities and provides leadership for the federal JAS. This office also coordinates the operations and activities of the five USDA Regional Aquaculture Centers.

Office of International Cooperation and Development

The **Office of International Cooperation and Development** (OICD) Research and Scientific Exchange Division seeks new knowledge and technology beneficial to the United States and cooperating countries through collaborative research in agriculture—including aquaculture—and forestry. Research programs are consistent with USDA's broad goals. They support the agricultural research system, increase net U.S. farm income, expand agricultural exports, conserve natural resources, and improve world agriculture.

Soil Conservation Service

Current **Soil Conservation Service** (SCS) policy recognizes that aquaculture is and will continue to be an important part of farming activities in some areas of the United States. Assistance for aquaculture may be a part of soil and water conservation activities in some areas as determined by soil conservation districts and reflected in their plans and priorities.

The primary objective of SCS assistance in fish farming is to ensure protection of the soil and water resource base. Careful resource assessment during conservation planning and application is a very important part of accomplishing this objective. SCS can assist with an initial resource assessment by furnishing data on—

- Water quality
- Water quantity
- Soils.

SCS may provide detailed assistance for planning an installation of projects with consideration for—

- Water quality
- Water quantity
- Appropriate fish species
- Soils investigation
- Site limitations
- Facility design and layout.

Contractors and consultants are encouraged to plan and install fish farming facilities, making use of guidance developed by SCS such as standards and technical notes. Where the fish farming workload is heavy, training sessions and field demonstrations are used.

U.S. DEPARTMENT OF THE ARMY

The U.S. Department of the Army includes all U.S. military land forces under the Department of Defense. This includes the U.S. Army Corps of Engineers.

Army Corps of Engineers

Agriculture projects that involve the discharge of dredged or fill material into waters of the United States, including marshes and other wetlands, or construction in the navigable waters of the United States must have a permit from the Army Corps of Engineers under Section 404 of the 1972 Clean Water Act and Section 10 of the 1899 Rivers and Harbors Act, respectively. The Corps makes its permit decisions based on a balancing of all the positive and negative impacts of the project. State and local approvals are frequently prerequisites to the Corps permit.

If a site is classified as a wetland, a permit is required from the Army Corps of Engineers before clearing or building. In some states, like Florida, permits are

FIGURE 10-7 The Army Corps of Engineers are involved when wetlands may be affected. (Photo courtesy J.V. Huner, Director, Crawfish Research Center, University of Southwestern Louisiana, Lafayette, LA)

needed from one or more state agencies before any clearing or building can take place on wetlands. Wetlands are those areas that are inundated or saturated by surface or ground water at a frequency and duration to support a prevalence of vegetation typically adapted for life in saturated soil conditions. Wetlands generally include swamps, marshes, bogs, and similar areas.

U.S. DEPARTMENT OF COMMERCE

The Department of Commerce is an executive department of the U.S. government and was established in 1913. This department is headed by the Secretary of Commerce and in general supervises transportation and shipping, the census, and food and drug laws. Six offices of the Department of Commerce play a role in aquaculture—

1. **Economic Development Administration**
2. **National Environmental Satellite, Data and Information Service** of the **National Oceanic and Atmospheric Administration** (NOAA)
3. **National Marine Fisheries Service** of NOAA
4. **National Sea Grant College Program** of NOAA
5. **National Technical Information Service** (NTIS)

Economic Development Administration

The programs of the Economic Development Administration (EDA) help alleviate conditions of substantial and persistent unemployment and underemployment in economically distressed areas and regions of the nation. While these programs do not expressly address efforts to generate new job opportunities and area income through aquaculture, EDA can consider funding aquaculture projects that meet the agency's regulations, applicant eligibility requirements, and project selection criteria.

National Environmental Satellite, Data and Information Service/National Oceanic and Atmospheric Administration

Four informational programs for aquaculture are sponsored by the National Environmental Satellite, Data and Information Service (NESDIS) of the National Oceanic and Atmospheric Administration (NOAA). These programs provide services and products to a wide range of clientele.

The **Aquatic Sciences and Fisheries Information System** (ASFIS) is a computerized system operated and maintained by several United Nations agencies and a network of information centers in member countries. A principal module of ASFIS is the **Aquatic Sciences and Fisheries Abstracts** (ASFA) database. This database covers international aquaculture and other topics in the marine and freshwater environmental sciences.

The **National Environmental Data Referral Service** (NEDRES) is available commercially as an on-line database containing over 22,000 descriptions and locations of data files on the environmental sciences held by people and organizations in the United States and Canada.

The NOAA **Earth System Data Directory** (NESDD) is an on-line guide to about 700 environmental data sets held by NOAA. NESDD provides data to NOAA centers and other organizational components, and provides the research community with the means to locate NOAA data.

The NOAA Library and Information Network consists of a central library in Rockville, Maryland, major branches in Miami and Seattle, and laboratory and information centers throughout the United States. Their collections encompass more than one million volumes on marine and atmospheric sciences, aquaculture and environmental disciplines.

National Marine Fisheries Service/National Oceanic and Atmospheric Administration

The **National Marine Fisheries Service** (NMFS) of the National Oceanic and Atmospheric Administration (NOAA) directs its aquaculture efforts toward managing common property resources and contributing to the restoration and protection of endangered species of stocks. NMFS conducts research at several laboratories in the United States. In addition, NMFS distributes aquaculture-related information and technological advances gained from its fisheries research. NMFS cooperates with federal and state agencies, international bodies and foreign governments, university and private interests, and promotes the development and expansion of domestic and international markets for products produced by the U.S. aquaculture industry.

The NMFS provides a comprehensive quality inspection program for fish and seafood production companies. This program is based on the Hazard Analysis and Critical Control Point (HACCP) testing method. The company develops criteria for acceptance and identifies any corrective actions needed. Then the product is inspected at each stage of production. The Food and Drug Administration proposed a mandatory HACCP regulation for seafood plants and the FSIS published its own requirements for meat and poultry plants. The HACCP responds to the consumer's concern for food safety.

National Sea Grant College Program/National Oceanic and Atmospheric Administration

The **National Sea Grant College Program** of the National Oceanic and Atmospheric Administration conducts research, extension, and educational programs with universities in all coastal and Great Lakes states. Sea Grant aquaculture research includes genetics, biotechnology, endocrinology, physiology, pathology, engineering, nutrition, policy economics, and others. The program chooses research projects on the basis of scientific merit, peer review, and the present needs of the aquaculture industry. It gives priority to projects that fit both the National Aquaculture and Sea Grant Aquaculture Development Plans. The extension aspect, called the Marine Advisory Service, uses a corps of area agents and specialists to provide public education, technology transfer, and demonstration projects in aquaculture. The program develops information generated by Sea Grant and other research for use by groups in the private sector to develop marine aquaculture. The program is operated through 30 Sea Grant programs in the coastal and Great Lakes states.

National Technical Information Service

The National Technical Information Service (NTIS), an agency of the U.S. Department of Commerce, is the central source for the archiving and public sale of U.S. government-sponsored research and development reports in all subjects, including those related to aquaculture.

U.S. DEPARTMENT OF ENERGY

Established in 1977 and headed by the Secretary of Energy, the Department of Energy controls energy use, research, and radioactive waste management.

Biofuels Systems Division

The U.S. Department of Energy supports aquatic species research as a part of the Biofuels Systems Division research program. This is one of the renewable energy options pursued by the department. The Solar Energy Research Institute (Golden, Colorado), universities, and private industry conduct aquatic species research to develop the technology base for production of liquid fuels from mass culture of microalgae. Research includes collecting, describing, and improving of microalgal strains that produce more fats in culture. The program also supports fundamental research on genetics, lipid (fat) physiology, and lipid biochemistry. Research on mass culture techniques includes studies of nutrient and carbon dioxide delivery, nutrient and water recycling, innovative harvesting, and lipid recovery and conversion.

U.S. DEPARTMENT OF HEALTH AND HUMAN SERVICES

Established in 1979, the U.S. Department of Health and Human Services is an executive department headed by the Secretary of Health and Human Services. This department administers social security, other welfare services, and supervises public health. The Food and Drug Administration falls under the Department of Health and Human Services.

Food and Drug Administration Center for Food Safety and Applied Nutrition

The Food and Drug Administration (FDA) Center for Food Safety and Applied Nutrition is the primary federal office with responsibility for the assurance of seafood safety. The center houses a wide range of programs devoted to the research and management of seafood, including aquaculture products. The FDA derives its authority for such programs primarily through two statutes, the Federal Food, Drug, and Cosmetic Act (FFDCA) and the Public Health Service Act (PHSA).

Under the FFDCA, the FDA is assigned responsibility to ensure that seafood shipped or received in interstate commerce is "safe, wholesome, and not misbranded or deceptively packaged." Under the PHSA, FDA is empowered to control the spread of communicable disease from one state, territory, or possession to another.

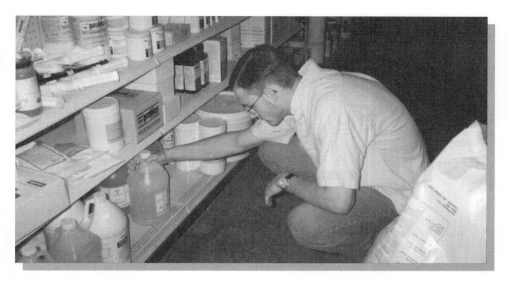

FIGURE 10-8 Drugs and feed additives require FDA approval.

Food and Drug Administration, Center for Veterinary Medicine

The **Center for Veterinary Medicine** (CVM) is responsible for the regulation of animal drugs, animal feeds, and veterinary medical devices. The center's involvement in aquaculture consists of four main areas—

1. **Approval of Animal Drugs and Feeds**—Under the provisions of the federal Food, Drug, and Cosmetic Act, animal drugs must ordinarily be approved before distribution and use. Manufacture of medicated feeds may require separate approval. Nondrug feed additives also fall under FDA regulation and ordinarily require approval prior to use.

2. **Oversight of Drug Distribution and Use**—CVM conducts surveillance and compliance programs relating to animal drugs, animal feeds, and other veterinary medical matters and coordinates the veterinary medical aspects of FDA inspections.

3. **Research**—CVM conducts basic drug metabolism and drug action research needed to support the development of analytical methods for detection of drug residues in aquaculture species.

4. **Educational Initiatives**—CVM works with the industry to develop quality assurance programs and educational materials to assist producers in using drugs and chemicals safely in aquaculture production systems.

U.S. DEPARTMENT OF THE INTERIOR

Congress established the U.S. Department of the Interior in 1849. The Secretary of the Interior heads this executive department, which includes the Bureau of Indian Affairs and agencies for the conservation of natural resources and supervision of power projects. The U.S. Fish and Wildlife Service and the U.S. Geological Survey come under its direction. These two agencies can be involved in aquaculture.

U.S. Fish and Wildlife Service

The U.S. **Fish and Wildlife Service** (FWS), Department of the Interior, restores depleted fish populations, preserves endangered species, moderates the impacts of federal water developments on fish populations, manages fish resources on federal lands, and provides scientific leadership in fishery resource management. A nationwide system of fish hatcheries, fisheries assistance offices, fish health centers, fish technology centers, fisheries research centers, and a training facility are operated to carry out these responsibilities. The service's fisheries activities involve research, management, and culture of

FIGURE 10-9 The U.S. Fish and Wildlife Service responsibilities extend to research, culture, and management of such things as a wildlife refuge.

freshwater, anadromous, estuarine, and exotic fishes of recreational, ecological, and commercial importance. The aquaculture mission of the FWS consists of two broad functions: (1) to encourage the development of private aquaculture in a manner that is compatible with responsible natural resource stewardship, and (2) to make service expertise, knowledge, and technical/scientific capabilities in fish culture and related disciplines available to the private aquaculture community.

U.S. Fish and Wildlife Service, Office of Extension and Publications

Through its Office of Extension and Publications, the U.S. Fish and Wildlife Service (FWS) continues a program of cooperation with the extension community to inform and educate the public concerning fish and wildlife resources. This program addresses the field of aquaculture both from the standpoint of extension education and 4-H. Memoranda of Understanding with the USDA Extension Service and NOAA enable the FWS to produce suitable extension-type material and to have them distributed effectively and efficiently through each agency's extensive communications network.

The Publications Unit distributes upon request the scientific and technical publications of the FWS. It also provides general interest information about fish and wildlife resources to the public.

U.S. Geological Survey

The **U.S. Geological Survey** (USGS) collects a large amount of information on the availability and quality of ground water and surface water supplies. Information such as **pollutant** levels, aquifer locations, lake levels, stream flows, and other data useful to aquaculture is available from the USGS in a variety of forms, including maps, reports, and computer output.

U.S. AGENCY FOR INTERNATIONAL DEVELOPMENT

The **Agency for International Development** (AID) supports aquacultural research and development projects where significant opportunities exist to use this form of food production to increase incomes and employment opportunities in developing countries. It supports applied research, usually involving U.S. scientists through the Bureau for Research and Development. The research addresses critical problem areas blocking the expansion of economically viable aquaculture in developing countries. AID assists in the education and training of aquaculturists, scientists, administrators, and extension specialists, and promotes the transfer of appropriate technology for aquacultural development in the tropics. Within the developing countries, AID supports aquacultural development projects designed to introduce and demonstrate technically sound aquacultural production methods and adapt them to local environmental, economic, and social circumstances. Special innovative and joint research projects are also funded through AID.

U.S. ENVIRONMENTAL PROTECTION AGENCY

Several programs of the U.S. Environmental Protection Agency (EPA) involve aquaculture. The Clean Water Act of 1972 designates fish hatcheries and farms as point sources of pollution of waters receiving their discharges. To comply with the Act, each state issues permits and monitors effluents from fish raising facilities. Proper management of effluents assures the protection of the environment. EPA's water quality programs are concerned with setting water quality criteria and effluent discharge standards for assuring the protection of the nation's waterways and water supplies.

FIGURE 10-10 In some states the Division of Environmental Quality (DEQ) works closely with the EPA.

TABLE 10-2 Issues Addressed by the Environmental Protection Agency (EPA)

Issues	Office
Water Quality and Effluent Guidelines	Office of Science and Technology
Discharge Permits	Office of Wastewater Enforcement and Compliance
Wastewater Treatment	Office of Wastewater Enforcement and Compliance
Wetlands	Office of Wetlands, Oceans, and Watersheds
Pesticide Registrations	Office of Pesticide Programs
Residual Wastes	Office of Solid Wastes

The **National Pollutant Discharge Elimination System** (NPDES) issues permits for the discharge of wastewaters to surface waters, including discharges from aquaculture systems in many cases. Permits are also available from this program to use wastes as nutrients in public waters for aquaculture purposes.

With the U.S. Army Corps of Engineers, the EPA also implements the Clean Water Act Section 404 Wetlands protection program, aimed at protecting natural wetlands from the impacts of dredging and filling.

NATIONAL SCIENCE FOUNDATION

At the **National Science Foundation** (NSF), the **Small Business Innovation Research Program** (SBIR) provides funding for aquaculture research. This program annually solicits high-quality research proposals from small business firms on important scientific or engineering problems that could lead to significant public benefit. Aquaculture research proposals generally concern marine or estaurine and freshwater problems.

SMALL BUSINESS ADMINISTRATION

The U.S. **Small Business Administration** (SBA) is an independent federal agency created to assist, counsel, and champion America's small businesses, including aquaculture firms. The agency provides prospective, new, and established small businesses with financial assistance, management, counseling, and training. The Small Business Administration also provides assistance to small firms interested in obtaining federal government contracts. The Office of Advocacy represents small business' interests before the U.S. Congress and other government agencies and promotes public awareness of the community's unique problems and needs.

TENNESSEE VALLEY AUTHORITY

Congress created the Tennessee Valley Authority (TVA) in 1933. Its chief purpose is the development of the Tennessee River system and land resources. Three programs of the TVA relate to aquaculture.

The National Fertilizer and Environmental Research Center is developing a research and development facility to evaluate artificial wetlands as biological filters for removing fertilizer nutrients from industrial waste streams.

The Valley Resource Center provides expertise to the private sector and other governmental organizations related to the design and operation of artificial wetlands for stabilizing industrial, municipal, or agricultural wastes.

The Agricultural Institute administers programs through the university extension service and a competitive grants program to encourage private sector adoption of new and emerging technologies in agriculture and aquaculture.

NATIONAL ASSOCIATION OF STATE AQUACULTURE COORDINATORS

The National Association of State Aquaculture Coordinators (NASAC) is an affiliate of the National Association of State Departments of Agriculture (NASDA) who are responsible for coordinating aquaculture programs at the state or territorial level.

The purpose of NASAC is to promote, encourage, and assist the development of aquaculture in the United States by enhancing communication among federal, state, local, and tribal governmental agencies, agricultural research and extension institutions, and trade and marketing organizations.

USDA Regional Aquaculture Centers

Congress established five regional aquaculture centers in Title XIV of the Agriculture and Food Act of 1980. Offices of the centers are located in Hawaii, Massachusetts, Michigan, Mississippi, and Washington.

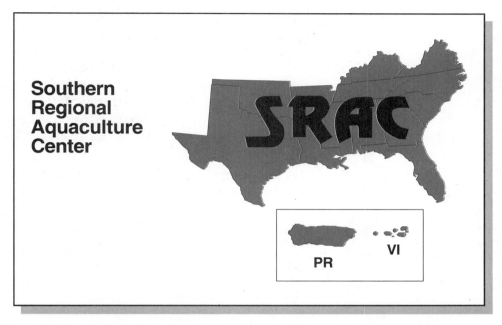

FIGURE 10-11 Logo of the Southern Regional Aquaculture Center.

The centers are jointly administered by USDA's Cooperative State Research Service and Extension Service. They are authorized to conduct research, extension education, and demonstration projects that have national or regional applications. These projects serve to implement the National Aquaculture Development Plan.

The primary mission of the five centers is to enhance viable, profitable commercial aquaculture production in the United States. The centers are organized to use the best scientific and educational skills and facilities among public and private institutions in the United States. Projects selected for funding address issues that cannot be adequately conducted by a single institution and problems of importance to commercial development in two or more states or territories.

INTERNATIONAL LAW

As aquaculture grows and migratory species become subject to ranching, international laws may become more important. International law is rooted in maritime laws. The first of these laws dates to 300 to 200 B.C., suggesting their importance.

In modern times, the concept of territorial waters began in 1945 with the Truman Proclamation, which stated that the continental shelf adjacent to the coast of the United States was part of the country's territory. In 1958, the conventions of modern maritime law were negotiated in Geneva, Switzerland. The **Geneva Conventions** were vague, deficient, and did not address the rights of landlocked countries. The Third **United Nations Conference on the Laws of the Sea** (UNCLOS III) convened in 1969. In 1982, a convention was adopted. Items covered in the new Laws of the Sea Treaty include—

■ A definition of the rights and duties of coastal countries in the **Exclusive Economic Zone** (EEZ). This includes the rights of "exploring and exploiting, conserving and managing the natural resources, whether living or nonliving."

■ Maximum size of an EEZ was established as being 200 nautical miles.

■ Conservation of living resources is called for by using the best scientific data available to establish allowable catches in the EEZ.

■ Coastal countries are required to promote best use of the EEZ living resources, and if the coastal countries cannot harvest an optimum catch, other countries have the right to harvest, as long as they come to an agreement with the coastal country.

■ Countries whose fishers harvest highly migratory species cooperate to maximize the catch, both within and out of the EEZs.

■ Countries from which anadromous fish originate have the primary responsibility for the stock.

■ All countries will reduce or prevent pollution of the seas.

Another form of international law is treaties between countries that may not mention aquaculture directly. These treaties include agreements on fishing, conservation, and ocean dumping of wastes. The United States has agreements with Mexico, Japan, and Canada.

FOOD AND AGRICULTURE ORGANIZATION OF THE UNITED NATIONS

In 1966, the **Food and Agriculture Organization of the United Nations** (FAO) organized the first world conference on aquaculture. Since then, the FAO established other smaller conferences and seminars. It also supports aquaculture projects in other countries and tracks world aquaculture production statistics.

FIGURE 10-12 Logo of the United Nations Food and Agriculture Organization.

STATE AND LOCAL LAWS

The business of aquaculture is often regulated in some way by laws of the state, province, county, or town. Some of these laws are fairly new and pertain directly to aquaculture while older laws relate only coincidentally. States with a long history of aquaculture have laws dealing more directly with the culture of aquatic species. Some of the permits that may be required include—

- Special vehicle license
- Water rights permit
- Water discharge permit
- Reservoir permit
- Game permit
- Harvesting permit
- Lease
- Hydraulics permit
- Aquaculture permit
- Live-fish transport permit
- Cultured fish processing license.

To coordinate the aquaculture activities in a state, some states have developed aquaculture plans. These plans suggest to aquaculturists the species and facilities to consider and recommend regulations and support to the legislature. The plans are written by a panel appointed by the state government or by working with a federal agency like NOAA or USDA.

INDIVIDUAL RESPONSIBILITY

A significant number of federal, state, and local government agencies are involved in the regulation of an aquaculture operation. This involvement includes site selection, facility design and construction, operations, species obtainment, production, processing, and marketing.

The regulatory environment is often a source of concern to individuals, investors, and corporations due to the possibility of unanticipated delays and increased capital and operating expenses. This concern is frequently based on a small percentage of proposed projects that encounter regulatory difficulties.

In most cases, regulatory difficulties arise because of inadequate planning, lack of knowledge of the process by the applicant, and incomplete information concerning the agencies' respective requirements. This is not to imply that improvements cannot be made in the regulatory environment, but to point out that information on government regulations is available and agency representatives are responsive to requests for assistance concerning their agency's jurisdiction. It is not the responsibility of an agency representative to be knowledgeable of the regulations of all other agencies that may have regulatory authority over some phase of a proposed project. This responsibility remains with the project applicant.

The applicant should start by asking—

■ Do state laws and regulations on aquaculture production and merchandising permit introduction, commercial rearing, and sale of the species of fish?

■ What permits or licenses are required?

■ Do any other federal or state regulations affect operations, such as those on interstate shipment, predator control, water rights, processing, and retailing?

■ Will an Environmental Impact Statement (EIS) be required?

■ Have all sources of information been checked?

SUMMARY

Many city, county, state, and federal regulations can affect aquaculture operations. Aquaculture often falls under catch-all legislation the creates unnecessary restrictions or demands. Existing land-use restrictions and environmental regulations reduce economic incentives to aquaculture by creating lengthy and uncertain permit processes, adding to the costs and delaying operation.

Aquaculture is a new industry in many areas. Old information and old laws and regulations applied in some cases hinder the development of aquaculture. A critical need exists to evaluate regulations and to separate valid ones from those that do not apply. Many services are available to the aquaculturalist to find information and meet regulations. Appendix Table A-11 provides addresses for all the services and agencies described in this chapter.

STUDY/REVIEW

Success in any career requires knowledge. Test your knowledge of this chapter by answering these questions or solving these problems.

True or False

1. The JAS was established to issue permits for aquaculture.

2. The Soil Conservation Service (SCS) seeks to improve the marketing of aquaculture products.

3. The Extension Service operates through the land grant universities.

4. The National Sea Grant College Program is administered by the USDA.

5. The Center for Veterinary Medicine (CVM) is responsible for wetlands.

Short Answer

6. Name five agencies or services provided through the Department of Agriculture.

7. Identify three states where the Agricultural Research Service conducts aquaculture research.

8. If an organization wished to import a new aquatic plant or animal, what government agency provides requirements for importation?

9. In cooperation with the Extension Service, what USDA agency administers the five regional aquaculture centers?

10. Name the five states that have regional aquaculture centers.

11. Which USDA agency might provide a loan for an aquaculture facility?

12. Which agency of the USDA provides assistance to aquaculture through the FSMIP?

13. Which agency of the USDA helps determine the economic feasibility of an aquaculture program?

14. Identify the following: TVA, FAO, FSMIP, APHIS, CSRS, SCS, EPA, CVM, JAS, MBTA.

15. Give the location of five national research programs in aquaculture.

16. Name three agencies that distribute information about aquaculture.

17. Indicate three types of assistance provided by the Soil Conservation Service to aquaculture.

18. List the two statutes that give the Food and Drug Administration authority for seafood products.

19. Which government service maintains a nationwide system of fish hatcheries?

20. Which government agencies will supply information about the quality and availability of groundwater and surface water?

21. Which government agency's prime responsibility affects aquaculture effluents?

22. Name the agency responsible for the National Fertilizer and Environmental Research Center and the Valley Resource Center.

Essay

23. What role can the Cooperative State Research Service play in aquaculture development?

24. Why would the Army Corps of Engineers be involved in an aquaculture facility?

25. What is the National Sea Grant College Program?

26. How may the Department of Energy become involved in an aquaculture project?

27. List the four main areas that involve the Center for Veterinary Medicine.

28. Describe how international law affects the development of aquaculture.

29. Indicate three things an individual must do before changing or starting an aquaculture facility.

KNOWLEDGE APPLIED

1. Using the addresses in Appendix Table A-11, contact the closest Regional Aquaculture Center. Request information on two species of fish that could be cultured in the area.

2. Using the address for the National Agricultural Library, request a bibliography on an aquaculture subject.

3. Contact the National Agricultural Statistic Service and request a copy of the most recent catfish and trout reports.

4. Visit a local SCS office. Ask for any information that they would supply an individual who wanted to start an aquaculture facility. Also, find out what other agricultural programs are involved in SCS.

5. Visit a local EPA office. Find out all aspects of aquaculture controlled by the EPA. Determine how the EPA would be involved in a start-up aquaculture facility.

6. Visit the offices of your local city or county government. During the visit, find out which individual should be contacted for local permits and licenses necessary to start an aquaculture business.

LEARNING/TEACHING AIDS

Books

DeVoe, M. R., & Whetstone, J. M. (1984). *An interim guide to aquaculture permitting in South Carolina.* Charleston, SC: South Carolina Sea Grant Consortium.

Hightower, M., Branton, C., & Granvil, T. (1990). *Summary tables of permits, licenses, certificates, and regulations affecting aquaculture operations.* Galveston, TX: Sea Grant College Program, Texas A&M University.

Hunt, J. W., & Pang, B. M. (1989). *Backyard aquaculture in Hawaii: A practical manual.* Kaneohe, HI: Windward Community College.

Jennings, F. D., Beeton, A. M., Davidson, J. R., Gaither, W. S., Gilmartin, M., & Crowder, B. (1982). *Sea grant aquaculture plan, 1983–1987.* Washington, DC:

Office of Sea Grant, National Oceanic and Atmospheric Administration, U.S. Department of Commerce.

Landau, M. (1992). *Introduction to aquaculture.* New York: John Wiley & Sons.

Pillay, T. V. R. (1992). *Aquaculture and the environment.* New York: Halstead Press.

Ziemann, D., Pruder, G., & Wang, J.-K. (1989). *Aquaculture effluent discharge program, Year 1, Final Report.* Honolulu, HI: Center for Tropical and Subtropical Aquaculture.

Career Opportunities in Aquaculture

One purpose of education and learning is to become employable and stay employable—to get and keep a job. People look for careers and careers look for people. Two broad categories of career opportunities in aquaculture are working for someone else and working for yourself. Success in any career requires some general skills and knowledge as well as some very specific skills and knowledge unique to a chosen occupation in aquaculture.

Learning Objectives

After completing this chapter, the student should be able to:

In Aquaculture

- List the basic skills and knowledge needed for successful employment and job advancement
- Describe the thinking skills needed for the workplace of today
- Identify the traits of an entrepreneur
- List six occupational areas of aquaculture
- Describe the general duties of the occupations in six areas of aquaculture

- Describe the education and experience needed to enter six areas of aquaculture
- List six general competencies needed in the workplace
- Describe five ways to identify potential jobs
- List eight guidelines for choosing a job
- List ten guidelines for filling out an application form
- Describe a letter of inquiry or application
- List the elements of a resume or data sheet
- Describe ten reasons an interview may fail

In Science

- Identify the careers that require a science background
- Discuss what research studies indicate about basic skills and thinking skills for the workplace

Understanding of this chapter will be enhanced if the following terms are known. Many are defined in the text and others are defined in the glossary.

KEY TERMS		
Competencies	Entrepreneur	Resources
Creative thinking	Follow-up letter	Resume
Cultural diversity	Forecast	Sociability
Curricular	Interpersonal	Systems
Data sheet	Letter of application	Visualization
Demographic	Letter of inquiry	

GENERAL SKILLS AND KNOWLEDGE

Over the past few years, study after study has indicated that potential employees seldom receive basic skills and knowledge. Without these basic skills and knowledge, the specific skills and knowledge for employment in aquaculture are of little value. Also, the new workplace demands a better prepared individual than in the past. Finally, those individuals working for themselves must develop a trait called entrepreneurship. This may also be a good trait for any employee.

Basic Skills

Success in the workplace requires that individuals possess skills in reading, writing, arithmetic, mathematics, listening, and speaking, at levels identified by employers nationwide.

Reading. An individual ready for the workplace of today and the future demonstrates reading with the following competencies:

- Locates, understands, and interprets written information, including manuals, graphs, and schedules to perform job tasks
- Learns from text by determining the main idea or essential message
- Identifies relevant details, facts, and specifications
- Infers or locates the meaning of unknown or technical vocabulary
- Judges the accuracy, appropriateness, style, and plausibility of reports, proposals, or theories of other writers.

Reading skills in aquaculture are necessary to keep up with new information and read directions for feeding or treating fish.

FIGURE 11-1 Reading skills are important to success.

Writing. Individuals ready for the workplace of today and the future demonstrate writing abilities with the following competencies:

- Communicates thoughts, ideas, information, and messages
- Records information completely and accurately
- Composes and creates documents such as letters, directions, manuals, reports, proposals, graphs, and flow charts with the appropriate language, style, organization, and format
- Checks, edits, and revises for correct information, emphasis, form, grammar, spelling, and punctuation.

In aquaculture, writing skills are necessary for such tasks as keeping pond or raceway records, describing disease conditions, or requesting a test.

Arithmetic and Mathematics. The workplace of today and the future requires individuals with competencies in arithmetic and mathematics. Arithmetic is computing with numbers by addition, subtraction, multiplication, and division. It is the application of mathematics. These important competencies are—

- Perform basic computations
- Use numerical concepts such as whole numbers, fractions, and percentages in practical situations
- Make reasonable estimates of arithmetic results without a calculator
- Use tables, graphs, diagrams, and charts to obtain or convey information
- Approach practical problems by choosing from a variety of mathematical techniques
- Use quantitative data to construct logical explanations of real-world situations
- Express mathematical ideas and concepts verbally and in writing
- Understand the role of chance in the occurrence and prediction of events.

Anyone not convinced of the value of arithmetic and mathematics to aquaculture should consider the skills required to figure pond volumes, treatment dosages, feed conversion ratios, and fish growth rates.

Listening. Individuals working today and in the future must demonstrate an ability to listen. This means to receive, attend to, and interpret verbal messages and other cues such as body language. Real listening means the individual comprehends, learns, evaluates, appreciates, or supports the speaker.

Speaking. Finally, individuals successful in the workplace of today and tomorrow must demonstrate these speaking competencies—

- Organize ideas and communicate oral messages appropriate to listeners and situations
- Participate in conversation, discussion, and group presentations
- Use verbal language, body language, style, tone, and level of complexity appropriate for audience and occasion
- Speak clearly and communicate the message
- Understand and respond to listener feedback
- Ask questions when needed.

Thinking Skills

Contrary to the old workplace, many research studies indicate that employers in the new workplace want workers who can think. Employers search for individuals showing competencies in creative thinking, decision making, problem solving, mental visualization, knowing how to learn, and reasoning.

FIGURE 11-2 Thinking skills are important for successful employment.

Creative Thinking. Creative thinkers generate new ideas by making non-linear or unusual connections or by changing or reshaping goals to imagine new possibilities. These individuals use imagination, freely combining ideas and information in new ways.

Decision Making. Individuals who use thinking skills to make decisions are able to specify goals and limitations to a problem. Next, they generate alternatives and consider the risks before choosing the best alternative.

Problem Solving. As silly as it sounds, the first step to problem solving is recognizing that a problem exists. After this, individuals with problem-solving skills identify possible reasons for the problem and then devise and begin a plan of action to resolve it. As the problem is being solved, problem solvers monitor the progress and fine-tune the plan. Being able to recognize a disease condition and look for solutions is a good example of problem solving in aquaculture.

Mental Visualization. This thinking skill is seeing things in the mind's eye by organizing and processing symbols, pictures, graphs, objects, or other information, for example, seeing a catfish pond from a diagram or a recirculating system's operation from a schematic.

Knowing How to Learn. Perhaps of all the thinking skills, this is most important because of rapid changes in technology. This skill is knowing how to learn techniques to apply and adjust existing and new knowledge and skills in familiar and changing situations. Knowing how to learn means awareness of personal learning styles and formal and informal learning strategies.

Reasoning. The individual who uses reasoning discovers the rule or principle connecting two or more objects and applies this to solving a problem. For example, chemistry teaches the theory of pH measurements, but the reasoning individual is able to use this information in understanding pH shifts in pond culture.

General Workplace Competencies

Besides the basic skills and the thinking skills, the workplace of today and tomorrow demands general competencies in the use of resources, interpersonal skills, information use, systems, and technology.

Resources. Resources of a business include time, money, materials, facilities, and people. Individuals in the workplace must know how to manage—

- Time through goals, priorities, and schedules
- Money with budgets and forecasts

FIGURE 11-3 Teams work together to solve problems and use resources efficiently.

■ Material and facility resources such as parts, equipment, space, and products

■ Human resources by determining knowledge, skills, and performance levels.

Interpersonal. More than ever, people cannot act in a vacuum. Most people are members of a team where they contribute to the group. They teach others in their workplace when new knowledge or skills are needed. More than ever and at all levels, individuals must remember to serve customers and satisfy their expectations. Through teams, individuals frequently exercise leadership to communicate, justify, encourage, persuade, or motivate individuals or groups. As part of employment teams, individuals negotiate resources or interests to arrive at a decision. Finally, all interpersonal skills require individuals to work with and use cultural diversity.

Information. The information age is here. Individuals in the workplace must cope with and use information. Successful individuals will identify the need for information and evaluate the information as it relates to a specific job. With the computer, individuals in the workplace must organize and process information in a systematic way. Also, with all this information available, individuals must interpret and communicate information to others using oral, written, or graphic methods. For example, production data of raceways must be summarized. To manage production information, computer skills are the key.

FIGURE 11-4 Computers open access to more information than ever before.

Systems. No longer can any aspect of a business or industry be viewed as standing alone. Every activity is part of a system, and individuals now seek to understand systems, whether these are social, organizational, technological, or biological. With an understanding of the systems in a business, trends can be identified and predictions and diagnosis can be made. Individuals then modify the system to improve the product or service. Aquaculturalists view the pond as a system; production is another system.

Technology. Technology makes life easier only for those who know how to select it, use it, maintain it, and troubleshoot it. Technology is complicated. Successful individuals learn to apply appropriate new technology through all the basic skills, the thinking skills, and general workplace competencies.

Personal Qualities

After all the training in basic skills, thinking skills, and general workplace competencies, individuals still fail for lack of some personal qualities. These include responsibility, self-esteem, sociability, self-management, and integrity or honesty.

Responsible individuals work hard at tasks even when the task is unpleasant. Responsibility shows in high standards of attendance, punctuality, enthusiasm, vitality, and optimism in starting and finishing tasks.

Those possessing self-esteem believe in themselves and maintain a positive view of themselves. These individuals know their skills, abilities, and emotional capacity. They feel good about themselves.

Successful individuals demonstrate understanding, friendliness, adaptability, empathy, and politeness to other people. These skills are demonstrated in familiar and unfamiliar social situations. The best examples are sincere individuals who take an interest in what others say and do.

Along with self-esteem is self-management. Individuals successful in business accurately assess their own knowledge, skills, and abilities while setting well-defined and realistic personal goals. Once goals are set, those who manage themselves monitor their progress and motivate themselves through the achievement of goals. Self-management also implies a person who exhibits self-control and responds to feedback unemotionally and nondefensively.

Finally, to be successful in aquaculture, an employee or entrepreneur requires good old-fashioned honesty and integrity. Good ethics are still a part of good business.

ENTREPRENEURSHIP

The most common view of an entrepreneur is one who takes risk and starts a new business. While this may be true for some in aquaculture, some traits of entrepreneurship are desirable at many levels of employment in aquaculture. Within any organization, an entrepreneur may—

- Find a better or higher use for resources
- Apply technology in a new way
- Develop a new market for an existing product
- Use technology to develop a new approach to serving an existing market
- Develop a new idea that creates a new business or diversifies an existing business.

Anyone can be an entrepreneur. The attitude of an entrepreneur includes—

- Risk-taking with clear expectations of the odds
- Focusing on opportunities and not problems
- Focusing on the customer
- Seeking constant improvement
- Emphasizing productivity over appearances
- Recognizing importance of example
- Keeping things simple
- Practicing open door and personal contact leadership

FIGURE 11-5 Entrepreneurship is responsible for Clear Springs Trout Farm, the world's largest commercial producers. (Photo courtesy Terry Patterson, College of Southern Idaho, Twin Falls, ID)

■ Encouraging flexibility

■ Being purposeful and communicating a vision.

Entrepreneurs are ready for the unexpected, differences, new needs, change, demographic shifts, changes in perception, and new knowledge. Entrepreneurs are good employees and good employers. Entrepreneurs keep the aquaculture industry growing.

JOBS IN AQUACULTURE

Some people consider the only jobs available in aquaculture to be those in the actual production or farm work. But the industry as a whole requires a large number of people to support the infrastructure of suppliers, producers, and marketers.

Specific jobs or employment opportunities in aquaculture can be grouped into general categories. These include supplies and services, training, production employment, marketing, inspection, and research and development. Each area requires some unique skills.

Supplies and Services

Occupations in the supplies and services area include those that support farm production and provide the inputs necessary for an operation to be productive. General areas of employment in supplies and services include financing, providing feed and other supplies, providing equipment, constructing facilities, and consulting.

Finance. Individuals involved in financing aquacultural operations provide money (capital) for the establishment and operation of facilities. They do this through loans and other types of financial assistance. Some examples of occupations in finance include—

- ■ Banker
- ■ Loan officer
- ■ Farm credit association employee.

Besides a good knowledge of finance, a good knowledge of aquaculture is necessary. Typical duties include helping fill out loan application forms, evaluation of loan application forms, locating credit sources, arranging loans, financial advisement, and collecting debts. The responsibilities of aquaculture may occupy only part of the job so other areas of agriculture could be part of the duties of an individual working in finance.

FIGURE 11-6 Bankers help arrange credit for the growth of aquaculture and nutrition.

Feed and Supplies. Providing feed and supplies includes manufacturing, hauling, selling, storing, and purchasing feed, feed ingredients, and other supplies. Obviously, more and different variations of employment will be available in areas where more aquaculture is found. Some types of occupations include—

- Feed salesperson
- Feed mill purchasing agent
- Feed truck driver
- Feed mill operator
- Feed mill worker
- Feed mill scales operator
- Nutritionist.

The knowledge and skills required for these jobs vary widely. For example, a feed salesperson and a feed mill purchasing agent require a strong background in nutrition, but a nutritionist requires the greatest knowledge and training in nutrition. Nutritionists must understand the nutritional requirements of the species being fed and what nutrients each feedstuff provides. Truck drivers and feed mill workers need some mechanical abilities and need to know how to operate equipment safely. The skills of a feed mill operator combine some nutritional, mechanical, and engineering knowledge and a greater level of responsibility.

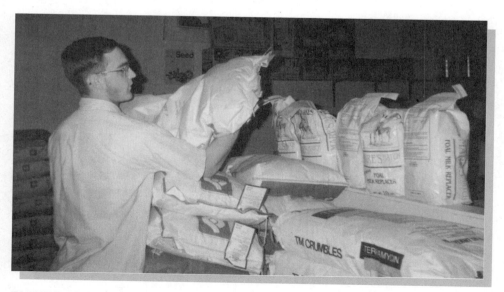

FIGURE 11-7 Selling feed requires an understanding of aquaculture and nutrition.

General duties of individuals providing feed and supplies include operating equipment, identifying feed ingredients, making careful measurements, product knowledge, following instructions, and consulting with farmers.

Equipment. Aquaculture requires some unique equipment and the development of new equipment. Equipment jobs include manufacturing, selling, hauling, and installing the equipment used in aquaculture facilities. Some examples are—

■ Equipment engineer

■ Manufacturing plant worker

■ Equipment installer

■ Equipment salesperson.

Again, the exact skills and knowledge vary considerably depending upon the job. For example, the mechanical knowledge of an engineer is greater than that of the manufacturing plant worker but the plant worker, is probably better skilled in welding. All these jobs require some mechanical aptitude with skills and knowledge in welding, electricity, and hydraulics. Sales positions require product knowledge with strong communication and interpersonal skills.

Duties for individuals who choose jobs in areas providing equipment to aquaculture facilities include working in plants to assemble equipment and traveling to farms to sell, deliver, install, and repair equipment.

Construction. Constructing aquaculture facilities involves designing, laying out designs, operating equipment, installing water facilities, and constructing buildings. Some examples of occupations include—

■ Architect

■ Carpenter

■ Heavy equipment operator

■ Electrician

■ Well driller

■ Surveyor.

These occupations are needed for many types of industries. In areas where aquaculture is prominent, these occupations will be more specific. For example, in Mississippi heavy equipment operators and surveyors develop skills for laying out and building catfish ponds. Electricians spend more time wiring pumps for aquaculture in Mississippi.

FIGURE 11-8 To construct new ponds requires the services of surveyors, heavy equipment operators, well drillers, and welders. (Photo courtesy Chuck Weirich, Delta Research and Extension Center, Stoneville, MS)

Typical duties for individuals in aquaculture construction include equipment operation, laying out facilities, installing pipe, making electrical connections, drawing and reading plans (blueprints), and working with all types of construction materials. Individuals interested in pursuing employment in construction areas need a mechanical aptitude.

Consulting. Consultants advise the aquaculturalist on how to establish, operate, and manage a business. A good consultant is worth good money. A bad consultant can cost lots of good money. Some examples of consultants include—

■ General aquaculture consultant

■ Extension aquaculture specialist

■ Veterinarian

■ Nutritionist.

The knowledge and skills required vary with the type of consultant. Most require at least a bachelor's degree, but a master's or doctorate is preferred in many cases. Also, the best consultants combine education with practical experience. Consultants stay current with new developments and can sift through information to discover trends and make judgment calls.

Typical duties of consultants, depending on the type, include making on-farm visits, holding meetings, providing information, developing reports, making suggestions, providing close examinations of an operation, and providing advice on a wide range of subjects.

Training

With all the interest in aquaculture and with all the potential that it holds, employment opportunities exist in the area of training and education. Some typical occupations include—

- High school agriculture instructor
- Postsecondary aquaculture instructor
- Fisheries and wildlife instructor
- Extension specialist.

To become an instructor requires at least a bachelor's degree. Teaching in a high school requires certification. The best instructors also possess some good practical experience. Instructors and trainers must have excellent communication skills.

Duties of instructors include finding or preparing curricular materials for the classes, preparing lesson plans, using a variety of methods to transfer information, and evaluating the effectiveness of the information transfer. The duties of instructors last for full days and weeks at a time. Extension specialists provide training but on a short-term basis and generally to those already involved in aquaculture.

FIGURE 11-9 Good instructors are needed for classroom and laboratory training in aquaculture programs.

Production Employment

Employment in production involves all of the occupations associated with reproducing, growing, and harvesting the aquacrop. These can be divided into management and worker categories.

Management. Management includes the planning organizing, staffing, directing, and controlling of farm activities. Management requires responsibility, the level of which should be reflected in the wage and the necessary skills and knowledge. Management occupations include—

- Farm manager
- Site manager
- Hatchery manager.

Managers possess a thorough knowledge of the production of the species being cultured. They know how to manage and supervise other people. Duties of managers can involve hiring and firing of workers, providing directions to workers, making decisions, meeting government regulations, selling the crop, buying feed and supplies, and keeping records. Managers need good communication skills to communicate with the owner and the workers.

Workers. Workers on an aquaculture facility can be very diverse. Workers can include unskilled as well as highly skilled and educated individuals. A few examples of some occupations include—

- Fishery technician
- Water technician
- Truck driver

FIGURE 11-10 Harvesting catfish requires seine operators.

- ■ Seine operator
- ■ Biologist
- ■ Skilled laborer
- ■ Unskilled laborer.

The type of worker depends on the size and type of facility. Some facilities may be small enough that a worker could do several different jobs. Large facilities may have individuals for each job. The knowledge and skills vary. Technicians and biologists require knowledge and skills beyond the high school level and even to the bachelor's degree. Seine operators, truck drivers, and skilled and unskilled laborers receive short-term training and on-the-job training. Because the work is around ponds, raceways, and tanks, most of these jobs require good physical skills.

General duties of workers on an aquaculture facility include such things as checking water quality, moving fish, harvesting fish, feeding fish, hauling seines, and operating a variety of equipment. In general, these individuals work with water and with fish.

Marketing

Employment in marketing includes jobs in all activities that connect the product from the farm with the consumer. This involves jobs in processing and promoting.

FIGURE 11-11 Truckers drive live haul trucks to catfish processing plants.

Processing. Processing prepares the aquaculture crop for the consumer. Jobs are often in processing plants, many in assembly line settings. Some of the possible jobs include—

■ Plant manager
■ Supervisor
■ Quality control specialist
■ Personnel officer
■ Skilled laborer
■ Unskilled laborer.

Some positions in processing plants require performance of repetitive activities that can be learned quickly, while other positions require considerable experience. Some of the typical duties in processing include working with the live and dead fish, operating and cleaning equipment, cleaning the facilities, working around water, working in refrigerated areas, and inspecting fish and fish products.

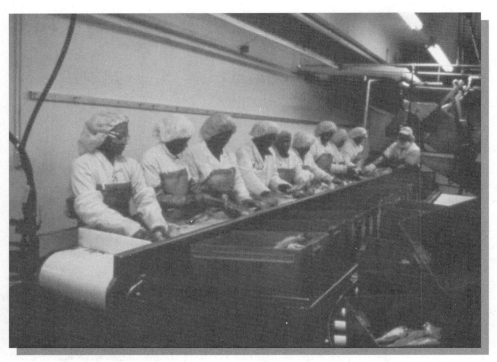

FIGURE 11-12   Many jobs in processing plants are in assembly line settings. (Photo courtesy Chuck Weirich, Delta Research and Extension Center, Stoneville, MS)

Promoting. Promotion convinces the consumer to purchase the product. Promotion includes advertising, public relations, market analyses, demonstrations, and selling. Some examples of occupations in promotion are—

- Salesperson
- Advertising account representative
- Writer
- Photographer
- Association executive.

Occupations in promotion can take place in a city far removed from the area of production. Still, the work requires a practical knowledge of aquaculture. The ability to talk the language of aquaculture is important for success. Additionally, occupations in promotion require education in marketing, communication, and advertising. Duties may include writing articles, writing advertising, making sales presentations, placing orders, planning advertising campaigns, reading reports, and preparing reports—anything that promotes the consumption of the products of aquaculture.

Inspection

Every industry witnesses more regulation by government, often initiated by consumer advocacy groups. This creates inspection or monitoring jobs to issue permits and check for compliance with regulations in such areas as water quality, chemical use, and processing facility standards. Examples of occupations in inspection and monitoring include—

- Soil conservation officer
- Game conservation officer
- Food inspector
- Water tester
- Grader
- Quality control officer
- Laboratory technician.

The primary focus of these occupations is on assuring quality products for the consumer while maintaining the environment. The work involves contact with farmers, processors, and others to enforce government regulations and laws. These occupations demand excellent human relations and communication skills.

Research and Development

Aquaculture progressed to the stage it is today because of a commitment to research and development. Occupations in research and development involve finding new and profitable ways to produce products in forms desired by consumers. Research may also identify new animals or plants for production. Some of the general occupations include—

- Research scientist
- Research technician
- Laboratory assistant
- Unskilled assistant.

Government agencies and private industries conduct research on new techniques and varieties. Depending upon the job, the duties include planning research, designing experiments or trials, selecting plants or animals for the trials, conducting the day-to-day steps of the experiment or trial, collecting data, laboratory analyses, analyzing data, reporting data, and writing final reports of the results. The work is often a mix of outdoor and indoor activities.

Depending on the job, the education required varies. The research scientist needs a doctorate in some related field—nutrition, fish biology, or water chemistry. Research technicians and laboratory assistants require some training beyond high school. Some of the training is often provided by the agency or industry that owns the laboratory.

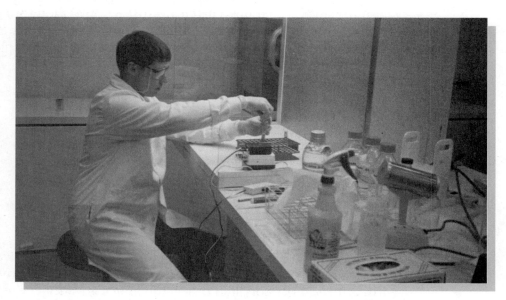

FIGURE 11-13 Research keeps aquaculture moving forward, and jobs in research areas are available on many levels.

EDUCATION AND EXPERIENCE

Requirements to begin working in aquaculture vary depending upon the level of work. One requirement common to all is practical work experience in aquaculture. Often to gain this practical experience, the new employee begins at an entry-level job and then is advanced through the organization. The advancement depends on the skills and knowledge the employee brings to the job, skills and knowledge gained on the job, and productivity.

Entry-level educational requirements vary. The basic skills, thinking skills, and general workplace competencies discussed in this chapter are important. These skills should be obtained in high school and reinforced during additional training and schooling. More specialized education in aquaculture is offered in some high schools, community colleges, and universities.

Many high school programs in aquaculture provide the education necessary for lower-level positions. Often high school programs in aquaculture provide students with supervised work experience in some aspect of aquaculture. This is invaluable in getting a job and in helping individuals determine if they wish to pursue additional education.

Some community colleges and other postsecondary schools provide specialized programs in aquaculture with practical experience as a part of the schooling. Programs at community colleges focus on entry-level technician jobs.

Universities and colleges offering bachelor's degrees, master's degrees, and doctorate programs provide highly specialized education in aquaculture. Depending on their location, they may specialize in warmwater culture, coldwater culture, genetics, nutrition, water chemistry, and other areas. The Appendix Table A-13 contains a list of colleges and universities with aquaculture programs.

IDENTIFYING A JOB

Finding that first job or finding a different job can be difficult. Books, videos, and seminars are available on finding jobs. What follows are some suggestions. The Learning/Teaching Aids section at the end of this chapter contains more information.

Sources for locating jobs include—

- ■ Classified advertisements of newspapers
- ■ Magazines or trade journals and publications
- ■ Personal contacts

- ■ Placement offices
- ■ Employment or personnel office of company
- ■ Public notices
- ■ Computerized on-line services.

Newspapers, magazines, trade journals, and publications can be good resources for locating a job. By reading the advertisements in these publications the potential employee can determine the demand for his or her job skills. Also, the potential employee can compare his or her skills and training with those listed in the advertisements.

Another kind of classified advertisement is the "situation wanted" section of newspapers, magazines, and trade journals. Many people secure excellent jobs by advertising their skills in these sections. An employer may read about an individual's skills and realize he or she is the answer to the employer's needs.

Personal contacts are still a top source of jobs. Employers do not like to make mistakes. Some feel that if a trusted acquaintance makes a recommendation this lessens the chances of making a mistake in hiring. Also, personal contacts may know of jobs opening up before being publicly announced. This gives the potential employee more time to prepare and research the job. Personal contacts include friends, relatives, teachers, guidance counselors, and employees of the company.

Placement offices provide vocational counseling, give aptitude and interest tests, locate jobs, and arrange job interviews. Three types of placement offices are available: public, private, and school. These agencies work to match employers with prospective employees. Often too, an agency knows how to help a potential employee prepare and present himself or herself.

Public placement offices are supported by federal and state funds. Their services are free. Private placement offices charge for services they provide. This usually is a percentage of the beginning salary. Individuals using private placement services sign a contract before services are provided. High schools, trade schools, and colleges may maintain a placement service for their students. They also provide help for individuals to identify their aptitude or interest for a job and help in preparation for job interviews.

Many companies support their own employment or personnel office. Individuals seeking employment can fill out application forms and leave resumes in case a job becomes available.

Finally, some companies seeking new employees may issue a public notice of some kind. This includes posters or fliers on bulletin boards around a community. Posters or fliers are sent to related businesses, which end up on their bulletin boards. Schools and colleges often receive public announcements of jobs.

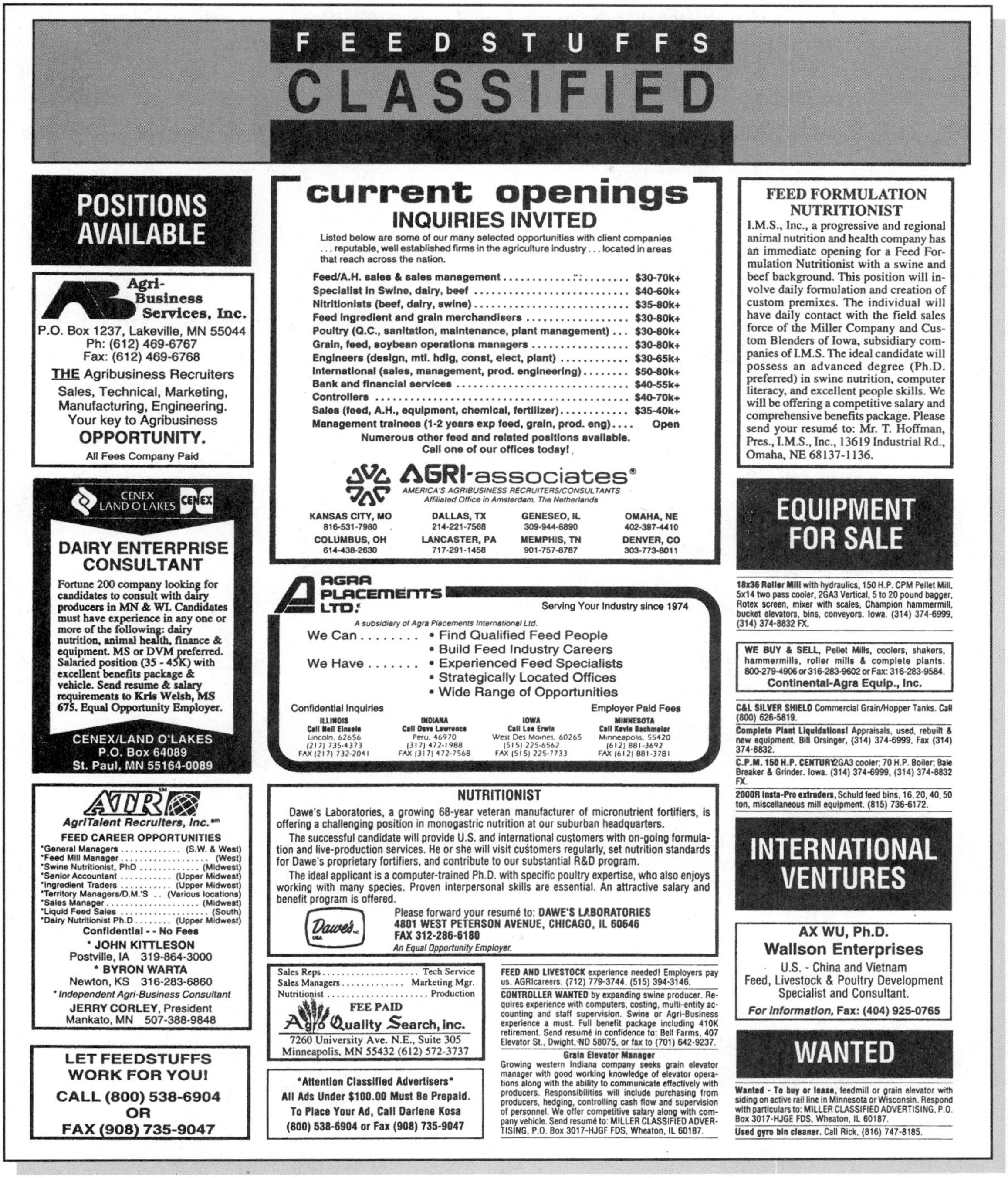

FEEDSTUFFS
CLASSIFIED

POSITIONS AVAILABLE

Agri-Business Services, Inc.
P.O. Box 1237, Lakeville, MN 55044
Ph: (612) 469-6767
Fax: (612) 469-6768
THE Agribusiness Recruiters
Sales, Technical, Marketing, Manufacturing, Engineering.
Your key to Agribusiness
OPPORTUNITY.
All Fees Company Paid

CENEX LAND O LAKES
DAIRY ENTERPRISE CONSULTANT
Fortune 200 company looking for candidates to consult with dairy producers in MN & WI. Candidates must have experience in any one or more of the following: dairy nutrition, animal health, finance & equipment. MS or DVM preferred. Salaried position (35 - 45K) with excellent benefits package & vehicle. Send resume & salary requirements to Kris Welsh, MS 675. Equal Opportunity Employer.
CENEX/LAND O'LAKES
P.O. Box 64089
St. Paul, MN 55164-0089

ATR
AgriTalent Recruiters, Inc.[sm]
FEED CAREER OPPORTUNITIES
*General Managers .... (S.W. & West)
*Feed Mill Manager .... (West)
*Swine Nutritionist, PhD .... (Midwest)
*Senior Accountant .... (Upper Midwest)
*Ingredient Traders .... (Upper Midwest)
*Territory Managers/D.M.'S .. (Various locations)
*Sales Manager .... (Midwest)
*Liquid Feed Sales .... (South)
*Dairy Nutritionist Ph.D .... (Upper Midwest)
Confidential - - No Fees
* JOHN KITTLESON
Postville, IA 319-864-3000
* BYRON WARTA
Newton, KS 316-283-6860
* Independent Agri-Business Consultant
JERRY CORLEY, President
Mankato, MN 507-388-9848

LET FEEDSTUFFS WORK FOR YOU!
CALL (800) 538-6904
OR
FAX (908) 735-9047

current openings
INQUIRIES INVITED
Listed below are some of our many selected opportunities with client companies . . . reputable, well established firms in the agriculture industry . . . located in areas that reach across the nation.

Feed/A.H. sales & sales management	$30-70k+
Specialist in Swine, dairy, beef	$40-60k+
Nitritionists (beef, dairy, swine)	$35-80k+
Feed ingredient and grain merchandisers	$30-80k+
Poultry (Q.C., sanitation, maintenance, plant management)	$30-80k+
Grain, feed, soybean operations managers	$30-80k+
Engineers (design, mtl. hdlg, const, elect, plant)	$30-65k+
International (sales, management, prod. engineering)	$50-80k+
Bank and financial services	$40-55k+
Controllers	$40-70k+
Sales (feed, A.H., equipment, chemical, fertilizer)	$35-40k+
Management trainees (1-2 years exp feed, grain, prod. eng)	Open

Numerous other feed and related positions available.
Call one of our offices today!

AGRI-associates®
AMERICA'S AGRIBUSINESS RECRUITERS/CONSULTANTS
Affiliated Office in Amsterdam, The Netherlands

KANSAS CITY, MO 816-531-7980	DALLAS, TX 214-221-7568	GENESEO, IL 309-944-8890	OMAHA, NE 402-397-4410
COLUMBUS, OH 614-438-2630	LANCASTER, PA 717-291-1458	MEMPHIS, TN 901-757-8787	DENVER, CO 303-773-8011

AGRA PLACEMENTS LTD.
Serving Your Industry since 1974
A subsidiary of Agra Placements International Ltd.
We Can .... • Find Qualified Feed People
• Build Feed Industry Careers
We Have .... • Experienced Feed Specialists
• Strategically Located Offices
• Wide Range of Opportunities
Confidential Inquiries
Employer Paid Fees

ILLINOIS Call Neil Einsele Lincoln, 62656 (217) 735-4373 FAX (217) 732-2041	INDIANA Call Dave Lawrence Peru, 46970 (317) 472-1988 FAX (317) 472-7568	IOWA Call Lee Erwin West Des Moines, 60265 (515) 225-6562 FAX (515) 225-7733	MINNESOTA Call Kevin Bachmeier Minneapolis, 55420 (612) 881-3692 FAX (612) 881-3781

NUTRITIONIST
Dawe's Laboratories, a growing 68-year veteran manufacturer of micronutrient fortifiers, is offering a challenging position in monogastric nutrition at our suburban headquarters.
The successful candidate will provide U.S. and international customers with on-going formulation and live-production services. He or she will visit customers regularly, set nutrition standards for Dawe's proprietary fortifiers, and contribute to our substantial R&D program.
The ideal applicant is a computer-trained Ph.D. with specific poultry expertise, who also enjoys working with many species. Proven interpersonal skills are essential. An attractive salary and benefit program is offered.
Dawe's USA
Please forward your resumé to: DAWE'S LABORATORIES
4801 WEST PETERSON AVENUE, CHICAGO, IL 60646
FAX 312-286-6180
An Equal Opportunity Employer.

Sales Reps .... Tech Service
Sales Managers .... Marketing Mgr.
Nutritionist .... Production
FEE PAID
Agro Quality Search, inc.
7260 University Ave. N.E., Suite 305
Minneapolis, MN 55432 (612) 572-3737

Attention Classified Advertisers
All Ads Under $100.00 Must Be Prepaid.
To Place Your Ad, Call Darlene Kosa
(800) 538-6904 or Fax (908) 735-9047

FEED AND LIVESTOCK experience needed! Employers pay us. AGRIcareers. (712) 779-3744. (515) 394-3146.

CONTROLLER WANTED by expanding swine producer. Requires experience with computers, costing, multi-entity accounting and staff supervision. Swine or Agri-Business experience a must. Full benefit package including 410K retirement. Send resumé in confidence to: Bell Farms, 407 Elevator St., Dwight, ND 58075, or fax to (701) 642-9237.

Grain Elevator Manager
Growing western Indiana company seeks grain elevator manager with good working knowledge of elevator operations along with the ability to communicate effectively with producers. Responsibilities will include purchasing from producers, hedging, controlling cash flow and supervision of personnel. We offer competitive salary along with company vehicle. Send resumé to: MILLER CLASSIFIED ADVERTISING, P.O. Box 3017-HJGF FDS, Wheaton, IL 60187.

FEED FORMULATION NUTRITIONIST
I.M.S., Inc., a progressive and regional animal nutrition and health company has an immediate opening for a Feed Formulation Nutritionist with a swine and beef background. This position will involve daily formulation and creation of custom premixes. The individual will have daily contact with the field sales force of the Miller Company and Custom Blenders of Iowa, subsidiary companies of I.M.S. The ideal candidate will possess an advanced degree (Ph.D. preferred) in swine nutrition, computer literacy, and excellent people skills. We will be offering a competitive salary and comprehensive benefits package. Please send your resumé to: Mr. T. Hoffman, Pres., I.M.S., Inc., 13619 Industrial Rd., Omaha, NE 68137-1136.

EQUIPMENT FOR SALE

18x36 Roller Mill with hydraulics, 150 H.P. CPM Pellet Mill, 5x14 two pass cooler, 2GA3 Vertical, 5 to 20 pound bagger, Rotex screen, mixer with scales, Champion hammermill, bucket elevators, bins, conveyors. Iowa. (314) 374-6999, (314) 374-8832 FX.

WE BUY & SELL, Pellet Mills, coolers, shakers, hammermills, roller mills & complete plants. 800-279-4906 or 316-283-9602 or Fax: 316-283-9584.
Continental-Agra Equip., Inc.

C&L SILVER SHIELD Commercial Grain/Hopper Tanks. Call (800) 626-5819.

Complete Plant Liquidations! Appraisals, used, rebuilt & new equipment. Bill Orsinger, (314) 374-6999, Fax (314) 374-8832.

C.P.M. 150 H.P. CENTURY2GA3 cooler; 70 H.P. Boiler; Bale Breaker & Grinder. Iowa. (314) 374-6999, (314) 374-8832 FX.

2000R Insta-Pro extruders, Schuld feed bins, 16, 20, 40, 50 ton, miscellaneous mill equipment. (815) 736-6172.

INTERNATIONAL VENTURES

AX WU, Ph.D.
Wallson Enterprises
U.S. - China and Vietnam
Feed, Livestock & Poultry Development Specialist and Consultant.
For information, Fax: (404) 925-0765

WANTED

Wanted - To buy or lease, feedmill or grain elevator with siding on active rail line in Minnesota or Wisconsin. Respond with particulars to: MILLER CLASSIFIED ADVERTISING, P.O. Box 3017-HJGE FDS, Wheaton, IL 60187.

Used gyro bin cleaner. Call Rick, (816) 747-8185.

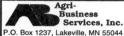

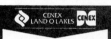

FIGURE 11-14 Advertisements from newspapers, magazines, or trade journals provide leads to jobs.

Computerized posting of jobs is another kind of a bulletin board. Some on-line information services maintain computerized databases of jobs. Interested individuals use a phone, computer, and modem to contact the computerized database to search for jobs that match their qualifications and desires. This type of job listing opens the door wide to potential jobs but often not local jobs.

GETTING A JOB

Once some job possibilities are identified the work begins. Getting the job is difficult and requires preparation. Again, books, videos, and seminars teach how to get a job. A few tips follow.

Once a job is identified, do a little research on the company and the job before applying. Know these things about the job and the company—

- Name of the company
- Name of personnel manager
- Company address and phone number
- Position available
- Requirements for the position
- Geographic scope of the company (local, county, state, regional, national)
- Company's product(s)
- Recent company developments
- Responsibilities of the position
- Demand for the company's product(s).

Before you get too far along in the application process, be certain the job is what you want to pursue. Money is not everything in a job. Compare the characteristics of the occupation with those you possess by answering these questions:

- Does the job description fit your interests?
- Is this the level of occupation in which you wish to engage?
- Does this type of work appeal to your interests?
- Are the working conditions suitable to you?
- Will you be satisfied with the salaries and benefits offered?
- Can you advance in this occupation as rapidly as you would like?
- Does the future outlook satisfy you?

■ Is there enough demand for this occupation that you should consider entering it?

■ Do you have or can you get the education needed for the occupation?

■ Can you get the finances needed to get into the occupation?

■ Can you meet the health and physical requirements?

■ Will you be able to meet the entry requirements?

■ Are there any other reasons you might not be able to enter this occupation?

■ Is the occupation available locally, or are you willing to move to a part of the country where it is available?

Application Forms

If the company requires an application form, remember you are trying to sell yourself by the information given. Review the entire application form before you begin. Pay particular attention to any special instructions to print or write in your own handwriting. When answering ads that require potential employees to apply in person, be prepared to complete an application form on the spot. Take an ink pen. Prepare a list of information you will need to complete the application form. The information may include: your social security number; the addresses of schools you have attended; names, phone numbers, and addresses of previous employers and supervisors; names, phone numbers, and addresses of references. The following guidelines will provide you with some direction when completing application forms.

■ Follow all instructions carefully and exactly.

■ If handwritten rather than typed, write neatly and legibly. Handwritten answers should be printed unless otherwise directed.

■ Application forms should be written in ink unless otherwise requested. If you make a mistake, mark through it with one neat line.

■ Be honest and realistic.

■ Give all the facts for each question.

■ Keep answers brief.

■ Fill in all the blanks. If the question does not pertain to you, write "not applicable" or "N/A." If there is no answer, write "none" or draw a short line through the blank.

■ Many application forms ask what salary you expect. If you are not sure what is appropriate, write "negotiable," "open," or "scale" in the blank. Before applying try to find out what the going rate for similar work is at other locations. Give a salary range rather than exact figure.

Letters of Inquiry and Application

The purpose of a **letter of inquiry** is to obtain information about possible job vacancies. The purpose of a **letter of application** is to apply for a specific position that has been publicly advertised. Both letters indicate your interest in working for a particular company, acquaint employers with your qualifications, and encourage the employer to invite you for a job interview.

Letters of inquiry and application represent you. They should be accurate, informative, and attractive. Your written communications should present a strong, positive, professional image both as a job seeker and future employee. The following list should be used as a guide when writing letters of inquiry and application.

- Be short and specific (one or two pages). Use $8\frac{1}{2} \times 11$ in white typing paper, not personal or fancy paper.
- Be sure the letter is neatly typed and error free.
- Use an attractive form, free from smudges.
- Write to a specific person. Use "To Whom It May Concern" if answering a blind ad.
- Write logically organized paragraphs that are to the point.
- Write carefully constructed sentences free from spelling or grammatical errors.
- Be positive in tone.
- Express ideas in a clear, concise, direct manner.
- Avoid slang words and expressions.
- Avoid excessive use of the word "I."
- Avoid mentioning salary and fringe benefits.
- Write a first draft, then make revisions.
- Proofread final letter yourself, and also have someone else proofread.
- Address and sign correctly. Type envelope addresses.

This information should be included in a letter of inquiry—

- Specify the reasons why you are interested in working for the company and ask if there are any positions available now or in the near future.
- Express your interest in being considered a candidate for a position when one becomes available.
- Since you are not applying for a particular position, you cannot relate your qualifications directly to job requirements. (You can explain how

your personal qualifications and work experience would help meet the needs of the company.)

■ Mention and include your resume.

■ State your willingness to meet with a company representative to discuss your background and qualifications. Include your address and phone number where you can be reached.

■ Address letters of inquiry to the "Personnel Manager" unless you know his or her name.

A letter of application should include—

■ Indicate your source of the job lead.

■ Specify the particular job you are applying for and the reason for your interest in the position and the company.

■ Explain how your personal qualifications meet the needs of the employer.

■ Explain how your work experience relates to job requirements.

■ Mention and include your resume.

■ Request an interview and state your willingness. Include your address and phone number where you can be reached.

Resume or Data Sheet

Some jobs require a resume or **data sheet**. The following information should be considered when writing a resume or data sheet—

■ Name, address, and phone number

■ Brief, specific statement of career objective

■ Educational background—names of schools, dates, major field of study, degrees, or diplomas—listed in reverse chronological order

■ Leadership activities, honors, and accomplishments

■ Work experience listed in reverse chronological order

■ Special technical skills and interests related to job

■ References

■ Limit to one page if possible

■ Neatly typed and error free

■ Logically organized

■ Honest statement of qualifications and experiences.

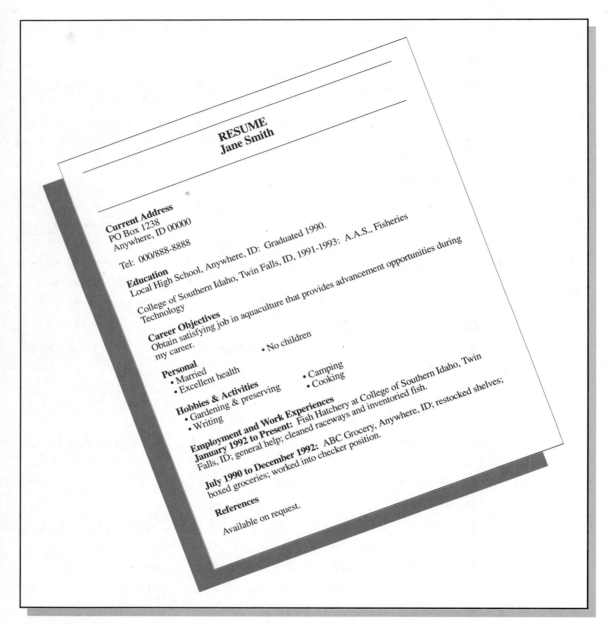

RESUME
Jane Smith

Current Address
PO Box 1238
Anywhere, ID 00000

Tel: 000/888-8888

Education
Local High School, Anywhere, ID: Graduated 1990.

College of Southern Idaho, Twin Falls, ID, 1991-1993: A.A.S., Fisheries
Technology

Career Objectives
Obtain satisfying job in aquaculture that provides advancement opportunities during
my career.

Personal
• Married   • No children
• Excellent health

Hobbies & Activities
• Gardening & preserving   • Camping
• Writing   • Cooking

Employment and Work Experiences
January 1992 to Present: Fish Hatchery at College of Southern Idaho, Twin
Falls, ID; general help; cleaned raceways and inventoried fish.

July 1990 to December 1992: ABC Grocery, Anywhere, ID; restocked shelves;
boxed groceries; worked into checker position.

References
Available on request.

FIGURE 11-15   Resumes are neat, specific, logical, and one page long.

Employers look for a quick overview of who you are and how you fit into their business. On the first reading, an employer may only spend 10 to 15 seconds reading a resume. Be sure to present relevant information clearly and concisely in an eye-catching format.

The Interview

The next step in the job-hunting process is the interview. While many do's and don'ts of an interview are available, perhaps the best advice comes from the interviewer's side of the desk. This list of items includes common reasons interviewers give for not being able to place applicants in a job.

- Poor attitude
- Unstable work record
- Bad references
- Lack of self-selling ability
- Lack of skill and experience
- Not really anxious to work
- "Bad-mouthing" former employers
- Too demanding (wanting too much money or to work only under certain conditions)
- Unable to be available for interviews
- Poor appearance
- Lack of manners and personal courtesy
- No attempt to establish rapport; not looking the interviewer in the eye
- Being interested only in the salary and benefits of the job
- Lack of confidence; being evasive
- Poor grammar, use of slang
- Not having any direction or goals.

Follow-up Letters

Follow-up letters are sent immediately after an interview. The follow-up letter demonstrates your knowledge of business etiquette and protocol. Always send a follow-up letter regardless of whether you had a good interviewing experience and regardless of whether you are interested in the position. When employers do not receive follow-up letters from job candidates, they often assume the candidate is not aware of professional protocol they will need to demonstrate on the job.

The major purpose of a follow-up letter is to thank those individuals who participated in your interview. In addition, a follow-up letter reinforces your name, application, and qualifications to the employer and indicates whether you are still interested in the job position.

WOW! MY FIRST JOB INTERVIEW

You just got your first job interview or maybe you just bombed two job interviews and you have a chance at another one. What should you do? Check over this list of do's. Do—

1. Find out about the company before you interview—its products, who its customers are, etc.

2. Be neat and well-groomed. Dress conservatively.

3. Be punctual—15 to 20 minutes early.

4. Have your resume and examples of your work available for quick reference.

5. Have a pen and note pad to take notes.

6. Have a prepared list of questions regarding the job. These may be answered by the interviewer during the course of the interview.

7. When meeting the receptionist, smile, introduce yourself, state you have an appointment, follow the receptionist's instructions, and wait patiently.

8. Greet the interviewer with a smile and by name.

9. If the interviewer offers his or her hand, shake it firmly.

10. Introduce yourself and state the purpose of your appointment.

11. Be seated only after the interviewer has asked you to do so.

12. Sit and stand erect.

13. Be polite and courteous.

14. Be sincere, enthusiastic, friendly, and honest.

15. Let the interviewer take the lead in the conversation.

16. Be alert. Sit slightly forward in the chair to give an alert appearance.

17. Be confident, look directly at the interviewer.

18. Make an effort to express yourself clearly and distinctly.

19. Speak correctly, use proper grammar, speak in clear, moderate tones.

20. Take time to think about your answer. Choose your words carefully.

21. Answer questions completely, but give only essential facts.

22. Convey positive answers.

23. Speak positively of former employers and associates.

24. Watch for signs that the interview is over, such as the interviewer shuffling papers, moving chair around, etc.

25. Thank the interviewer for his or her time.

26. Shake hands with the interviewer and leave promptly at the completion of the interview.

27. Write a follow-up letter to express your interest in the job and your appreciation for the opportunity to interview.

Some don'ts of interviewing include the following. Don't—

1. Take others with you to the interview—parents, friends, etc.

2. Put your hat or coat on the interviewer's desk.

3. Use a limp or overpowering handshake.

4. Lean against a wall, chair, or desk.

5. Interrupt the interviewer.

6. Chew gum, smoke, eat candy, etc.

7. Giggle, squirm in your chair, tap your fingers, swing a crossed leg, etc.

8. Use slang or swear.

9. Talk too long.

10. Try to flatter the interviewer.

11. Give all yes or no answers.

12. Talk about personal problems.

13. Press for a decision on being hired.

Employers have good and bad interviews, but they remember bad interviews longer!

SUMMARY

A primary goal of education and training is to become employable and stay employable—to get and keep a job or run a successful business. The world of work still requires people who can read, write, do math, and communicate. Rapidly advancing technology makes this even more apparent. Also, the modern workplace now looks for people who possess thinking skills. With a solid set of basic skills, future employees also need to relate to other people, to use information, and to understand the concept of systems and use technology. Old-fashioned ideas like responsibility, self-esteem, sociability, self-management, and integrity are not out of date either.

Aquaculture jobs range from those very closely tied to aquaculture to those that support aquaculture. In general, potential job areas include supplies and services, training, production, marketing, inspection, and research and development. Education and training for jobs in aquaculture vary from on-the-job training to high school to college degrees.

After training and education, finding and getting the right job may still be a challenge. Several good resources exist for locating a job. Still, the best one is personal contact. Letters of inquiry, letters of application, a resume, and being prepared for the job interview help secure a job.

STUDY/REVIEW

Success in any career requires knowledge. Test your knowledge of this chapter by answering these questions or solving these problems.

True False

1. Technology makes the basic skills of reading, writing, and math less necessary for today's workplace.

2. The only jobs in aquaculture are at trout and catfish facilities.

3. Jobs in aquaculture range from unskilled to highly skilled and highly educated.

4. Employers in the modern work force want employees who will take orders and suppress thinking skills.

5. Entrepreneurs are risk takers only involved in the development of a new aquabusiness.

Short Answer

6. List five basic skills necessary for employment.

7. List five thinking skills desired by today's employers.

8. List three personal qualities found in successful employees.

9. List five occupations in supplies and services for aquaculture.

10. An individual who wants to become an educator in aquaculture would look for jobs in what organizations?

11. Name two general areas in aquaculture that are apt to rely on unskilled labor.

12. Name the general occupation in aquaculture where the actual work can take place in a city far removed from the area of production.

13. Give three examples of occupations in the area of inspection.

14. If an individual wants to work in research and development in aquaculture, what occupations are available?

15. List five ways to find a job in aquaculture.

16. The following competency demonstrates an individual's writing ability:
    a. Identifies relevant details, facts, and specifications
    b. Checks for correct grammar and spelling
    c. Understands the role of change in the prediction of events
    d. Uses tables and graphs to obtain information

17. List three guidelines for filling out an application form.

18. What is the purpose of a follow-up letter?

Essay

19. Give an example of why basic skills would be important to each of the following occupations in aquaculture: loan officer, feed mill salesperson, equipment salesperson, and nutritionist.

20. Why are thinking skills important in any occupation in aquaculture?

21. Describe interpersonal skills.

22. Define an entrepreneur.

23. Describe how architects, carpenters, electricians, surveyors, and heavy equipment operators could be involved in occupations related to aquaculture.

24. Define a consultant for aquaculture and give examples of the duties of a consultant.

25. Briefly outline the process of getting a job in aquaculture.

26. Describe ten reasons an interview may fail.

27. What should a resume contain and how long should it be?

28. What duties and responsibilities might be a part of working in a fish processing plant?

KNOWLEDGE APPLIED

1. Gather sample resumes and develop your own resume or data sheet.

2. Collect position announcements and classified ads for jobs in aquaculture. Write a letter of job inquiry and a letter of job application for a selected job using this information.

3. Develop a list of questions frequently asked during an interview. Use the questions in role-playing job interviews and videotape the interviews.

4. Organize a field trip to a public or private placement office. Following the field trip, discuss the office's policies and how they affect job searchers and employers. Alternatively, invite a representative from a state employment agency to explain how employment agencies can help students gain employment.

5. Hold an aquaculture career field day. Invite individuals currently employed in aquaculture to present a panel discussion on career opportunities. For example, invite representatives from production, aquaculture, research, education, and government.

6. Select one career in aquaculture of interest and prepare a research paper on the career using a computer and word processing software. The paper should identify the knowledge and skills required and the employment opportunities.

7. Collect pictures or photographs of people engaged in various careers in aquaculture and prepare a bulletin board collage.

8. Invite a resource person such as a business owner or personnel manager to discuss what he or she looks for in resumes, application letters and forms, and during interviews.

9. Invite a panel of local agribusiness people to discuss the importance of employee work habits, basic skills and attitudes, and how they affect the entire business.

LEARNING/TEACHING AIDS

Books

Benson, C., White, W. C., & Collins, D. (1988). *Opportunities in agriculture careers.* Lincolnwood, IL: National Textbook Company.

Business Council for Effective Literacy Bulletin. (1987). *Job-related basic skills: A guide for planners of employee programs.* New York: Business Council for Effective Literacy (BCEL).

Campbell R., & Thompson, M. J. (1987). *Working: Today and tomorrow.* St. Paul, MN: The Changing Times Education Service Division, EMC Publishing.

Department of Adult Vocational and Technical Education. (1986). *Find it; Get it; Keep it: A pre-employment skills curriculum for the special needs individual.* Macomb, IL: Research and Development Section, Curriculum Publications Clearinghouse, Western Illinois University.

Harrington, T. F., & O'Shea, A. J. (1984). *Guide for occupational exploration.* Circle Pines, MN: National Forum Foundation.

Littrell, J. J. (1984). *From school to work.* Homewood, IL: National Farm Book Company. Goodheart-Willcox.

Smith, M., Underwood, J. M., & Bultmann, M. (1991). *Careers in agribusiness and industry* (4th ed.). Danville, IL: Interstate Publishers.

U.S. Department of Education, U.S. Department of Labor. (1988). *The bottom line: Basic skills in the workplace.* Washington, DC.

U.S. Department of Education. (1990) *Policy perspectives.* Workplace Competencies: The Need to Improve Literacy and Employment Readiness. Washington, DC.

U.S. Department of Education. (1991). *America 2000: An education strategy.* Sourcebook. Washington, DC.

U.S. Department of Labor, Employment and Training Administration. (1978). *Dictionary of occupational titles.* Washington, DC: American Association for Vocational Instructional Materials.

U.S. Department of Labor. (1991). *What work requires of schools.* A SCANS Report for America 2000. Washington, DC.

U.S. Department of Labor. (1992). *Learning a living: A blueprint for high performance.* A SCANS Report for America 2000. Washington, DC.

U.S. Department of Labor, Bureau of Labor Statistics. *Occupational outlook handbook.* Washington, DC.

Vocational Agriculture Service (1986). *Applying for a job.* Urbana, IL: College of Agriculture, University of Illinois.

Weymann, C. J. (N.D.). *Work in the new economy: Careers and job seeking in the 21st century.* Macomb, IL: AAVIM, Curriculum Publications Clearinghouse, Western Illinois University.

Videos

Choosing careers series. (1986). Charleston, WV: Cambridge Career Products.

Career planning. (1988). Shawnee Mission, KS: RMI Media Productions, Inc.

Agricultural careers. (1990). Santa Monica, CA: Ready Reference Press.

Appendix

TABLE A-1 Miscellaneous Conversion Factors for Aquaculture Use

1 acre-foot	=	43,560 cubic feet
1 acre-foot	=	325,850 gallons
1 acre-foot of water	=	2,718,144 pounds
1 cubic foot of water	=	62.4 pounds
1 gallon of water	=	8.34 pounds
1 gallon of water	=	3,785 grams
1 liter of water	=	1,000 grams
1 fluid ounce	=	29.57 grams
1 fluid ounce	=	1.043 ounces
1 grain per gallon	=	17.1 milligrams/liter
1 milliliter of water	=	1 gram
1 cubic meter of water	=	1 metric ton
1 quart of water	=	946 grams
1 teaspoon	=	4.9 milliliters
1 tablespoon	=	14.8 milliliters
1 cup	=	8 fluid ounces

(continued)

TABLE A-1 Miscellaneous Conversion Factors for Aquaculture Use
(concluded)

1 acre-foot/day of water	=	226.3 gallons/minute
1 acre-inch/day of water	=	18.9 gallons/minute
1 acre-inch/hour of water	=	452.6 gallons/minute
1 second foot of water	=	448.8 gallons/minute
1 cubic foot/second of water	=	448.8 gallons/minute
1 foot of water	=	0.43 pound/square inch
1 foot of water	=	0.88 inch of mercury (Hg)
1 horsepower	=	550 foot-pounds/second
1 horsepower	=	745.7 watts
1 kilowatt	=	1,000 watts
1 kilowatt	=	1.34 horsepower
1 hectare	=	10,000 square meters
1 hectare	=	2.47 acres
1 acre	=	4,048 square meters

TABLE A-2 Conversion Factors (C.F.): The Weight of a Chemical That Must Be Added to One Unit Volume of Water to Give One Part Per Million (ppm)

Amount of Chemical	To Equal
2.72 pounds per acre-foot	1 ppm
1,233 grams per acre-foot	1 ppm
0.0283 gram per cubic foot	1 ppm
0.0000624 pound per cubic foot	1 ppm
0.0038 gram per gallon	1 ppm
0.0584 grain per gallon	1 ppm
1 milligram per liter	1 ppm
0.001 gram per liter	1 ppm
8.34 pounds per million gallons of water	1 ppm
1 gram per cubic meter	1 ppm
1 milligram per kilogram	1 ppm
10 kilograms per hectare-meter	1 ppm

TABLE A-3   Conversion of Units of Volume

FROM	TO							
	cm^3	liter	m^3	ft^3	fl oz	fl pt	fl qt	gal
cm^3	1	0.001	1×10^{-6}	3.53×10^{-5}	0.0338	0.00211	0.00106	2.64×10^{-4}
liter	1,000	1	0.001	0.0353	33.81	2.113	1.057	0.2642
m^3	1×10^6	1,000	1	5.31	3.38×10^4	2,113	1,057	264.2
ft^3	2.83×10^4	28.32	0.0283	1	957.5	59.84	29.92	7.481
fl oz	29.57	0.0296	2.96×10^{-5}	0.00104	1	0.0625	0.0313	0.0078
fl pt	473.2	0.4732	4.73×10^{-4}	0.0167	16	1	0.5000	0.1250
fl qt	946.4	0.9463	9.46×10^{-4}	0.0334	32	2	1	0.2500
gal	3,785	3.785	0.0038	0.1337	128	8	4	1

cm^3 = cubic centimeter = milliliter = ml; m^3 = cubic meter; ft^3 = cubic foot; fl oz = fluid ounce;
fl pt = fluid pint; fl qt = fluid quart; gal = gallon

TABLE A-4   Conversion for Units of Weight

FROM	TO				
	GM	kg	gr	oz	lb
gm	1	0.001	15.43	0.0353	0.0022
kg	1,000	1	1.54×10^4	35.27	2.205
gr	0.0648	6.48×10^{-5}	1	0.0023	1.43×10^{-4}
oz	28.35	0.0284	437.5	1	0.0625
lb	453.6	0.4536	7,000	16	1

gm = gram; kg = kilogram; gr = grain; oz = ounce; lb = pound

TABLE A-5 Conversion for Parts Per Million, Proportion, and Percent

Parts Per Million	Proportion	Percent
0.1	1:10,000,000	0.000010
0.25	1:4,000,000	0.000025
1.0	1:1,000,000	0.0001
2.0	1:500,000	0.0002
3.0	1:333,333	0.0003
4.0	1:250,000	0.0004
5.0	1:200,000	0.0005
8.4	1:119,047	0.00084
10.0	1:100,000	0.001
15.0	1:66,667	0.0015
20.0	1:50,000	0.002
25.0	1:40,000	0.0025
50.0	1:20,000	0.005
100.0	1:10,000	0.01
150.0	1:6,667	0.015
167.0	1:6,000	0.0167
200.0	1:5,000	0.02
250.0	1:4,000	0.025
500.0	1:2,000	0.05
1,667.0	1:600	0.1667
5,000.0	1:200	0.5
6,667.0	1:150	0.667
30,000.0	1:33	3.0

TABLE A-6   Preparation of Standard Solutions

Concentration of Standard Solution (ppm)	Amount of Concentrated Standard Solution (mL)	Volume Diluted to mL	Concentration of Final Solution (ppm)
1,000	1	1,000	1
1,000	5	1,000	5
1,000	10	1,000	10
500	2	1,000	1
500	10	1,000	5
500	20	1,000	10
250	4	1,000	1
250	20	1,000	5
250	40	1,000	10
100	1	100	1
100	5	100	5
100	10	100	10
50	2	100	1
50	10	100	5
50	20	100	10
10	10	100	1
10	50	100	5
1	1	10	0.1
1	5	10	0.5

TABLE A-7 Solubility of Oxygen in Parts Per Million (ppm) in Freshwater at Various Temperatures and at a Pressure of 760 mm Hg (Sea Level)

Temperature		Oxygen Concentration (ppm)	Temperature		Oxygen Concentration (ppm)
°F	°C		°F	°C	
32	0	14.6	69.8	21	9.0
33.8	1	14.2	71.6	22	8.8
35.6	2	13.8	73.4	23	8.7
37.4	3	13.5	75.2	24	8.5
39.2	4	13.1	77	25	8.4
41	5	12.8	78.8	26	8.2
42.8	6	12.5	80.6	27	8.1
44.6	7	12.2	82.4	28	7.9
46.4	8	11.9	84.2	29	7.8
48.2	9	11.6	86	30	7.6
50	10	11.3	87.8	31	7.5
51.8	11	11.1	89.6	32	7.4
53.6	12	10.8	91.4	33	7.3
55.4	13	10.6	93.2	34	7.2
57.2	14	10.4	95.0	35	7.1
59	15	10.2	96.8	36	7.0
60.8	16	10.0	98.6	37	6.8
62.6	17	9.7	100.4	38	6.7
64.4	18	9.5	102.2	39	6.6
66.2	19	9.4	104.0	40	6.5
68	20	9.2			

TABLE A-8 Altitude Correction Factor for the Solubility of Oxygen in Freshwater

Atmospheric Pressure (mm Hg)	or	Equivalent Altitude (Ft)	=	Correction Factor
775		540		1.02
760		0		1.00
745		542		.98
730		1,094		.96
714		1,688		.94
699		2,274		.92
684		2,864		.90
669		3,466		.88
654		4,082		.86
638		4,756		.84
623		5,403		.82
608		6,065		.80
593		6,744		.78
578		7,440		.76
562		8,204		.74
547		8,939		.72
532		9,694		.70
517		10,472		.68
502		11,273		.66

Example: Solubility of oxygen at sea level (760 mm Hg) at 20°C is 9.2 ppm. The solubility of oxygen at an altitude of 1,688 feet in 20°C water is 9.2 ppm × 0.94 (Correction Factor) = 8.65 ppm

TABLE A-9 Number of Water Samples Required from Ponds to Estimate the Averages of Water Quality Variables with 95 Percent Certainty that Errors Will Not Exceed the Specified Values

Water Quality Variable	Number of Water Samples Per Determination
Dissolved Oxygen	
± 0.5 ppm	6
± 1.0 ppm	2
pH	
± 0.5 unit	1
± 1.0 unit	1
Temperature	
± 0.5°C	2
± 1.0°C	1
Total Hardness	
± 1.0 ppm	1
Secchi disk (underwater) visibility	
± 5 cm	7
± 10 cm	2

TABLE A-10 Finfish Propagated in Aquaculture

Scientific Family Name	Common Scientific Name	*Scientific Name*	Common Name
Acinpenseridae	Sturgeons	*Acipenser transmontanus*	White sturgeon
		Acipenser fulvescens	Lake sturgeon
		Acipenser ruthensus	Sterlet
		Acipenser guldenstadti	Russian sturgeon
		Acipenser nudiventris	Thorn sturgeon
		Acipenser stellatus	Starred sturgeon
		Huso huso	Beluga sturgeon
Amiidae	Bowfins	*Amia calva*	Bowfin
Anguillidae	Eels	*Anguilla anguilla*	European eel
		Anguilla rostrata	American eel
		Anguilla japonica	Japanese eel
Clupeidae	Herrings	*Dorsoma petenense*	Threadfin shad
		Dorsoma cepedianum	Gizzard shad
		Alosa sapidissima	American shad
Salmonidae	Trout, Salmon, Chars	*Salmo salar*	Atlantic salmon
		Salmo clarki	Cutthroat trout
		Salmo trutta	Brown trout
		Salmo aguabonita	Golden trout
		Salvelinus fontinalis	Brook trout
		Salvelinus namaycush	Lake trout
		Salvelinus malma	Western brook char (or Dolly Varden)
		Oncorhynchus mykiss	Rainbow trout (many strains)
		Oncorhynchus nerka	Sockeye salmon
		Oncorhynchus gorbuscha	Pink salmon
		Oncorhynchus keta	Chum salmon
		Oncorhynchus tshawytscha	Chinook salmon
		Oncorhynchus kisutch	Coho salmon
		Oncorhynchus masu	Masu salmon
		Thymallus arcticus	Arctic grayling
		Coregonus sp.	Cisco and Whitefish
		Stenodus leucichthys	Sheerfish
Esocidae	Pikes	*Esox lucius*	Northern pike
		Esox masquinongy	Muskellunge
		Esox niger	Chain pickerel
Chanidae	Milkfishes	*Chanos shanos*	Milkfish

(continued)

TABLE A-10   Finfish Propogated in Aquaculture *(continued)*

Scientific Family Name	Common Scientific Name	*Scientific Name*	Common Name
Cyprinidae	Minnows and Carps	*Carassius auratus*	Goldfish
		Cyprinus carpio	Common Carp
		Notemegonus crysoleucas	Golden shiner
		Notropis sp.	Shiners
		Pimephales sp.	Minnows
		Hypophthalmicthys molitrix	Silver carp
		Ctenopharyngodon idella	Grass carp
		Aristichthys nobilis	Bighead carp
		Abramis brama	Bream
		Tinca tinca	Tench
		Carassius carassius	Crucian carp
		Rutilus rutilus	Roach
		Cirrhinus molitorella	Mud carp
		Mylopharyngodon piceus	Black carp
		Labeo rohita	Rohu
		Catla catla	Catla or Bhakur
		Cirrhina mrigala	Mrigal or Naini
		Osteochilus hasselti	Nilem
		Puntius gonionotus	Puntius carp
Catastomidae	Suckers	*Ictobius niger*	Buffalofish
Ictaluridae	Catfishes	*Ictalurus puctatus*	Channel catfish
		Ictalurus furcatus	Blue catfish
		Ictalurus nebulosis	Brown bullhead
		Ictalurus catus	White catfish
Claridae	Walking catfishes	*Clarias batrachus*	Walking catifish
		Clarias macrocephalus	Walking catifish
		Clarias fusais	Walking catifish
		Clarias largera	Walking catifish
		Clarias gariepinus	Sharp-toothed catfish
Siluridae	Freshwater sharks	*Wallagonia attu*	Freshwater shark
		Siluris glanis	Sheatfish
Pangasidae	River catfishes	*Pangasius sutchi*	River catfish
		Pangasius laurnaudi	River catfish
Atherinidae	Whitefishes	*Odonthectes basilichthys*	Pejerrey
		Chirostoma sp.	Mexican whitefishes
Cyprinodontidae	Killifishes	*Fundulus sp.*	Topminnows and Killfishes

(continued)

TABLE A-10 Finfish Propogated in Aquaculture *(concluded)*

Scientific Family Name	Common Scientific Name	*Scientific Name*	Common Name
Poeciliidae	Livebearers	*Gambusia sp.*	Gambusias
Percichthyidae	Temperate basses	*Morone saxatilis*	Striped bass
Centrarchidae	Sunfishes	*Lepomis sp.* *Micropterus sp.* *Pomoxix sp.*	Sunfishes Basses Crappie
Percidae	Perches	*Perca flavescens* *Stizostedion vitreum vitreum*	Yellow perch Walleye
Carangidae	Jacks and Pompano	*Trachinotus carolinus*	Florida pompano
Sciaenidae	Drums	*Aplodinotus grunniens* *Cynosion nebulosus* *Sciaenops ocellata*	Freshwater drum Spotted seatrout Red drum
Mugilidae	Mullets	*Mugil cephalus*	Striped mullet
Cichlidae	Cichlids	*Tilapia sp.*	Tilapia
Scombridae	Tunas and Mackerels	*Euthynnus sp.* *Thunnus albacares*	Skipjacks Yellowfin tuna
Anoplopomatidae	Sablefishes	*Anoplopoma fimbria*	Sablefish
Bothidae	Lefteye flounders	*Paralichthys lethostigma*	Southern flounder
Pleuronectidae	Righteye flounders	*Hippoglossus hippoglossus* *Hippoglossus stenolepis* *Pleuronichthys sp.*	Atlantic halibut Pacific halibut Turbot
Ophicephalidae	Snakeheads	*Ophicephalus sp.*	Snakeheads
Osphronemidae	Gouramis	*Osphronemus goramy* *Trichogaster pectoralis*	Giant gourmi Siamese gourami

TABLE A-11 Addresses of Agencies Providing Information or Regulations to Aquaculture

U.S. Department of Agriculture

Federal-State Marketing Improvement Program
Transportation and Marketing Division
Agricultural Marketing Service
U.S. Department of Agriculture
Room 4007, South Building
PO Box 96456
Washington, DC 20090-6456
(202) 720-2704
FAX (202) 690-0338

Chief, Commodity Procurement Branch
Livestock and Seed Division
Agricultural Marketing Service
U.S. Department of Agriculture
2610 South Building
PO Box 96456
Washington, DC 20090-6456
(202) 720-2650
FAX (202) 720-7271

Director, Transportation and Marketing Division
Distribution Services Branch
Agricultural Marketing Service
U.S. Department of Agriculture
2945 South Building
PO Box 96456
Washington, DC 20090-6456
(202) 720-8357
FAX (202) 690-0031

Director, Policy and Program Development
Animal and Plant Health Inspection Service
U.S. Department of Agriculture
Room 305E, Administration Building
14th and Independence Avenue, S.W.
Washington, DC 20250
(202) 720-5283
FAX (202) 690-2251

Director, Office of Aquaculture
Cooperative State Research Service
U.S. Department of Agriculture
14th and Independence Avenue, S.W.
Aerospace Building, Suite 342
Washington, DC 20250-2200
(202) 401-4929
FAX (202) 401-5179

Principal Aquacultural Scientist
Cooperative State Research Service
U.S. Department of Agriculture
14th and Independence Avenue, S.W.
Aerospace Building, Suite 330
Washington, DC 20250-2200
(202) 401-4061
FAX (202) 401-5179

Agricultural Economist-Aquaculture
Economic Research Service
U.S. Department of Agriculture
1301 New York Avenue, N.W., Room 1237-B
Washington, DC 20005-4788
(202) 219-0888
FAX (202) 219-0042

The National Program Leader, Aquaculture
Extension Service
U.S. Department of Agriculture
Room 3863, South Building
Washington, DC 20250-0900
(202) 720-5004
FAX (202) 690-4869

(continued)

TABLE A-11 Addresses of Agencies Providing Information or Regulations to Aquaculture *(continued)*

Trade Assistance and Planning Office
Foreign Agricultural Service
U.S. Department of Agriculture
3101 Park Center Drive, Suite 1103
Alexandria, VA 22302
(703) 305-2916
FAX (703) 305-2788

Dairy, Livestock, and Poultry Division
Foreign Agricultural Service
U.S. Department of Agriculture
Room 6616, South Building
Washington, DC 20250
(202) 720-8031
FAX (202) 720-0617

Legislative Affairs and Public Information Staff
Farmers Home Administration
U.S. Department of Agriculture
14th and Independence Avenue, S.W.,
    Room 5037-S
Washington, DC 20250
(202) 720-4323
FAX (202) 690-0311

Director, Office of Aquaculture
U.S. Department of Agriculture
14th and Independence Avenue, S.W.
Aerospace Building, Suite 342
Washington, DC 20250-2200
(202) 401-4929
FAX (202) 401-5179

Director, Research and Scientific Exchange Division
Office of International Cooperation and Development
U.S. Department of Agriculture
Room 3220, South Building
Washington, DC 20250-4300
(202) 690-4872
FAX (202) 690-0892

National Biologist-Aquaculture
Ecological Sciences Division
Soil Conservation Service
U.S. Department of Agriculture
South Building, Room 6151
Washington, DC 20250
(202) 720-5991
FAX (202) 720-2646

National Agricultural Library
Aquaculture Information Center
U.S. Department of Agriculture
10301 Baltimore Blvd., Room 304
Beltsville, MD 20705-2351
(301) 504-5558
FAX (301) 504-5472

Livestock, Dairy, and Poultry Branch
National Agricultural Statistics Service
U.S. Department of Agriculture
Room 5906, South Building
Washington, DC 20250-2000
(202) 720-6147
FAX (202) 690-0675

U.S. Department of the Army

Director, Research and Development Directorate
    (Civil Works)
U.S. Headquarters
Army Corps of Engineers (CERD-C)
U.S. Department of the Army
20 Massachusetts Avenue, NW
Washington, DC 20314-1000
(202) 272-0257
FAX (202) 272-0907

Regulatory Branch
Operations and Readiness Division
Civil Works Directorate
Army Corps of Engineers
U.S. Department of the Army
20 Massachusetts Avenue, N.W.
Washington, DC 20314-1000
(202) 272-1785
FAX (202) 504-5096

(continued)

TABLE A-11 Addresses of Agencies Providing Information or Regulations to Aquaculture *(continued)*

U.S. Department of Commerce

Director, Technical Assistance and Research
  Division
Economic Development Administration
U.S. Department of Commerce
14th and Pennsylvania Ave., N.W., Room 7810
Washington, DC 20230
(202) 377-4085
FAX (202) 377-0995

Aquaculture Specialist
Office of Research and Environmental
  Information
National Marine Fisheries Service
U.S. Department of Commerce
1335 East-West Highway, Room 6314
Silver Spring, MD 20910
(301) 713-2363
FAX (301) 588-4853

Program Director, Marine Advisory Service
National Sea Grant College Program
U.S. Department of Commerce
1335 East-West Highway
Silver Spring, MD 20910
(301) 713-2431
FAX (301) 713-0799

National Oceanographic Data Center
NOAA/NESDIS (E/OCx7)
User Services Division
U.S. Department of Commerce
1825 Connecticut Avenue, NW
Washington, DC 20235
(202) 606-4549
FAX (202) 606-4586

Associate Program Director, Aquaculture
National Sea Grant College Program
U.S. Department of Commerce
1335 East-West Highway, Room 5492
Silver Spring, MD 20910
(301) 713-2451
FAX (301) 713-0799

National Technical Information Service
U.S. Department of Commerce
5285 Port Royal Road
Springfield, VA 22161
(703) 487-4650
FAX (703) 321-8547

U.S. Department of Energy

Biofuels Systems Division
CE-331
U.S. Department of Energy
1000 Independence Avenue, S.W.
Washington, DC 20585
(202) 586-8078
FAX (202) 586-7114

(continued)

TABLE A-11 Addresses of Agencies Providing Information or Regulations to Aquaculture *(continued)*

U.S. Department of Health and Human Services

Associate Director
Office of Seafood
Center for Food Safety and Applied Nutrition
Food and Drug Administration
U.S. Department of Health and Human Services
200 C Street, S.W., HFF 503
Washington, DC 20204
(202) 254-3888
FAX (202) 254-3984

Center for Veterinary Medicine
Food and Drug Administration
U.S. Department of Health and Human Services
7500 Standish Place
Rockville, MD 20855
(301) 295-8761
FAX (301) 295-8807

U.S. Department of the Interior

National Aquaculture Coordinator
ARLSQ 820
U.S. Fish and Wildlife Service
U.S. Department of the Interior
Mail Stop III, Arlington Square
Washington, DC 20240
(703) 358-1715
FAX (703) 358-2210

Chief, Office of Extension and Publications
U.S. Fish and Wildlife Service
U.S. Department of the Interior
Arlington Square Building, Mail Stop 725
Washington, DC 20240
(703) 358-1706
FAX (703) 358-2202

Hydrologic Information Unit
U.S. Geological Survey
U.S. Department of the Interior
12201 Sunrise Valley Drive
Reston, VA 22092
(703) 648-4000
FAX (703) 648-4250

Independent Agencies

Senior Fisheries Advisor
Agency for International Development
Bureau for Research and Development
Office of Agriculture
Washington, DC 20523-1809
(703) 875-4098
FAX (703) 875-4186

SBIR Program
National Science Foundation
1800 G Street, N.W., Room V-502
Washington, DC 20550
(202) 653-5335
FAX (202) 653-7699

Division of Ocean Sciences
National Science Foundation
1800 G Street, N.W., Room 609
Washington, DC 20550
(202) 357-9639
FAX (202) 357-7621

Division of Environmental Biology
National Science Foundation
1800 G Street, N.W., Room 215
Washington, DC 20550
(202) 357-7332
FAX (202) 357-1191

(continued)

TABLE A-11 Addresses of Agencies Providing Information or Regulations to Aquaculture *(concluded)*

Small Business Administration
Small Business Answer Desk
409 Third Street, S.W.
Washington, DC 20416
(202) 205-6600
FAX (202) 205-7064

U.S. Environmental Protection Agency
401 M Street, S.W.
Washington, DC 20460
FAX (202) 260-7883

Regional Manager
Agricultural Institute
Tennessee Valley Authority
National Fertilizer and Environmental Research Center
PO Box 1010
Muscle Shoals, AL 35660-1010
(205) 386-3488
FAX (205) 386-2284

Regional and State Networks

Secretary/Treasurer
National Association of State Aquaculture Coordinator
PO Box 1163
Richmond, VA 23209
(804) 371-6094
FAX (804) 371-7786

Director, North Central Regional Aquaculture Center
Department of Fisheries and Wildlife
13 Natural Resources Building
East Lansing, MI 48824-1222
(517) 353-1962
FAX (517) 353-7181

Director, Southern Regional Aquaculture Center
Delta Branch Experiment Station
PO Box 197
Stoneville, MS 38776
(601) 686-9311
FAX (601) 686-9744

Director, Tropical and Subtropical Regional Aquaculture Center
Administrative Center
The Oceanic Institute
Makapuu Point
Waimanalo, HI 96795
(808) 259-7951
FAX (808) 259-5971

Director, Northeastern Regional Aquaculture Center
University of Massachusetts, Dartmouth
North Dartmouth, MA 02747
(508) 999-8157
FAX (508) 999-8590

Director, Western Regional Aquaculture Center
School of Fisheries
College of Ocean and Fishery Sciences
University of Washington
Seattle, WA 98195
(206) 543-4290
FAX (206) 685-4674

TABLE A-12 Aquaculture Journals and Related Periodicals

American Fish Farmer and World Aquaculture News
Aquaculture
Aqua-cultura (in Spanish)
Aquaculture Digest
Aquaculture and Fisheries Management
Aquaculture Ireland
Aquaculture Magazine
Bamidgeh (Journal of Israeli Fish Breeders' Association)
Bulletin of the Japanese Society of Scientific Fisheries (Nippon Suisan Gakkaishi; English
    abstracts)
California Department of Fish and Game (Fish Bulletin)
Canadian Journal of Fisheries and Aquatic Sciences
Commercial Fish Farmer
European Aquaculture Society (quarterly newsletter)
Fish Farmer
Fish Farming International
Freshwater and Aquaculture Contents Tables
Il Pesce (in Italian and English)
Infofish Marketing Digest
Informationen für die Fischwirtschaft (in German)
Journal of Aquaculture Engineering
Journal of Fish Biology
Journal of Fish Diseases
Journal of the World Aquaculture Society
NAGA (ICLARM Newsletter)
Progressive Fish-Culturist
Recent Advances in Aquaculture
Revista Latinoamericana de Acuicultura (in Spanish)
FDEC Asian Aquaculture
Transactions of the American Fisheries Soceity
Trout Cultivation

Note: The above list excludes certain scientific journals (e.g. *Comparative Biochemistry
    and Physiology; Marine Biology*) as well as less specialist publications (e.g. *Scientific
    American; New Scientist*), which frequently contain articles relating to aquaculture.

TABLE A-13 Aquaculture Training and Education

Institutions with specialized courses or training in aquaculture include—

Auburn University
Clemson University
Texas A & M University
Louisiana State University
University of California at Davis
Florida Institute of Technology
University of Hawaii at Manoa
University of Washington
Rutgers State University
University of Delaware
Virginia Institute of Marine Science
Southern Illinois University at Carbondale
Oregon State University
Hofstra University
Colorado State University
University of South Carolina
University of Arkansas at Pine Bluff
University of Hawaii
University of Michigan
Michigan State University
Oregon State University

One- and two-year aquaculture training programs for technicians are available from a number of junior or community colleges. Some of these include—

College of Southern Idaho, Twin Falls, Idaho
Lummi Indian College of Fisheries, Washington
Windward Community College, Hawaii
Grays Harbor College, Washington
Haywood Technical College, North Carolina

Many universities offer specialized M.S. and Ph.D. degrees in aquaculture and have excellent research and experimental facilities. Land grant colleges and universities have a tradition of freshwater aquaculture (mainly biology) and good extension and continuing education services. The Sea Grant-funded universities and colleges deal mainly with brackishwater and marine aquaculture.

TABLE A-14 Dichotomous Key for Pond Water

1. Cells single or, if undergoing cell division, found in pairs 2
1. Cells numerous, multicellular; arranged in chains, filaments or other
     multicellular organism 9
2(1) Cells or cultures of cells green in color 3
2(1) Cells or cultures of cells not green or a shade of green; generally
     yellow to brown in color 5
3(2) Cells motile by cilia .... ciliated protozoan
3(2) Cells non-motile by flagellum.. 4
4(3) Cells with bright green (e.g. grass green) chloroplast, larger (10 to 15
     microns in diameter), oval or flattened shape .... *Tetraselmis*
4(3) Cells a shade of green, generally small (2 to 5 microns in diameter),
     spherical or oval in shape .... *Nannochloropsis*
5(2) Cells with radial symmetry .... (centric diatoms) 6
5(2) Cells not radially symmetrical, with somewhat bilateral
     symmetry .... (pennate diatoms) 7
6(5) Cells large, spherical or oval, usually connected in chains of two or
     more cells, spines not present .... *Melosira*
6(5) Cells smaller; square, rectangular, or oval; single or in short chains,
     spines originating at the corner of each cell .... *Chaetoceros*
7(5) Cells tapering to a long, thin spine .... *Nitzschia closterium*
7(5) Cell endings rounded or in a point, but not tapering to a long spine 8
8(7) Cells bilaterally symmetrical in all views .... *Navicula* or other navicula-type diatom
8(7) Cells asymmetrical in some views .... *Achnanthes*
9(1) Cells photosynthetic, arranged in a filament, variable in color 10
9(1) Cells non-photosynthetic, arranged into tissues and organs, generally
     clear or may be pigmented 12
10(1) Cells arranged as a branching filament, green in color .... *Cladophora*
10(1) Cells arranged in a non-branching filament, brown in color 11
11(10) Cells generally in long chains of large, spherical cells, spines not
     present .... *Melosira*
11(10) Cells in short or long chains or smaller, square, or rectangular cells
     possessing spines which originate at the corner of each cell........... *Chaetoceros*
12(9) Body with segmented body parts, appearing shrimp-like, lacking
     cilia around the mouth .... copepod
12(9) Body smooth, non-segmented, appearing sac-like, with cilia around
     the mouth .... rotifer

Glossary

Like a foreign language, terms unique to aquaculture can be baffling to the newcomer. An individual who traveled to a foreign country to do business would be expected to know the language of the country. The same is true for the individual wanting to learn aquaculture. Successful individuals use the glossary and learn the language. Words not found in the glossary may be listed in the index and defined within the book.

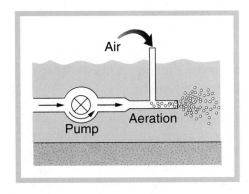

A

Abdomen—Belly; the ventral side of the fish surrounding the digestive and reproductive organs.

Abdominal—Pertaining to the belly.

Abrasion—A spot scraped of skin, mucous membrane, or superficial epithelium.

Abscess—A localized collection of necrotic (dead) debris and white blood cells surrounded by inflamed tissue.

Absorption—The process by which water and dissolved substances pass into the cells.

Acclimated—Gradually introduced to changes in water temperature and quality.

Acclimatization—The adaptation of fishes to a new environment or habitat or to different climatic conditions.

Accounting—A system of recording, classifying, and summarizing commercial transactions in terms of money.

Accrual—Expenses are considered expenses when they are accrued (or committed) and income is counted as income when it is earned. This includes changes in inventories.

Acid—A compound that yields hydrogen ions when dissolved in an ionizing solvent.

Acidosis—Metabolic condition caused by low pH.

Acre-foot—A water volume equivalent to that covering a surface area of one acre to a depth of one foot; equal to 325,850 gal or 2,718,144 lbs of water.

Activated sludge process—A system in which organic waste is continually circulated in the presence of oxygen and digested by aerobic bacteria.

Acute—Having a short and relatively severe course; for example, acute inflammation.

Acute catarrhal enteritis—See Infectious pancreatic necrosis.

Acute toxicity—Causing death or severe damage to an organism by poisoning during a brief exposure period, normally 96 hours or less. See Chronic.

Adaptation—The process by which individuals, or parts of individuals, populations, or species change in form or function in order to better survive under given or changed environmental conditions. Also the result of this process.

Adductor—Term used to describe muscle that draws toward the axis.

Adipose fin—A small fleshy appendage located posterior to the main dorsal fin; present in Salmonidae and Ictaluridae.

Adipose tissue—Tissue capable of storing large amounts of neutral fats.

Advertising—The act or practice of attracting public notice to create an interest or induce to purchase.

Aerated lagoon—A waste treatment pond in which the oxygen required for biological oxidation is supplied by mechanical aerators.

Aeration—The mixing of air and water by wind action or by air forced through water; generally refers to a process by which oxygen is added to water.

Aerobes—Organisms that can live and grow only where free oxygen is present.

Aerobic—Referring to a process, for example, respiration; or organism, for example, a bacterium that requires oxygen.

Affiliate—To associate or unite a branch or member to the larger group.

Aggregate—Soil made of a mixture of mineral particles.

Agitator—Mechanism for stirring up and thus aerating water in hatching tanks and troughs.

Agriculture—The art, science, and business of producing every kind of plant and animal useful to humans.

Air—The gases surrounding the earth; consists of approximately 78 percent nitrogen, 21 percent oxygen, 0.9 percent argon, 0.03 percent carbon dioxide, and minute quantities of helium, krypton, neon, and xenon, plus water vapor.

Air bladder (swim bladder)—An internal, inflatable gas bladder that enables a fish to regulate its buoyancy.

Air stripping—Removal of dissolved gases from water to air by agitation of the water to increase the area of air-water contact.

Alevin—A life stage of salmonid fish between hatching and feeding when the yolk sac is still present. Equivalent to sac fry in other fishes.

Alevins (sac fry)—Fry that obtain nourishment from an attached yolk sac.

Algal bloom—A high density or rapid increase in abundance of algae.

Algal toxicosis—A poisoning resulting from the uptake or ingestion of toxins or toxin-producing algae; usually associated with blue-green algae or dinoflagellate blooms in fresh or marine water.

Alimentary tract—The digestive tract, including all organs from the mouth to the anal opening.

Aliquot—An equal part or sample of a larger quantity.

Alkalinity—The power of a mineral solution to neutralize hydrogen ions; usually expressed as equivalents of calcium carbonate. Measure of pH buffering capacity.

Alkalosis—Metabolic condition caused by a high pH.

Amino acid—A building block for proteins; an organic acid containing one or more amino groups ($-NH_2$) and at least one carboxylic acid group ($-COOH$).

Ammonia—The gas NH_3; highly soluble in water; toxic to fish in the unionized form, especially at low oxygen levels.

Ammonia nitrogen—Also called total ammonia. The summed weight of nitrogen in both the ionized (ammonium, NH_4^+) and molecular (NH_3) forms of dissolved ammonia (NH_4-N plus NH_3-N). Ammonia values are reported as N, the hydrogen being ignored in analyses.

Ammonium—The ionized form of ammonia, NH_4^+.

Anabolism—Constructive metabolic processes in living organisms—tissue building and growth.

Anadromous fish—Fish that leave the sea and migrate up freshwater rivers to spawn.

Anaerobes—Organisms that can live and grow where there is no free oxygen.

Anaerobic—Referring to a process or organism not requiring oxygen.

Anal—Pertaining to the anus or vent.

Anal fin—The fin on the ventral median line behind the anus.

Anal papilla—A protuberance in front of the genital pore and behind the vent in certain groups of fishes.

Anatomy—The structure of an organism or any of its parts.

Anchor ice—Ice that forms from the bottom up in moving water.

Anemia—A condition characterized by a deficiency of hemoglobin, packed cell volume, or erythrocytes (red blood cells). Anemias in fish include normocytic anemia, microcytic anemia, and macrocytic anemia.

Anesthetics—Chemicals used to relax fish and facilitate the handling and spawning of fish. Commonly used agents include tricane methane sulfonate (MS-222), benzocain, quinaldine, and carbon dioxide.

Anions—Negative ions

Annulus—A yearly mark formed on fish scales when rapid growth resumes after a period (usually in winter) of slow or no growth.

Anomalies—Data values that deviate from the normal range but are accepted as accurate.

Anoxia—Reduction of oxygen in the body to levels that can result in tissue damage.

Antennae—Paired, lateral, moveable, jointed appendages on the head of crustaceans.

Anterior—In front of or toward the head end.

Anthelmintic—An agent that destroys or expels worm parasites.

Antibiotic—A chemical produced by living organisms, usually molds or bacteria, capable of controlling or inhibiting the growth of other bacteria or microorganisms.

Antibody—A specific protein produced by an organism in response to a foreign chemical (antigen) with which it reacts.

Antigen—A large protein or complex sugar that stimulates the formation of an antibody. Generally, pathogens produce antigens and the host protects itself by producing antibodies.

Antigenicity—A measure of a substance's ability to cause the development of antibodies.

Antimicrobial—Chemical that inhibits microorganisms.

Antinutrients—Substances that occur naturally in plant or animal feedstuffs that adversely affect the performance of the animal.

Antioxidant—A substance that chemically protects other compounds against oxidation, for example, ethoxyquin, BHT, BHA, and propyl gallate prevent oxidation and rancidity of fats.

Antiseptic—A compound that kills or inhibits microorganisms, especially those infecting living tissues.

Antivitamin—Substance chemically similar to a vitamin that can replace the vitamin or an essential compound but cannot perform its role.

Anus—The external posterior opening of the alimentary tract; the vent.

Appendages—Any part jointed to or diverging from the main body.

Aquaculture—The art, science, and business of cultivating plants and animals in water.

Aquatic—Growing or living in or upon water.

Aquifer—Underground supply of water.

Archive—Historical records.

Artery—A blood vessel carrying blood away from the heart.

Ascites—The accumulation of serum-like fluid in the abdomen.

Ascorbic acid (vitamin C)—A water-soluble vitamin important for the production of connective tissue; deficiencies cause spinal abnormalities and reduce wound-healing capabilities.

Asexual—Reproduction without eggs and sperm.

Asphyxia—Suffocation caused by too little oxygen or too much carbon dioxide in the blood.

Assembling—Fitting together parts to form the final product.

Assets—The property or resources owned and controlled by a business.

Assimilation—The transformation of digested foods into an intregal and homogenous part of the solids or fluids of the organism.

Asymptomatic carrier—An individual that shows no signs of a disease but harbors and transmits it to others.

Atmosphere—The envelope of gases surrounding the earth; also, pressure equal to air pressure at sea level, approximately 14.7 pounds per square inch (psi).

Atrophy—A degeneration or diminution of a cell or body part due to disuse, defect, or nutritional deficiency.

Auditory—Referring to the ear or to hearing.

Automatic feeder—Feeder that is time controlled providing the correct amount of feed at set intervals of the day.

Autopsy—A medical examination after death to ascertain the cause of death.

Available energy—Energy available from nutrients after food is digested and absorbed.

Available oxygen—Oxygen present in the water in excess of the amount required for minimum maintenance of a species and that can be used for metabolism and growth.

Avirulent—Not capable of producing disease.

Avitaminosis (hypovitaminosis)—A disease caused by deficiency of one or more vitamins in the diet.

Axilla—The region just behind the pectoral fin base.

B

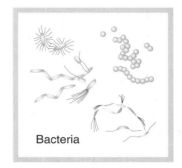

Bacteria

B.O.D.—See Biochemical oxygen demand.

Bacteremia—The presence of living bacteria in the blood with or without significant response by the host.

Bacteria—See Bacterium.

Bacterial gill disease—A disease usually associated with unfavorable environmental conditions followed by secondary invasion of opportunist bacteria. See Environmental gill disease.

Bacterial hemorrhagic septicemia—A disease caused by many of the gram-negative rod-shaped bacteria, usually of the genera *Aeromonas* or *Pseudomonas*, that invade all tissues and blood of the fish. Other names include infectious dropsy, red pest, freshwater eel disease, redmouth disease, motile aeromonad septicemia (MAS).

Bacterial kidney disease—An acute to chronic disease of salmonids caused by *Renibacterium salmoninarum*. Other names include corynebacterial kidney disease, Dee's disease, kidney disease.

Bacterin—A vaccine prepared from bacteria and inactivated by heat or chemicals in a manner that does not alter the cell antigens.

Bacteriocidal—Having the ability to kill bacteria.

Bacteriostat—Chemical that stops the growth or multiplication of bacteria.

Bacteriostatic—Having the ability to inhibit or retard the growth or reproduction of bacteria.

Bacterium (plural: bacteria)—One of a large, widely distributed group of typically one-celled microorganisms, often parasitic or pathogenic.

Baffle—Device such as a screen that interferes with water flow, thus stirring up and aerating the water.

Balance sheet—A statement of the assets owned and liabilities owed in dollars; it shows equity or networth at a specific point in time.

Balanced diet (feed)—A diet that provides adequate nutrients for normal growth and reproduction.

Bar marks—Vertical color marks on fishes.

Barbel—An elongated fleshy projection, usually of the lips.

Basal metabolic rate—The oxygen consumed by a completely resting animal per unit weight and time.

Basal metabolism—Minimum energy requirements to maintain vital body processes.

Bath—A solution of therapeutic or prophylactic chemicals in which fish are immersed. See Dip; Short bath; Flush; Long bath; Constant-flow treatment.

Beer's Law—The light passing through a colored liquid decreases as the concentration of the substance dissolved in the liquid increases.

Benign—Not endangering life or health.

Benthos—Organisms living on or in the bottom sediment of a pond.

Best management practices—For agricultural activities, land use practices or management styles that maintain land productivity and decrease pesticide and fertilizer expenditures.

Binders—A substance promoting the cohesion of particles.

Bioassay—Any test in which organisms are used to detect or measure the presence or effect of a chemical or condition.

Biochemical oxygen demand (BOD)—The quantity of dissolved oxygen taken up by nonliving organic matter in the water.

Biofilter—A water filtration device that uses bacteria to convert harmful substances such as ammonia into harmless substances.

Biological control (biocontrol)—Control of undesirable animals or plants by means of predators, parasites, pathogens, or genetic diseases (including sterilization).

Biological filtration—Removal of ammonia from a recirculating system.

Biological oxidation—Oxidation of organic matter by organisms in the presence of oxygen.

Biologics—A drug obtained from animal tissue or some other organic source.

Biomass—Used to express total living weight.

Biotin (vitamin H)—One of the water-soluble B-complex vitamins.

Bivalve—Animals with two sides of a shell hinged together.

Black grub—Black spots in the skin of fishes caused by metacercaria (larval stages) of the trematodes *Uvilifer ambloplitis*, *Cryptocotyle lingu*, and others. Black-spot disease is another name.

Black spot—Usually refers to black cysts of intermediate stages of trematodes in fish. See Black grub.

Black-spot disease—See Black grub.

Black-tail disease—See Whirling disease.

Blank egg—An unfertilized egg.

Blastopore—Channel leading into a cavity in the egg where fertilization takes place and early cell division begins.

Blastula—A hollow ball of cells, one of the early stages in embryological development.

Blood flagellates—Flagellated whip-like tail protozoan parasites of the blood.

Bloom—Used to describe flourishing algae in a pond.

Blue-sac disease—A disease of sac fry characterized by opalescence and distension of the yolk sac with fluid and caused by previous partial asphyxia.

Blue slime—Excessive mucus accumulation on fish, usually caused by skin irritation due to ectoparasites or malnutrition.

Blue-slime disease—A skin condition associated with a deficiency of biotin in the diet.

Blue stone—See Copper sulfate.

Boil—A localized infection of skin and subcutaneous tissue developing into a solitary abscess that drains externally.

Borings—Soil samples taken by drilling.

Brackish water—A mixture of fresh and sea water; or water with total salt concentrations between 0.05 percent and 3.0 percent, or a salinity of 1 to 10 parts per thousand.

Branchiae (singular: branchia)—Gills, the respiratory organs of fishes.

Branchiocranium—The bony skeleton supporting the gill arches.

Branchiomycosis—A fungal infection of the gills caused by *Branchiaomyces sp.* Other names include gill rot, European gill rot.

Branded—Refers to unique products often identifying a single producer.

Breakdown—Metabolic oxidation into component parts.

Break-even point—The point where income is equal to the total of the fixed costs and variable costs of doing business.

Broadcast—To scatter feed or seed over a wide area.

Broodstock—Adult fish retained for spawning.

Buccal cavity—Mouth cavity.

Buccal incubation—Incubation of eggs in the mouth; oral incubation.

Budget—A formal plan that projects the use of assets for a future time; a schedule of expected returns or costs.

Buffer—Any substance in a solution that tends to resist pH change by neutralizing any added acid or alkali. A chemical that by taking up or giving up hydrogen ions sustains pH within a narrow range.

Byssus—A bunch of silky threads secreted by certain mollusks and serving as a means of attachment to an object.

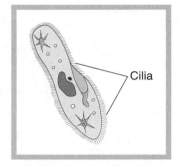

Cilia

C

CCVD—Channel catfish virus disease.

Cfs—Cubic feet per second.

Cage—Encloses the fish in a container or basket that allows water to pass freely between the fish and the water source.

Calcareous—Composed of, containing, or like limestone or calcium carbonate.

Calcinosis—The deposition of calcium salts in the tissues without detectable injury to the affected parts.

Calcium carbonate (CaCO₃)—A relatively insoluble salt, the primary constituent of limestone and a common constituent of hard water.

Calcium cyanamide (CaCN₂)—Used as a pond disinfectant, also known as lime nitrogen.

Calcium oxide—See Lime.

Calorie—The amount of heat required to raise the temperature of one gram of water one degree centigrade; a measure of energy.

Capital—The amount of money that can be obtained through borrowing or selling of assets that is used to promote the production of other goods.

Carbohydrate—Any of the various neutral compounds of carbon, hydrogen, and oxygen, such as sugars, starches, and celluloses, most of which can be used as an energy source by animals.

Carbon dioxide—A colorless, odorless gas, CO_2, resulting from the oxidation of carbon-containing substances; highly soluble in water. Toxic to fish at high levels. Toxicity to fish increases at low levels of oxygen. May be used as an anesthetic.

Carbonate—The CO_3^- ion, or any salt formed with it, such as calcium carbonate, $CaCO_3$.

Carcinogen—Any agent or substance that produces cancer or accelerates the development of cancer.

Carnivore—Animals feeding or preying on animals, eating only animal food.

Carotenoid—A form of vitamin A responsible for coloration of flesh in fish.

Carrier—An individual harboring a pathogen without indicating signs of the disease.

Carrier host (transport host)—An animal in which the larval stage of a parasite will live but not develop.

Carrying capacity—The population, number, or weight of a species that a given environment can support for a given time.

Cartilage—A substance more flexible than bone but serving the same purpose.

Cash basis—Income is recorded as income when it is received and expenses are recorded as expenses when they are paid.

Cash-flow projection—An estimate of monthly cash inflows and outflows over a period of time, usually one year.

Cash-flow summary—A list or record of actual monthly cash levels for a business.

Catabolism—The metabolic breakdown of materials with a resultant release of energy.

Catadromous—Fish that leave freshwater and migrate to the sea to spawn.

Catalyst—A substance that speeds up the rate of chemical reaction but is not itself used up in the reaction.

Cataract—Partial or complete opacity of the crystalline lens of the eye or its capsule.

Cations—Positive ions.

Caudal—Pertaining to the posterior end; toward the tail.

Caudal fin—The tail fin of fish.

Caudal peduncle—The relatively thin posterior section of the body to which the caudal fin is attached; region between base of the caudal fin and base of the last ray of the anal fin.

Cecum (plural: ceca)—A blind sac of the digestive tract, such as a pyloric cecum at the posterior end of the stomach.

Cellular—Of, pertaining to, or like a cell or cells.

Chemical coagulation—A process in which chemical coagulants are put into water causing suspended colloidal solids to flocculate and settle out.

Chemical oxygen demand (COD)—A measure of the chemically oxidizable components in water, determined by the quantity of oxygen consumed.

Chemotherapy—Cure or control of a disease by the use of chemicals (drugs).

Chinook salmon virus disease—See Infectious hematopoietic necrosis.

Chitin—A structural carbohydrate (sugar) found in the exoskeletons of crustaceans, and also contains some tightly bound noncarbohydrate material, including proteins and inorganic salts.

Chitinoverous—Deriving nutrition from chitin, the support substance of the exoskeleton.

Chlorophyll—Green pigment essential to photosynthesis in plants.

Chromatophores—Colored pigment cells.

Chromosomes—Structural units of heredity in the nuclei of cells.

Chronic—Occurring or recurring over a long time.

Chronic inflammation—Long-lasting inflammation.

Cilia—Movable organelles (small organs) that project from some cells, used for locomotion of one-celled organisms or to create fluid currents past attached cells.

Ciliate protozoan—One-celled animal bearing motile cilia.

Ciliated—Having short, fine, hairlike growths that aid in movement, as found on many adult protozoans.

Circuli—The more or less concentric growth marks in a fish scale.

Clarification—Process of removing impurities or making clear.

Clean cropping—Harvesting all fish at one time.

Clinical infection—An infection or disease generating obvious symptoms and signs of pathology.

Cloaca—The common cavity into which rectal, urinary, and genital ducts open. Common opening of intestine and reproductive system of male nematodes.

Closed-formula feed (proprietary feed)—A diet for which the formula is known only to the manufacturer.

Clutch—Eggs laid at one time.

Cock—Male trout.

Coelomic cavity—The body cavity containing the internal organs.

Coelomic fluid—Fluid inside the body cavity.

Coelozoic—Living in a cavity, usually of the urinary tract or gall bladder.

Coffee can—Essential measuring device used by some fish culturalists in lieu of a graduated cylinder.

Coherent—Sticking together, as with soil particles.

Cold water disease—See Peduncle disease or Fin rot disease.

Coldwater species—Generally, fish that spawn in water temperatures below 55°F (12.8°C). The main cultured species are trout and salmon. See Coolwater species; Warmwater species.

Collateral—Property, savings, stocks, etc., deposited as security additional to one's personal or contractual obligations.

Colloid—A substance so finely divided that it stays in suspension in water but does not pass through animal membranes.

Colorimetric—Determining the quantity of a substance by the measurement of the intensity of light transmitted by a solution of the substance.

Columnaris disease—An infection, usually of the skin and gills, by *Flexibacter columnaris*, a myxobacterium.

Commensal—A plant or animal that lives in association with a host organism but is not injurious to it.

Communicable disease—A disease that naturally is transmitted directly or indirectly from one individual to another.

Community—Group of animal and plant populations living together in the same environment.

Compensation point—That depth at which incident light penetration is just sufficient for plankton to photosynthetically produce enough oxygen to balance their respiration requirements.

Competencies—Abilities or capabilities of employees.

Complete diet (complete feed)—See Balanced diet.

Complete feed—Feed that supplies 100 percent of the dietary requirements of the fish; used when there is little or no access to natural food.

Complicating disease—An additional disease during the course of an already existing ailment.

Compressed—Applied to fish, flattened from side to side, as in the case of a sunfish. See Depressed.

Conditioned response—Behavior that is the result of experience or training.

Conductance—The ability of a substance to allow the passage of electrical current.

Congenital disease—A disease that is present at birth; may be infectious, nutritional, genetic, or developmental.

Congestion—Unusual accumulation of blood in tissue; may be active (often called hyperemia) or passive. Passive congestion is the result of abnormal venus return and is characterized by dark cyanotic blood.

Connective tissue—Type of tissue which lies between groups of nerve, gland, and muscle cells and beneath the skin cells.

Constant-flow treatment—Continuous automatic metering of a chemical to flowing water.

Contamination—The presence of material or microorganisms making something impure or unclean.

Control (disease)—Reduction of mortality or morbidity in a population, usually by use of drugs.

Control (experimental)—Similar test specimens subjected to the same conditions as the experimental specimens except for the treatment variable under study.

Coolwater species—Generally, fish that spawn in temperatures between 40° and 60°F (4.4° and 15.6°C). The main cultured coolwater species are northern pike, muskellunge, walleye, sauger, and yellow perch. See Coldwater species; Warmwater species.

Copepods—Small, free-swimming, freshwater and marine crustaceans.

Copper sulfate (blue stone)—Blue stone is copper sulfate pentahydrate ($CuSO_4 \cdot 5H_2O$); effective in the prevention and control of external protozoan parasites, fungal infections, and external bacterial diseases, highly toxic to fish.

Cornea—Outer covering of the eye.

Corporation—A body of people recognized by law as an individual person, having a name, rights, privileges, and liabilities distinct from the individual members.

Corynebacterial kidney disease—See Bacterial kidney disease.

Costiasis—An infection of the skin, fins, and gills by flagellated protozoans of the genus *Costia*.

Cranium—The part of the skull enclosing the brain.

Creative thinking—Ability to generate new ideas by making nonlinear or unusual connections or by changing or reshaping goals to imagine new possibilities; using imagination freely, combining ideas and information in new ways.

Crossbreeding—The mating of unrelated strains of the same species to avoid inbreeding.

Crustacean—Class in arthropod phylum containing crayfish, crabs, lobsters, shrimp, prawn, and others.

Cryogenically frozen—Frozen at very low temperatures.

Cultural diversity—Term used to describe the American workplace representing people from different backgrounds.

Culture—The business of producing, propagating, transporting, possessing, and selling fish or shellfish raised in a private pond, raceway, or tank.

Cultured fish—Farm-raised fish or shellfish.

Curricular—Having to do with a course of study.

Cyanocobalamin (vitamin B$_{12}$)—One of the water-soluble B-complex vitamins that is involved with folic acid in blood-cell production in fish; enhances growth in many animals.

Cyst—Round, thick membrane with which some parasites are surrounded when in the resting state.

Cyst (host)—A connective tissue capsule, liquid or semi-solid, produced around a parasite by the host.

Cyst (parasite)—A noncellular capsule secreted by a parasite.

Cyst (protozoa)—A resistant resting or reproductive stage of protozoa.

Cytoplasm—The contents of a cell, exclusive of the nucleus.

D

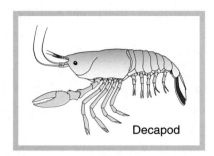

Decapod

Daily temperature unit (DTU)—Equal to one degree Fahrenheit above freezing (32°F) for a 24-hour period.

Data sheet—Similar to a resume; contains pertinent information about potential employee.

Database—Computerized collection of information that can be sorted and retrieved for various reports.

Decapod—An animal with ten legs, such as a crayfish.

Dechlorination—Removal of the residual hypochlorite or chloramine from water to allow its use in fish culture. Charcoal is used frequently because it removes much of the hypochlorite and fluoride. Charcoal is inadequate for removing chloramine.

Dee's disease—See Bacterial kidney disease.

Deficiency—A shortage of a substance necessary for health.

Deficiency disease—A disease resulting from the lack of one or more essential constituents of the diet.

Deheading—Removal of the head.

Demand—The desire to possess a product combined with the ability to purchase.

Demand feeder—Provides feed as animals desire.

Demography—The study of vital and social statistics.

Denitrification—A biochemical reaction in which nitrate (NO_3) is reduced to NO_2, N_2O, and nitrogen gas.

Density index—The relationship of fish size to the water volume of a rearing unit; calculated by the formula: Density index = (weight of fish) / (fish length × volume of rearing unit).

Dentary bones—The principal or anterior bones of the lower jaw or mandible. They usually bear teeth.

Depreciation—The decrease in value resulting from the wear and tear of use, accident, destructive weather, poor management, and the obsolescence of equipment and processes.

Depressed—Flattened in the vertical direction, as a flounder.

Depth of fish—The greatest vertical dimension; usually measured just in front of the dorsal fin.

Dermal—Pertaining to the skin.

Dermatomycosis—Any fungus infection of the skin.

Detritus—Debris from plants and animals.

Diagnosis—The process of recognizing diseases by their characteristic signs.

Diarrhea—Profuse discharge of fluid feces.

Diel—Involving a 24-hour day that includes a day and the adjoining night.

Diet—Food regularly provided and consumed.

Dietary fiber—Nondigestible carbohydrate.

Diffusion—The spreading out of molecules in a given space.

Digestion—The breakdown of foods in the digestive tract to simple substances that may be absorbed by the body.

Diluent—A substance used to dissolve and dilute another substance.

Dilution water—Refers to the water used to dilute toxicants in aquatic toxicity studies.

Dip—Brief immersion of fish into a concentrated solution of a treatment, usually for one minute or less.

Diplostomiasis—An infection involving larvae of any species of the genus *Diplostomum*, Trematoda.

Discharge—To release or to remove by unloading.

Disease—Any departure from health; a particular destructive process in an organ or organism with a specific cause and symptoms.

Disease agent—A physical, chemical, or biological factor that causes disease. Sometimes called etiologic agent; pathogenic agent.

Disease prevention—Steps taken to stop a disease outbreak before it occurs; may include environmental manipulation, immunization, administration of drugs, etc.

Disinfectant—An agent that destroys infective agents.

Disinfection—Destruction of pathogenic microorganisms or their toxins.

Dissolved oxygen (DO)—The amount of elemental oxygen, O_2, in solution under existing atmospheric pressure and temperature.

Dissolved solids—The residue of all dissolved materials when water is evaporated to dryness. See Salinity.

Distal—The remote or extreme end of a structure.

Distributors—An agent that sells merchandise.

Diurnal—Relating to daylight; opposite of nocturnal.

Domestic—Tame; bred and raised in captivity.

Dorsal—Top side of the body, opposite of the ventral; the back.

Dorsal fin—The fin on the back or dorsal side, in front of the adipose fin if the latter is present.

Dose—A quantity of medication administered at one time.

Drainage—May refer to methods of draining a pond or to surface water runoff.

Drawdown—Process of lowering the water level in a pond, or completely draining a pond.

Dredges—A large powerful scoop or suction apparatus for removing mud and gravel.

Dress—To clean and eviscerate for marketing or consumption.

Dressed—Killed and prepared for food market.

Drip treatment—See Constant-flow treatment.

Dropsy—See Edema.

Drug resistant—A microorganism, usually a bacterium, that cannot be controlled (inhibited) or killed by a drug.

Drug sensitive—A microorganism, usually a bacterium, that can be controlled (inhibited) or killed by use of a drug.

Dry feed—A diet prepared from air-dried ingredients, formed into distinct particles and fed to fish.

Dysentery—Disease characterized by passage of liquid feces containing blood and mucus; inflammation of the colon.

E

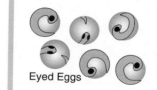
Eyed Eggs

ERM—See Enteric redmouth disease.

Economic—Development and management of the material wealth of a government, community or business.

Ecosystem—A system of interrelated organisms and their physical-chemical environment.

Ectocommensal—A commensal that lives on the surface of the host's body.

Ectoderm—The outer layer of cells in an embryo that gives rise to various organs.

Ectoparasite—Parasite that lives on the surface of the host.

Edema—Excessive accumulation of fluid in tissue spaces.

Efficacy—Ability to produce effects or intended results.

Effluent—Water discharge from a rearing facility, treatment plant, or industry.

Egg—The mature female germ cell, ovum.

Egtved disease—See Viral hemorrhagic septicemia.

Emaciation—Wasting of the body.

Emarginate fin—Fin with the margin containing a shallow notch, as in the caudal fin of the rock bass.

Emboli—Abnormal materials carried by the blood stream, such as blood clots, air bubbles, cancers or other tissue cells, fat, clumps of bacteria or foreign bodies, until they lodge in a blood vessel and obstruct it.

Embryo—Developing organism before it is hatched or born.

Encyst—To enclose or become enclosed in a cyst, capsule, or sac.

Endocrine—A ductless gland or the hormone produced therein.

Endoparasite—A parasite that lives in the host.

Endoskeleton—The skeleton proper; the inner bony and cartilaginous framework of vertebrates.

Energy—Capacity to do work.

Enteric redmouth disease (ERM)—A disease, primarily of salmonids, characterized by general bacteremia. Caused by an enteric bacterium, *Yersinia ruckeri.* Synonym is Hagerman redmouth disease.

Enteritis—Any inflammation of the intestinal tract.

Enterprise—A specific process or activity that requires a certain amount of risk to make a profit.

Enterprise budget—A look at the costs and risks involved with producing one commodity or making one product.

Entrepreneur—One who starts and conducts a business assuming full control and risk.

Environment—All of the external conditions that affect growth and development of an organism.

Environmental gill disease—Rapid growth of gill tissue (hyperplasia) caused by presence of a pollutant in the water that is a gill irritant. See Bacterial gill disease.

Enzootic—A disease that is present in an animal population at all times but occurs in few individuals at any given time.

Enzyme—A protein that catalyzes biochemical reactions in living organisms.

Epidermis—The outer layer of the skin.

Epizootic—A disease attacking many animals in a population at the same time—widely diffused and rapidly spreading.

Epizootiology—The study of epizootics—the field of science dealing with relationships of various factors that determine the frequencies and distributions of diseases among animals.

Equity—The value remaining in a business in excess of any liability or mortgage.

Equivalents per million (epm)—A measure of ionized salts.

Eradication—Removal of all recognizable units of an infecting agent from the environment.

Erosion—The wearing away of land surface by wind and water. Erosion can occur naturally and by land use activities.

Esophagus—The gullet; a muscular, membranous tube between the pharynx and the stomach.

Essential amino acids—Those amino acids that must be supplied by the diet and cannot be synthesized within the body. In fish, these include: arginine, histidine, isoleucine, leucine, lysine, methionine, phenylalanine, threonine, tryptophan, and valine.

Essential fatty acids—A fatty acid that must be supplied by the diet, for example, linoleic acid or linolenic acid.

Estate—One's entire property and possessions.

Estuary—The part of the mouth or lower course of a river in which the river's current meets the ocean's tide to create a mixing of fresh and salt water.

Etiologic agent—See Disease agent.

Etiology—The study of the causes of a disease, both direct and predisposing, and the mode of their operation; not synonymous with cause or pathogenesis of disease, but often used to mean pathogenesis.

European gill rot—See Branchiomycosis.

Eutrophic—Bodies of water that have excessive concentrations of plant nutrients causing excessive algal production and low transparency.

Evaluation—Determining worth; appraisal.

Eviscerate—To gut a fish; to remove the viscera.

Eviscerated—Gutted; with internal organs removed.

Exclusive economic zone—An area 200 nautical miles off the coast of a country reserved for exploring and exploiting, conserving, and managing the natural resources, whether living or nonliving.

Excretion—The process of getting rid or throwing off metabolic waste products by an organism.

Exophthalmos—Abnormal protrusion of the eyeball from the orbit.

Exoskeleton—Hard outer shell that protects the body of an organism such as a crawfish, crab, or lobster.

Exotic—A fish or animal that is not native to the state or locale.

Export—To send merchandise or raw materials to other countries for sale or trade.

Extended aeration system—A modification of the activated-sludge process in which the retention time is longer than in the conventional process.

Extensive culture—Rearing of fish in ponds with low water exchange and at low densities; the fish utilize primarily natural foods.

Extensive production—Raising of fish in low densities in ponds where the fish feed primarily on natural feeds.

Extruded—Pushed through a die to give a certain shape; method of producing floating fish food.

Eyed egg—Egg in which two black spots—the developing eyes of the embryo—can easily be seen.

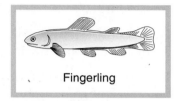

Fingerling

F

Fps—Feet per second.

F1—The first generation of a cross.

F2—The second filial generation obtained by random crossing of F1 (first filial) individuals.

Facultative—Capable of living under varying conditions.

Farming—Breeding and raising stock under controlled conditions.

Fat—An ester composed of fatty acid(s) and glycerol.

Fatty acid—Organic acid present in lipids, varying in carbon content from 2 to 34 atoms (C_2 to C_{34}).

Fauna—The animals inhabiting any region, taken collectively.

Fecundity—Number of eggs in a female spawner.

Feed conversion ratio (FCR)—The average number of pounds of feed needed to gain 1 pound in weight.

Feeding chart—Guidelines provided by feed manufacturer.

Feeding level—The amount of feed offered to fish over a unit time, usually given as percent of fish body weight per day.

Feeding ring—Part of a cage that keeps feed from floating out of the cage.

Fertility—Ability to produce viable offspring.

Fertilization—(1) The union of sperm and egg; (2) addition of nutrients to a pond to stimulate natural food production.

Fillet—Boneless sides of fish cut lengthwise away from the backbone.

Fin ray—One of the cartilaginous rods that support the membranes of the fin.

Fin rot disease—A chronic, necrotic disease of the fins caused by invasion of a myxobacterium into the fin tissue of an unhealthy fish.

Fines—Small particles of feed.

Fingerling—The stage in a fish's life between 1 inch (2.5 cm) and the length at 1 year of age.

Fish farm—The property including private ponds, raceways, or tanks from which fish or shellfish are produced, propagated, transported, or sold.

Fish farmer—Any person engaged in fish farming.

Fish farming—The business of producing, propagating, transporting, possessing, and selling cultured fish or shellfish raised in a private pond, raceway, or tank.

Fistula—An abnormal tube-like passage from an abscess or hollow organ to the skin.

Fixative—A chemical agent chosen to penetrate tissues very soon after death and preserve the cellular components in an insoluble state as nearly life-like as possible.

Fixed costs—Costs that usually do not fluctuate with an increase or decrease in production.

Flagellum (plural: flagella)—Whip-like locomotion organelle of single (usually free-living) cells.

Flashing—Quick turning movements of fish, especially when fish are annoyed by external parasites, causing a momentary reflection of light from their sides and bellies. When flashing, fish often scrape themselves against objects to rid themselves of the parasites.

Flatland—Area with not more than 3 percent slope.

Flow index—The relationship of fish size to water inflow (flow rate) of a rearing unit; calculated by the formula: (fish weight) / (fish length × water inflow).

Flow rate—The volume of water moving past a given point in a unit of time, usually expressed as cubic feet per second (cfs) or gallons per minute (gpm).

Flush—A short bath in which the flow of water is not stopped, but a high concentration of chemical is added at the inlet and passed through the system as a pulse.

Folic acid (folacin)—A vitamin of the water-soluble B complex that is necessary for maturation of red blood cells and synthesis of nucleoproteins; deficiency results in anemia.

Follow-up letter—Letter written immediately after a job interview.

Fomites—Inanimate objects (brushes or dipnets) that may be contaminated with and transmit infectious organisms. See Vector.

Food chain—Transfer of energy from one living thing to another in the form of food.

Food conversion—A ratio of food intake to body weight gain; more generally, the total weight of all feed given to a lot of fish divided by the total weight gain of the fish lot. The units of weight and the time interval over which they are measured must be the same. The better the conversion, the lower the ratio.

Foot—Drain end of holding trough.

Forage—Food for animals taken by browsing or grazing, or the act of browsing or grazing to obtain food.

Forecast—To calculate before hand.

Fork length—The distance from the tip of the snout to the fork of the caudal fin.

Formalin—Solution of approximately 37 percent by weight of formaldehyde gas in water. Effective in the control of external parasites and fungal infections on fish and eggs; also used as a tissue fixative.

Formulated feed—A combination of ingredients that provides specific amounts of nutrients per weight of feed.

Fortification—Addition of nutrients to feeds.

Free living—Not dependent on a host organism.

Freeboard—Distance between pond surface and top of levees or dam—generally between 1 and 2 feet (30 to 61 cm).

Freshwater—Water containing less than 0.05 percent total dissolved salts by weight.

Friable—Easily crumbled or crushed into powder.

Fry—The stage in a fish's life from the time it hatches until it reaches 1 inch (2.5 cm) in length.

Fungus—Any of a group of primitive plants lacking chlorophyll, including molds, rusts, mildews, smuts, and mushrooms. Some kinds are parasitic on fishes. Reproduces by spores.

Fungus disease—See Saprolegniasis.

Furuncle—A localized infection of skin or subcutaneous tissue which develops a solitary abscess that may or may not drain externally.

Furunculosis—A bacterial disease caused by *Aeromonas salmonicida* and characterized by the appearance of furuncles.

Fusiform—Long and tapered toward the ends like a torpedo.

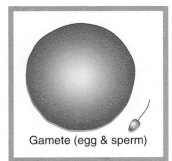

Gamete (egg & sperm)

G

Gpm—Gallons per minute.

Gall bladder—The internal organ containing bile.

Gametes—Sexual cells; eggs and sperm.

Gape—The opening of the mouth.

Gas bubble disease—Gas embolism in various organs and cavities of the fish, caused by supersaturation of gas (mainly nitrogen) in the blood.

Gas bladder—See Air bladder.

Gastric—Relating to the stomach.

Gastritis—Inflammation of the stomach.

Gastroenteritis—Inflammation of the mucosa of the stomach and intestines.

Gastrointestinal—Referring to stomach and intestines, the digestive system.

Gastroliths—Two small stones in the crayfish's stomach in which calcium carbonate is stored; used for hardening the shell after molting.

Gastropod—Marine, freshwater, and land mollusks with one shell.

Gene—The unit of inheritance. Genes are located at fixed loci in chromosomes and can exist in a series of alternative forms called alleles.

Genetic dominant—Character donated by one parent that masks in the progeny the recessive character derived from the other parent.

Genetics—The science of heredity and variation.

Genital—Pertaining to the reproductive organs.

Genital papilla—Small nipple-like projection of tissue on male catfish.

Genus—A unit of scientific classification that includes one or several closely related species. The scientific name for each organism includes designations for genus and species.

Geographic distribution—The geographic areas in which a condition or organism is known to occur.

Germinal disc—The disc-like area of an egg yolk on which cell segmentation first appears.

Gill arch—The U-shaped cartilage that supports the gill filaments.

Gill clefts (gill slits)—Spaces between the gills connecting the pharyngeal cavity with the gill chamber.

Gill cover—The flap-like cover of the gill and gill chamber; the operculum.

Gill disease—See Bacterial gill disease and Environmental gill disease.

Gill filament—The slender, delicate, fringe-like structure composing the gill.

Gill lamellae—The subdivisions of a gill filament where most gas and some mineral exchanges occur between blood and the outside water.

Gill openings—The external openings of the gill chambers, defined by the operculum.

Gill rakers—A series of bony appendages, variously arranged along the anterior and often the posterior edges of the gill arches.

Gill rot—See Branchiomycosis.

Gills—The highly vascular, fleshy filaments used in aquatic respiration and excretion.

Glair—Blue-like substance secreted by the female crayfish; used to attach laid eggs to her swimmerets.

Globulin—One of a group of proteins insoluble in water, but soluble in dilute solutions of neutral salts.

Glycogen—Animal starch, a carbohydrate storage product of animals.

Goals—The end objectives or terminal points of a business.

Gonadotrophin—Hormone produced by pituitary glands to stimulate sexual maturation.

Gonads—The reproductive organs; testes or ovaries.

Grading—Sorting of fish by size, usually by some mechanical device.

Gram-negative bacteria—Bacteria that lose the purple stain of crystal violet and retain the counterstain, in the gram staining process.

Gram-positive bacteria—Bacteria that retain the purple stain of crystal violet in the gram staining process.

Grants—Money offered through the process of a proposal to an organization or a business for research.

Grazers—Animals that eat continuously in well-defined bites.

Gregarine—Name for a group of parasitic protozoans that live in insects, crustaceans, earthworms, and several other types of invertebrate animals.

Gross pathology—Pathology apparent from the naked-eye appearance of tissues.

Groundwater—The supply of water found beneath the Earth's surface.

Group immunity—Immunity enjoyed by a susceptible individual by virtue of membership in a population with enough immune individuals to prevent a disease outbreak.

Grow-out—Facilities that produce crops (fish) from the seed.

Gullet—The esophagus.

Gyro infection—An infection of any of the monogenetic trematodes or, more specifically, of *Gyrodactylus* species.

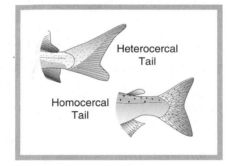

H

HRM—See Enteric redmouth disease.

Habitat—Those plants, animals, and physical components of the environment that constitute the natural food, physical-chemical conditions, and cover requirements of an organism.

Hagerman redmouth disease—See Enteric redmouth disease.

Haptor—Posterior attachment organ of monogenetic trematodes.

Hardness—The power of water to neutralize soap, due to the presence of cations such as calcium and magnesium; usually expressed as parts per million equivalents of calcium carbonate. Refers to the calcium and magnesium ion concentration in water on a scale of very soft (0-20 ppm as $CaCO_3$), soft (20-50 ppm), hard (50-500 ppm), and very hard (500+ ppm).

Harvesting—Involves the gathering or capturing of the fish for marketing and processing. Aquaculture harvesting is typically topping or partial and total harvest.

Hatchery—Produce the seed or young fish.

Hatchery constant—A single value derived by combining the factors in the numerator of the feeding rate formula: Percent body weight fed daily = (3 × food conversion × daily length increase × 100) / length of fish. This value may be used to estimate feeding rates when water temperature, food conversion, and growth rate remain constant.

Head—Inflow end of holding trough.

Headwaters—Place of origin (where groundwater first surfaces) of a creek, stream, or river.

Heat capacity—Characteristic of water making it resistant to temperature changes.

Heavy metals—Metals such as Cadmium (Cd), Cobalt (Co), Chromium (Cr), Copper (Cu), Lead (Pb), Lithium (Li), Manganese (Mn), Mercury (Hg), Nickel (Ni), Zinc (Zn), or Iron (Fe).

Heme—The iron-bearing constituent of hemoglobin.

Hematocrit—Percent of total blood volume that consists of cells; packed cell volume.

Hematoma—A tumor-like enlargement in the tissue caused by blood escaping the vascular system.

Hematopoiesis—The formation of blood or blood cells in the living body. The major hematopoietic tissue in fish is located in the anterior kidney.

Hematopoietic kidney—The anterior portion of the kidney, involved in the production of blood cells.

Hemocoel—The enclosed space around the organs of crustaceans which contain the animal's blood.

Hemoglobin—The respiratory pigment of red blood cells that takes up oxygen at the gills or lungs and releases it at the tissues.

Hemorrhage—An escape of blood from its vessels, through either intact or ruptured walls; bleeding.

Hen—Female trout.

Hepatic—Pertaining to the liver.

Hepatitis—Inflammation of the liver.

Hepatoma—A tumor with cells resembling those of liver; includes any tumor of the liver. Hepatoma is associated with mold toxins in feed eaten by cultured fishes. The toxin having the greatest affect on fishes is aflatoxin B_1, from *Aspergillus flavus*.

Herbivore—Animals that subsist primarily on the available vegetation and decayed organic material in the environment.

Hermaphroditic—Possess gonads, testes and ovaries, for both sexes and can release eggs and sperm.

Heterocercal—Forked tail fin.

Heterotrophic bacteria—Bacteria that oxidize organic material (carbohydrate, protein, fats) to CO_2, NH_4-N, and water (H_2O) for their energy source.

Histology—Microscopic study of cells, tissues, and organs.

Histopathology—The study of microscopically visible changes in diseased tissues.

Homing—Return of fish to their stream or lake of origin to spawn.

Homocercal—Single-lobed tail fin.

Horizontal transmission—Any transfer of a disease agent between individuals except for the special case of parent-to-progeny transfer via reproductive processes.

Hormone—A chemical product of endocrine gland cells affecting organs that do not secrete it.

Host—Animal on or in which a parasite lives.

Humectants—Prevents bacterial growth.

Husbandry—The occupation or business of farming.

Hyamine—See Quaternary ammonium compounds.

Hybrid—Crossbreeding fish of different varieties, races, or species.

Hybrid vigor—Condition in which the offspring perform better than the parents. Sometimes called heterosis.

Hybridization—The crossing of different species.

Hydrate—To combine with water.

Hydrogen ion concentration (activity)—The cause of acidity in water. See pH.

Hydrogen sulfide—An odorous, soluble gas, H_2S, resulting from anaerobic decomposition of sulfur-containing compounds, especially proteins.

Hydrologic cycle—Circular flow of water between the atmosphere and the Earth; precipitation, runoff, surface water, groundwater, evaporation, and transpiration.

Hydroponics—The cultivation of land plants without soil, in a water solution.

Hyoid—Bones in the floor of the mouth supporting the tongue.

Hyper—A prefix denoting excessive, above normal, or situated above.

Hyperemia—Increased blood resulting in distension of the blood vessels.

Hyperplasia—Increased, abnormal tissue growth.

Hypo—A prefix denoting deficiency, lack, below, beneath.

Incubator (upwelling)

IHN—See Infectious hematopoietic necrosis.

IPN—See Infectious pancreatic necrosis.

Ich—A protozoan disease caused by the ciliate *Ichthyophtherius multifilis*; sometimes called white-spot disease.

Immune—Unsusceptible to a disease.

Immunity—Lack of susceptibility; resistance; an inherited or acquired status.

Immunization—Process or procedure by which an individual is made resistant to disease, specifically infectious disease.

Import—To receive merchandise or raw materials from other countries for sale or trade.

Impound—To gather and enclose water for fish pond or irrigation.

Impoundment—A dam, dike, floodgate, or other barrier confining a body of water.

Imprinting—The imposition of a behavior pattern in a young animal by exposure to stimuli.

In berry—Female crayfish carrying eggs.

In vitro—Used in reference to tests or experiments conducted in an artificial environment, including cell or tissue culture.

In vivo—Used in reference to tests or experiments conducted in or on intact, living organisms.

Inbred line—A line produced by continued matings of brothers to sisters and progeny to parents over several generations.

Inbreeding—Mating of closely related animals.

Incidence—The number of new cases of a particular disease occurring within a specified period in a group of organisms.

Income—Amount of money received periodically in return for goods, labor, or services.

Income statement—Financial record that reflects the profitability of the business over a specified period of time; also known as a profit and loss statement or an operating statement.

Incubation—Process by which eggs are placed in a favorable environment for hatching.

Incubation (disease)—Period of time between the exposure of an individual to a pathogen and the appearance of the disease it causes.

Incubation (eggs)—Period from fertilization of the egg until it hatches.

Incubator—Device for artificial rearing of fertilized fish eggs and newly hatched fry.

Indigenous—Refers to a species of fish, shellfish, or aquatic plant usually found in the public waters of the state.

Indispensable amino acid—See Essential amino acids.

Inert gases—Those gases in the atmosphere that are inert or nearly inert; nitrogen, argon, helium, xenon, krypton, and others. See Gas bubble disease.

Infection—Contamination (external or internal) with a disease-causing organism or material, whether or not overt disease results.

Infection, focal—A well circumscribed or localized infection in or on a host.

Infection, secondary—Infection of a host that already is infected by a different pathogen.

Infection, terminal—An infection, often secondary, that leads to death of the host.

Infectious catarrhal enteritis—See Infectious pancreatic necrosis.

Infectious disease—A disease that can be transmitted between hosts.

Infectious pancreatic necrosis (IPN)—A disease caused by an infectious pancreatic necrosis virus that presently has not been placed into a group. Sometimes called infectious catarrhal enteritis, chinook salmon virus disease, Oregon sockeye salmon virus, Sacramento River chinook disease.

Inferior mouth—Mouth on the under side of the head, opening downward.

Infiltration—To filter through small gaps or passages.

Inflammation—The reaction of the tissues to injury, characterized clinically by heat, swelling, redness, and pain.

Ingest—To eat or take into the body.

Injection—Method of introducing a drug or vaccine into the muscle or body cavity.

Inoculation—The introduction of an organism into the tissues of a living organism or into a culture medium.

Inorganic—Not characterized by life processes; not containing carbon.

Input costs—Money required to begin production.

Inspection—Careful or critical examination of a product.

Instinct—Inherited behavioral response.

Intensive culture—Rearing of fish at densities greater than can be supported in the natural environment; utilizes high water flow or exchange rates and requires the feeding of formulated feeds.

Intensive production—Raising of fish in densities higher than could be supported in the natural environment; requires feeding of formulated feeds.

Interest—Payment for the use of money or credit.

Interpersonal—Between people.

Interspinals—Bones to which the rays of the fins are attached.

Intestine—The lower part of the alimentary tract from the pyloric end of the stomach to the anus.

Intragravel water—Water occupying interstitial spaces within gravel.

Intramuscular injection—Administration of a substance into the muscles of an animal.

Intraperitoneal injection—Administration of a substance into the body cavity (peritoneal cavity).

Inventory—The value of goods or stock of a business.

Invertebrate—Organism with a hard outer skeleton and lacking a spinal column.

Ion—Electrically charged atom, radical, or molecule.

Ion exchange—A process of exchanging certain cations or anions in water for sodium, hydrogen, or hydroxyl (OH^-) ions in a resinous material.

Isotonic—No osmotic difference; one solution having the same osmotic pressure as another.

Isthmus—The region just anterior to the breast of a fish where the gill membranes converge; the fleshy interspace between gill openings.

K

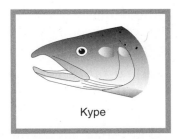

Kype

Kidney—One of the pair of glandular organs in the abdominal cavity that produces urine.

Kidney disease—See Bacterial kidney disease.

Kilogram calorie—The amount of heat required to raise the temperature of one kilogram of water one degree centigrade, also called kilocalorie (kcal), or large calorie—a measure of energy.

Kype—Upward curving hook of lower jaw that occurs at spawning.

L

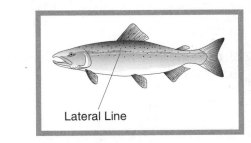

Lateral Line

LDV—See Lymphocystis disease.

Larva (plural: larvae)—An immature form that must undergo change of appearance or pass through a metamorphic stage to reach the adult state.

Lateral—Side of the body.

Lateral band—A horizontal pigmented band along the sides of a fish.

Lateral line—A series of sensory pores, sensitive to low-frequency vibrations, located laterally along both sides of the body.

Least-cost—Method of feed formulation where the formula varies, within limits, as ingredient prices change.

Length—May refer to the total length, fork length, or standard length.

Lesion—Any visible alteration in the normal structure of organs, tissues, or cells.

Lethargy—A state of sluggishness, inaction, indifference, or dullness.

Letter of application—Sent with resume or data sheet when applying for a job.

Letter of inquiry—Sent to potential employer requesting possibility of employment.

Leucocyte—A white blood corpuscle.

Levee—Earth dike used to enclose water.

Liabilities—Just or legal responsibilities.

Lime (calcium oxide, quicklime, burnt lime)—CaO; used as a disinfectant for fish-holding facilities. Produces heat and extreme alkaline conditions.

Line breeding—Mating individuals so that their descendants will be kept closely related to an ancestor that is regarded as unusually desirable.

Linolenic acid—An 18-carbon fatty acid with two double bonds; certain members of the series are essential for health, growth, and survival of some, if not most, fishes.

Lipid—Any of a group of organic compounds consisting of the fats and other substances of similar properties. They are insoluble in water but soluble in fat solvents and alcohol.

Liquidity—A business's ability to meet short-run obligations when due without disrupting the normal operation of the business.

Live-car—Net attached to harvesting seine and used to crowd, grade, and hold fish in the pond.

Live-haulers—Individuals or business who specialize in trucking live fish.

Logarithm—The exponent to which a base (usually 10) must be raised to produce a given number.

Long bath—A type of bath frequently used in ponds. Low concentrations of chemicals are applied and allowed to disperse by natural processes.

Lymphocystis disease—A virus disease of the skin and fins affecting many freshwater and marine fishes of the world; caused by the lymphocystis virus of the *Iridovirus* group.

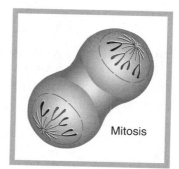

Mitosis

M

MAS—See Motile aeromonas septicemia.

Macrominerals—Minerals in the body in large quantities.

Macrophyte—Vascular plants with true roots, stems, and leaves.

Malignant—Progressive growth of certain tumors that may spread to distant sites or invade surrounding tissue and kill the host.

Malnutrition—Faulty or inadequate nutrition.

Mandible—Lower jaw.

Mantle—Soft covering over the organs of a mollusk.

Mariculture—Raising of organisms in the ocean.

Marine—Of the sea or ocean.

Maritime—Pertaining to the sea, its navigation and commerce.

Market—Buyer of product; place to sell a product.

Marketing—The process of getting the product from the producer to the consumer. It is the final step in food production.

Mass selection—Selection of individuals from a general population for use as parents in the next generation.

Matching funds—Money provided on a prearranged split.

Mating system—Any of a number of schemes by which individuals are assorted in pairs leading to sexual reproduction.

Maxilla or maxillary—The hindmost bone of the upper jaw.

Mean—The arithmetic average of a series of observations.

Mechanical damage—Extensive connective tissue proliferation, leading to impaired growth and reproductive processes, caused by parasites migrating through tissue.

Median—A value in a series below and above which there are an equal number of values.

Melanophore—A black pigment cell; large numbers of these give fish a dark color.

Menadione—A fat-soluble vitamin; a form of vitamin K.

Meristic characters—Body parts that can be counted, such as scales, gill rakers, and vertebrae; useful in species identifications.

Metabolic rate—The amount of oxygen used for total metabolism per unit of time per unit of body weight.

Metabolism—Vital processes involved in the release of body energy, the building and repair of body tissue, and the excretion of waste materials; combination of anabolism and catabolism.

Metamorphose—To change from one form to another, as from a tadpole to a frog.

Methylene blue—A quinoneimine dye effective against external protozoans and superficial bacterial infections.

Microbe—Microorganism, such as a virus, bacterium, fungus, or protozoan.

Microencapsulation—The coating of small feed particles with a substance that is insoluble in water but digestible by the enzymes in the digestive tract of the fish.

Microminerals—Minerals in the body in small quantities.

Microohms—Measure of electrical resistance; one-thousandth of an ohm.

Micropyle—Opening in egg that allows entrance of the sperm.

Microscopic—Small enough to be invisible or obscure except when observed through a microscope.

Microsporidean—A small spore-producing organism.

Migration—Movement of fish populations.

Milt—Secretion that contains sperm produced by a male fish.

Mitosis—The process by which the nucleus is divided into two daughter nuclei with equivalent chromosome complements.

Molt—For crustaceans, the shedding of the exoskeleton; molting occurs at intervals during a crustacens's life and allows for expansion in size.

Monoculture—Raising a single species in a pond or enclosure.

Monthly temperature unit (MTU)—Equal to one degree Fahrenheit above freezing (32°F) based on the average monthly water temperature (30 days).

Morbid—Caused by disease; unhealthy, diseased.

Morbidity—The condition of being diseased.

Morbidity rate—The proportion of individuals with a specific disease during a given time in a population.

Moribund—Obviously progressing towards death; nearly dead.

Morphology—The science of the form and structure of animals and plants.

Mortality—Death, particularly death from disease or on a large scale.

Mortality rate—The number of deaths per unit of population during a specified period. May be called death rate, crude mortality rate, or fatality rate.

Motile aeromonas septicemia (MAS)—An acute to chronic infectious disease caused by any motile bacteria belonging to the genus *Aeromonas*, primarily *Aeromonas hydrophila* or *Aeromonas punctate*. Sometimes called bacterial hemorrhagic septicemia or pike pest.

Mottled—Blotched; color spots running together.

Mouth fungus—See Columnaris disease.

Mouthbrooders—Fish that hold young or eggs in the mouth.

Mucking (egg)—The addition of an inert substance such as clay or starch to adhesive eggs to prevent them from sticking together during spawn taking, commonly used with esocid and walleye eggs.

Mucus (mucous)—A viscid or slimy substance secreted by the mucus glands of fish.

Mud line—Bottom of a pond seine that keeps fish from escaping from under the seine.

Multiple harvest (partial harvest)—Harvesting strategy that requires numerous yearly seinings or trappings and complete draining of a pond each 6 to 8 years.

Mutation—A sudden heritable variation in a gene or in a chromosome structure.

Mycology—The study of fungi.

Mycosis—Any disease caused by an infectious fungus.

Mycotoxin—Poisons derived from fungus.

Myomere—An embryonic muscular segment that later becomes a section of the side muscle of a fish.

Myotome—Muscle segment.

Myxobacteriosis—A disease caused by any member of the Myxobacterales group of bacteria, for example, peduncle disease, coldwater disease, fin rot disease, or columnaris disease.

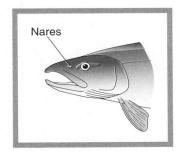

Nares

N

Nares—The openings of the nasal cavity.

Native fish—All fish documented to live, spawn, or reproduce in public waters of a state, and whose first documented occurrence in public waters was not the result of direct or indirect importation by people.

Natural foods—Plants and animals normally found in a pond or other water source, sometimes their production is enhanced by fertilization.

Necropsy—A medical examination of the body after death to ascertain the cause of death; an autopsy in humans.

Necrosis—Dying of cells or tissues within the living body.

Nematoda—A diverse phylum of roundworms, many of which are plant or animal parasites.

Nephrocalcinosis—A condition of kidney (renal) insufficiency due to the precipitation of calcium phosphate ($CaPO_4$) in the tubules of the kidney.

Nerve fibers—Anatomical structures carrying nerve impulses.

Net pens—Large cages used for raising fish, for example, salmon.

Net worth statement—Financial condition of a business at a definite point in time; it lists all assets, values of assets, and liabilities of a business; also known as a balance sheet, financial statement, or statement of financial condition.

Niacin—One of the water-soluble B-complex vitamins, essential for maintenance of the health of skin and other epithelial tissues in fishes.

Nicotinic acid—See Niacin.

Nitrification—A method through which ammonia is biologically oxidized to nitrite and then nitrate.

Nitrite—The NO_2^- ion.

Nitrogen (N_2)—An odorless, gaseous element that makes up 78 percent of the earth's atmosphere and is a constituent of all living tissue. It is almost inert in its gaseous form.

Nitrogenous wastes—Simple nitrogen compounds produced by the metabolism of proteins, such as urea and uric acid.

Nodule—Small knot, knob, or lump of tissue.

Nonindigenous—Refers to a species of fish, shellfish, or aquatic plant not usually found in public waters of the state.

Noninfectious—Refers to diseases that are not contagious and that usually cannot be cured by medications.

Nonpathogenic—Refers to an organism that may infect but causes no disease.

Nonpoint source pollution—Pollution from a diffuse source.

Nostril—See Nares.

Noxious—Harmful or undesirable.

Nursery—Ponds or tanks for newly hatched fry.

Nutrient—A chemical used for growth and maintenance of an organism.

Nutrition—All of the processes in which an animal takes in and uses food.

Nutritional gill disease—Gill hyperplasia caused by deficiency of pantothenic acid in the diet.

Nymph—The larva of various insects, especially dragonfly and mayfly larvae.

Operculum

O

Ocean ranching—Type of aquaculture involving the release of juvenile aquatic ani-

mals into marine waters to grow on natural foods to harvestable size.

Offal—Waste parts of a slaughtered animal.

Off-flavor—Musty- or muddy-tasting fish flesh.

Omnivore—Eating both vegetable and animal food.

On-line—Refers to connecting computers via telephone lines for the purpose of transferring information.

Open-formula feed—A diet in which all the ingredients and their proportions are public (nonproprietary).

Operculum—A bony flap-like protective gill covering.

Opportunistic—Waiting for a combination of favorable circumstances.

Optic—Referring to the eye.

Organic—Related to or derived from living organisms, contains carbon.

Osmoregulation—The process by which organisms maintain stable osmotic pressures in their blood, tissues, and cells in the face of differing chemical properties among tissues and cells, and between the organism and the external environments.

Osmosis—The diffusion of liquid that takes place through a semipermeable membrane between solutions starting at different osmotic pressures, and that tends to equalize those pressures. Water always will move toward the more concentrated solution, regardless of the substances dissolved, until the concentration of dissolved particles is equalized, regardless of electric charge.

Osmotic pressure—The pressure needed to prevent water from flowing into a more concentrated solution from a less concentrated one across a semipermeable membrane.

Outfall—Wastewater at its point of effluence or its entry into a river or other body of water.

Outliers—Data values that lie outside the normal range and may be false (unacceptable) or true (acceptable anomaly).

Ovarian—Having to do with ovaries, the female egg producing glands.

Ovarian fluid—Fluid surrounding eggs inside the female's body.

Ovaries—The female reproductive organs producing eggs or ova.

Overflow pipe—Vertical pipe placed in a tank so the top is at desired water height; water above this height drains from the tank.

Overt disease—A disease, not necessarily infectious, that is apparent or obvious by gross inspection; a disease exhibiting clinical signs.

Oviduct—The tube that carries eggs from the ovary to the exterior.

Oviparous—Producing eggs that are fertilized, develop, and hatch outside the female body.

Ovoviviparous—Producing eggs, usually with much yolk, that are fertilized internally. Little or no nourishment is furnished by the mother during development; hatching may occur before or after expulsion.

Ovulate—Process of producing mature eggs (ova) capable of being fertilized.

Ovum (plural: ova)—Egg cell or single egg.

Oxidation—Combination with oxygen; removal of electrons to increase positive charge.

Oxygen transfer efficiency—A measure of the percent of the total oxygen used that a device is able to put into solution.

Oxytetracycline (terramycin)—One of the tetracycline antibiotics produced by *Streptomyces rimosus* and effective against a wide variety of bacteria pathogenic to fishes.

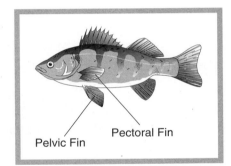

Pelvic Fin   Pectoral Fin

P

Paddlewheel—Type of aeration device.

Pancreas—The organ that functions as both an endocrine gland secreting insulin and an exocrine gland secreting digestive enzymes.

Pantothenic acid—One of the essential water-soluble B-complex vitamins.

Para-aminobenzoic acid (PABA)—A vitamin-like substance thought to be essential in the diet for maintenance of health of certain fishes.

Parasite—An organism that lives in or on another organism (the host) and that depends on the host for its food, has a higher reproductive potential than the host, and may harm the host when present in large numbers.

Parasite, obligate—An organism that cannot lead an independent, nonparasitic existence.

Parasiticide—Antiparasite chemical (added to water) or drug (fed or injected).

Parasitology—The study of parasites.

Parr—A life stage of salmonid fishes that extends from the time feeding begins until the fish become sufficiently pigmented to obliterate the parr marks, usually ending during the first year.

Parr mark—One of the vertical color bars found on young salmonids and certain other fishes.

Part per billion (ppb)—A concentration at which 1 unit is contained in a total of 1 billion units. Equivalent to 1 microgram per kilogram (1 mcg/kg).

Part per million (ppm)—A concentration at which 1 unit is contained in a total of 1 million units. Equivalent to 1 milligram per kilogram (1 ml/kg) or 1 microliter per liter (1 ml/liter).

Part per thousand (ppt)—A concentration at which 1 unit is contained in a total of 1,000 units. Equivalent to 1 gram per kilogram (1 g/kg) or 1 milliliter per liter (1 ml/liter). Normally, this term is used to specify salinity.

Partnership—A form of business organization with multiple owners.

Pathogen—Disease-causing organism.

Pathology—The study of diseases and the structural and functional changes produced by them.

Payback—Number of years it takes to recover the initial investment.

Pectoral fins—The anterior and ventrally located fins whose principle function is locomotor maneuvering.

Peduncle disease—A chronic, necrotic disease of the fins, primarily the caudal fin, caused by invasion of a myxobacterium (commonly *Cytophaga psychrophilia*) into fin and caudal peduncle tissue of an unhealthy fish. Other names include fin rot disease or cold water disease.

Peeling plant—Crayfish processing plant.

Pelvic fins—Paired fins corresponding to the posterior limbs of the higher vertebrates (sometimes called ventral fins), located below or behind the pectoral fins.

Peptide bond—Chemical connection between amino acids when forming proteins.

Percolation—The process of a liquid passing through a filter.

Peritoneum—The membrane lining the abdominal cavity.

Perivitelline fluid—Fluid lying between the yolk and outer shell (chorion) of an egg.

Perivitelline space—Area between yolk and chorion of an egg where embryo expansion occurs.

Permeability—Rate of penetration by liquids.

Peroxide—Containing oxygen.

Pesticide—Broad name for chemicals that control or kill insects, fungi, parasites, and other pests.

Petechia—A minute rounded spot of hemorrhage on a surface, usually less than one millimeter in diameter.

pH—An expression of the acid-base relationship designated as the logarithm of the reciprocal of the hydrogen-ion activity; the value of 7.0 expresses neutral solutions; values below 7.0 represent increasing acidity; those above 7.0 represent increasingly basic solutions.

Pharynx—The cavity between the mouth and esophagus.

Phenotype—Appearance of an individual as contrasted with its genetic makeup or genotype. Also used to designate a group of individuals with similar appearance but not necessarily identical genotypes.

Photoperiod—The number of daylight hours best suited to the growth and maturation of an organism.

Photosynthesis—The formation of carbohydrates from carbon dioxide and water that takes place in the chlorophyll-containing tissues of plants exposed to light. Oxygen is produced as a by-product.

Phycocolloid—Gelatin-like substance obtained from seaweed.

Physiology—Study of body functions.

Phytoplankton—Minute plants suspended in water with little or no capability for controlling their position in the water mass; frequently referred to as algae.

Pig trough—See Von Bayer trough.

Pigment—The coloring matter in the cells of plants and animals.

Pigmentation—Disposition of coloring matter in an organ or tissue.

Pituitary—Small endocrine organ located near the brain.

Plankton—Microscopic plants and animals.

Planting of fish—The act of releasing fish from a hatchery into a specific lake or river. Sometimes called distribution or stocking.

Plasma—The fluid fraction of the blood, as distinguished from corpuscles. Plasma contains dissolved salts and proteins.

Plasticity—Capacity of soil to be bent without breaking and to remain bent after force is removed.

Poikilothermic—Having a body temperature that fluctuates with that of the environment.

Point source pollution—Pollution from any single identifiable source.

Pollutant—A term referring to a wide range of toxic chemicals and organic materials introduced into waterways from industrial plants and sewage wastes.

Pollution—The addition of any substance not normally found in or occurring in a material or ecosystem.

Polyculture—Raising two or more species in the same pond or enclosure.

Polysaccharide—Class of carbohydrates of high molecular weight formed by the union of three or more monosaccharide molecules (sugar). Examples include starch, cellulose, dextrin, and glycogen.

Polytrophic—Eating a wide variety of material, both plant and animal.

Pond run—Ungraded by size or sex.

Population—A coexisting and interbreeding group of individuals of the same species in a particular locality.

Population density—The number of individuals of one population in a given area or volume.

Portal of entry—The pathway by which pathogens or parasites enter the host.

Portal of exit—The pathway by which pathogens or parasites leave or are shed by the host.

Posterior—The back side.

Posttreatment—Treatment of hatchery wastewater before it is discharged into the receiving water; pollution abatement.

Potassium permanganate—$KMnO_4$, a strong oxidizing agent used as a disinfectant and to control external parasites.

Pox—A disease sign in which eruptive lesions are observed primarily on the skin and mucous membranes.

Pox disease—A common disease of freshwater fishes, primarily minnows, characterized by small, flat epithelial growths and caused by a virus as yet unidentified. Sometimes called carp pox or papilloma.

Precipitate—To separate out from a solution.

Predator—Animal that preys on, destroys, or eats other animals.

Premix—Feed additive that contains vitamins and minerals.

Pretreatment—Treatment of water before it enters the hatchery.

Primary producers—Lowest link on the food chain.

Principal—The property or capital.

Processor—Business that takes the raw product to its final form.

Product pull—Creating consumer demand.

Product push—Providing product shelf space.

Production ponds—Ponds used for the final growing stages.

Profit—The money that remains after all fixed and variable costs are deducted from income.

Progeny—Offspring.

Progeny test—A test of the value of an individual based on the performance of its offspring produced in some definite system of mating.

Prognosis—Expected outcome of the disease.

Prolific—Producing many young.

Promotion—To encourage the growth or development of a business or industry.

Propagation—Reproduction, raising or breeding.

Prophylactic—Activity or agent that prevents the occurrence of disease.

Protandrous—Changing sex one or more times during life.

Protein—Any of the numerous naturally occurring complex combinations of amino acids that contain the elements carbon, hydrogen, nitrogen, oxygen, and occasionally sulfur, phosphorus, or other elements.

Protozoa—The phylum of mostly microscopic animals made up of a single cell or a group of more or less identical cells and living chiefly in water; includes many parasitic forms.

Proximate composition—Analysis of protein, fat, and ash content.

Pseudobranch—The remnant of the first gill arch that often does not have a respiratory function and is thought to be involved in hormone activation or secretion.

Pseudomonas septicemia—A hemorrhagic, septicemic disease of fishes caused by infection of a member of the genus *Pseudomonas*. This is a stress-mediated disease that usually occurs as a generalized septicemia. See Bacterial hemorrhagic septicemia.

Purging—Process of cleansing crayfish systems by not feeding and changing water often during the holding period.

Pus—Yellowish-white liquid produced in certain infections.

Pyloric cecum—See Cecum.

Pyridoxine (vitamin B$_6$)—One of the water-soluble B-complex vitamins involved in fat metabolism, but playing a more important role in protein metabolism; carnivorous fish have stringent requirements for this vitamin.

Q

Quality assurance—The total integrated program for assuring reliability of monitoring and measurement data.

Quality assurance project plan—A plan that details the monitoring objectives, program scope, methods, field and lab procedures and the quality assurance and control activities necessary to meet the stated data quality objectives.

Quality control—Routine application and procedures for obtaining prescribed standards of performance and for controlling the measurement process.

Quaternary ammonium compounds—Several of the cationic surface-active agents and germicides, each with a quaternary ammonium structure; bactericidal but will not kill external parasites of fish; used for controlling external bacterial pathogens and disinfecting hatching equipment.

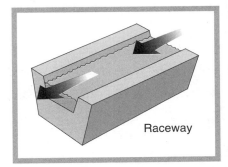

Raceway

R

Raceway—An aquaculture rearing unit in which the water flows through it.

Radii of scale—Lines on the proximal part of a scale, radiating from near center to the edge.

Ranching—Obtaining wild-bred stock and raising it to marketable size under controlled conditions.

Random mating—Matings without consideration of definable characteristics of the broodfish; nonselective mating.

Ration—A fixed allowance of food for a day or other unit of time.

Ray—A supporting rod for a fin; two kinds: hard (spines) and soft rays.

Rearing unit—Any facility in which fish are held during the rearing process, such as rectangular raceways, circular ponds, circulation raceways, and earth ponds.

Recessive—Character possessed by one parent that is masked in the progeny by the corresponding alternative or dominant character derived from the other parent.

Reciprocal mating (crosses)—Paired crosses in which both males and females of one parental line are mated with the other parental line.

Reconditioning treatment—Treatment of water to allow its reuse for fish rearing.

Recruitment—Production of young.

Rectum—Most distal part of the intestine; repository for the feces.

Recycle/reuse—The use of water more than one time for fish propagation. There may or may not be water treatment between uses, and different rearing units may be involved.

Red pest—See Motile aeromonas disease.

Red sore disease—See Vibriosis.

Redd—Area of stream or lake bottom excavated by a female salmonid during spawning.

Redmouth disease—An original name for bacterial hemorrhagic septicemia caused by an infection of *Aeromonas hydrophila* specifi-

cally. Synonyms: motile aeromonas disease; bacterial hemorrhagic septicemia.

Regenerate—The ability to regrow a lost body part, such as a claw.

Relift pump—Moves water from one source to another.

Resistance—The natural ability of an organism to withstand the effects of various physical, chemical, and biological agents that potentially are harmful to the organism.

Resources—Available means or property; a supply that can be drawn on.

Respiration—The process by which an animal or plant takes in oxygen from the air and gives off carbon dioxide and other products of oxidation.

Resume—A summary of an individual's employment record.

Retail sales—Sales of fish in small quantities directly to the consumer.

Return—The money available after production expenses are subtracted from total income.

Riboflavin—An essential water-soluble vitamin of the B-complex group (B_2).

Riparian area—The vegetated area constituting a buffer zone between the stream bank and the beginning of land use.

Ripe—Containing fully developed eggs; ready to spawn.

Risk—A chance of encountering harm or loss.

Roe—The eggs of fishes.

Rotational line-crossing—System for maintaining broodstocks while preventing inbreeding.

Rotifers—Many-celled, microscopic aquatic organisms having rings of cilia that in motion resemble revolving wheels.

Roundworm—See Nematoda.

Runoff—Rain that does not infiltrate the soil and so flows to ponds, streams, and depressions.

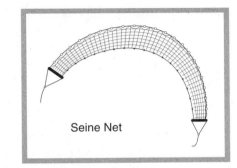

Seine Net

S

Sac fry—A fish with an external yolk sac.

Safe concentration—The maximum concentration of a material that produces no adverse sublethal or chronic effect.

Salinity—Concentration of sodium, potassium, magnesium, calcium, bicarbonate, carbonate, sulfate, and halides (chloride, fluoride, bromide) in water. See Dissolved solids.

Salmonids—Refers to trout and salmon.

Salt—Compound resulting from an acid and a base.

Saltwater—Water with a salinity of 30 to 35 parts per thousand.

Sample—A part, piece, item, or observation taken or shown as representative of a total population.

Sample count—A method of estimating fish population weight from individual weights of a small portion of the population.

Sanitizer—A chemical that reduces microbial contamination on equipment.

Saprolegniasis—An infection by fungi of the genus *Saprolegnia*, usually on the external surfaces of a fish body or on dead or dying fish eggs.

Saprophyte—Organism that lives on dead or decaying organic matter.

Saturation—In solutions, the maximum amount of a substance that can be dissolved in a liquid without it being precipitated or released into the air.

Scale formula—A conventional formula used in identifying fishes. "Scales 7 + 65 + 12," for example, indicates 7 scales above the lateral line, 65 along the lateral line, and 12 below it.

Scales—Horny or bony plate-like outgrowth from skin of fish.

Scales above the lateral line—Usually, the number of scales counted along an oblique row beginning with the first scale above the lateral line and running anteriorly to the base of the dorsal fin.

Scales below the lateral line—The number of scales counted along a row beginning at the origin of the anal fin and running obliquely dorsally either forward or backward, to the lateral line. For certain species this count is made from the base of the pelvic fin.

Secchi disk—A circular metal plate with the upper surface divided into four quadrants, two painted white and two painted black; lowered into the water on a graduated line, and the depth at which it disappears is noted as the limit of visibility.

Second dorsal fin—The posterior of two dorsal fins, usually the soft-rayed dorsal fin of spiny-rayed fishes.

Secondary invader—An opportunist pathogen that obtains entrance to a host following breakdown of the first line of defense.

Sedentary—Remaining in one place; attached or fixed.

Sediment—Solids settling out to form bottom deposits.

Sedimentation pond (settling basin)—A wastewater treatment facility in which solids settle out removing them from the hatchery effluent.

Seed stock—Larval fish or crustaceans, fry, small fingerlings.

Seeding—Pumping plankton from a pond with bloom to a pond without bloom to promote plankton growth.

Seepage—To flow out slowly through the pond bottom material.

Seine—Harvesting net.

Selective breeding—Selection of mates in a breeding program to produce offspring possessing certain defined characteristics.

Self-feeders—Device that allows animals the choice of when to receive feed.

Semipermeable—Permeable to different substances to different degrees.

Sensory receptors—Organs or receptors that receive stimuli and convey these by the nerve fibers to the brain or spinal cord where they are interpreted.

Septicemia—A clinical sign characterized by a severe bacteremic infection, generally involving the significant invasion of the blood stream by microorganisms.

Serum—The fluid portion of blood that remains after the blood is allowed to clot and the cells are removed.

Settleable solids—That fraction of the suspended solids that will settle out of suspension under quiescent conditions.

Settling pond—Area where the solids settle to the bottom and the top water is released into the environment or a stream.

Sex determination—Genetic process leading to an organism being male or female.

Sexing—Identifying males and females.

Shareholders—Owner of a share of a company or a stockholder.

Shelf life—Length of time a product maintains quality before being sold or used.

Shocking—Act of mechanically agitating eggs, which ruptures the perivitelline membranes and turns infertile eggs white.

Short bath—A type of bath most useful in facilities having a controllable rapid exchange of water. The water flow is stopped, and a relatively high concentration of chemical is thoroughly mixed in and retained for about 1 hour.

Side effect—An effect of a chemical or treatment other than that intended.

Sign—Any manifestation of disease, such as an aberration in structure, physiology, or behavior, as interpreted by an observer. (Note the term "symptom" is only appropriate for human medicine because it includes the patient's feelings and sensations about the disease.)

Silo—Deep cylindrical tanks that are similar in operation to horizontal raceways.

Silt—Soil particles carried or deposited by moving water.

Single-pass system—A system in which water is passed through fish rearing units without being recycled and then discharged from the hatchery.

Sinuses—Opening, hollow cavities.

Siphon—Tubular structure for drawing in or expelling liquids.

Slope—Incline from the level; slant.

Sludge—The mixture of solids and water that is drawn off a settling chamber.

Slurry—Thin, watery mixture of feed.

Smolt—Juvenile salmonid at the time of physiological adaptation to life in the marine environment.

Snatch block—Pulley.

Snout—The portion of the head in front of the eyes. The snout is measured from its most anterior tip to the anterior margin of the eye socket.

Sociability—The quality or character of being agreeable in company.

Sock—Same as live car.

Soft-egg disease—Pathological softening of fish eggs during incubation, the etiological agent(s) being unknown but possibly a bacterium.

Soft fins—Fins with soft rays only, designated as soft dorsal, etc.

Soft rays—Fin rays that are cross-striated or articulated, like a bamboo fishing pole.

Sole proprietorship—Form of business organization where one individual owns the business.

Solubility—The degree to which a substance can be dissolved in a liquid; usually expressed as milligrams per liter or percent.

Solvency—Having sufficient means to pay all debts.

Sp. and spp.—Singular and plural abbreviations for species, respectively. The singular abbreviation is often used when identity of the genus of the organism is known but the exact species is not known.

Spat—Spawn of the oyster; a young oyster.

Spawning (hatchery context)—Act of obtaining eggs from female fish and sperm from male fish.

Spawning net—Artificial nest, generally of Spanish moss or a synthetic material such as spandex, on which fish lay eggs.

Species—The largest group of similar individuals that actually or potentially can successfully interbreed with one another but not with other such groups; a systematic unit including geographic races and varieties, and included in a genus.

Specific drug—A drug that has therapeutic effect on one disease but not on others.

Spent—Spawned out.

Spermatozoon—A male reproductive cell, consisting usually of head, middle piece, and locomotory flagellum.

Spinal cord—The cylindrical structure within the spinal canal, a part of the central nervous system.

Spines—Unsegmented rays, commonly hard and pointed.

Spiny rays—Stiff or noncross-striated fin rays.

Spleen—The site of red blood cell, thrombocyte, lymphocyte, and granulocyte production.

Sporadic disease—A disease that occurs only occasionally and usually as a single case.

Spore—Singe-cell reproductive unit capable of creating a new adult individual.

Stabilization pond—A simple waste-water treatment facility in which organic matter is oxidized and stabilized—converted to inert residue.

Standard environmental temperature (SET)—The temperature at which all of the species physiological systems operate optimally.

Standard length—The distance from the most anterior portion of the body to the junction of the caudal peduncle and anal fin.

Standard metabolic rate—The metabolic rate of poikilothermic animals under conditions of minimum activity, measured per unit time and body weight at a particular temperature. Close to basal metabolic rate, but animals rarely are at complete rest. See Basal metabolism.

Standing crop weight—Total weight of all fish in a pond.

Stenohaline marine—Fish unable to withstand a wide variation in water salinity.

Sterilant—An agent that kills all microorganisms.

Sterilize—To destroy all microorganisms and their spores in or about an object.

Stimuli—Any factor or environmental change producing activity or response.

Stock—Group of fish that share a common environment and gene pool.

Stocker—Fish 8 in (20.3 cm) or over.

Stocking rate—The number of fish per unit of water.

Stomach—The expansion of the alimentary tract between the esophagus and the pyloric valve.

Strainers—Fish that select food primarily by size rather than by type and strain water through gill rakers to remove food.

Strains—Group of fish with presumed common ancestry.

Strategic planning—Analyzing the business and the environment in which it operates to create a broad plan for the future.

Stratification—Separation into distinct layers.

Stress—A state manifested by a syndrome or bodily change caused by some force, condition, or circumstance—a stressor—in or on an organism or on one of its physiological or anatomical systems. Any condition that forces an organism to expend more energy to maintain stability.

Stressor—Any stimulus, or succession of stimuli, that tends to disrupt the normal stability of an animal.

Stripping—Manually releasing eggs and milt from broodfish.

Stun—To render unconscious or incapable of action; performed on fish before entering processing plant.

Subacute—Not lethal; between acute and chronic.

Substrata—Subsoils.

Substrate—A underlying substance on which something takes hold or takes root.

Suckers—Fish that feed primarily on the bottom of their habitat—sucking in mud, filtering and extracting digestible material.

Sulfadimethoxine sulfonamide—Drug effective against certain bacterial pathogens of fishes.

Sulfaguanidine—Sulfonamide drug used in combination with sulfamerazine to control certain bacterial pathogens of fishes.

Sulfamerazine—Sulfonamide drug effective against certain bacterial pathogens of fish.

Sulfamethazine (sulmet)—Sulfonamide drug effective against certain bacterial pathogens of fishes.

Sulfate—Any salt of sulfuric acid; any salt containing the radical SO_4.

Sulfisoxasole—Sulfonamide drug effective against certain bacterial pathogens of fishes.

Sulfonamides—Antimicrobial compounds, for example, sulfamerazine or sulfamethazine.

Superior—As applied to the mouth, opening in an upward direction.

Supersaturation—Greater than normal solubility of a chemical—oxygen and nitrogen, for example—as a result of unusual temperatures or pressures.

Supplemental diet—A diet used to augment available natural foods; used in extensive fish culture.

Supplier—An individual or business that furnishes what is needed.

Surface runoff—Any surface water from precipitation, snowmelt, or irrigation that runs off the land into any surface waterbody.

Susceptible—Having little resistance to disease or to injurious agents.

Suspended solids—Particles retained in suspension in the water column.

Swim bladder—See Air bladder.

Swimmerets—Appendages on the abdomen of a crustacean.

Swim-up—Term used to describe fry when they begin active swimming in search of food.

Swim-up fry—Fry that have lost their yolk sac and are ready for food.

Syndrome—A group of signs that together characterize a disease.

Synthesize—Process of assembling parts into a whole.

Systems—Orderly combinations or arrangements of parts, elements, etc., into a whole, especially such combinations according to some rational principle.

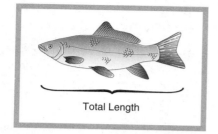

Total Length

T

Temper—To allow fish to adjust to different water chemistry and temperature.

Temperature shock—Physiological stress induced by sudden or rapid changes in temperature, defined by some as any change greater than 3 degrees per hour.

Tempering—Gradually acclimating (accustoming) fish to changes in water chemistry and temperature.

Tender stage—Period of early development, from a few hours after fertilization to the time pigmentation of the eyes becomes evident, during which the embryo is highly sensitive to shock. Also called green-egg stage, sensitive stage.

Terramycin—See Oxytetracycline.

Terrestrial—Existing on land.

Testes—The male reproductive organs producing sperm cells and hormones.

Therapeutic—Serving to heal or cure.

Thermal stress—Stress caused by rapid temperature change or stress caused by extreme high or low temperature.

Thermocline—Zone separating waters of varying densities.

Thiamine—An essential water-soluble B-complex vitamin that maintains normal carbohydrate metabolism and essential for certain other metabolic processes.

Thiosulfate, sodium (sodium hyposulfite, hypo, antichlor) ($Na_2S_2O_3$)—Used to remove chlorine from solution or as a titrant for determination of dissolved oxygen by the Winkler method.

Thorax—Middle region of the body between the head and abdomen.

Tissue residue—Quantity of a drug or other chemical remaining in body tissues after treatment or exposure is stopped.

Titrant—The reagent or standard solution used in titration.

Titration—A method of determining the strength (concentration) of a solution by adding known amounts of a reacting chemical until a color change is detected.

Titrimetric—Analyses using a solution of known strength—the titrant—which is added to a known or specific volume of sample in the presence of an indicator. The indicator produces a color change indicating the titration is complete.

Tocopherol—Vitamin E, an essential vitamin that acts as a biological antioxidant.

Toggle—A pin or rod inserted through a loop of haul line to attach it to the seine.

Tolerance—Residue levels of a drug or chemical that are permitted by regulatory agencies in food eaten by humans.

Topical—Local application of concentrated treatment directly onto a lesion.

Topography—Surface features of a region; the lay of the land.

Topping—Harvesting only those fish that have grown to marketable size.

Total dissolved solids (TDS)—See Dissolved solids.

Total harvest—Harvesting strategy that involves one-time seining or trapping and annual draining of a pond.

Total length—The distance from the most anterior point to the most posterior tip of the fish tail.

Total solids—All of the solids in the water, including dissolved, suspended, and settleable components.

Toxicity—A relative measure of the ability of a chemical to be poisonous. Usually refers to the ability of a substance to kill or cause an adverse effect. High toxicity means that small amounts are capable of causing death or ill health.

Toxicology—The study of the interactions between organisms and a toxicant.

Toxin—A particular class of poisons, usually albuminous proteins of high molecular weight produced by animals or plants, to which the body may respond by the production of antitoxins.

Trademark—The name or design officially registered and used by a merchant or manufacturer to identify goods and distinguish from those made by others.

Trammel seine—A seine with two course outer nets that support a fine-mesh inner net that fish swim into and force through the course layers, trapping themselves as the fine-mesh net is forced in around them.

Transmission—The transfer of a disease agent from one individual to another.

Transplanting—The moving of shellfish from one growing area to another.

Trauma—An injury caused by a mechanical or physical agent.

Treaty—Formal agreement or compact duly concluded and ratified between two or more states or countries.

Trematoda—The flukes from the subclass Monogenea; ectoparasitic in general, one host; from the subclass Digenea; endoparasitic in general, two hosts or more.

Tubercles—Hornlike projections on the head of breeding fathead minnows.

Tumor—An abnormal mass of tissue, the growth of which exceeds and is uncoordinated with that of the tissues and persists in the same excessive manner after the disappearance of the stimuli that evoked the change.

Turbidity—Presence of suspended or colloidal matter or planktonic organisms that reduces light penetration of water.

Turbulence—Agitation of liquids by currents, jetting actions, winds, or stirring forces.

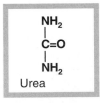

UDN—See Ulcerative dermal necrosis.

Ubiquitous—Existing everywhere at the same time.

Ulcer—A break in the skin or mucous membrane with loss of surface tissue; disintegration and necrosis of epithelial tissue.

Ulcer disease—An infectious disease of eastern brook trout caused by the bacterium *Hemophilus piscium*.

Ulcerative dermal necrosis (UDN)—A disease of unknown etiology occurring in older fishes, usually during spawning, and primarily involving salmonids.

Understocking—Periodically harvesting marketable fish from the rearing unit and stocking smaller fish in their place.

Unisex culture—Raising only one sex—usually male—of a species.

United States Pharmacopeia (USP)—An authoritative treatise on drugs, products used in medicine, formulas for mixtures, and chemical tests used for identity and purity of the above.

Urea—One of the compounds in which nitrogen is excreted from fish in the urine. Most nitrogen is eliminated as ammonia through the gills.

Uremia—The condition caused by faulty renal function and resulting in excessive nitrogenous compounds in the blood.

Urinary bladder—The bladder attached to the kidneys; the kidneys drain into it.

Urinary ducts—Tube for conveying urine.

Urogenital pore (urinary opening)—External outlet for the urinary and genital ducts.

Uropod—Last swimmeret on a crustacean that develops into a flipper.

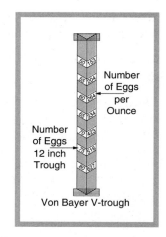

Number of Eggs per Ounce

Number of Eggs 12 inch Trough

Von Bayer V-trough

V

VHS—See Viral hemorrhagic septicemia.

Vaccine—A preparation of nonvirulent disease organisms (dead or alive) that retains the capacity to stimulate production of antibodies against it. See Antigen.

Value-added—Further processing to increase the value of a product.

Variable costs—Costs that increase or decrease in relation to an increase or decrease in production.

Vector—A living organism that carries an infectious agent from an infected individual to another, directly or indirectly.

Vein—A tubular vessel that carries blood to the heart.

Vent—The external posterior opening of the alimentary canal; the anus.

Ventral—Underside of the body where the belly is located.

Ventral fins—Pelvic fins.

Vertebrate—Organism with an inner skeleton and a segmented spinal column.

Vertical transmission—The parent-to-progeny transfer of disease agents via eggs or sperm.

Viable—Alive.

Vibriosis—An infectious disease caused by the bacterium *Vibrio anguillarium*. Also called pike pest, eel pest, or red sore.

Viral hemorrhagic septicemia (VHS)—A severe disease of trout caused by a virus of the Rhabdovirus group. Sometimes called egtved disease, infectious kidney swelling and liver degeneration (INUL), or trout pest.

Viremia—The presence of virus in the blood stream.

Virulence—The relative capacity of a pathogen to produce disease.

Virus—Ultramicroscopic infective agent that is capable of multiplying in connection with living cells. Normally, viruses are many times smaller than bacteria.

Viscera—The internal organs of the body, especially of the abdominal cavity.

Visualization—Being able to see things in the mind's eye.

Vitamin—An organic compound occurring in minute amounts in foods and essential for numerous metabolic reactions.

Vitamin D—A radiated form of ergosterol that has not been proved essential for fish.

Vitamin K—An essential, fat-soluble vitamin necessary for formation of prothrombin; deficiency causes reduced blood clotting.

Vitamin premix—A mixture of crystalline vitamins or concentrates used to fortify a formulated feed.

Viviparous—Bringing forth living young; the mother contributes food toward the development of the embryos.

Volumetric—Measurement of substances by comparison of volume.

Vomer—Bone of the anterior part of the roof of the mouth, commonly triangular and often with teeth.

Von Bayer trough—A 12-inch (30.5 cm) V-shaped trough used to count eggs.

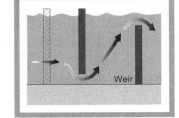

W

Warmwater species—Generally, fish that spawn at temperatures above 60°F (15.6°C). The chief cultured warmwater species are basses, sunfish, catfish, and minnows. See Coldwater species; Coolwater species.

Wastewater—Water leaving a processing plant or production facility.

Water column—Vertical pattern of water from the surface to the bottom of a waterbody.

Water hardening—Process by which an egg absorbs water that accumulates in the perivitelline space.

Water hardness—Measure of the total concentration of primarily calcium and magnesium expressed in milligrams per liter (ppm) of equivalent calcium carbonate ($CaCO_3$).

Water quality—As it relates to fish nutrition, involves dissolved mineral needs of fishes inhabiting that water (ionic strength).

Water table—Level below which the ground is saturated with water.

Water treatment—Primary: removal of a substantial amount of suspended matter, but little or no removal of colloidal and dissolved matter. Secondary: biological treatment methods, for example, by contact stabilization, extended aeration. Tertiary (advanced): removal of chemicals and dissolved solids.

Watershed—The whole region from which a river receives its supply of water.

Weir—A structure for measuring water flow.

Western gill disease—See Nutritional gill disease.

Wetland—Area that is covered with standing water or is saturated most of the year, and that supports mainly water-loving plants.

Whirling disease—A disease of trout caused by *Myxosoma cerebalis.*

White grub—An infestation of *Neodiplostomum multicellulata* in the liver of many freshwater fishes.

White spot disease—A noninfectious malady of incubating eggs or on the yolk sac of alevins. The cause of the disease is thought to be mechanical damage. Also see Ich.

Wholesale sales—Sales of fish in large quantities to buyer who then sells to distributor or retail market.

Withdrawal time—Period of time that must pass after drug, chemical, or pesticide treatment before an animal can be eaten.

X

Xanthophyll—Yellow to orange color; derivative of carotene. High levels in diet impart undesirable yellow color to light-fleshed fish.

Y

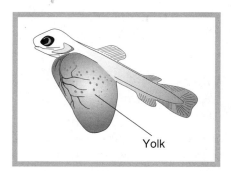

Yolk

Yellow grub—An infestation of *Clinostomum marginatum.*

Yolk—The food part of an egg.

Yolk sac—Source of nutrition for fish immediately after hatching.

Z

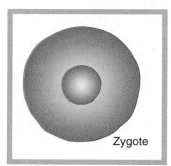

Zygote

Zooplankton—Minute animals in water, chiefly rotifers and crustaceans, that depend upon water movement to carry them about, having only weak capabilities for movement; important prey for young fish.

Zoospores—Motile spores of fungi.

Zygote—Cell formed by the union of the male and female gametes—the sperm and egg—and the individual developing from this cell.

Index